Food Security Handbook

Food Security Handbook

Edited by Alexis Crum

MURPHY & MOORE
www.murphy-moorepublishing.com

Published by Murphy & Moore Publishing,
1 Rockefeller Plaza,
New York City, NY 10020, USA
www.murphy-moorepublishing.com

Food Security Handbook
Edited by Alexis Crum

International Standard Book Number: 978-1-63987-233-6 (Hardback)

Cataloging-in-Publication Data

Food security handbook / edited by Alexis Crum.
      p. cm.
Includes bibliographical references and index.
ISBN 978-1-63987-233-6
1. Food security. 2. Food supply. 3. Food. I. Crum, Alexis.
HD9000.5 .F66 2022
338.19--dc23

TABLE OF CONTENTS

PREFACE

The availability of food that is nourishing, diverse and balanced to sustain the food consumption of individuals is known as food security. It is concerned with the accessibility that individuals have to nutritious food to fulfil their dietary needs to lead a healthy and active lifestyle. The main components in this field are access, utilization, availability and stability. To measure these components, various scales such as Household Food Insecurity Access Scale, Household Hunger Scale and Coping Strategies Index have been developed. Some of the adverse effects of food shortage are famine, hunger, physical and psychological deficiencies. Factors such as global water crisis, climate change, food waste, politics, agricultural diseases and land degradation directly impact food security. Various approaches such as large-scale food stockpiling, agricultural insurances and use of genetically modified crops are aimed at achieving food security across the world. The book studies, analyses and upholds the pillars of food security and its utmost significance in modern times. It presents researches and studies performed by experts across the globe. Coherent flow of topics, student-friendly language and extensive use of examples make this book an invaluable source of knowledge.

The information contained in this book is the result of intensive hard work done by researchers in this field. All due efforts have been made to make this book serve as a complete guiding source for students and researchers. The topics in this book have been comprehensively explained to help readers understand the growing trends in the field.

I would like to thank the entire group of writers who made sincere efforts in this book and my family who supported me in my efforts of working on this book. I take this opportunity to thank all those who have been a guiding force throughout my life.

Editor

Short rotation plantations policy history in Europe: lessons from the past and recommendations for the future

Kevin N. Lindegaard[1], Paul W. R. Adams[2], Martin Holley[3], Annette Lamley[3], Annika Henriksson[4], Stig Larsson[5], Hans-Georg von Engelbrechten[6], Gonzalo Esteban Lopez[7] & Marcin Pisarek[8]

[1]Crops for Energy Ltd, 15 Sylvia Avenue, Knowle, Bristol BS3 5BX, UK
[2]Department of Mechanical Engineering, Faculty of Engineering and Design, Institute for Sustainable Energy and Environment (I·SEE), University of Bath, Claverton Down, Bath BA2 7AY, UK
[3]Centre for Sustainable Energy, 3 St Peter's Court, Bedminster Parade, Bristol BS3 4AQ, UK
[4]Henriksson Salix AB, Almhög, 241 92 Eslöv, Sweden
[5]EWB, Spannmålsgatan 28, SE-268 32 SVALÖV, Sweden
[6]Agraligna GmbH, Oststrasse 7, 38315 Schladen/OT Beuchte, Germany
[7]Agencia Provincial de la Energía de Granada, Edificio CIE - 1°Planta. Avda. Andalucía s/n., 18015 Granada, Spain
[8]PGNiG TERMIKA SA, Siedziba główna - Elektrociepłownia Żerań, ul. Modlińska 15, 03-216 Warszawa, Poland

Keywords
Biomass, energy crop, policy, short rotation coppice, short rotation plantations.

Correspondence
Kevin N. Lindegaard, Crops for Energy Ltd, 15 Sylvia Avenue, Knowle, Bristol BS3 5BX, UK.

E-mail: kevin@crops4energy.co.uk
and
Paul W. R. Adams, Institute for Sustainable Energy and Environment (I·SEE), Department of Mechanical Engineering, Faculty of Engineering and Design, University of Bath, Claverton Down, Bath BA2 7AY, UK.

E-mail: paul.adams@bath.edu

Funding Information
No funding information provided.

Abstract

Short rotation plantations (SRPs) are fast-growing trees (such as willow (*Salix* spp.), poplar (*Populus* spp.) and *Eucalyptus*) grown closely together and harvested in periods of 2–20 years. There are around 50,000 hectares of SRPs in Europe, a relatively small area considering that there have been supportive policy measures in many countries for 30 years. This paper looks at the effect that the policy measures used in different EU countries have had, and how other external factors have impacted on the development of the industry. Rokwood was a 3-year European funded project which attempted to understand the obstacles and barriers facing the woody energy crops sector using well established methods of SWOT and PESTLE analysis. Stakeholder groups were formed in six different European regions to analyze the market drivers and barriers for SRP and propose ways that the industry could make progress through targeted research and development and an improved policy framework. Based upon the outcomes of the SWOT and PESTLE analysis, each region produced a series of recommendations for policymakers, public authorities, and government agencies to support the development, production, and use of SRP-derived wood fuel in each of the partner countries. This study provides details of the SRP policy analysis and reveals that each region shared a number of similarities with broad themes emerging. There is a need to educate farmers and policymakers about the multifunctional benefits of SRPs. Greater financial support from regional and/or national government is required in order to grow the SRP market. Introducing targeted subsidies as an incentive for growers could address lack of local supply chains. Long-term policy initiatives should be developed while increasing clarity within Government departments. Research funding should enable closer working between universities and industry with positive research findings developed into supportive policy measures.

Introduction

Short rotation plantations (SRPs) offer an opportunity to increase energy security by providing a local source of low carbon renewable biomass fuel. Bioenergy offers an alternative to fossil fuels, reductions in greenhouse gas emissions, and assists in the economic development of rural communities (Defra, 2007). Policies have therefore been implemented across Europe to promote bioenergy and the domestic planting of perennial energy crops

(Mangan 1997; Thornley and Cooper 2008; EC 2009; Adams 2011; Natural England, 2013; Adams & Lindegaard, 2016). Despite the various policy instruments, grants, and incentives implemented, the cultivation of woody energy crops has proceeded at a low rate across the EU (IEE 2009; Aebiom 2015; EurObserv'ER, 2015). The research is part of an EU-funded project that is exploring ways to increase the cultivation of SRPs throughout Europe; hence, the hypothesis is that the positive benefits of SRP outweigh potential negative impacts. This paper provides an overview of the historical development of SRPs in Europe drawing on specific policy and market examples from six EU countries. An assessment is conducted of the current state of play for the SRP market in each country, with an analysis performed to provide policy recommendations for the future development of SRPs in Europe.

History of SRPs in Europe

Short rotation plantations (SRPs) are fast-growing trees grown closely together and harvested in periods of 2–20 years. Trees that are cultivated this way include willow (*Salix* spp.), poplar (*Populus* spp.), *Eucalyptus*, and *Robinia* (Rokwood, 2015a). SRPs have been considered as an option in modern agriculture for biomass energy and fiber production for over 40 years, although the historical use can be traced back to centuries (Bergendorff and Emanuelsson 1996). Initially, interest was sparked in the early 1970s by the potential shortage in pulp wood used for paper and cardboard production (Anon, 1980; Richards 1987). This potential land use also received significant attention in the wake of the OPEC oil crisis of 1973, with the oil embargo increasing oil prices and leading to supply shortages (Ross 2013). Countries like Sweden and the Netherlands with low levels of indigenous fossil fuels were particularly exposed to this issue and endured energy rationing (Chitadze 2012; Verwijst et al. 2013). In light of this incident, the need for greater security of energy supply became important and research on willow for biomass energy began in Sweden and United Kingdom (Dawson 1992; Mangan 1997; Lindegaard et al. 2001).

Initial research efforts suggested that high yields could be achieved on marginal land and an industry started to develop. The first commercial willow plantings took place in Sweden in 1981 and cuttings suppliers were offering large volumes of material from 1985; the first mechanized planter (the Step Planter) was developed in 1986 and Svalöf-Weibull AB began commercial willow breeding in 1987 (Larsson 1998; Verwijst et al. 2013).

The industry began to grow with the introduction of set-aside in 1988 under the Common Agricultural Policy (CAP) (EC 1988). This program imposed production quotas and forced farmers to take a proportion of their land out of food production in order to control the over-supply of agricultural commodities such as milk and grain. There were suggestions at the time that 6 million hectares of UK farmland would need to be removed from food production; SRPs therefore emerged as an attractive diversification option (Dawson 1992).

Other geo-political factors also stimulated the industry. The realization that over reliance on fossil fuels was causing global warming, led to the Earth Summit in Rio in 1992 and the signing of Kyoto Protocol in 1997 (UNCED, 1992; Keating 1993; UNFCCC, 1997). The need for large-scale carbon emissions reduction led some countries like Sweden and Denmark to adopt carbon taxes (McCormick and Kåberger 2005). This gave a favorable advantage to renewable energy and home grown biomass production (ETSU, 1999; Mola-Yudego and Pelkonen 2008).

Throughout the history of the SRP sector there have been key breakthroughs in research and technology development (Verwijst et al. 2013). For instance, there are numerous breeding and selection programs for SRPs in Sweden, UK, Italy, Belgium, Germany, Poland, and Spain (Lindegaard and Barker 1997; Larsson 1998; Karp et al. 2011; Isebrands and Richardson 2014). From these efforts there are some exceptional, high- yielding and disease-resistant varieties (Danfors et al. 1997; Lindegaard et al. 2001, 2011; Caslin et al. 2012). In addition, planting and harvesting technology has been developed, making it easier to ensure a good establishment and efficient harvesting (PAMI, 2003; Spinelli et al. 2008, 2009; Schweier and Becker 2012; Henriksson and Henriksson 2015).

Nonetheless, despite this promising start and 30 years of supportive policy measures in many countries, the industry has faltered and there are currently estimated to be 50,000 hectares of SRPs in the EU28 (Aebiom 2015). In many countries, there has been a similar trend with relatively large areas of SRPs being established in a short time followed by a rapid decline (See section Short Rotation Plantations (SRP) policy review).

Rokwood

Rokwood was a six-country study which aimed to make the regionally based production of woody biomass economically attractive, technically feasible, and environmentally sustainable (Rokwood, 2015b). SRPs provide a quick and efficient way of producing large volumes of woody biomass where there is a local market or need. Besides their high productivity, SRPs offer further advantages such as providing landscape diversity, increased biodiversity compared to annual crops, and numerous ecosystem services such as reduction in soil erosion, reductions in nutrient leaching, and a possible approach to flood mitigation (Johnston et al. 2015; Adams and Lindegaard 2016; Styles et al. 2016). These promising attributes are not being fully exploited due to a variety of obstacles and barriers hindering or even preventing the further development of the SRP sector (Adams et al. 2011; Lindegaard 2013a,b). These obstacles and barriers comprise missing or unfavorable legal framework conditions, missing financial support, and various technical and nontechnical barriers.

Rokwood as a trans-European research project attempted to confront these issues to find innovative ways to increase the market penetration of woody energy crops. The project involved a large consortia of 20 partners from six European regions (Northern Germany, South West England, Mazovia in Poland, Skåne in Sweden, Andalusia in Spain, and the Midlands and Western Region of Ireland) as well as EUBIA, the European Biomass Industries Association. Each region is represented by three partners, respecting the triple-helix concept (a business entity, a research entity, and a local or a regional authority) (Lindegaard et al. 2015). The six regions, in spite of their structural differences and levels of SRP engagement, face similar challenges in terms of developing the SRP market. Rokwood was intended to enforce the cooperation between these countries through a collective Joint Action Plan (JAP) for tackling the most important obstacles and barriers on the European level. By connecting these regions, Rokwood has striven to promote the exchange of established best practices and thus improve the economic growth of SRPs (Rokwood, 2015c).

Aims and objectives

The first aim of this study is to briefly review the past history of SRP policy development in Europe and critique the main policies and incentives that were designed and implemented to increase SRP supply. The focus of the policy review is on those policies that specifically incentivized either production or use of SRP, or are beneficial to the farmed environment. The secondary aim of the research is to develop policy recommendations for each country, comparing and contrasting key themes and differences between countries. Policy recommendations are intended for policy makers, public authorities, and governmental agencies to support the development, production, and use of SRP-derived wood fuel in each of the partner countries and beyond. Specific objectives required to address this aim include:

- Review key policies that have supported the development of SRP in Germany, Ireland, Poland, Spain, Sweden, and UK;
- Perform a SWOT (Strengths, Weaknesses, Opportunities, and Threats) and PESTLE (Political, Economic, Social,

Technological, Legal, and Environmental) analysis to identify factors influencing the market for SRP in each of the six countries;
- Present the SRP policy recommendations produced as part of the Rokwood project.

Methodology

This section describes the research methods that were followed to critique and assess different policies implemented to incentivize the cultivation of SRPs, analyze the factors influencing SRP markets, and present policy recommendations. Figure 1 provides a visual summary of the research methodology, and the following subsections describe the main research stages.

Review of the key SRP policies for each country

The research commenced with a review of the policy measures that have been used in the past in Germany, Ireland, Poland, Spain, Sweden, and UK and the effect these had on the uptake of SRP. Consideration has also been given as to how other external factors have impacted on the development of the industry. For each country considered, a literature review was performed to identify policies that promoted the planting of SRPs, the use of SRPs or are beneficial to the farmed environment. The review focused on national and local policies and strategies, industry and research publications, and other literature in the public domain. Through project workshops, these policies were further identified and assessed for their relevance to the development of the SRP industry. Workshops consisted of varied stakeholders in each region with a knowledge and interest in SRP to provide different perspectives from policy, industry, agriculture, and research. A workshop was held in each region with the details of attendees provided in the Rokwood publications. The policies described in section Short Rotation Plantations (SRP) policy review provide a summary of some key policy examples in each partner country but do not provide an exhaustive list of all the policies identified.

PESTLE and SWOT analysis

The Rokwood partners performed PESTLE (Political, Economic, Social, Technological, Legal, and Environmental factors) and SWOT (Strengths, Weaknesses, Opportunities, and Threats) analyses to identify all the factors that influence the SRP sector within their countries in order to prioritize and select those which could be best targeted by policymakers to help the industry expand.

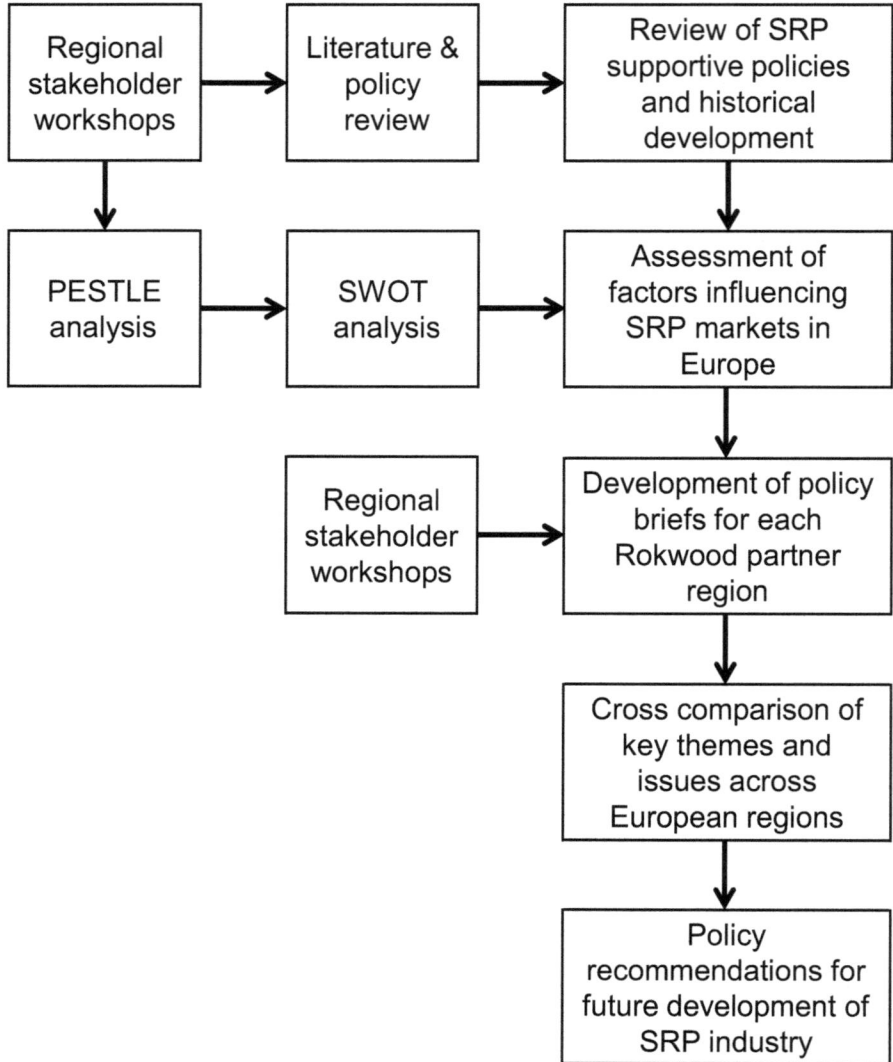

Figure 1. Overview of key stages of the research conducted for this paper.

The PESTLE analysis was performed first and represents a checklist of factors currently affecting the production and use of SRP, and those which are likely to affect it in the future (CPID, 2015a). It gives an overview of the whole environment from many different perspectives that need to be considered in policy development. Key questions that were considered for the PESTLE analysis were:

- What is the political situation of the country and how can it affect the SRP industry?
- What are the prevalent economic factors and market conditions for SRP?
- How much importance do social aspects have in the market for SRP?
- What technological innovations could arise that may affect the market structure?

- Which current legislations impact on the SRP industry and what influence can stakeholders have on future development of policy?
- What are the environmental considerations for the SRP industry?

The PESTLE analysis was undertaken by each cluster focusing on their corresponding region and developed in workshops (Parra-López et al. 2015). These regional workshops consisted of different actors in the SRP market and included academics, agronomists, farmers, funders, plant breeders, and policymakers who have a specialist knowledge or influence on the development of SRP. To recruit workshop participants the Rokwood partners consulted expert knowledge outside the cluster group to list and rank external PESTLE factors. The method used involved key actors identified as:

- Experts in their field
- Able to look beyond the borders of their specialism
- Well-renowned in the cluster for their ideas and opinion

Experts identified in the first step of the analysis were used. The key actors were not limited to persons active, in the current situation, within the SRP sector. Key actors may also be persons with prior experience from the SRP sector, representatives of organizations planning to expand within this sector, and other biomass fuel actors that as of today for different reasons refrain to involve in the SRP sector (e.g. operators of biomass plants not using SRP today). The workshop participants totaled 4–23 stakeholders in each region to ensure a good coverage of specialisms while still allowing active participation. The delegate's names remain anonymous as agreed at the onset of the research. Further details are available from the Rokwood representatives for each region. The Spanish cluster was the only project members to formally publish the PESTLE results (Sayadi et al. 2014; Parra-López et al. 2015).

The first objective of the PESTLE analysis was to identify impediments and factors of success of regional regions through the evaluation of current markets and ascertain the barriers to growing SRP. A second objective was to identify potential policy mechanisms which stimulate growers and end users to develop the industry (Rokwood, 2014a).

The follow-up SWOT analysis drew on the PESTLE outputs. A SWOT analysis is an established method of evaluating a situation or a market in a structured appraisal to help plan for the future (CPID, 2015b). In this context, it helps to identify strengths and weaknesses of the SRP market and map these to external opportunities and threats so that the most effective policies can be formulated to achieve market success. Each region populated a SWOT chart made up of four quadrants to identify "internal" strengths and weaknesses within the SRP industry alongside external opportunities and threats (Rokwood, 2014a). The most important factors in each quadrant (up to a maximum of 10) were then recorded. Factors that could make up a "common" SWOT across all the regions were then decided at a consortium meeting workshop. Results from the PESTLE and SWOT analyses are presented in section Summary outcomes of the PESTLE and SWOT analysis with further analysis provided in section Policy recommendations.

Policy recommendations

Findings from the policy and market review were consolidated and summarized to ensure all opportunities

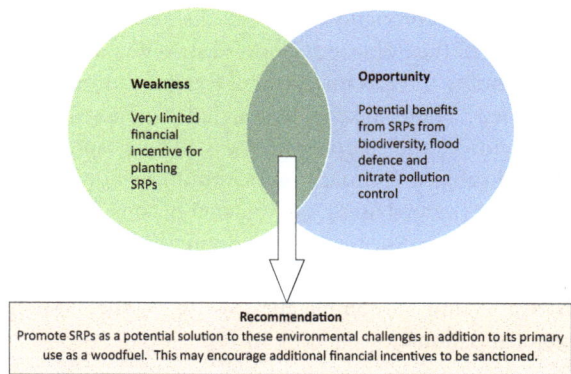

Figure 2. Example of using identified opportunities to negate weaknesses to SRPs in the UK.

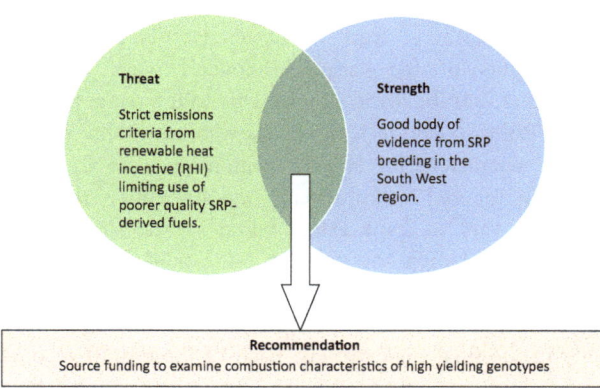

Figure 3. Example of using identified strengths to overcome threats to SRPs in the UK.

for and threats against SRP development were fully considered. This included an evaluation of the SWOT and PESTLE analyses as well as a detailed appraisal of the policy outcomes identified for each region (see section Policy recommendations). The SWOT analysis, in particular, was heavily used both to identify issues and to find potential solutions, with most of the issues being derived from weaknesses and threats in the SWOT analysis. Solutions were sought by looking for relationships, for example, a strength that could overcome a threat or an opportunity that could negate a weakness. Examples from the UK region are shown in Figs 2 and 3.

Stakeholder consultation exercise to validate policy measures

In order to validate the policy recommendations, a stakeholder consultation exercise was performed. To ensure that the final policy brief documents were suitably targeted and that they thoroughly addressed the key barriers identified, each of the Rokwood regions distributed draft copies of their briefs to relevant stakeholders for comment and peer review. These stakeholders included policymakers at local, regional, and national levels, academics specializing in related subjects (primarily agricultural, economic development, or energy-based), and bioenergy consultants. The consultation exercise was an important final stage of the methodology to ensure the accuracy, appropriateness, validity, and completeness of the policy recommendations.

Policy briefs development for each of the six regions

Following the consultation exercise, each of the six regions produced a "policy brief" that drew upon the robust evidence base gathered from the stages described above and the outputs of the various Rokwood work packages (Rokwood, 2014b, 2015d,e,f,g,h). Due to the regional focus of the Rokwood project, the briefs were inevitably shaped by the characteristics of each region and are therefore primarily focused on influencing regional policy, although this does vary to some extent based on the structure of governance in each country. Once the policy recommendations were agreed following consultation, each region chose a number of recommendations to elaborate on more fully in the final policy briefs. Each of these draft recommendations was presented in line with the following four headings:

- What is the problem or issue?
- How could the problem or issue be addressed?
- Which stakeholders should take this forward?
- What are the benefits and potential outcomes?

Section Policy recommendations presents the key aspects of the policy briefs by assessing the policy mechanisms required to develop the SRP industry in the different European countries. A discussion of the main issues and solutions is presented within the policy recommendations to identify key themes and differences between regions.

Short Rotation Plantations (SRP) Policy Review

This section provides a high-level summary of the key policies and market situation for SRPs in each of the six European regions considered. Table 1 provides a summary of the different characteristics of each region, and the following subsections describe each region.

Germany

The area of SRP in Germany has grown from <500 hectares in 2004 to 5969 hectares in 2014 (DBFZ, 2015). Poplar is by far the most common SRP option and planting peaked in 2010 with around 1400 hectares planted (BMELV, 2012; DBFZ, 2015), falling to 600 hectares in 2015 (von Engelbrechten 2015). The largest area of SRC is in the Brandenburg region with around 50% of the total (Murach et al. 2013). The largest market for SRP poplar is the 5 MW_e Märkisches Viertel CHP plant (Vattenfall, 2014).

In 2010, there was a grant available in the Federal State of Saxony paying 30% of the establishment costs of SRC (Faasch and Patenaude 2012). There was also a grant for €1200 for planting SRPs in Saarland in 2012, but this brought about limited planting. (von Engelbrechten 2015). Currently, there is a planting grant for SRC under the Gemeinschaftsaufgabe Agrarstruktur und Küstenschutz (GAK) or Joint Task Agricultural Structures and Coastal Protection Framework Plan (FNR, 2015). This is not a national scheme and only five of the 15 German regions took part in the program: Baden-Wurttemberg, Brandenburg -Berlin, Mecklenburg-Western Pomerania, North Rhine-Westphalia, and Thuringia.

Republic of Ireland

The Republic of Ireland is 85% dependent on fossil fuel imports costing the nation around €6.5 billion per year (TEAGASC, 2014). Energy crops are viewed as one of the ways that Ireland could help itself meet the target for 12% renewable heat by 2020. Over 900 hectares of SRC willow was planted in Ireland between 2006 and 2013 (DCENR, 2014). The increase in farmer interest was boosted by the introduction of the Bioenergy Scheme in 2007 which covered SRC and Miscanthus establishment costs (DAFM, 2015a). Since 2012, the scheme has been launched annually with narrow application windows (less than 2 months).

The peak planting year was 2010 when over 200 hectares were planted (see Fig. 4). Planting levels have fallen rapidly since. The largest market for energy crops in Ireland is co-firing with peat in the Edenderry plant (Bord na Móna, 2015). The main reason for the fall in planting of SRC seems to be the bad press associated with energy crops as a result of the lack of markets for Miscanthus (Independent, 2013; Irish Examiner, 2013). The majority of Miscanthus growers were located in County Cork and County Limerick, too far away from Edenderry site in County Offaly to make it a viable market. This coupled with the low price offered for the fuel led the majority

Table 1. Summary information regarding the six Rokwood regions (Lindegaard et al. 2015).

	Northern Germany	Midlands & Western Ireland	Mazovia, Poland	Andalusia, Spain	Skåne, Sweden	South West England
Population (millions)	19.5	1.1	5.3	8.4	1.3	5.3
Area (m ha)	13.77	3.25	3.56	8.76	1.09	2.38
Area of SRPs today (ha)	3600	117	1100	150–170	2042	93
Forest cover (m ha)	2.37	0.34	0.85	2.54	0.39	0.25
% of land cover that is forest	17.2	10.5	23.8	29.0	35.7	10.5
Installed capacity of biomass (MW$_{th}$)	approx. 500	94	2480	1555	1840	280.3
Number of biomass heating & CHP installations	7500	951	32,262	23,431 heating and 18 power plants	33,140 heating and 33 district heating and CHP plants	3414
Area of agricultural land (m ha)	6.91	2.05	2.31	3.85	0.51	1.91
% of land cover that is agricultural	50.2	63.1	65.0	43.9	46.3	80.4
Predominant agricultural land use	Cereal farming and cultivated pasture	Pasture/ grassland for livestock	Fruit, vegetables, potatoes, cereals, canola, berries	Olive plantations	Livestock farming and arable crops cultivation	Livestock farming, particularly, dairy cows and sheep

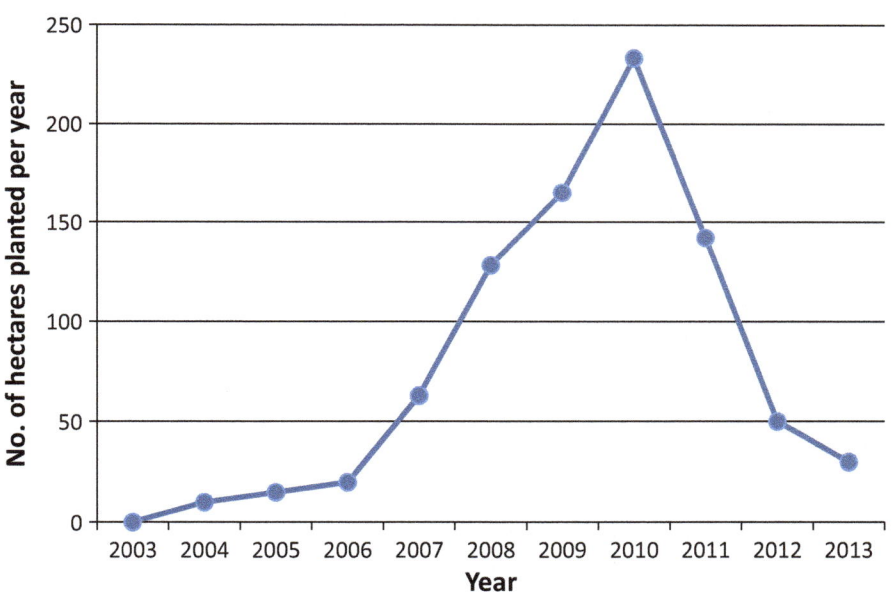

Figure 4. SRC plantings in the Republic of Ireland 2003–2013.

of Miscanthus growers to grub out their crops (Irish Times, 2013).

The Bioenergy Scheme in 2015 now covers just SRC willow with the maximum area that can be planted, increased substantially; but, the maximum grant and percentage of eligible costs covered, both reduced (DAFM, 2015a). It is expected that a Renewable Heat Incentive (RHI) scheme will be introduced in 2016 (DCENR, 2014) which could see an increase in SRC willow planting.

Poland

Poland is the largest coal producer in EU and its dependency on fossil fuels imports is among the lowest in the EU (Wisniewski and Oniszk-Popławska 2009). The renewable energy market is dominated by co-firing biomass with coal in CHP plants. In between 2000 and 2006, regional environmental funds supported local initiatives for SRP introduction, but that resulted in limited planting. Establishment grants at national level for various energy crops (SRC willow and poplar, Miscanthus and *Sida hermaphrodita*) were introduced in 2007–2008 and supported 1300 ha of planting until the scheme was withdrawn in 2009 (Szymańska and Chodkowska-Miszczuk 2011). Despite the absence of further support, the planted area of energy crops continued to grow from 6193 ha in 2010 to 11,509 ha in 2013, with SRC willow making up the majority of this area (Aebiom 2015; Gajewski 2015).

In 2005, Poland introduced Tradable Green Certificates (TGC) to support renewable electricity production (Heinzel and Winkler 2010). This led to a surge in biomass co-firing with coal as well as dedicated 100%-fuelled biomass systems. By 2012, co-firing alone had increased the biomass consumption to 12 M tons (9.5 TWH electricity generated) at 50 plants. However, a fall in the price of TGCs from the peak of around €62/MWH (275 PLN/MWh) in quarter 3 of 2011 to €23/MWH (100 PLN/MWh) in quarter 1 of 2013 resulted in a large decrease in total biomass consumption to around 8.5–9 million tons in 2014–2015 (Polenergia, 2015).

Several energy utilities including MONDI, KGHM SA, Fortum Poland, and PGNiG TERMIKA SA have provided incentives for planting SRPs either through providing support toward planting and maintenance costs or land-lease arrangements. As an example PGNiG TERMIKA is currently offering farmers €570/ha (2500 PLN/ha) to plant and maintain SRC willow and a guarantee to buy the fuel produced for 17 years or 5 harvests (PGNiG TERMIKA, 2015).

Despite this, the planted area of SRC has seen a slight decrease since 2013. Furthermore, SRC is not explicitly supported under the Polish CAP scheme for 2015–2020 (Pisarek, 2015).

Spain

Spain has 18.4 million hectares of forest and is the EU's third most wooded country behind Sweden and Finland (Eurostat, 2015). Estimates suggest that there are 500,000 hectares of Eucalyptus and 100,000 hectares of poplar planted (FAO, 2008; Ruiz and Lopez 2010; Isebrands and Richardson 2014). Most of this is planted as single-stem trees and used for industrial uses such as pulp wood for the paper industry and veneer production (Parra-López et al. 2015).

There is a great deal of interest in local supplies of woody energy crops from bioenergy project developers and also policymakers seeking to increase production from marginal farmland. Despite this, the area devoted to SRP for biomass production is mainly restricted to experimental plots (Pérez-Cruzado et al. 2014). There are only around 150–170 hectares of SRP trials in Andalusia but evidence

suggest that high yields are achievable (Kauter et al. 2003; Durán Zuazo et al. 2013). The main reason for the lack of commercial planting is the price competition from abundant waste biomass, especially olive trees and forestry residues (Parra-López et al. 2015).

The production costs of SRPs are much higher and the price paid by power plants is too low to attract more farmers to plant SRPs. Estimates suggest that the production costs for poplar are €20–40/ton while the price paid by the user is €40–55/ton at the farm gate (equivalent to €70/ton for dry chips), green electricity tariffs would therefore need to rise for the power companies to offer farmers a better deal (Carrasco and Sixto 2007). There have never been any establishment grants to support planting of SRPs in Spain.

Ence is Spain's market leader in the production of renewable energy using forest biomass and energy crop. The company currently runs four biomass power stations at Huelva, Navia, Merida, and Pontevedra with a total installed biomass generation capacity of 220 MW (Ence, 2016).

Sweden

The introduction of a generous planting grant in Sweden led to a mini boom in planting in the mid-1990s; at its peak, there were 18,000 hectares planted and over 1250 growers with numerous harvesting machines developed (Rosenqvist et al. 2000; Mola-Yudego and Pelkonen 2008). However, the reduction of compulsory set-aside from 15% to 10% in 1996/97 brought about a huge slump with planting levels falling from 2000 to 200 hectares in the space of a year with 10–15 cuttings producers leaving the market (Larsson and Lindegaard 2003).

During the years that followed, the Swedish market shrunk due to the removal of crops. Some plantations had been established on poor land hundreds of kilometers from heating plants and were not economical (Helby et al. 2006). Additionally, the price paid to farmers has reduced due to competition from imported biomass and increased incineration of waste. Since the peak planting year of 1996 (see Fig. 5), there has only been two years (2001 and 2008) when a greater area was planted than removed; currently, Sweden has around 12,000 ha of SRC remaining (see Fig. 6) (Swedish Board of Agriculture, 2015).

The main market for SRC is district heating and CHP. In Skåne in 2010, willow contributed only 0.4% of the fuel used in these plants (Nylander 2014). The opening of the 110 MW CHP plant at Örtofta in 2014 and the reopening of the 55 MW Flintrännan heat plant in 2015 could improve the market situation for SRPs (Henriksson 2014).

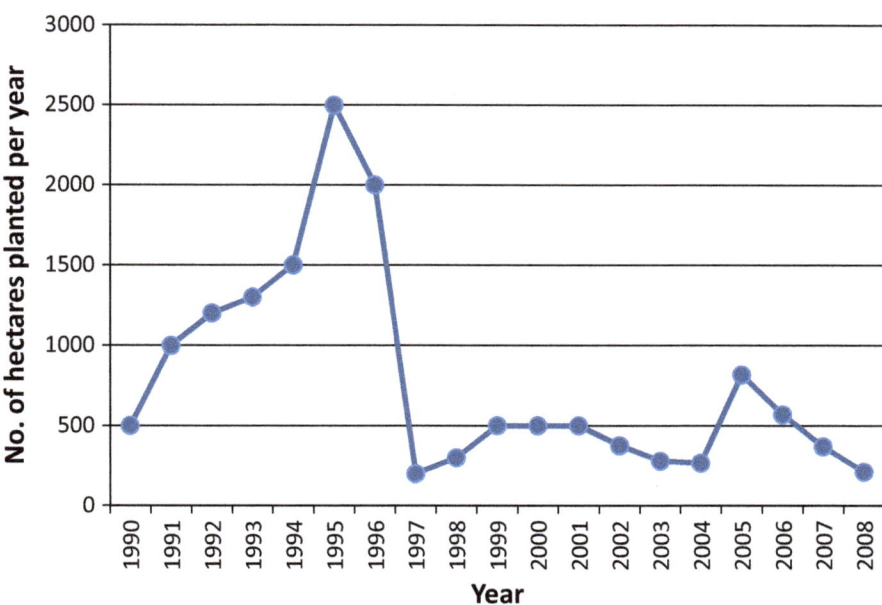

Figure 5. SRC plantings in Sweden 1990–2008 (Larsson 2015).

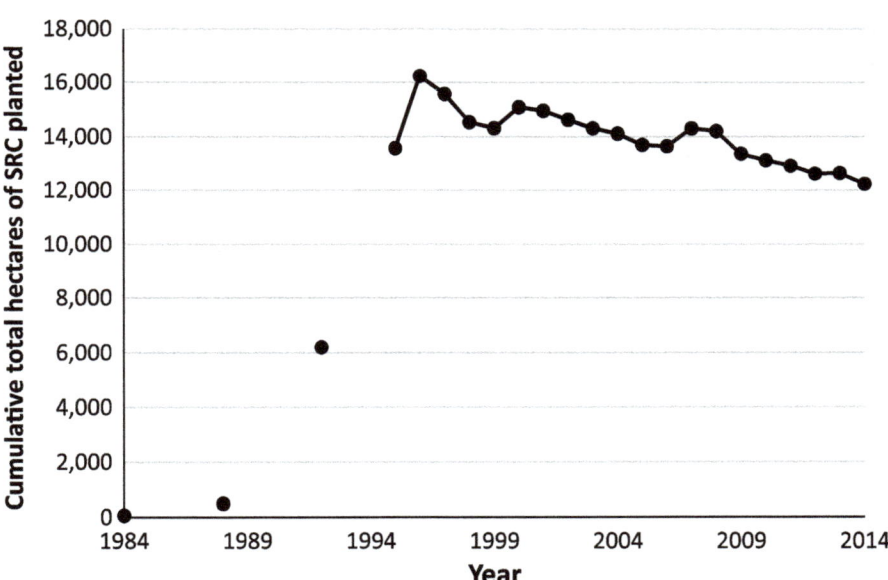

Figure 6. Cumulative SRC area in Sweden 1984–2014 (Sydkraft 1987; Jonsson 1992; Åström and Ramstedt 1993; Swedish Board of Agriculture, 2001, 2003, 2004, 2008, 2011, 2013, 2015; Andersson 2005).

There is currently an establishment grant worth of €615/ha (5800 SEK/ha) for planting SRC willow.

United Kingdom

UK energy crops policy from 1990 to 2015 is assessed in detail by Adams and Lindegaard 2016. Of the four countries making up the UK, only significant amounts of SRC have been planted in England and Northern Ireland. The majority of planting in England is in Yorkshire, East Midlands, and Cumbria. There were two boom and busts experienced in the English SRC industry. The Arbre Energy project was supported by the UK Government's Non Fossil Fuel Obligation (NFFO) and European development funds created a market for around 1500 hectares of SRC willow. The plant was built but never became fully operational and was closed in 2002 (Piterou et al. 2008). Despite the introduction of an establishment grant, farmer confidence was badly affected and planting levels fell from a peak of 422 ha in 2000 to just 65 ha in 2002 (Lindegaard 2013b). The introduction of policy favoring the co-firing of energy crops with coal led to a gradual increase in planting, peaking at 502 ha in 2007 (see Fig. 7), but this again plummeted due to uncertainty because of the cessation of the Energy Crops Scheme for 18 months, the abandonment of set-aside, and the sudden increase in cereal prices at this time.

The main market for most SRC growers has been Drax Power Station but they are pulling out from the contract in 2017. There has been a recent small increase in planting due to the completion of the Iggesund Paperboard CHP plant in Workington Cumbria with 50–100 hectares of willow planted for this project in 2015 (Iggesund, 2016).

The majority of SRC planted in Northern Ireland took place between 2005 and 2007 (see Fig. 8) when a scheme called the SRC Challenge Fund was in place (NIDOE, 2016). Most SRC produced in Northern Ireland is used for small scale heat supply (Farming Futures, 2015). A Renewable Heat Incentive (RHI) covering England, Wales, and Scotland was introduced in 2011 (DECC, 2011). This provided a role for some SRC that was already planted but did not lead to significant new plantings. The Northern Ireland RHI which was introduced in 2014 similarly enabled markets for existing plantings; this scheme closed to new applicants in February 2016 (NI Direct, 2016).

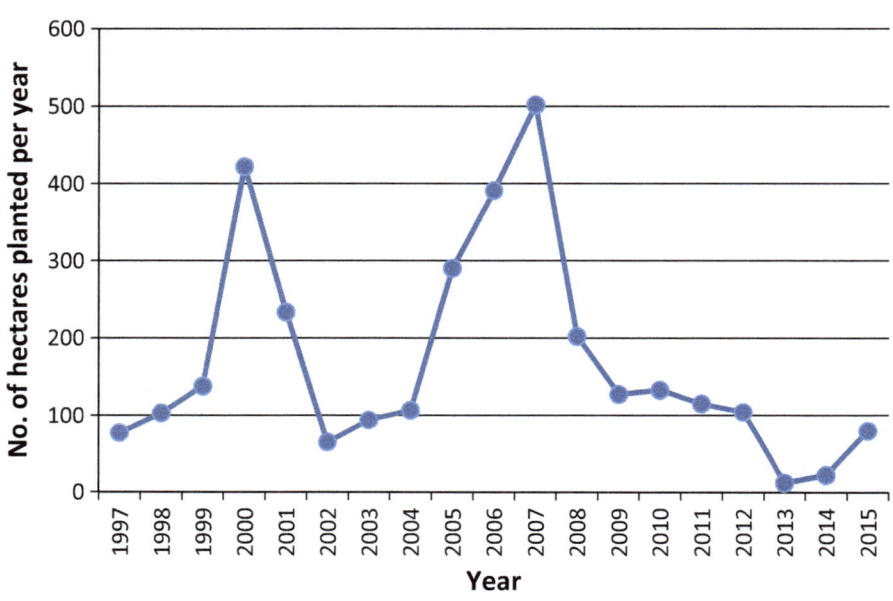

Figure 7. SRC plantings in England 1997–2015.

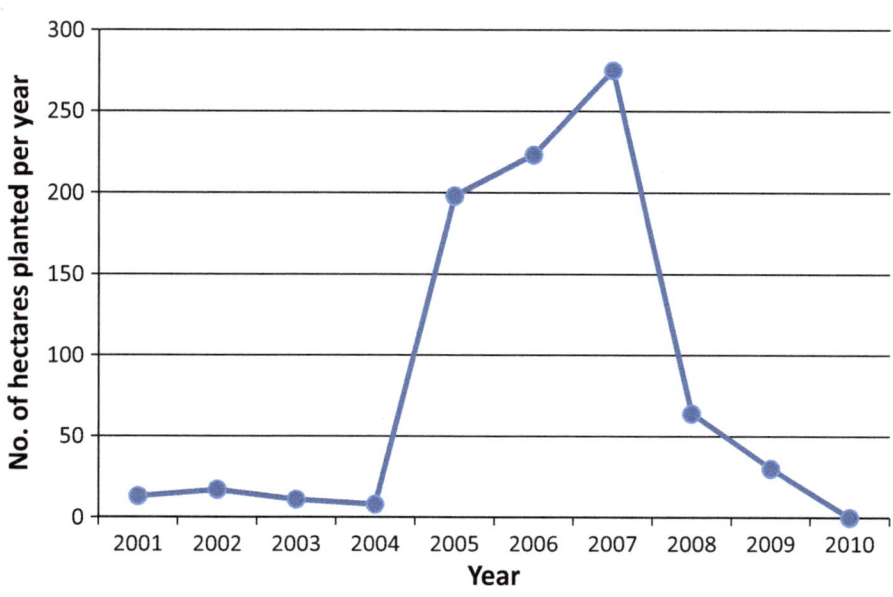

Figure 8. SRC plantings in Northern Ireland 2001–2010.

Comparative summary of EU SRP establishment schemes

A comparative summary of the different SRP establishment schemes from the different EU countries is provided in Table 2.

Current SRP market situation

In most of these western European countries, the current SRP market is static with very little planting taking place. The current areas of growth for SRP are in Eastern Europe. Large plantations have been planted in Lithuania, Latvia, and Ukraine over the last 3–4 years and this looks set to continue (Rokwood, 2015c).

The recent introduction of short rotation coppice (SRC) as an Ecological Focus Area (EFA) option under the current CAP so called "greening" measures was envisaged as possibly having some impact on planting levels (Andersons, 2015 Lindegaard 2013b; Larsson and Henriksson 2015).

Any farm in the EU28 with more than 15 hectares of arable land must have 5% of the land in EFAs. This could rise to 7% after 2017. EFA measures can be thought of as land set aside for environmental benefits. The long list of EFA measures proposed by the European Commission (EC) to member states included SRC. This has been adopted by Germany, Ireland, Poland, and Sweden. In the UK, agriculture is a devolved matter and individual countries make the decisions. SRC has been adopted as an EFA measure in Wales and Northern Ireland, but not in England or Scotland. Neither are likely to have much effect on planting levels although both regions are dominated by grassland (<5% arable).

Each EFA measure is given an EFA weighting that is used to transform the lengths/areas of the EFA measures into equivalent land areas depending on how environmentally friendly they are deemed to be. Where SRC has been included as an EFA measure, it has been assigned one of the lowest weightings (0.3) because it is considered to have a relatively low ecological benefit per unit area. This assessment seems to disagree with numerous studies that suggest that SRC can provide significant biodiversity benefits and ecosystem services (Sage et al. 2006; Rowe et al. 2009, 2011, 2013; Nisbet et al. 2011; Environment Agency, 2015). The low weighting compared to other options (see Tables 3 and 4) discriminates against SRC as it requires as much as 2–7 times the amount of land to be taken up compared to other measures. The weighting of 0.7 awarded to nitrogen-fixing crops would suggest that these crops are deemed to be more environmentally friendly than SRC. Originally, the European Commission had also proposed a weighting

of 0.3 for these crops but this was increased following strong lobbying from the European Parliament (Hart et al. 2016).

Summary Outcomes of the PESTLE and SWOT Analysis

Findings from the PESTLE analysis have been merged into the SWOT analysis which is described in a Rokwood publication (Rokwood, 2014a); hence, only summarized results are presented here. Developing and prioritizing a combined SWOT for all regions was challenging due to the differing characteristics, circumstances, and priorities for each region but a summary of the SWOT outputs by theme is shown in Fig. 9. Table 5 presents the identified strengths, weaknesses, opportunities, and threats.

Strengths and weaknesses

The hypothesis used in this study is that the positive benefits outweigh possible negative impacts of SRP, which is supported by several studies (Rowe et al. 2009; Langeveld et al. 2012; Adams and Lindegaard 2016). Key environmental benefits include increased carbon sequestration, reduced GHG emissions, reduced soil erosion, and groundwater nitrate and surface runoff. Also, SRP can be used in phytoremediation of contaminated land and can lead to an increase in biodiversity. Furthermore, most of the unfavorable impacts can be managed and mitigated. Land use can be restricted to suitable sites such as marginal land, contaminated land, buffer strips, and lower quality agricultural areas (ADAS, 2008; Styles et al. 2016). For hydrology, guidelines on catchment management can be enforced to ensure detrimental effects do not occur to hydrological resources (Rowe et al. 2009).

Fuel security is a key strength of SRP due to local production and reduced imports which highlights some regional benefits. Local supply is good for employment and farm diversification as it allows farmers an alternative income stream from a different enterprise. Despite these strengths, SRP currently offers a poor cash flow for farmers and there is a lack of political will to support SRP which contributes to uncertainty in supply and demand. A lack of incentives has limited uptake of SRP which is also compounded by a lack of skills and infrastructure. SRP requires significant land take and a long-term contractual commitment with the landowner which is recognized as a weakness. A general lack of public awareness of the industry, the supply chain and end-user benefits also featured highly as a weakness.

Table 2. Summary of SRP establishment grants offered in five European countries.

Country	Name of scheme	Type (N = national, R = regional)	Years running	Crops covered	Grant value €/ha	% of total eligible costs	Min. area (ha)	Max. area (ha)	Min. no. of plants/ha	Min. length of plant-ation lifetime	Notes	Source
Germany		R	2008–2013	SRC poplar, willow, Robinia		30					Saxony only	Faasch and Patenaude (2012)
		R	2015–2018	SRC poplar, willow, Robinia	1200	40	3.75	10	3000	12	5 German regions	FNR (2015)
Ireland	Bioenergy Scheme	N	2007–2014	SRC willow, miscanthus	1300	50	3	30	13,300		900 ha planted	DAFM (2015a)
			2015	SRC willow	1040	40	3	50	13,300			
Poland		N	2007–2008	SRC willow	980	50	/	100			4300 PLN/ha. 1300 ha planted 2520 PLN/ha	Szymańska and Chodkowska-Miszczuk (2011)
Sweden	Omställning 90 (Deregulation -90)	N	1991–1996	SRC poplar	575	30	/	100			10,000 SEK/ha. 15,500 ha planted	Blomquist (2006), Mola-Yudego and Pelkonen (2008)
		N	1997–1998	SRC willow, poplar, hybrid aspen	1065		0.5	n.a.			approx 3,000 SEK/ha. 242 ha planted.	Mola-Yudego & Gonza'lez-Olabarria (2010)
		N	1999	SRC willow, poplar, hybrid aspen	535			n.a.			5000 SEK/ha. 358 ha planted.	Mola-Yudego & Gonza'lez-Olabarria (2010)
	Miljö- och landsbygd-sprogrammet 2000–2006 (Env and Rur Dev Progr)	N	2000–2006	Willow, poplar, hybrid aspen	321	40		n.a.			5000 SEK/ha. 1653 ha planted during 2000–2003.	Helby et al. (2006)
	Landsbygdsprogram för Sverige 2007–2013 (Rural Dev Progr)	N	2007–2013	Willow, poplar, hybrid aspen	513	40	1.0 ha willow 0.1 ha poplar, aspen	n.a.			5000 SEK/ha. 1500 ha planted.	Swedish Board of Agriculture (2014).
	Landsbygdsprogram för Sverige 2014–2020 (Rural Dev Program)	N	2014–2020	Willow, poplar, hybrid aspen	615	40	3.45	n.a.	14,500		5800 SEK/ha	Swedish Board of Agriculture (2014)

(Continues)

Table 2. (Continued)

Country	Name of scheme	Type (N = national, R = regional)	Years running	Crops covered	Grant value €/ha	% of total eligible costs	Min. area (ha)	Max. area (ha)	Min. no. of plants/ha	Min. length of plant-ation lifetime	Notes	Source
UK England	Energy Crops Scheme	N	2000–2007	SRC willow & poplar	1285 / 2056	/ /	3	n.a.	n.a.	5	All land types. £1000/ha Ex forage land only. £1600/ha 1815 ha planted.	Lindegaard (2013a)
			2008–2013	SRC (Willow, Poplar, Ash, Alder, Hazel, Silver Birch, Sycamore, Sweet Chestnut and Lime)	/	50	3	n.a.	n.a.	5	Ca. 900 ha planted	
Northern Ireland	SRC Challenge Fund	N	2004-2006	SRC willow	/	n.a.					Competitive scheme. No minimum grant. 696 ha planted.	NIA (2008)

Table 3. Ecological Focus Area (EFA) options and their weightings (DARD 2014; The Scottish Government 2015, Welsh Government 2015).

Measure	Weighting
Fallow land	1.0
Hedges/Wooded strips	2.0
Buffer strips	1.5
Catch crops or green cover	0.3
Nitrogen fixing crops	0.7
Field margins	1.5
Ditches	2.0
Traditional stone walls	1.0
Archaeological features	1.0
Earth banks	1.0
Agroforestry	1.0
Short rotation coppice	0.3
Afforested areas	1.0

A number of factors were found to be specific to each country or region due to variations in market advancement, existing national/local policies, and the local characteristics of the area. For example, the possibility of using SRPs to assist flood mitigation was included as a strength by the UK region, which reflects the high incidents of flooding experienced in the South West of England. The UK, Irish, and Spanish regions noted weaknesses in the lack of harvesting infrastructure and supply chain logistics, while the Swedish, Polish, and German regions identified a lack of profitable specialized machinery for SRP and lack of technological development to address this.

Opportunities and threats

While the results varied across the regions, there was a noted dominance of political and economic issues that could either be viewed as opportunities or threats depending on policymakers' decisions. Common Agricultural Policy (CAP) reform and the role of SRC in Ecological Focus Areas (EFAs), government national policy, and the extent to which SRP is prioritized and supported, and EU/national targets for renewables and emission reductions – all featured highly in this respect, with most being viewed as opportunities. This made clear that with the appropriate political encouragement, backed up by the right economic incentives, the SRP market could be kick-started to ultimately compete on an equal footing with other feedstocks in the sustainable heat market. The wider issue of increasing fossil fuel costs was noted by all regions as a significant opportunity in this respect.

Local heat networks offer a substantial opportunity for SRP as it builds on the strengths of fuel security and regional benefits which benefits rural regeneration. Diversification of agriculture and farming offers multi-functional benefits such as economic potential and environmental enhancement. A key opportunity for SRP is to combine the energy production with other opportunities to improve ecosystems services such as flood mitigation, water treatment, and reduced runoff.

Common threats included a lack of local markets, with the more advanced regions also highlighting the risk of local markets being affected by an increased import of cheap biomass fuel and the low prices attracted

Table 4. Tree species permitted to be grown as SRC and their management practices in the EFA options for six European countries (DAFM, 2015b; Hart 2015).

Species covered	Germany	Ireland	Poland	Sweden	UK (Wales)	UK (Northern Ireland)
Salix spp	✓		✓	✓	✓	
Populus spp	✓	✓	✓		✓	
Alnus spp	✓	✓			✓	
Betula pendula	✓	✓	✓		✓	
Fraxinus excelsior	✓	✓			✓	
Acer spp		✓			✓	
Quercus spp	✓	✓			✓	
Tilia spp		✓			✓	
Castanea sativa		✓			✓	
Corylus spp		✓			✓	
Max harvest cycle (years)	6	30	3	10	20	5
Mineral fertilizers	Not allowed	Not allowed	Allowed with limits	Only the first year	Not allowed	
Plant protection products	Not allowed	Not allowed	Not allowed	Only the first year	No use of plant protection products, except for spot treatment of invasive non-native species within the first 2 years of planting	Allowed until end of year 2

Figure 9. Main outcomes from the SWOT and PESTLE analysis performed for the SRP market.

by the biomass power sector. Regulations around landscape protection and nature conservation were viewed as threats by the German and Swedish regions. Market competition and technical issues were identified as threats in all regions due to alternative renewable energy technologies, competition for land, and problems such as air pollution, combustion efficiency, cheap imports, and limited species optimization for SRP. Awareness of SRP and relevant information in the public domain is a threat to obtaining future support for market development.

Potential negative aspects of SRP cultivation warrant further assessment and consideration before making policy recommendations to support SRP. It is important that SRP is grown in appropriate locations as competition for land and concerns over land use change is a crucial issue (EEA 2006; ADAS, 2008; DECC, 2012), for example, cultivating SRP on permanent unimproved grassland can lead to soil carbon loss (Rowe et al. 2009). Indirect land use change (ILUC) and "food versus fuel" is an ongoing debate which is a limitation restricting the development of SRPs in many countries with limited land availability (Berndes et al. 2011; Ecofys, 2015). Some trees managed as SRPs

are planted as monocultures and exotic species such as Eucalyptus provide few biodiversity benefits. There are also concerns around water availability due to considerable water requirements, landscape change due to visual impacts, and potential for deep roots to affect archaeological remains (Finch et al. 2009).

In summary, the SWOT analysis highlighted the key benefits and risks associated with further support for SRPs. The key barriers are assessed in section Policy recommendations along with policy recommendations to increase the uptake of SRP.

Policy Recommendations

In this section, the results of the regional policy briefs are amalgamated and assessed as there was a high degree of alignment in the issues identified by the countries. Six broad requirements emerged, with each identified by more than one region as an area where appropriate policy change was required. A summary of the problems and potential options toward a solution are summarized in Table 6 with a more detailed description provided in the subsections below.

Table 5. Strengths, weaknesses, opportunities, and threats identified for the SRP market in each region (N.B. these apply to all regions unless otherwise stated).

Strengths	Weaknesses

1. Carbon Reduction
  a. Reduced CO_2 emissions and greenhouse gasses
  b. Implementation of international commitments to reduce emissions e.g. EU 2020
  c. Increasing the role of renewable energy in national and regional energy policies.
  d. Building Regulations – RES requirements and Carbon compliance targets.
2. Fuel Security
  a. Stable energy supply to meet demand
  b. Advances in technical development make SRP a good long-term fuel option
  c. Increase in energy coming from renewable sources which most governments support.
3. Regional Benefits
  a. Provides sustainable rural development
  b. Flexible in scale to fit a particular area
  c. Employment potential for the local area
  d. Suitable climate for growing SRP (Sweden/Ireland/UK in particular).
4. Economic
  a. Stimulates the national economy, particularly the agriculture sector
  b. In the long term will lead to cheaper heating
  c. Some countries have grants/funding from governments to support SRP (see section Short Rotation Plantations (SRP) policy review)
  d. Existing expertise – willow breeding (UK), whole supply chain (Sweden).
5. Biodiversity
  a. Promotes flora and fauna better than traditional mono-crops
  b. Willow in particular supports invertebrate species
  c. Reduces soil erosion.
6. Added Benefits
  a. Using sludge as fertilizer
  b. Multifunctional crop
  c. Water treatment option
  d. Provides a natural windbreak
  e. Acts as a flood defence.
	1. Land
	  a. Limited land availability
	  b. Not all land is suitable for SRP
	  c. Established traditions of land use are difficult to change
	  d. Protection of the landscape is an issue in some areas.
	2. Lack of Political Will
	  a. Certain agriculture rules and regulations may impede the process
	  b. Lack of joined up thinking from policymakers for how SRP can achieve multiple goals
	  c. Lack of subsidies/grants to establish industry (in some cases)
	  d. Focus on other alternative energy sources, e.g., wind, biogas.
	3. Lack of support
	  a. Lack of awareness by most of society
	  b. Skepticism of the technology and opposition to change
	  c. Resistance to change by producers, including peer pressure to maintain existing practices
	  d. Minimal lobbying for change.
	4. Lack of Skills and Infrastructure
	  a. Need to further develop the technical infrastructure, e.g., combustion systems
	  b. Harvesting infrastructure is limited (UK, Ireland and Spain)
	  c. Lack of profitable specialized machinery for SRP (Germany, Poland and Sweden)
	  d. Lack of training for best practice in both agricultural and business
	  e. Lack of working examples demonstrating the possibilities and as a way of knowledge sharing.
	5. SRP Market
	  a. Lack of established market for SRP
	  b. Higher price compared to some other fuels
	  c. Lack of collaboration between relevant stakeholders to develop market
	  d. All of the above creates uncertainty for potential investors.
	6. High Costs
	  a. Establishment – long-term investment is required
	  b. Poor cashflow for farmers, does not provide a good return in the short term
	  c. Transport – potential long distances involved
	  d. Combustion – new area requiring investment
	  e. Grid connections are expensive where they do not exist.
	7. Operational Issues
	  a. Storage and drying of high moisture content woodchips
	  b. No guarantee that heating plants will accept SRP
	  c. Drainage issues of the land while growing
	  d. Transprt logistics need to be developed.

Opportunities	**Threats**

1. Political
  a. Potential to make required legal changes, e.g., making SRP a subsector of forestry
  b. Possibility of being included in beneficial CAP policies
  c. Taxes on fossil fuels could further advance economic advantage
  d. Positive environmental impact such as carbon reduction is good for national/council targets.
	1. Political Barriers
	  a. Agricultural reform may prove negative
	  b. New emissions criteria targeting NOx emissions could pose a problem for nitrogen rich willow
	  c. Tax issues for energy crops
	  d. Conservation laws and regulations (Germany and Sweden)
	  e. Bureaucracy creates complexity
	  f. Isolated local authorities lack leadership
	  g. Inconsistent policy and regular changes leading to uncertainty.

(continues)

Table 5. (Continued)

Opportunities	Threats

2. Regional
  a. Good for rural areas where gas use is low and alternative heating sources are expensive
  b. Potential to reduce fuel poverty (UK)
  c. Use marginal land that is not currently being utilized
  d. Opportunity to engage local government in energy matters
  e. Reduce logistical issues by promoting local use
  f. Create more local jobs.
3. Economic
  a. Potential to provide good value heating in the long term
  b. Possibility of government funding
  c. Increasing price of fossil fuel internationally makes biomass more attractive
  d. International trading possibilities including "high grade" SRP.
4. Promotion of SRP
  a. Need to challenge negative public opinion
  b. Promotion of the use of pellets/wood chips to stimulate demand
  c. Target young farmers and farm sectors most likely to adopt SRPs.
5. Possible Benefits
  a. Sewage water for irrigation
  b. Remediation of brown-field sites
  c. Development of ecosystems.
6. Technical Improvements
  a. Better quality due to improved SRP varieties and harvesting techniques
  b. Possibility of linking to heat networks and CHP.
	2. Technical Issues
	  a. High levels of particulate matter (air pollution) possible in urban areas due to large-scale domestic biomass
	  b. Need for improved air pollution mitigation measures, e.g., filtering technology
	  c. Lack of a plan to change existing power generation (locally and nationally) to biomass
	  d. Tree diseases resulting in a glut of wood fuel.
	3. Economic
	  a. Difficult to compete with sources of waste biomass
	  b. Possible reduction in government funding
	  c. Tenant farmers have insufficient funds to invest
	  d. Competition from other fuels, e.g., gas, coal, oil, kerosene
	  e. Immature market and limited current development
	  f. Subsidies have tended to promote energy generation, not feed-stock supply.
	4. Market
	  a. Competion for land from crops and other uses (land use change)
	  b. No competitive advantage over imported biomass
	  c. Expansion of gas infrastructure
	  d. Competition from other renewable energy sources
	  e. High food prices lead to an unwillingness to use land for energy crops
	  f. In some countries Miscanthus is a more popular energy crop with farmers (UK)
	  g. Changing policy leads to market uncertainty and reduces investor confidence.
	5. Limited R&D
	  a. Future funding of R&D
	  b. Bioenergy funding often more focused on technical solutions overlooking feedstock supply.

Better dissemination of information regarding the benefits of SRP

The Problem

All of the regions identified the need to educate relevant groups about the benefits of SRP, particularly farmers and policymakers. The UK region identified a lack of knowledge regarding the potential positive social and economic impacts of SRP such as reducing fuel poverty by providing cheaper fuel and job creation from a new industry. Spain recognized the potential of SRP to provide cheaper fuel to rural communities with no access to mains gas. The Swedish region also highlighted a lack of awareness of SRP's potential to provide energy security to the nations in which it is grown.

Despite there being multiple environmental benefits of SRP, a coherent body of evidence in one place is currently lacking. This undermines efforts to assess these benefits in a holistic way. The German and Swedish regions state that the positive ecological effects of SRP need a much greater focus. This would include the ability of

SRP to: regulate groundwater levels, clean wastewater, prevent ground erosion, and increase biodiversity (Langeveld et al. 2012). These benefits are significant when compared to alternative energy crops such as maize (Farnworth and Melchett 2015), however, they frequently go unnoticed. This is in part due to a lack of practical examples and evidence (in particular, as a potential source of flood mitigation) of the value of these multifunctional benefits.

The lack of clarity on the environmental benefits of SRP has been pinpointed by the Swedish region as a large contributory factor to the halting of their progression in Sweden, a country which originally led the way in developing the production of SRP. The very low weighting for SRC as an EFA option under the CAP 2015–2020 means that if farmers choose to grow SRC, the proportion of land they must turn over to EFAs increases (Hart 2015). This sends the signal that SRC is less environmentally friendly than other crops, such as peas and beans which have a higher weighting. This has played a part in the slowing of development of

Table 6. Summary of the policy recommendations, problems identified and potential options toward a solution.

The Problem	Toward a Solution
Better dissemination of information regarding the benefits of SRP	
Need to educate relevant groups about the benefits of SRP	Provide courses, disseminate information via literature and workshops/events.
A coherent body of evidence of benefits of SRP in one place is currently lacking.	Increase weighting factor of SRC to 1.0 in Ecological Focus Areas (EFA).
	Conduct a full evidence-based review of SRP including a cost benefit analysis.
	Further research into the multifunctional benefits of SRP to society.
Increased financial support to foster the SRP market	
Need for greater financial support to grow the nascent SRP market.	Additional funding from regional or national government to kick-start industry.
	Regional establishment grants, interim payments during the establishment period, interest-free loans and subsidy payments.
Developing the supply chain at the local level	
Lack of local supply chains is a barrier to the uptake of SRP. This leaves growers isolated and lacking adequate infrastructure.	Provide subsidies in areas where the SRP market is able to grow. Grants for crucial infrastructure could be made available.
Often supply is not linked to end-user demand leading to imbalances.	Establish pilot projects that connect growers to end users. Create a strong demand for biomass through taxes and perception.
Improved clarity regarding SRP funding and land use	
Broadening definitions to include SRP in environmental stewardship, biomass, forestry, and agricultural support schemes.	Improve legislation so that SRP can be incorporated into land sector support schemes to increase competitiveness.
Issues over the suitability of different land use for SRP, e.g., forestry or agricultural.	Improved classifications and clarifications of land use so that farmers can make informed decisions about SRP.
More research and development in SRP leading to better resources	
Continued R&D on specific aspects of SRP cultivation to increase commercial viability.	Appropriate funding for research programs in EU countries.
	Increase pilot projects and field trials and work closely with policy-makers, industry, and researchers to maximize value of R&D.
Formation of a policy development group	
Lack of lobby groups supporting SRP and limited policy development has hindered development.	Formation of lobby groups to improve the way that Government deals with energy crops policy.
More political support is required as policy-making often falls between different Government departments.	Potentially push for an interdepartmental body for energy policy to ensure that different Department's objectives are aligned.

SRC in Sweden and could have similar implications in other EU countries (Hart et al. 2016). For instance, in Germany, the area of SRC as an EFA measure in 2015 was just 2200 ha out of a total area of 1,367,400 ha. In contrast, nitrogen-fixing crops covered an area of 161,800 ha and catch crops and cover crops covered an area of 930,200 ha.

The Spanish region states that Spain suffers from a lack of knowledge regarding heat production in general and of SRP in particular (Parra-López et al. 2015). This is at multiple levels, from farmers all the way to the general public. Similarly, the Irish region identified a lack of understanding of the sector by prospective growers.

Results from the Polish workshop suggest the lack of knowledge on SRP is also widespread; even those in charge of making energy and heat production policies are largely unaware of its existence. The lack of locally available infrastructure has resulted in some producers trying to adapt machinery to the biomass process or to build their own systems, resulting in an inefficient and suboptimal production process. This is not a conducive environment to promoting the benefits of biomass production.

Toward a solution

The methods suggested for better targeting of information to relevant stakeholders varied, with various regions suggesting different strategies. The proposed solution in the UK and Germany is to conduct a full evidence-based review of SRP including a thorough cost benefit analysis. An up-to-date evidence base would facilitate a process of clarifying opportunities which in turn could be used to raise awareness of the range of additional benefits around biodiversity and ecosystem services that SRP can provide (Lindegaard 2013a,b) and potentially lead to greater funding through environmental schemes. The Swedish region found that more research on SRPs in general is required, and in particular, on the potential economic value of SRP multifunctionality to society. Both researchers and industry should push for funding for well-documented research and experimental field trials to show the benefits to industry stakeholders and policymakers.

If firm conclusions are drawn from this research about the strong environmental benefits of SRP, then it is important that a review is made of the weighting given to SRC in the EFA guidance. Increasing the weighting factor of SRC to 1.0 would send an important signal to farmers that it is an environmentally beneficial crop that is worth cultivating. This might encourage countries like England, who have not adopted SRC as an EFA option to reconsider.

Spain suggested a wider range of options to tackle this problem through the provision of information to all stakeholders (politicians, energy generators, farmers, and the general population) through courses, networking, campaigns, and technical visits. This would be done at all policy levels, from national to local. As with Germany, the hope is that more informed policymakers would increase the levels of funding available. Ireland recognizes the important need of countering not only the lack of knowledge in some cases but also some of the popular misconceptions regarding SRP. They advocate a program of training and education but also focusing on farmers involving the development of factsheets, workshops, and seminars. This would cover the whole wood fuel production process including: which varieties are suitable to the local climate and soil conditions, the management techniques required, harvesting and drying processes, and applications for end use.

Poland suggested that the best way to increase knowledge of the SRP process was to provide working examples of the stages in the biomass production chain. For instance, EU or national funding could be sourced in order to build biomass heat production facilities in local communities which would serve as a prototype for those hoping to invest in similar facilities. It is also suggested that building contacts with institutes dealing in the biomass market abroad might help to disseminate the necessary knowledge.

Increased financial support to foster the SRP market

The Problem

The regions identified a need for greater financial support as an issue that needs to be addressed in order to grow the SRP market. The UK region recognizes the need to lower the investment risk for SRP growers. This is largely due to the view that SRP is a high-risk, long-term commitment that most farmers are unwilling to undertake. Sweden also recognizes this issue; it cites the example of *Salix* grown as SRC which needs an investment time of

20–25 years before it turns a profit, with the first 8–10 years likely to result in negative cash flow. Financial support may be necessary in order to support farmers in taking the economic risk of turning land over to SRP production. The German region also highlighted the lack of funding for SRP growers; this is particularly significant as currently funding is available for other, more traditional crops making them more attractive than SRP. Ireland, similarly, noted a funding discrepancy, as relatively generous grants are available for forestry but this is not currently defined as including SRP. The result is similar to Germany whereby farmers are unlikely to grow SRP without a subsidy when funding is available for other crops. Spain does not specifically focus on financial incentives as the subject of any of their policies. However, they do include additional funding as a necessary part of the solution in broad terms regarding all of their other policies.

In Poland, extra funding was also observed to be necessary, in this case not to compete with other crops but with cheaper fossil fuel energy production. In Poland, the energy market is dominated by cheap coal and currently the same amount of energy can be produced from a much smaller tonnage of coal at a substantially lower price. Thus, if biomass is to get a foot in the door, then financial incentives are critical.

Toward a solution

All of the regions are in broad agreement that additional funding should come from regional or national government in addition to other funding sources. The UK recommends that government at both the national and local levels work together to improve finance, with the local element creating opportunities for targeting of schemes in appropriate areas that create the most benefit. The complete package of support could include a variety of attractive options such as crop regional establishment grants, interim payments during the establishment period, interest-free loans, and subsidy payments. The latter two options would be in the event of the SRP planting aiding in flood defence or water treatment. Germany envisages the motivation coming primarily from local government, through engagement with national farmers' unions to make use of existing CAP funding infrastructure, specifically, allowances for greening (Hart 2015). They argue that by increasing the EFA weighting factor for SRP, the level of funding available would increase and make these crops a more attractive financial proposition.

Ireland proposes two funding options, one possibility would be to make use of the current grant that supports forestry by extending it to include SRP thereby allowing

SRP to compete financially for land use. The Irish part-ners also called for a renewable heat incentive (RHI) which would lead to a market pull for energy crops (as a result of the country's low indigenous woodland cover). It is hoped that such a measure could help Ireland in achieving its 2016 RES-H targets; a target which if not achieved may have large financial implications (TEAGASC, 2014).

In Poland, there is precedence of EU funding being used to realize the development of community projects. At present, however, funding is prioritized toward pro-jects focused on other renewable energy sources, such as wind and solar. The biomass sector must therefore present a strong case for accessing this support mechanism.

Developing the supply chain at the local level

The Problem

Three of the regions, the UK, Germany, and Spain identi-fied the lack of local supply chains as a barrier to the uptake of SRP. Poland and Sweden highlighted, in par-ticular, that a lack of retail market for SRP was likely to hinder its success. They all saw the development of such supply chains via their policies to be a potential opportunity.

There are only two instances of established supply chains in the UK, serving the Iggesund Paperboard CHP plant and Drax Power Station (Iggesund, 2016, REGRO, 2016). Outside these regions, the supply chain is struggling to develop as growers are relatively isolated and infrastructure and specialized machinery required is not sufficient for economies of scale. The investment risk means there is still little incentive to invest in the necessary infrastructure. The workshop identified a pressing need for locally avail-able infrastructure, so that smaller growers can join the SRP market and produce the properly prepared wood fuel which will allow access to the lucrative domestic energy market.

The German workshop recognized the importance of developing the supply chain as a possible means of regional development, citing that some parts of the country such as the Saxony-Anhalt region are quite economically deprived.

The Spanish region identified a barrier to SRP – over-capacity in installed power due to the large growth over the last decade of wind farms and combined cycle gas turbine (CCGT) plants (STORE, 2013). This is likely to mean a very competitive electricity market and low prices paid to SRP growers. By focusing on local district heating systems, this barrier to SRP uptake could be overcome.

They also see local biomass networks as a way to increase energy security and independence.

In Sweden, there is quite a well-developed production and supply chain. However, the lack of the final link in the chain, a home market to support their biomass pro-duction, is threatening to cripple the industry. Poland has a similar and serious lack of a retail market for any biomass produced as the country's infrastructure is set up to use coal, from large coal-fired power stations to the coal-burnings stoves in individual houses.

Toward a solution

The broad solution from each of the regions is that sub-sidies are required to stabilize the market and that this should be focused in specific areas where the SRP market is able to grow. The solution to this issue proposed by the UK is to establish regional pilot projects that connect growers to end users. This would involve setting up one or more biomass heating, CHP, or district heating systems. The Rokwood case studies showcased other projects such as the Beuchte Bioenergy Village in Germany (a district heating scheme linking 65 households) that could be rep-licated elsewhere (Rokwood, 2015c). It is suggested that the required funding could come from the Rural Development Program, with targeted loans and leasing arrangements to help growers establish SRP (Adams and Lindegaard 2016).

In order to facilitate the move to biomass district heat-ing, the Spanish region propose that grants and subsidies should be more favorable to locally sourced biomass including SRPs because of their lower lifecycle GHG emis-sions compared to imported biomass (Adams et al. 2013).

The expansion of SRP is already happening in specific parts of Germany such as the Achental region where the Biomassehof project is connecting producers and consum-ers of biomass locally (SRCPlus, 2014; Rokwood, 2015c). It is plausible that this could extend to other parts of the Saxony-Anhalt region, and that the creation of local supply chains could be an opportunity to increase the economic prosperity of struggling regions. As such, the German policy proposal advocates extending current net-works, especially in rural areas with economic difficulties (Rokwood, 2015d).

There is a belief from Polish stakeholders that focusing on creating a strong demand for biomass will help the rest of the chain to fall into place. Legal and financial measures will be needed in order to create this buyers' market, and this would entail external funding from the state or the EU. The current 2014–2020 Development Strategy (which aims to build the capacity in each com-munity for renewable energy production) could offer an opportunity to build on this aim (Rakowski 2015).

Improved clarity regarding SRP funding and land use

The problem

Three of the regions recognized opportunities for improved clarity that would support the expansion of SRP. This ambiguity exists in different areas; for Ireland, it affects funding, in Spain and Germany, there are land use issues.

Ireland notes there is a need to examine the specific definitions of biomass currently used in the country. The region advocates the possibility of broadening these definitions in order to bring additional funding streams to different varieties of SRP. This would mean including other crops in the existing bioenergy schemes as opposed to just willow and previously Miscanthus. Additionally, there is also a pressing need to identify the lifetime gap between grants which support forestry and those which support SRP to determine new potential support structure.

In Spain, there are issues with regard to which areas of land are suitable for SRP: agricultural or forest. This confusion partly arises due to the different SRP species which can be planted. There are similar land use uncertainties in Germany. The legal status of the land used for SRP production has caused a great deal of confusion as it was unclear whether farmland with SRP remained farmland or became woodland after a certain period of cultivation. An amendment to the German Federal Forests Act in 2011 deemed that SRC plantations are not considered to be forests as long as the harvest is performed within 20 years (BWaldG, 2011). This should have enabled German farmers to claim farming subsidies but has not happened in practice. An additional problem is unused land which SRP could help bring back into use but which is currently being underexploited.

Toward a solution

The regions are in broad agreement that policymakers need to do more to foster SRP through improved classifications and clarifications of land use so that farmers can make informed decisions about SRP.

In Ireland, this would be the responsibility of the relevant government departments and state agencies to re-examine the current Bioenergy Scheme. They would need to look at the support offered to forestry growers and develop a plan to address the disparity between this and what is offered to SRP growers. Ireland envisages that the policy development group (discussed in section Formation of a policy development group) would liaise with the relevant government departments to re-evaluate the definition of energy crops in a positive way.

Spain, very clearly advocates the need for new national legislation as a way to address this issue. This is required to redress the financial disparity between SRP and forestry in terms of financial support as currently the energy crops industry cannot compete with forestry. By legislating for increased subsidies for SRP, interest and participation in growing energy crops is expected to increase which should help Spain reach their RES-H target. The benefit of re-classifying energy crops and the widening of the Bioenergy Scheme in Ireland, would allow for the development and promotion of native species and royalty-free species into the market. Spain is also looking for clearer criteria on land use with regard to agricultural and forest land and specific parameter on SRPs in CAP regulation.

Germany proposes a different solution; in recognition of the large amount of fallow, unused land available, they envisage local authorities promoting SRP as a way to bring this land back into use. The potential plots owned by local authorities could be made available to farmers to plant SRPs. In a similar vein to Ireland, Germany sees a role for their national farmers union to advocate for SRP, though current organizations are seen as sufficient without the need for a new lobby group as suggested by Ireland.

More research and development in SRP leading to better resources

The problem

Three of the regions, Spain, Sweden, and to a lesser extent the UK, note the importance of continued research on SRP as part of developing better resources. The Spanish region identified the need for research to identify SRP species and varieties more adapted to the warmer climate as well as suitable farming methods and technologies.

The UK recognizes that in addition to consolidating the current body of evidence into a consistent format to share more widely (as discussed in section Better dissemination of information regarding the benefits of SRP), further research also needs to be undertaken to ensure benefits and drawbacks are understood in terms of a holistic whole. There is also support for this in the Swedish region, who suggest that more research is needed into the multifunctional benefits and ecosystem services provided by SRP. Without furthering the knowledge base, incentives for SRP production are not being implemented and opportunities are therefore not being realized. Through consolidating existing evidence, research gaps would be identified which in turn would help prioritize future actions and priorities for SRP.

Toward a solution

The issue of increased research will necessarily work in tandem with other policy goals, particularly the key point of wider dissemination of information outlined in section Better dissemination of information regarding the benefits of SRP. Achieving this goal will rely on necessary funding for research programs and working closely with relevant institutions.

Spain recommends a range of policies to help solve this issue. One suggestion is closer working with universities and research institutions to develop new species and machinery to ensure better yields. Another component of this is to conduct further research into methods that use less fertilizer and weed control agents as well as further work examining more efficient use of water and wastewater. In a similar vein to the UK, Spain also suggests starting pilot projects to work on improving the adaptability of SRPs to marginal lands. The Swedish partners are in broad agreement that more field trials are required to demonstrate the environmental benefits of SRP in practice. Such field trials could inform best practice protocols for planting SRPs for environmental benefits and reduce the risk of poorly sited crops that have detrimental effects.

The UK states that identifying and addressing research gaps that increase the overall knowledge and understanding of SRP will in turn trigger constructive and coordinated policy initiatives and incentives. This could work in tandem with the pilot-based schemes advocated by the UK (see section Developing the supply chain at the local level) to demonstrate possibilities. An example project could involve the use of SRP in flood mitigation in addition to energy supply.

Formation of a policy development group

The problem

The Irish and Polish regions both identified that the lack of lobby groups supporting an increased uptake of SRP was hindering its progress. The Irish region identified the lack of policy development for SRP in Ireland as an issue. They attribute this to the lack of lobbyists in this sector compared to other players in the energy market (oil, gas, electricity, and wind). The need for a cohesive group representing the energy crop industry has been identified by stakeholders.

This view was also presented by Poland, who believed that that marginalization of biomass in legislation and in business decisions was due to a lack of political backing. The coal industry is well established, with all the country's infrastructure set up to serve it and massive lobbying power behind it. Political backing and lobbying will be necessary if biomass is to be recognized in laws of regional and national importance. The Polish region is clear that until local and national authorities begin to be lobbied to include biomass in their development strategies, the achievement of more specific goals will not be possible.

Toward a solution

The Irish group believes that support must be provided to help grow a fledgling group who would start to lobby for improvements in the way SRP and energy crops are dealt with by government. An initial goal of this group might be to push for an interdepartmental body for energy policy. This is because there is a lack of continuity in governmental responsibility for biomass with multiple departments having responsibility for different aspects. The Polish region also suggests formation of political support groups to lobby local and national authorities and to make a stand against the large lobbying power of the coal industry in the country.

Concluding Remarks

The issues that must be tackled in order to ensure the creation of a successful path to market for SRP are numerous and complex, but not insurmountable. However, there is a lack of awareness of the multifaceted benefits of SRP at the level of both farmers and policymakers and thus a coordinated top-down and bottom-up approach is needed in order to promote the widespread uptake of SRP. The strategic planting of perennial biomass crops in arable farmland to increase landscape heterogeneity and enhance ecosystem function is recommended to strike a balance between energy and food security (Haughton et al. 2015).

At present, there is a lack of knowledge of SRP as both a feasible crop choice for farmers and as an energy source for heat producers. This is due both to a lack of dissemination of knowledge and the absence of recognition of SRP in governmental policy. SRP is unlikely to be looked on favorably by farmers and producers unless it is afforded the same benefits, subsidies and support that other crops and fuel sources receive from the government. Change in policy is crucial and the support of bureaucrats at the district, regional, and national level is vital for this to be effected. Researchers will also play a part in ensuring there is clear and concrete evidence in the field of the environmental and socioeconomic benefits of SRP. The simultaneous dissemination of this knowledge upwards to policymakers and downwards to producers and farmers is critical in the success of SRP. The formation and development of groups to lobby for the uptake and support of SRP and bioenergy is also of great importance.

The summation of the policy briefs of the six regions of the Rokwood project highlights that there are common obstacles to the wider uptake of SRP across each of the different countries. It also highlights, however, that there are some issues that are unique to one region that are the result of specific circumstances within that country, for example, a structure of governance or the characteristics of existing energy markets. There are similarities in the solutions offered by the regions, but again the variations between them highlight the importance of specific policy changes which are locally relevant.

Acknowledgements

The work reported here is based on research performed as part of the EU project Rokwood and the Supergen Bioenergy Hub. Rokwood an ambitious 3-year study involving partners from six European countries (Germany, Ireland, Poland, Spain, Sweden, and UK). It exists to encourage interaction between biomass research, industry, policy, and business in order to fulfill the potential of woody energy crops like SRC. It is funded by the European Commission's Seventh Framework Programme under call FP7-REGIONS-2012-2013-1 "Regions of Knowledge" and involves 20 partner organizations. The Rokwood policy briefs for the six partner countries can be found at: www.rokwood.eu/public-library/policy-briefs.html

Bioenergy research at the University of Bath is supported by the EPSRC Supergen Bioenergy Hub [Grant Ref: EP/J017302/1] (http://www.supergen-bioenergy.net/) and BBSRC BSBEC Sustainable Bioenergy Centre [Grant Ref: BB/G01616X/1].

These are large interdisciplinary program and the views expressed in this paper are those of the authors alone, and do not necessarily reflect the views of the collaborators or the policies of the funding bodies.

We thank Tobias Markensten of the Swedish Board of Agriculture and Professor Håkan Rosenqvist who assisted in finding historical planting records in Sweden.

Conflict of Interest

None declared.

References

Adams, P. W. R. 2011. An Assessment of UK Bioenergy Production, Resource Availability, Biomass Gasification, and Life Cycle Environmental Impacts, PhD Thesis. University of Bath, Bath. Available at http://opus.bath.ac.uk/27930/ (accessed 5 January 2016).

Adams, P. W. R., and K. Lindegaard. 2016. A critical appraisal of the effectiveness of UK perennial energy crops policy since 1990. Renew. Sustain. Energy Rev. 55:188–202.

Adams, P. W., G. P. Hammond, M. C. McManus, and W. G. Mezzullo. 2011. Barriers to and drivers for UK bioenergy development. Renew. Sustain. Energy Rev. 15:1217–1227.

Adams, P., A. Gilbert, J. Hammond, D. Howard, N. McNamara, P. Thornley, et al. 2013. Understanding the greenhouse gas balances of bioenergy systems. Supergen Bioenergy Hub, Tyndall Centre for Climate Change Research, University of Manchester, UK. Available at http://epsassets.manchester.ac.uk/medialand/supergen/Publications/GHG_balances.pdf (accessed 3 June 2016).

ADAS. 2008. Addressing the land use issues for non-food crops, in response to increasing fuel and energy generation opportunities. NNFCC project 08-004 funded by DEFRA. ADAS, Hereford.

Aebiom, A.. 2015. Aebiom statistical report 2014. Available at www.aebiom.org/blog/aebiom-statisticalreport-2014-2/ (accessed 13 January 2016).

Andersons. 2015. Greening factsheet. The Andersons Centre, Melton Mowbray, Leicestershire, U.K.

Andersson, F. C. A.. 2005. The Swedish 1990 Agricultural Reform - Adjustments of the Use of Land. Paper prepared for presentation at the XIth International Congress of the European Association of Agricultural Economists - EAAE. Copenhagen. 2005. Swedish Institute for Food and Agricultural Economics. Lund.

Anon. 1980. The outlook for energy forestry in France and in the European Economic Community. Pp. 172–180 In W. Palz, P. Chartier, D. O. Hall, eds. Energy from Biomass. 1st EC Conference Proceedings. Applied Science Publishers, Brighton, U.K. ISBN 0-85334-970-3

Åström, B., M. Ramstedt. 1993. Willow-a new crop with new disease problems. Swedish University of Agricultural Sciences 1993 (Salix-en ny gröda med nya sjukdomsproblem. Växtskyddsnotiser. Sveriges Lantbruksuniversitet.) http://www.vaxteko.nu/html/sll/slu/vaxtskyddsnotiser/VSN93-1/VSN93-1D.HTM

Bergendorff, C., and U. Emanuelsson. 1996. History and traces of coppicing and pollarding in Scania, South Sweden. Pp. 235–304 in H. Slotte, H. Göransson, eds. Lövtäkt ochstubbskottsbruk II, Kungl. Skogs-och lantbruksakademien. Stockholm.

Berndes, G., N. Bird, and A. Cowie. 2011. Bioenergy, land use change and climate change mitigation. Background Technical Report. IEA Bioenergy: ExCo:2011:04

Blomquist, A. 2006. Uppföljning av plantering på nedlagd åkermark i Skåne 1991-1996. Follow-up of forest plantation on former agriculture land in southernmost Sweden 1991-1996. Examensarbete nr 76. Institutionen för sydsvensk skogsvetenskap. Sveriges Lantbruksuniversitet. Alnarp.

BMELV. 2012. Poplars and Willows in Germany: Report of the National Poplar Commission. Available at http://www.bmel.de/SharedDocs/Downloads/EN/Publications/PoplarsReport.pdf?__blob=publicationFile (accessed 18 January 2016).

Bord na Móna. 2015. Suppliers/Biomass Growers. Available at http://www.bordnamona.ie/our-company/our-businesses/feedstock/biomass-growers/ (accessed 28 January 2016).

BWaldG. 2011. Bundesministerium der Justiz und für Verbraucherschutz, Gesetz zur Erhaltung des Waldes und zur Förderung der Forstwirtschaft, amendment 2011. Available at http://www.gesetze-im-internet.de/bwaldg/ (accessed 1 June 2016).

Carrasco, J., and H. Sixto (2007). Short Rotation Forestry, Short Rotation Coppice and perennial grasses in the European Union: Agro-environmental aspects, present use and perspectives. in J. F. Dallemand, J. E. Petersen, A. Karp, eds. Harpenden, U.K. Available at http://publications.jrc.ec.europa.eu/repository/bitstream/JRC47547/eur%2023569%20proceedings%20srf-src%20final.pdf (accessed 29 March 2016).

Caslin, B., J. Finnan, and A. McCracken. 2012. Short Rotation Coppice Willow – Best Practice Guidelines. Teagasc, Crops Research Centre, Oak Park, Carlow, Ireland. 72 p. Available at http://www.teagasc.ie/forestry/docs/grants/WillowBestPracticeManual_2012.pdf (accessed 12 January 2016).

Chitadze, N.. 2012. The role of the OPEC in the international energy market. J. Soc. Sci. 1:5–12.

CPID. 2015a. PESTLE analysis. Chartered Institute of Personnel and Development (CPID), London.

CPID. 2015b. SWOT analysis. Chartered Institute of Personnel and Development (CPID), London.

DAFM. 2015a. Bioenergy scheme. Department of Agriculture, Food and the Marine (DAFM). Available at http://www.agriculture.gov.ie/bioenergyscheme/ (accessed 19 January 2016).

DAFM. 2015b. A guide to greening 2015. Department of Agriculture, Food and the Marine (DAFM). Available at http://www.agriculture.gov.ie/media/migration/farmingschemesandpayments/basicpaymentscheme/greeningdocuments/Greeningmanual200215.pdf (accessed 29 February 2016).

Danfors, B., S. Ledin, and H. Rosenqvist. 1997. Short Rotation Willow Coppice – Growers Manual. Swedish Institute of Agricultural Engineering. JTI-informerar No. 1. 40 p.

DARD. 2014. Greening: ecological focus areas. Department of Agriculture and Rural Development (DARD), Belfast, U.K.

Dawson, M.. 1992. Some aspects of the development of short-rotation coppice willow for biomass in Northern Ireland. Proc. R. Soc. Edinb. Biol. 98:193–205.

DBFZ. 2015. Schnellwachsende Baumarten in Deutschland und deren Einsatz zur Wärmebereitstellung. Deutsches Biomasseforschungszentrum gemeinnützige (DBFZ), Leipzig.

DCENR. 2014. Draft bioenergy plan. Department of Communications, Energy and Natural Resources (DCENR), Dublin.

DECC, 2011. Renewable Heat Incentive. Department for Energy & Climate Change (DECC), Available at https://www.gov.uk/government/uploads/system/uploads/attachment_data/file/48041/1387-renewable-heat-incentive.pdf (accessed 5 January 2016).

DECC. 2012. UK bioenergy strategy. Department of Energy and Climate Change (DECC), HMSO, London.

DEFRA. 2007. UK biomass strategy. Department of Environment, Food & Rural Affairs (DEFRA), HMSO, London.

Durán Zuazo, V. H., J. A. Jiménez, F. Perea, C. R. Rodríguez, and J. R. Francia. 2013. Biomass yield potential of paulownia trees in a semi-arid Mediterranean environment (S Spain). Int. J. Renew. Energy Res. 3:789–793.

Ecofys. 2015. The land use change impact of biofuels consumed in the EU: Quantification of area and greenhouse gas impacts. Ecofys, IIASA and E4tech, EC Project number: BIENL13120. Available at https://ec.europa.eu/energy/sites/ener/files/documents/Final%20Report_GLOBIOM_publication.pdf (accessed 1 June 2016).

ENCE. 2016. Ence is the market leader in biomass-fuelled renewable energy. Available at http://www.ence.es/index.php/en/energy.html (accessed 24 February 2016).

von Engelbrechten, H. G. (2015) Questionnaire on Poplars and Willows 2012-2015. Completed for the 25th Session of the International Poplar Commission (IPC) on request of the Food and Agriculture Organization of the United Nations.

Environment Agency. 2015. Energy crops and floodplain flows. Report - SC060092/R2. Available at https://www.gov.uk/government/uploads/system/uploads/attachment_data/file/480799/Energy_crops_and_floodplain_flows_report.pdf (accessed 24 February 2016).

ETSU. 1999. New and renewable energy: prospects in the UK for the 21st century. Report for the DTI. Energy Technologies Support Unit (ETSU), London.

EurObserv'ER. 2015. Solid biomass barometer 2015. Available at http://www.eurobserv-er.org/solid-biomass-barometer-2015/ (accessed 13 January 2016).

European Commission. 1988. The future of rural society. Luxembourg: Office for Official Publications of the European Communities. ISBN: 92-825-9073-9. Available from: http://ec.europa.eu/agriculture/cap-history/crisis-years-1980s/com88-501_en.pdf (accessed 13 January 2016).

European Commission. 2009. Directive 2009/28/EC of The European Parliament and of The Council of 23 April 2009 on the promotion of the use of energy from renewable sources. Official Journal of the European Union 2009, 16-62, Brussels: European Commission.

European Environment Agency (EEA). 2006. How much bioenergy can Europe produce without harming the environment? European Environment Agency (EEA) Report No 7/2006. Copenhagen: EEA.

Eurostat. 2015. Forest area and ownership 2010 and 2015. Available at http://ec.europa.eu/eurostat/statistics-explained/images/8/80/T1_Forest_area_and_ownership%2C_2010_and_2015.png (accessed 29 February 2016).

Faasch, R. J., and G. Patenaude. 2012. The economics of short rotation coppice in Germany. Biomass Bioenergy 45:27–40.

FAO. 2008. Poplars, willows and people's wellbeing. Synthesis of Country Progress Reports, prepared for 23rd Session of the International Poplar Commission, Beijing, China, 27–30 October 2008.

Farming Futures. 2015. Effective approach to drying willow stems. Available at http://www.farmingfutures.org.uk/blog/effective-approach-drying-willow-stems (accessed 22 February 2016).

Farnworth, G., and P. Melchett. 2015. Runaway maize: Subsidised soil destruction. Soil Association. Available at https://www.soilassociation.org/media/4671/runaway-maize-June-2015.pdf (accessed 1 June 2016).

Finch, J. W., A. Karp, D. P. M. McCabe, S. Nixon, A. B Riche, and A. P. Whitmore. 2009. Miscanthus, short-rotation coppice and the historic environment. English Heritage. Available at https://core.ac.uk/download/files/79/60358.pdf (accessed 1 June 2016).

FNR. 2015. Mögliche KUP-Förderung nach GAK ab 2015. Die Fachagentur Nachwachsende Rohstoffe e. V. (FNR). Available at http://energiepflanzen.fnr.de/energiepflanzen/mehrjaehrige-energiepflanzen/energieholz/kup-foerderung/ (accessed 29 January 2016).

Gajewski, R. 2015. Bioenergy as the key to economic growth of the regions-EO Based Service Supporting Energy Crops Cultivation (SERENE). European Space Agency. Available at https://www.nauka.gov.pl/g2/oryginal/2015_03/be9a18f168a4a7aa3e62eb6b2ab43123.pdf (accessed 22 February 2016).

Hart, K. 2015. Green direct payments: implementation choices of nine Member States and their environmental implications. Institute for European Environmental Policy (IEEP), London.

Hart, K., A. Buckwell, and D. Baldock 2016. Learning the lessons of the Greening of the CAP. LUPG The UK Statutory Conservation, Countryside and Environment Agencies In collaboration with The European Nature Conservation Agencies Network (ENCA-net). Available at http://www.ieep.eu/assets/2028/Learning_the_lessons_from_CAP_greening_-_April_2016_-_final.pdf (accessed 26 May 2016).

Haughton, A. J., D. A. Bohan, S. J. Clark, M. D. Mallott, V. Mallott, R. Sage, et al. 2015. Dedicated biomass crops can enhance biodiversity in the arable landscape. GCB Bioenergy. doi: 10.1111/gcbb.12312.

Heinzel, C., and T. Winkler. 2010. Tradable Green Certificates as a Policy Instrument? A Discussion on the Case of Poland. Environmental Economics Research Hub Research Reports, Crawford School of Economics and Government, ISSN 1835-9728. Available at http://ageconsearch.umn.edu/bitstream/95068/2/Tradable%20Green%20Certificates%20as%20a%20Policy%20Instrument%20A%20Discussion%20on%20the%20Case%20of%20Poland.pdf (accessed 22 February 2016).

Helby, P., H. Rosenqvist, and A. Roos. 2006. Retreat from Salix—Swedish experience with energy crops in the 1990s. Biomass Bioenergy 30:422–427.

Henriksson, A. 2014. Willow in Sweden. Salix Energi Europa AB. Available at http://3d3a514068.url-de-test.ws/wp-content/uploads/2013/10/Annika-Presentaion-Brussel-19-March-2014-short.pdf (accessed 22 February 2016).

Henriksson, A., and G. Henriksson. 2015. Rokwood: Development of direct chipping harvesters for short rotation plantations by the Swedish family enterprise Henriksson Salix AB. Asp. Appl. Biol. 131: Biomass and Energy Crops V 45–52.

Iggesund. 2016. A new cash crop. Iggesund Holmen Group, Workington, Cumbria, U.K.

Independent. 2013. Promised market for miscanthus crop fails to emerge. Available at http://www.independent.ie/business/farming/promised-market-for-miscanthus-crop-fails-to-emerge-29783160.html (accessed 28 January 2016)

Intelligent Energy for Europe. 2009. BAP DRIVER—European Best Practice Report: towards national biomass action plans, Contract No. EIE/07/118/SI2.467614. IEE. Available at http://ec.europa.eu/energy/intelligent/projects/sites/iee-projects/files/projects/documents/bap_driver_european_best_practice_report.pdf (accessed 13 January 2016).

Irish Examiner. 2013. Miscanthus — a hot product for farmers or just a load of hot air? Available at http://www.irishexaminer.com/farming/news/miscanthus-a-hot-product-for-farmers-or-just-a-load-of-hot-air-238492.html (accessed 28 January 2016).

Irish Times. 2013. Energy crop faces uncertain future in price dispute. Available at http://www.irishtimes.com/news/environment/energy-crop-faces-uncertain-future-in-price-dispute-1.1337267 (accessed 28 January 2016).

Isebrands, J. G., and J. Richardson 2014. Poplars and willows: trees for society and the environment. CABI, Rome.

Johnston, C. R., L. Walsh, and A. R. McCracken. 2015. Biomass production – exploiting short rotation coppice willow plantation multifunctionality to achieve the joint goals of biomass production and waste water management. Asp. Appl. Biol. 131: Biomass and Energy Crops V 89–96.

Jonsson, H. 1992. Summary evaluation of Swedish experimental plantations of willow 1986-1991. ramprogram Coppice NUTEK.: R, ISSN 1102-2574

(Sammanfattande utvärdering av svenska försöksodlingar med salix 1986-1991. Malmöhus läns hushållningssällskap. 1992: ramprogram Energiskog)

Karp, A., et al. 2011. Genetic improvement of willow for bioenergy and biofuels. J. Integr. Plant Biol. 53:151–165.

Kauter, D., I. Lewandowski, and W. Claupein. 2003. Quantity and quality of harvestable biomass from *Populus* short rotation coppice for solid fuel use—a review of the physiological basis and management influences. Biomass Bioenergy 24:411–427.

Keating, M. 1993. Agenda for change: a plain language version of agenda 21 and other rio agreements. Centre for Our Common Future, Geneva, Switzerland.

Langeveld, H., et al. 2012. Assessing environmental impacts of short rotation coppice (SRC) expansion: model definition and preliminary results. Bioenerg. Res. 5:621–635.

Larsson, S.. 1998. Genetic improvement of willow for short-rotation coppice. Biomass Bioenergy 15:23–26.

Larsson, S. 2015. Planted *Salix* areas 199-2008 from own production in Sweden and Poland.

Larsson, S., and A. Henriksson. 2015. Salix on arable land - now and in the future. Position Paper. European Willow Breeding.

Larsson, S., and K. Lindegaard 2003. Full scale implementation of short rotation willow coppice, SRC, in Sweden. Agrobränsle AB, Örebro, Sweden.

Lindegaard, K. 2013a. Why we need any energy crops scheme 3. Crops4Energy, Bristol.

Lindegaard, K. 2013b. CAP reform consultation response. Response from a broad coalition supporting short rotation coppice and the energy crops sector. Crops4Energy, Bristol.

Lindegaard, K., and J. H. Barker. 1997. Breeding willows for biomass. Asp. Appl. Biol., Biomass and Energy Crops 49:1–9.

Lindegaard, K., R. I. Parfitt, G. Donaldson, T. Hunter, W. M. Dawson, E. G. A. Forbes, et al. 2001. Comparative Trials of Elite Swedish and UK Biomass Willow Varieties. Asp. Appl. Biol. 65:183–198.

Lindegaard, K., M. M. Carter, A. McCracken, I. Shield, W. MacAlpine, M. Hinton Jones, et al. 2011. Comparative trials of elite Swedish and UK biomass willow varieties 2001–2010. Asp. Appl. Biol. 112: Biomass and Energy Crops IV 57–66.

Lindegaard, K. N., M. Holley, A. Lamley, P. Zapata Aranda, S. Sayadi, C. Parra-López, et al. 2015. Fuelling dialogue between biomass research, industry, policy and business. Asp. Appl. Biol. 131: Biomass and Energy Crops V 33–42.

Mangan, C.. 1997. Overview of EU energy crop policy. Asp. Appl. Biol. 49:11–15.

McCormick, K., and T. Kåberger. 2005. Exploring a pioneering bioenergy system: the case of Enköping in Sweden". J. Clean. Prod. 13:1003–1014.

Mola-Yudego, B., and P. Pelkonen. 2008. The effects of policy incentives in the adoption of willow short rotation coppice for bioenergy in Sweden. Energy Pol. 36:3062–3068.

Mola-Yudego, B. and González-Olabarria, J. R. 2010. Mapping the expansion and distribution of willow plantations for bioenergy in Sweden: Lessons to be learned about the spread of energy crops. Biomass and Bioenergy, 34(4):442–448. DOI: 10.1016/j.biombioe.2009.12.008

Murach, D., H. Hartmann, N. Koim, C. Mollnau, P. Rademacher, R. Schlepphorst, et al. 2013. Recent experiences with agrowood production in Brandenburg/ Germany. Kongress – Agrarholz 2013 19/20 February 2013. Available at http://veranstaltungen.fnr.de/ fileadmin/veranstaltungen/Agrarholz2013/Murach__FH_ Eberswalde_Agrarholz_2013.pdf (accessed 12 January 2016).

Natural England. 2013. Energy crops scheme: establishment grants handbook, 3rd edn. Natural England, Worcester, U.K.

NI Direct, 2016. RHI for non-domestic customers. Available at http://www.nidirect.gov.uk/rhi-for-non-domestic-customers (accessed 24 February 2016)

NIA. 2008. The Committee for Agriculture and Rural Development Report into Renewable Energy and Alternative Land Use. Northern Ireland Assembly. Available at http://archive.niassembly.gov.uk/ agriculture/2007mandate/reports/390708R.htm (accessed 1 March 2016).

NIDOE. 2016. Draft PPS18: Renewable Energy Annex 1 Technology: Energy Crops. Northern Ireland Dept. of Environment (NIDOE) Planning, Belfast. Available at http://www.planningni.gov.uk/index/policy/policy_ publications/planning_statements/pps18/pps18_annex1/ pps18_annex1_biomass/pps18_annex1_biomasstechnology/ pps18_annex1_energy.htm (accessed 28 February 2016).

Forest Research 2011. Woodland for Water: Woodland measures for meeting Water Framework Directive Objectives. Forest Research for the Forestry Commission and the Environment Agency. July 2011. Available at www.forestry.gov.uk/pdf/FRMG004_Woodland4Water. pdf/$file/FRMG004_Woodland4Water.pdf (accessed 2 March 2016).

Nylander, A. 2014. Key SRP Policies in Skåne, Sweden. Powerpoint presentation as part of 3rd Rokwood partner meeting.

PAMI. 2003. Agroforestry Planting Equipment Guide Prairie Agricultural Machinery Institute (PAMI), Saskatchewan Forest Centre; Prince Alberta. Available at http://pami.ca/ pdfs/reports_research_updates/Agroforestry/agroforestry_ planting_equipment_guide.pdf (accessed 16 February 2016).

Parra-López, C., S. Sayadi, and V. H. Duran-Zuzáo. 2015. Production and use of biomass from short-rotation plantations in Andalusia, southern Spain: limitations and opportunities. New Medit. 14:40–49.

Pérez-Cruzado, C., D. Sanchez-Ron, R. Rodríguez-Soalleiro, R. José Hernández, M. Sánchez-Martín, I. Cañellas, et al. 2014. Biomass production assessment from *Populus* spp. short-rotation irrigated crops in Spain. GCB Bioenergy 6:312–326.

PGNiG TERMIKA, 2015. Kontraktacja wieloletnich plantacji drzewnych (wierzby i topoli). Available at http://termika. pgnig.pl/biomasa/kontraktacja-plantacji/ (accessed 28 January 2016).

Pisarek, M. 2015. PGNIG TERMIKA's practices outlook in contracting of SRP willow production for energy (Agro-biomass supply building to Warsaw CHP plants: EC Żerań, EC Siekierki K1 biomass boiler). Presentation as part of Rokwood project 6th partner meeting.

Piterou, A., S. Shackley, and P. Upham. 2008. Project ARBRE: Lessons for bio-energy developers and policy-makers. Energy Pol. 36:2044–2050.

Polenergia. 2015. Warchoł "Miejsce producentów i dostawców biomasy w obliczu nowej Ustawy OZE" conference presentation Powermeetings SCC "Forum Biomasy" Połaniec. 26.03.2015

Rakowski, D.. 2015. Renewable energy: a development opportunity for rural regions in Poland. Barometer Regionalny 13:21–25. Available at http://br.wszia.edu.pl/ zeszyty/pdfs/br39_03rakowski.pdf (accessed 1 June 2016).

Regro. 2016. Available at www.energycrop.co.uk (accessed 28 March 2016).

Richards, E. G. 1987. Forestry and the forest industries: past and future. Martinus Nijhoff Publishers, Dordrecht. United Nations.

Rokwood. 2014a. Findings of the SWOT analysis. Available at http://www.rokwood.eu/public-library/ public-project-reports/send/5-public-project-reports/19-findings-of-the-swot-analysis-rokwood.html (accessed 25 March 2016).

Rokwood. 2014b. Short rotation coppice in the UK – A briefing for policymakers. Available at http://www. rokwood.eu/public-library/policy-briefs/send/20-policy-briefs/17-rokwood-uk-cluster-policy-briefs.html (accessed 25 March 2016).

Rokwood. 2015a. Resource efficient production and utilization of woody biomass from SRPs: European Best Practice and Key Findings. Available at http://www. rokwood.eu/public-library/final-publication/send/29-final-publication/57-rokwood-final-publication.html (accessed 25 March 2016).

Rokwood. 2015b. Joint Action Plan. Available at http://www. rokwood.eu/public-library/joint-action-plan/send/26-joint-action-plan/44-rokwood-joint-action-plan.html (accessed 25 March 2016).

Rokwood. 2015c. Energy crops in Europe: Best practice in SRP biomass from Germany, Ireland, Poland, Spain, Sweden & UK. Available at http://www.rokwood.eu/ public-library/best-practice-booklet/send/27-best-practice-booklet/45-best-practice-booklet.html (accessed 25 March 2016).

Rokwood. 2015d. Short rotation coppice in Germany – A briefing for policymakers. Available at http://www. rokwood.eu/public-library/policy-briefs/send/20-policy-briefs/52-rokwood-german-cluster-policy-briefs.html (accessed 25 March 2016).

Rokwood 2015e. Short rotation coppice in Ireland – A briefing for policymakers. Available from: http://www. rokwood.eu/public-library/policy-briefs/send/20-policy-briefs/53-rokwood-irish-cluster-policy-briefs.html (accessed 25 March 2016).

Rokwood. 2015f. Short rotation coppice in Poland – A briefing for policymakers. Available at http://www. rokwou.eu/public-library/policy-briefs/send/20-policy-briefs/56-rokwood-polish-cluster-policy-briefs.html (accessed 25 March 2016).

Rokwood. 2015g. Short rotation coppice in Spain – A briefing for policymakers. Available at http://www. rokwood.eu/public-library/policy-briefs/send/20-policy-briefs/58-rokwood-spanish-cluster-policy-briefs.html (accessed 25 March 2016)

Rokwood. 2015h. Short rotation coppice in Sweden – A briefing for policymakers. Available at http://www. rokwood.eu/public-library/policy-briefs/send/20-policy-briefs/55-rokwood-swedish-cluster-policy-briefs.html (accessed 25 February 2016).

Rosenqvist, H., A. Roos, E. Ling, and B. Hektor, 2000. Willow Growers in Sweden. Biomass & Bioenergy, 18(2):137–145. DOI:10.1016/S0961-9534(99)00081-1

Ross, M. L. 2013. How the 1973 Oil Embargo Saved the Planet. Foreign Affairs, 15 October 2013. Available at https://www.foreignaffairs.com/articles/north-america/2013-10-15/how-1973-oil-embargo-saved-planet (accessed 28 January 2016).

Rowe, R. L., N. R. Street, and G. Taylor. 2009. Identifying potential environmental impacts of large-scale deployment of dedicated bioenergy crops in the UK. Renew. Sustain. Energy Rev. 13:271–290.

Rowe, R. L., M. E. Hanley, D. Goulson, D. J. Clarke, C. P. Doncaster, and G. Taylor. 2011. Potential benefits of commercial willow Short Rotation Coppice (SRC) for farm- scale plant and invertebrate communities in the agri-environment. Biomass Bioenergy 35:325–336.

Rowe, R. L., et al. 2013. Evaluating ecosystem processes in willow short rotation coppice bioenergy plantations. Glob. Change Biol. Bioenergy 5:257–266.

Ruiz, F., and G. Lopez 2010. Review of cultivation, History and Uses of Eucalypts in Spain. Conference of Eucalyptus species management, history, status and trends in Ethiopia Addis Ababa (Ethiopia) 15th-17th September 2010.

Sage, R., et al. 2006. Birds in willow short-rotation coppice compared to other arable crops in central England and a review of bird census data from energy crops in the UK. The Ibis 148(Suppl S1):184–197.

Sayadi, S., C. Parra-López, V. H. Durán-Zuazo, and V. Magnolfi. 2014. Short-Rotation Wooden Biomass Production: Barriers and Opportunities in Andalusia, Spain. Proceedings of 22nd European Biomass Conference and Exhibition (EU B&C 2014). ISBN: 978-88-89407-52-3, pp: 1640–1643. Hamburg, June. (Germany) Available at http://www.etaflorence.it/proceedings/index.asp?detail=10130&mode=keyword&categories=0&items=barriers&keywords=t (accessed 1 June 2016).

Schweier, J., and G. Becker. 2012. Harvesting of short rotation coppice – harvesting trials with a cut and storage system in Germany. Silva Fenn. 46:298–299.

Spinelli, R., C. Nati, and N. Magagnotti. 2008. Harvesting short-rotation-poplar plantations for biomass production. Croatian J. Forest Eng. 29:129–139.

Spinelli, R., C. Nati, and N. Magagnotti. 2009. Using modified foragers to harvest short rotation poplar plantations. Biomass Bioenergy 33:817–821.

SRCPlus. 2014. Short Rotation Woody Crops (SRC) plantations for local supply chains and heat use. Best practice examples on sustainable local supply chains of SRC. Available at http://www.srcplus.eu/images/SRCBestPractice_EN.pdf (accessed 1 June 2016).

STORE. 2013. Energy Storage Needs in Spain. Available at http://www.store-project.eu/en_GB/current-situation-in-the-target-countries-spain (accessed 1 June 2016).

Styles, D., P. Börjesson, T. D'Hertefeldt, K. Birkhofer, J. Dauber, and P. Adams, et al. 2016. Climate regulation, energy provisioning and water purification: quantifying ecosystem service delivery of bioenergy willow grown on riparian buffer zones using life cycle assessment. Ambio. doi: 10.1007/s13280-016-0790-9.

Swedish Board of Agriculture. 2001. Agriculture Statistical Yearbook 2001. (Jordbruksstatistisk årsbok 2001)

Swedish Board of Agriculture. 2003. Agriculture Statistical Yearbook 2001. (Jordbruksstatistisk årsbok 2003)

Swedish Board of Agriculture. 2004. Agriculture Statistical Yearbook 2001. (Jordbruksstatistisk årsbok 2004)

Swedish Board of Agriculture. 2008. Agriculture Statistical Yearbook 2001. (Jordbruksstatistisk årsbok 2008)

Swedish Board of Agriculture. 2011. Agriculture Statistical Yearbook 2001. (Jordbruksstatistisk årsbok 2011)

Swedish Board of Agriculture. 2013. Agriculture Statistical Yearbook 2001. (Jordbruksstatistisk årsbok 2013)

Swedish Board of Agriculture. 2014. Jordbruksverket. Rapport 215:18. Landsbygdsprogram för Sverige år 2007-2013. Årsrapport 2014.

Swedish Board of Agriculture. 2015. Agriculture Statistical Yearbook compilation 2015 (Jorbruksstatistisk sammanställning 2015). Available at http://www.jordbruksverket.se/swedishboardofagriculture/statistics/agriculturalstatistics.4.2d224fd51239d5ffbf780001098.html) (accessed 2 February 2016).

Sydkraft. 1987. Energy forestry in southern Skåne. Progress report Phase 5 (Storförsök Syd. Energiskogsodling i södra Skåne. Lägesrapport etapp 5).

Szymańska, D., and J. Chodkowska-Miszczuk. 2011. Endogenous resources utilization of rural areas in shaping sustainable development in Poland. Renew. Sustain. Energy Rev. 15:1497–1501.

TEAGASC 2014. Tillage Sectoral Energy Crop Development Group. Available at http://www.teagasc.ie/energy/Policies/Tillage_Sectoral_Energy_Crop_Development_GroupPlan2014.pdf (accessed 29 January 2016).

The Scottish Government 2015. Basic payments scheme: greening. Agriculture, Food and Rural Communities Directorate, Edinburgh, U.K.

Thornley, P., and D. Cooper. 2008. The effectiveness of policy instruments in promoting bioenergy. Biomass Bioenergy 32:903–913.

UNCED 1992. Non-Legally binding authoritative statement of principles for a global consensus on the management, conservation and sustainable development of all types of forests. United Nations, Rio de Janeiro. Available from: http://www.un.org/documents/ga/conf151/aconf15126-3annex3.htm (accessed 12 January 2015).

UNFCCC 1997. Kyoto Protocol, United Nations Framework Convention on Climate Change (UNFCCC); 1997. Available from: http://unfccc.int/kyoto_protocol/items/2830.php (accessed 12 January 2015).

Vattenfall 2014. Biomasse-Heizkraftwerk im Märkischen Viertel startet Test mit KUP-Hölzern. Available from: http://corporate.vattenfall.de/newsroom/pressemeldungen/2014/biomasse-heizkraftwerk-im-markischen-viertel-startet-test-mit-kup-holzern/ (accessed 12 March 2016).

Verwijst, T., A. Lundkvist, S. Edelfeldt, and J. Albertsson 2013. Development of Sustainable Willow Short Rotation Forestry in Northern Europe. Biomass Now - Sustainable Growth and Use in M. Darko Matovic, ed. InTech. ISBN 978-953-51-1105-4, Available at http://cdn.intechopen.com/pdfs-wm/44392.pdf (accessed 28 January 2016) DOI: 10.5772/55072

Welsh Government 2015. The common agricultural policy reform: 2016 greening booklet. Welsh Government, Cardiff.

Wisniewski, G., and A. Oniszk-Popławska. 2009. Fast growing renewable energy sector in Poland. Int. Sustain. Energy Rev. 1:6–8. Available at http://www.ieo.pl/pl/raporty/doc_download/259-international-sustainable-energy-review-nr-12009.html (accessed 11 March 2016).

2

GplusE: beyond genomic selection

Ian Mackay[1], Eric Ober[1] & John Hickey[2]

[1]John Bingham Laboratory, NIAB, Huntingdon Road, Cambridge CB3 0LE, UK
[2]The Roslin Institute and Royal (Dick) School of Veterinary Studies, University of Edinburgh, Easter Bush Research Centre, Midlothian EH25 9RG, UK

Keywords
Genomic selection, high throughput phenotyping, phenomics, selection index, wheatwheat.

Correspondence
Ian Mackay, John Bingham Laboratory, NIAB, Huntingdon Road, Cambridge CB3 0LE, UK

E-mail: ian.mackay@niab.com

Funding Information
We thank Alison Bentley from NIAB and John Wolliams from the Roslin Institute for helpful discussions. BBSRC funding for the GplusE project started in February 2015.

Abstract

GplusE is a strategy for genomic selection in which the accuracy of assessment in the reference population for a primary trait such as yield is increased by the incorporation of data from high-throughput field phenotyping platforms. This increase in precision comes from both exploiting genetic relationships between traits and reducing the effect of environmental influences upon them. We describe a collaborative project among researchers and breeders to develop a large reference population of elite UK wheat lines. This will be used to test the method, to study the design of the reference population, and to test genotyping strategies and imputation methods. Finally, it will provide data to pump-prime the application of genomic selection to UK winter wheat breeding.

Introduction

In this paper, we argue that the collection of high-throughput phenotype data from field trials (HTP) (Montes et al. 2007; Araus and Cairns 2014) can be integrated directly into plant breeding programs in a manner analogous to the use of high-density marker data in genomic selection (GS) (Meuwissen et al. 2001; Jannink et al. 2010). We focus on our own plans to integrate HTP with genomic prediction for UK winter wheat. To this end, we first give a brief overview of GS, then describe how HTP and GS can be brought together to improve response to selection. This combined approach we refer to as GplusE: in acknowledgment of the work of Johannsen (1857–1927) who first described phenotype as a function of genotype and environmental factors: an insight which still underpins plant breeding.

We suggest that this approach can be extended to incorporate other sources of high throughput data into genomic prediction, for example, from metabolomics or gene expression experiments, and can also be used to improve the accuracy of field trials in general.

The rate of genetic improvement depends on four factors: (1) the accuracy of selection; (2) the time taken to generate new lines; (3) the number of selection candidates and proportion selected (selection intensity); (4) the genetic variability among the candidates for selection (Hallauer and Darrah 1985; Falconer and Mackay 1996). In wheat, most recent research effort has been placed on the fourth of these through projects that are curating and exploiting novel sources of genetic variation from landraces and wild species. For example, in the United Kingdom, £12 m is being invested by the BBSRC over 6 years from 2011 in a large collaborative project to introgress novel sources of germplasm into hexaploid wheat from crosses between adapted lines, 4× and 2× related species, and landraces (Galushko and Gray 2014; www.wheatisp.org). Similarly, the Mexican government invested $70 m into "Seeds of Discovery" at CIMMYT (Wenzl 2013; www.seedsofdiscovery. org), a 7-year project to characterize wheat and maize germplasm. In recent years, however, the development of high throughput systems to record large quantities of genetic and phenotypic information cheaply presents other opportunities for increasing the rate of genetic improvement. In the case of genetics, high-density genetic markers can be used directly to predict traits without recourse to initial QTL mapping experiments, leading to genomic selection (Meuwissen et al. 2001). We propose that in an analogous manner, measurements from high- throughput phenotyping systems can also be used in trait prediction without recourse to physiological or biochemical hypothesis testing or interpretation of those measurements (though these can help).

Genomic Selection

In breeding programs, a major application for high-density genetic markers in plant and animal breeding is to predict traits for individuals (Scutari et al., 2013; Lin et al. 2014). These predicted traits can then be used in place of direct phenotyping to select among individuals. Since selection is no longer constrained by the time required to develop lines and to bulk up seed for phenotyping, rates of response to selection can be greatly increased. The optimum method of implementing genomic selection in any breeding program is an active area of research and varies with species and target traits (Jonas and de Koning 2013) but its use in commercial animal breeding is now well established (Hayes et al. 2013).

The principle of genomic selection is straight forward: a large population of individuals (the reference or training population) is phenotyped accurately and genotyped with a high-density of genetic markers. Traits are regressed against markers to generate a prediction equation for the traits from the markers. Candidates for selection, which are not part of the reference population (RP), are genotyped and selected on the basis of their predicted trait values alone. Selected individuals can then be crossed and their progeny immediately genotyped in turn to enable selection to be performed amongst them. GS thus enables high intensities of selection (because large populations can be raised and genotyped at lower costs than when phenotyping is required) and a rapid cycle time (because candidates for selection do not need to be phenotyped). However, the devil is in the detail: in particular, in development of appropriate prediction algorithms, the number of markers required, and the composition and size of the RP. We comment briefly on each of these in turn.

Prediction algorithms

Algorithm development for trait prediction is no longer viewed as critical because most methods give similar prediction accuracies in most situations (e.g., Daetwyler et al. 2013). The core statistical problem is that, with more markers than individuals, least squares regression methods do not work and statistical models that treat markers as random effects are needed (Whittaker et al. 2000; Meuwissen et al. 2001). Several of these methods (e.g., GBLUP, BayesA, BayesB, Bayes Lasso, Elastic Net) are implemented in freely available software (e.g., BGLR: Pérez and de los Campos 2014; AlphaBayes: Hickey and Tier 2009; GenSel: Fernando and Garrick 2009; ASReml: Gilmour et al. 2009) but perhaps the simplest method "GBLUP" is most widely used in practice because it is easy to implement, fast and gives results that are generally as good as those obtained with other methods (e.g., Daetwyler et al. 2013).

Marker numbers

GS can require genotyping of thousands of selection candidates in each generation, with potentially several generations per year. Consequentially, even though cheap high-throughput genotyping platforms are available for the major crops, cost can still be a constraint. There are two approaches to its control.

Firstly, GS can be limited to cases for which only low densities of markers are required. For example, cycles of genomic prediction and selection among individuals within a single cross require only a small reference population of lines from the same cross and only small numbers of markers. In this case, linkage disequilibrium (correlation between pairs of loci) extends over large genetic distances along the chromosomes and accurate predictions can be made from very small numbers of markers, potentially <100 (Hickey et al. 2014). However, making repeated cycles of selection within a cross is not the best breeding strategy; greater progress is made by crossing selected individuals from different crosses.

As relationships between the RP and the selection candidates decrease, marker density must also increase. There may be no recent pedigree relationship between candidates and the RP and in these cases high marker densities are required, of the order of tens of thousands (Hickey et al. 2014). A second approach in these circumstances is to control genotyping costs by marker imputation. Here, the RP is genotyped with a full set of markers but selection candidates are genotyped with a smaller selected subset. The pattern of linkage disequilibrium among markers in the RP is then used to predict genotypes for the missing markers for the candidates. This is standard practice in animal breeding where, for example, as many as 600,000 markers or whole genome sequence may be genotyped in the RP but as few as 384 markers on the candidates (e.g., Huang et al. 2012; Hickey et al., 2011; Habier et al. 2009; VanRaden et al. 2010). The cost reduction is therefore substantial. The process is less well developed in plant breeding; methods and software may need to be developed to account for complexity arising from large repetitive genomes with multiple polymorphic chromosome rearrangements, polyploidy, selfing, and the absence of a genome sequence which allows markers to be easily ordered. There is therefore a requirement for software which works well for plants and data sets in which some markers may be mapped and some not.

Reference population design

The biggest problem in implementing GS is the design of the RP. Simulations (Clark et al. 2011; Hickey et al. 2014) and empirical studies (VanRaden et al. 2009; Habier et al. 2010; Clark et al. 2012) have demonstrated that the RP should be large; several thousand individuals ideally for an inbreeding cereal (Hickey et al. 2014), though much smaller populations can be used when candidates and RP are very closely related. Initial implementation of GS in a breeding program may, therefore, be best within a very closely related genetic pool, where linkage disequilibrium is extensive, allowing accurate predictions from small numbers of markers and individuals. However, to exploit the full potential of GS, larger populations and markers densities are required.

Examples of Trait Prediction in Wheat

We give some examples of the accuracy of genomic prediction from work in United Kingdom and European wheat in Table 1. Each example illustrates the success of the process, but also the diverse ways in which reduced relationships between test and RP can greatly reduce the accuracy of prediction.

Example 1 uses historical records of wheat yields from 1948 to 2007, reanalyzed by Mackay et al. 2011). High accuracies are achieved when lines are partitioned into

test and reference sets with equal representation of lines from the whole time series (case 1). However, when lines are partitioned into an older set in the RP and modern lines in the test set (case 2), the correlation drops from 0.8 to 0.2, though for lines released within 10 years of the youngest line in the reference set the correlation is 0.4 (Mackay et al. 2011). Case 2 mimics more closely what breeders desire by predicting forward over generations. In this data set, variety yields have increased over time and marker allele frequencies have changed over time. In essence, markers give a good prediction of age and age gives a good prediction of yield, but this fails if the full age range is not represented in the RP.

In the second example, yield and marker data from the publically available doubled haploid mapping population Avalon x Cadenza (http://www.wgin.org.uk/) were partitioned at random into test and RPs (case 1). The cross-validation correlation of 0.5 is acceptable. However, if the lower yielding lines are used as the reference set to predict the yield of the higher yielding set the correlation drops (case 2). Once more, case 2 mimics more accurately breeders' requirements: to predict varieties with better performance than the best currently available.

The final example uses the TriticeaeGenome association mapping panel (Bentley et al. 2014, http://www.triticaeegenome.eu/ ie. no hypen) of 384 French, German, and UK winter wheat lines, genotyped with DArT markers. If test and reference sets are created without reference to country of origin, then the cross validation correlation is acceptable (case 1). However, if lines from one country of origin are used to predict performance of lines from another (case 2), the correlation drops. Predicting across a greater genetic distance, measured here by country of origin, is more in line with the needs of the plant breeder.

These three examples show that differences between test and RPs in age of varieties, in yield, or in country of origin can all have a major effect on the accuracy of predictions. In all examples, the underlying cause of the reduction in prediction accuracy is the increased genetic distance between lines in the test and RPs. All three also give reasonable examples of what breeders would like to do: predict forward in time, predict transgressive segregation and predict into differing pools of germplasm. In every case, the composition of the RP is key.

Table 1. Genomic prediction in wheat.

Data source	Case 1	Case 2	RP size	No. markers
1. UK NL/RL 1948–2007	0.8	0.2	80	217
2. AxC yield data	0.5	0.2	100	351
3. Triticae Genome	0.5	0.3	376	1804

Data source: See text for details.
Case 1: Data partitioned at random into test and reference populations.
Case 2: Data partitioned selectively: for details see text.

High Throughput Field Phenotyping

Ideally, RPs should be large, closely related to the candidates for selection and accurately phenotyped. These requirements are antagonistic. Accurate phenotyping generally requires multiple replicates of large plots. However, large RPs limit replicate number, and the requirement to multiply seed over several years for trials may cause selection candidates to be several generations removed from the RP. There is a requirement, therefore, to increase the precision with which yields are estimated from low replicate, large entry number yield trials.

Trial design and analysis

To date, the principle mechanism for increasing precision has been through improvements in trial design. Sophisticated designs are now readily available (e.g., http:/ www.expdesigns.co.uk/co.uk/, http://www.austatgen. org/software/). These have largely removed the constraints on permissible combinations of block size, variety number and replicate number present in old published catalogs of designs in classic texts such as Cochran and Cox (1957). Augmented (Federer 1956), p-rep (Cullis et al. 2006), and augmented p-rep trial designs (Williams et al. 2011) warrant mention as they are targeted at very low replicate experiments such as early generation breeders' trials. They are therefore also highly suitable for testing large RPs for GS. Trials which incorporate known genetic relationships among varieties in their arrangement in the field will also help improve precision (Moehring et al. 2014). However, there is a limit to the increased precision that trial design alone can deliver.

An old but little used method of increasing precision is through the analysis of covariance (Fisher 1934; Wishart 1950). Here, adjustment for field environmental effects (on yield, say) is made through the correlation between yield and some other measurement made on the plots or plants. The additional measurement should have no genetic link to yield, but can still be a phenotype. For example, adjustment could be made for plant vigour if it acted as a surrogate for plot fertility. This approach has parallels with genomic selection, where any relationship between trait and markers is purely genetic in origin so selection on marker genotype is indirect selection on G: the trait genotype. In the analysis of covariance, any relationship between the covariate and trait is treated as purely environmental in origin so selection on the covariate is indirect selection on E: the environmental effect. Selection to reduce E then acts to increase the accuracy of G. However, in many cases, covariates also have a genetic correlation with yield. In the case of plant vigour, the analysis of covariance could result in the selection of less vigorous lines. Partly for this reason, analysis of covariance has been little used in plant breeding, and mainly in disaster recovery: when the covariate could be poor field emergence for example. Moreover, routine scoring of additional traits for potential use as covariates is expensive and time consuming, particularly in breeders' large early generation trials. However, the recent introduction of automated methods and platforms to score very large numbers of traits on field plots (Montes et al. 2007; Araus and Cairns 2014) makes their collection cost effective and we believe the routine application of these methods to variety trials should be re-examined.

Using currently available technology for field phenotyping and for precision agriculture, large numbers of covariates can be recorded quickly at plot level. Many, if not most of these, will correlate with both environmental and genetic determinants of yield so cannot be incorporated into trait prediction in the same way as markers in GS or as environmental covariates in the analysis of covariance. However, using methods originally employed in creating "selection indices" (Smith 1936; Hazel and Lush 1942; Henderson 1963) to select optimally across multiple correlated traits of varying economic importance, it is possible to include these in estimation of yield by taking into account their environmental and genetic covariances with each other and with yield.

Selection index

A selection index is a linear combination of variables that is used to compute, for each individual or line, a criterion for selection (Henderson 1963). To compute such an index, estimates of genetic and environmental variances and covariances are required among all traits. These estimates are possible from all well designed field trials provided there is some, possibly incomplete, replication, and/or by taking into account genetic relationships among lines estimated by pedigree or by genetic markers (Astle and Balding 2009; Lee et al. 2010). The relative importance of each trait is also required. This is referred to as the trait's "economic value" and is measured as the cash value of a unit increase in the trait. Estimating economic values across multiple traits can be complex, but with only a single-target trait for selection (say yield) it is simple: the economic value of the trait is 1 (or −1 if we are selecting for decreasing values) and the economic value of all other traits is 0. For completeness, we give the equation to compute the coefficients of the selection index below:

$$\mathbf{b} = \mathbf{G}\mathbf{P}^{-1}\mathbf{e} \qquad (1)$$

where $\mathbf{b}$ = the vector of regression coefficients to be estimated (including for yield itself), $\mathbf{e}$ = the vector of economic values (1 for yield and 0 for all other traits and covariates), $\mathbf{G}$ = the genetic (co)variance matrix, $\mathbf{P}$ = the phenotypic (co)variance matrix. Excellent accounts of the theory of selection indices are given in Falconer and Mackay (1996) and Bulmer (1980). Once the coefficients are estimated, then for each line each trait in turn is multiplied by its corresponding value in $\mathbf{b}$ and these are summed to give the value of the selection index for that line. Selecting on this index will give a greater response than selecting on the trait alone and this increase

Table 2. Selection indices to increase trait Y incorporating data from a second trait X.

Phenotypic correlation	Cause of correlation	Selection index	Relative response[1]
0.5	All genetic	Select on Y + X	1.155
0.5	All environmental	Select on Y − X/2	1.155
0.5	Genetic = environmental	Select on Y	1.000
0.0	No cause; genetic = environmental	Select on Y	1.000
0.0	Genetic = - environmental = 0.5	Select on Y + X/2	1.118

Heritability of X = heritability of Y = 0.5 in all cases. Genetic variance = 1.
[1]Expected response to selection on the index relative to direct selection on Y.

can be predicted (Falconer and Mackay 1996). In Table 2, we give an idealized example with one target trait, Y, and one additional trait, X. The heritability of each trait is fixed at 0.5 and, for ease of interpretation, each trait has the same variance. The correlation between the traits may be all genetic, all environmental or made up of both environmental and genetic effects.

The relative strengths and signs of the genetic and environmental correlation determine the nature of the index: X is selected for if the correlation is all genetic but against if it is all environmental. If the two correlations are equal, then there is no gain to be made by including the second trait. The expected gain in response from selection on the index is substantial for the other examples. Note that the phenotypic correlation may be 0 because there is neither genetic nor environmental correlation, or because the genetic and environmental correlations are of opposite sign and cancel. In the first case, the best selection strategy is to select on Y alone, but in the second some weight is given to X. This emphasizes the importance of decomposing phenotypic correlations into genetic and environmental components. The Excel spreadsheet used to compute the coefficients given in Table S1 is given in the supporting information and can be used to compare coefficients for other values of the parameters, given in equation 1.

Differences in sign of genetic and environmental correlations (r_g and r_e respectively) do occur. Table 3 gives an example from the TriticeaeGenome dataset (Bentley et al. 2014) of correlations between height and flowering time in five trials. The correlations vary from site to site, but in general the environmental correlation is positive and the genetic correlation is negative. A positive environmental correlation between height and yield is most simply interpreted as an indicator of soil fertility: fertile plots yield more and the plants grow taller. The negative genetic correlation is most likely due to the presence in this association mapping panel of older lines, which tend to be taller, carrying no semi-dwarfing alleles, and modern lines which are shorter and yield more. Table 3 also lists the heritabilities of the two traits and the relative merit of selecting on an index to improve yield by taking into account the additional information provided by height. In this example, the improvement is slight except for data from France in 2010 and the United Kingdom in 2011. At the 2011 United Kingdom trial, seedling establishment, flowering time and tiller number were also scored. If these traits are incorporated into the index, selection is predicted to be 11% more efficient than selecting on yield alone. In breeders' trials, the heritabilities for yield are commonly lower than the high values given here and the scope for improved precision from use of an index is therefore greater.

Table 3. TriticeaeGenome trials 2010–2011: 387 wheat varieties of French, German and UK origin.

Correlations between yield and height					
Trial location	r_e	r_g	h^2y	h^2x	RM
Fr 2010	0.03	−0.45	0.43	0.91	1.074
Ge 2010	−0.07	−0.18	0.59	0.91	1.003
Ge 2011	0.36	−0.13	0.61	0.91	1.009
UK 2010	0.24	−0.03	0.30	0.82	1.008
UK 2011	0.48	−0.24	0.44	0.69	1.066
Average	0.2	−0.2	0.5	0.9	1.03

r_e: Environmental correlation coefficient.
r_g: Genetic correlation coefficient.
h^2y: Heritability of yield.
h^2x: Heritability of height.
RM: Relative merit of index selection.

Combining information from multiple sources

Additional trait and phenotype information collected on the RP can be used to create a selection index for any key traits(s). Genetic markers will be used to generate a prediction equation for that index. Subsequently, candidates for selection will be genotyped and their selection index, say for yield, predicted from that score. The additional gain to come from the use of an index needs to be tested empirically and must compensate for the cost of collection. For our own trials, our current best estimate is that the additional phenotyping adds 28% to the cost of a trial. This estimate is based on actual costs within the GplusE project (described below) in which a 2 × 6 m plot for assessing yield is £18.67, including drilling, harvesting and land rental, and the cost of gathering multiple spectrophotometric readings from six unmanned aerial flights together with ground-based readings for calibration is £5.25 per plot. Statistical methods are needed to incorporate very many plot covariates into a selection index: the classical selection index equation will only work if there are more varieties in trial than there are additional measurements and the measured traits are not very highly correlated (i.e., there is no colinearity among the measurements). This is analogous to the position with genomic prediction prior to the introduction of statistical methods to circumvent the problem (Whittaker et al. 2000; Meuwissen et al. 2001). Some methods are already available to account for these problems (Hayes and Hill 1980, 1981; Tai 1989) but borrowing and adapting methods from genomic selection may be more effective.

The GplusE Project

In our study, with the collaboration of four breeders – Elsoms, KWS, Limagrain and RAGT – we are creating a large population of at least 3000 wheat lines from up to 44 elite crosses. Representative views of the linked pedigree are given in Figure 1, and illustrate the consanguineous, unstructured and intermeshed nature of the UK winter wheat pedigree. Allele frequencies at genetic markers have changed over time in the United Kingdom (White et al. 2008), but genetic variation has not declined; it increases in periods when more independent breeding programs contribute varieties. The 44 crosses have been selected to cover the diversity of parents that breeders are currently using. The lines we generate will be phenotyped over 2 years and genotyped with the 35 k UK Affymetrix SNP chip (http://www.affymetrix.com/). The large size of the population, and the known pedigree structure, will allow its partition into RPs and test populations with varying degrees of relationship between the two to assess how accuracy of prediction varies with genetic distance and marker density.

This approach has been simulated for wheat by Hickey et al. (2014) and the results of those simulations have been used in the experimental design of GplusE. For example, at one extreme, the reference and training populations can be constructed so that for every cross, there are representatives in both. Alternatively, the two populations could be selected to minimize the pedigree relationship between them. This might involve ensuring that they contain no common grandparents or more distant relationships. In between these extremes are partitions where, say, if two crosses have a single parent in common, then lines from one are allocated to the test population only and lines from the other to the reference population only, while individuals from the same cross are never represented in both. There are many other possible partitions, and selection can also take place using marker-based estimates of kinship rather than those from the pedigree. Superimposed on these alternatives, the size of the reference population can be varied. In comparing results from these alternatives, we hope to understand how to construct the most effective RP for the germplasm which UK breeders are currently using.

Marker imputation algorithms and software will be developed specifically for use in crops, building on early work which is already in routine use in commercial animal breeding programs (Hickey et al. 2012). The contemporary and highly commercially relevant composition of this population will assist in establishing GS for UK wheat as a cost effective strategy for our commercial partners. This collaboration is unique in the level of cooperation required among competitor companies and the degree of goodwill shown in releasing pedigree information of elite crosses to academic collaborators: a measure of the importance with which GS is viewed in the breeding community.

In addition to the study of RP composition for predicting yield, additional covariate and trait information will be collected. Data on soil composition at the plot level will be collected by ground-based electromagnetic induction (EMI) by SOYL Precision Farming. (http://www.soyl.com/). EMI is used in precision agriculture to measure various soil properties, including topsoil depth, soil texture and moisture content (Doolittle and Brevik 2014). Soil penetrometer measurements will be used as a proxy for root activity in surface soil layers (Whalley et al. 2008). Airborne multispectral reflectance signatures (visible and infra red) will be captured at the plot level by Ursula Agriculture (http://www.ursula-agriculture.com/). Ground-based spectral reflectance measurements will be used in combination and to inform targeted airborne data captured through the use of the UAS (Unmanned Aerial System). We plan to score the trial from the ground three times and from the air six times between sowing and harvest, though this partition may vary subject to initial results. UA have proprietary algorithms to predict biomass,

Figure 1. Partial pedigree of 44 crosses selected for GplusE project. Crosses are represented by "X". Upper panel: descendents of the variety Moulin (listed 1984), the most recent common ancestor of 43 out of 44 of the selected crosses. Lower panel: ancestors of an arbitrarily selected cross. The positions of Moulin and Robigus, a recent ancestor of many of the lines on the current UK recommended list, are identified. Plots were created using Pedigree Viewer (Kinghorn 1994).

lodging potential, establishment, vigour, crop cover, pests, diseases, weeds, and stress. These will be included, but it is possible that the raw reflectance scores function better as covariates in creating an index for selection: there is no requirement in the construction of a selection index for the use of these covariates to have a biological interpretation any more than there is a requirement in genomic prediction to select subsets of markers which tag statistically significant QTL. This will be tested: First selection indices will be constructed in the reference population from yield and either the raw reflectance scores or from yield and the traits derived from those scores. Then, within the RP, yield and the selection indices with be regressed on the markers to derive prediction equations for each. The data in the test population will then be used to compare observed yield with yield predicted from the three different prediction equations (recalling that the selection indices are merely predictors of yield). A simplified schematic of the analysis pipeline is given in Figure 2.

Our population will be grown in augmented p-rep trials at two sites in each year (though additional testing may occur outside the scope of the project). The trials protocol will otherwise follow standard practice in the United Kingdom for fungicide-treated trials, as described for running the UK recommended list system (http://www.hgca.com/). Using standard breeders' trial plot sizes of 2 × 6 m, this means

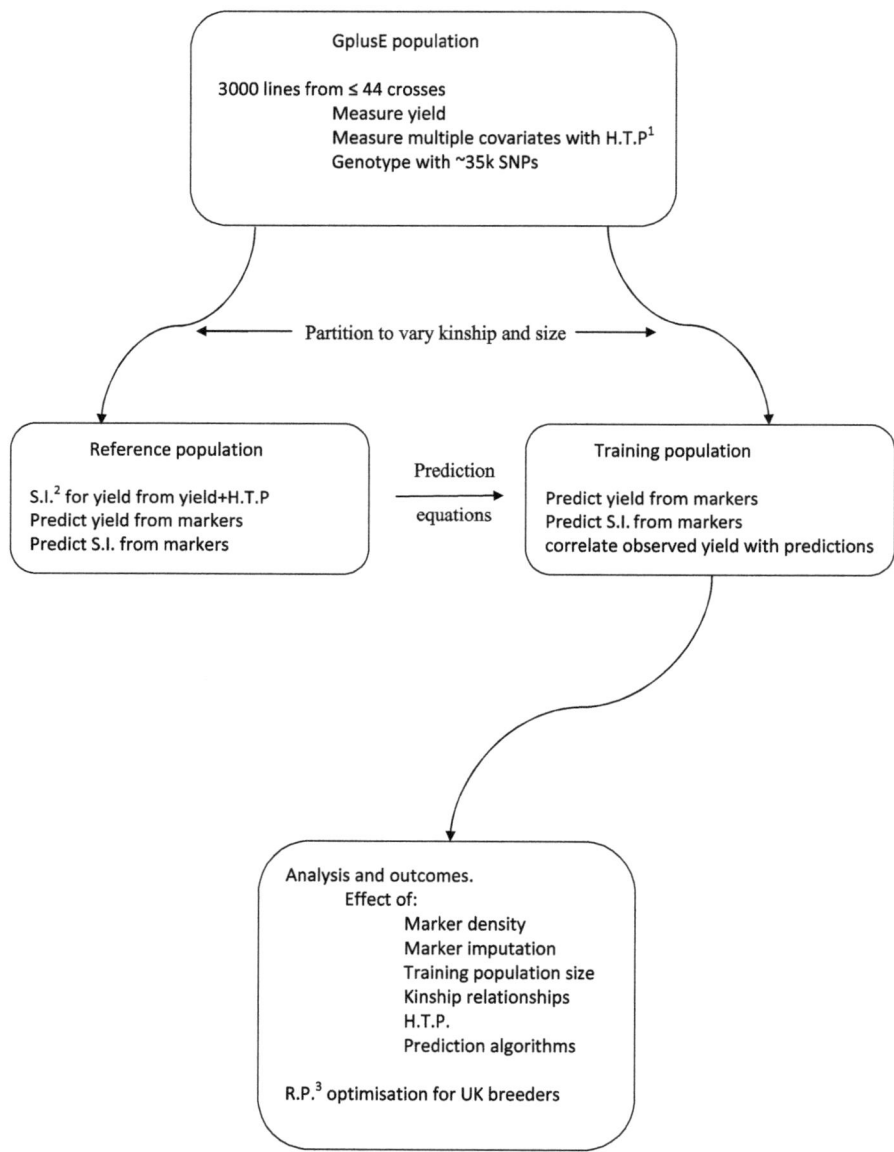

Figure 2. Outline workflow for GplusE. Marker and trait data on the full population are partitioned into a reference population of variable size and a test population with a minimum size of ~300 lines. The reference population can be varied in size and selected to alter kinship relationships among its members and with members of the training population. One or more selection indices are constructed for yield by incorporation of data from high throughput phenotyping. These are then regressed onto a genome wide marker set to create prediction equations which are tested in the training population by comparing observed yield with predicted yield and with the predicted selection indices. All combinations of parameters will be tested in replicated cross-validations. [1]H.T.P.: high thoughput phenotyping. [2]S.I.: selection index. [3]R.P.:reference population.

that each trial is ~4 ha. No fixed field-phenotyping platform could cover this area: those currently available commercially are prohibitively expensive and might contain 80 plots, at best. For GplusE, the requirement is to gather additional trait and covariate information cheaply on a large scale. Our choice of field phenotyping platforms was dictated by cost, throughput and availability. The potential of automated phenotyping to add value to large field experiments is great and several high throughput ground-based and aerial systems now exist or are in development (Araus and Cairns 2014). Although GplusE is focussed on developing and enhancing GS of UK wheat, the data the project generates will have other uses. The 44 crosses in the RP are linked by pedigree and can be used in linkage analysis and join linkage disequilibrium linkage analysis (LDLA) (Meuwissen and Goddard 2001) to detect QTL of larger size. The field phenotyping data can be used directly to develop and test physiological hypotheses. Moreover, if the methods and algorithms we describe work, these can be applied to any field experiment to improve the precision of the assessment of the primary outcome (commonly yield). We look forward to a time when additional covariate and trait information is collected routinely in any field experiment; it can be incorporated as easily into fertilizer or pesticide trials as into variety trials.

Although we have focused on yield, bread making quality or any other quantitative trait is amenable to the GplusE approach. Limited progress in improving the quality of UK wheat can be regarded as a "market failure" (Galushko and Gray 2014): the return to the commercial breeder from improving this trait does not warrant the investment required. It is possible that the cost of breeding could be reduced substantially if marker-based prediction of quality was substituted for direct phenotypic assessment. The cost of phenotyping the reference population cannot be avoided, but its accuracy may be increased by applying the selection index methods described above to combine the currently used array of predictive tests of grain quality (e.g., Cavanagh et al. 2010) into an index. Genomic selection could then be based on this index. A similar approach has been described by Heffner et al. (2011). Nevertheless, genomic selection for improved quality may require restriction to selection among individuals very closely related to the RP – in the extreme the candidates and the RP could all come from a single cross. In these circumstances, the size of the training population can be reduced substantially.

Conclusion

Crop genetics is becoming data rich as the technologies underpinning the various new 'omics disciplines (genomics, metabolomics, phenomics etc.) are applied. Many new biological insights will emerge. Crop improvement through plant breeding, however, will remain the major route by which targets of sustainable intensification will be achieved. Currently, GS is the most promising way in which improvements in polygenic traits such as yield will be made. This process of translating high throughput genetic marker systems into breeding practice is well developed in animal breeding and is beginning to be applied in crops. GplusE will assist in this process for UK wheat by increasing understanding of the dynamics of RP design and by testing if data from high throughput field phenotyping can be integrated directly into the process. Success would pump-prime the routine application of GS for UK wheat breeding, which could also improve the precision of all future field experimentation.

Acknowledgments

The population of doubled-haploid (DH) individuals, derived from F1 progeny of a cross between cvs Avalon and Cadenza, was developed by Clare Ellerbrook, Liz Sayers and the late Tony Worland (John Innes Centre), as part of a Defra funded project led by ADAS. The parents were originally chosen (to contrast for canopy architecture traits) by Steve Parker (CSL), Tony Worland and Darren Lovell (Rothamsted Research). We thank Alison Bentley from NIAB and John Wolliams from the Roslin Institute for helpful discussions. BBSRC funding for the GplusE project started in February 2015.

Conflict of Interest

None declared.

References

Araus, J. L., and J. E. Cairns. 2014. Field high-throughput phenotyping: the new crop breeding frontier. Trends Plant Sci. 19:52–61.

Astle, W., and D. J. Balding. 2009. Population structure and cryptic relatedness in genetic association studies. Stat. Sci. 24:451–471.

Bentley, A. R., M. Scutari, N. Gosman, S. Faure, F. Bedford, P. Howell, et al. 2014. Association mapping and genomic selection allow dissection of the genetic control of key traits in elite European wheat. Theor. Appl. Genet. 127:2619–2633.

Bulmer, M. G. 1980. The mathematical theory of quantitative genetics. Clarendon Press, Oxford, U.K.

Cavanagh, C. R., J. Taylor, O. Larroque, N. Coombes, A. P. Verbyla, Z. Nath, et al. 2010. Sponge and dough bread making: genetic and phenotypic relationships with wheat quality traits. Theor. Appl. Genet. 121:815–828.

Clark, S. A., J. M. Hickey, and J. H. Van der Werf. 2011. Different models of genetic variation and their effect on genomic evaluation. Genet. Sel. Evol. 43:18.

Clark, S. A., J. M. Hickey, H. D. Daetwyler, and J. H. van der Werf. 2012. The importance of information on relatives for the prediction of genomic breeding values and the implications for the makeup of reference data sets in livestock breeding schemes. Genet. Sel. Evol. 44:4.

Cochran, W. G., and G. M. Cox. 1957. Experimental designs, 2nd ed. Wiley Classics Library Edition 1992, John Wiley and Sons Inc. USA.

Cullis, B. R., A. B. Smith, and N. E. Coombes. 2006. On the design of early generation variety trials with correlated data. J. Agric. Biol. Environ. Stat. 11:381–393.

Daetwyler, H. D., M. P. Calus, R. Pong-Wong, G. de los Campos, and J. M. Hickey. 2013. Genomic prediction in animals and plants: simulation of data, validation, reporting, and benchmarking. Genetics 193:347–365.

Doolittle, J. A., and E. C. Brevik. 2014. The use of electromagnetic induction techniques in soils studies. Geoderma 223:33–45.

Falconer, D. S., and T. F. C. Mackay. 1996. Introduction to quantitative genetics, 4th ed. Pearson, Harlow, U.K.

Federer, W. T. 1956. Augmented (or Hoonuiaku) designs. Hawaiian Planters' Rec. 55:191–208.

Fernando, R. L., and D. J. Garrick. 2009. GenSel–user manual of genomic selection related analyses. Iowa State University, Ames, Iowa.

Fisher, R. A. 1934. Statistical methods for research workers, 5th ed. Oliver and Boyd, Edinburgh, U.K.

Galushko, V., and R. Gray. 2014. Twenty five years of private wheat breeding in the UK: lessons for other countries. Sci. Public Policy 41:765–779. doi:10.1093/scipol/scu004.

Gilmour, A. R., B. J. Gogel, B. R. Cullis, and R. Thompson. 2009. ASReml user guide release 3.0. VSN International Ltd, Hemel Hempstead, U.K.

Habier, D., R. L. Fernando, and J. C. Dekkers. 2009. Genomic selection using low-density marker panels. Genetics 182:343–353.

Habier, D., J. Tetens, F. R. Seefried, P. Lichtner, and G. Thaller. 2010. The impact of genetic relationship information on genomic breeding values in German Holstein cattle. Genet. Sel. Evol. 42:5.

Hallauer, A. R., and L. L. Darrah. 1985. Compendium of recurrent selection methods and their application. Crit. Rev. Plant Sci. 3:1–33.

Hayes, J. F., and W. G. Hill. 1980. A reparameterization of a genetic selection index to locate its sampling properties. Biometrics 36:237–248.

Hayes, J. F., and W. G. Hill. 1981. Modification of estimates of parameters in the construction of genetic selection indices ('bending'). Biometrics 37:483–493.

Hayes, B. J., H. A. Lewin, and M. E. Goddard. 2013. The future of livestock breeding: genomic selection for efficiency, reduced emissions intensity, and adaptation. Trends Genet. 29:206–214.

Hazel, L. N., and J. L. Lush. 1942. The efficiency of three methods of selection. J. Hered. 33:393–399.

Heffner, E. L., J. L. Jannink, and M. E. Sorrells. 2011. Genomic selection accuracy using multifamily prediction models in a wheat breeding program. Plant Genome 4:65–75.

Henderson, C. R. 1963. Selection index and expected genetic advance. Stat. Genet. Plant Breed. 982:141–163.

Hickey, J. M., and B. Tier. 2009. AlphaBayes: user manual. UNE, Australia.

Hickey, J. M., B. P. Kinghorn, B. Tier, J. H. van der Werf, and M. A. Cleveland. 2012. A phasing and imputation method for pedigreed populations that results in a single-stage genomic evaluation. Genet. Sel. Evol. 44:9.

Hickey, J. M., B. P. Kinghorn, B. Tier, J. F. Wilson, N. Dunstan, and J. H. van der Werf. 2011. A combined long-range phasing and long haplotype imputation method to impute phase for SNP genotypes. Genet. Sel. Evol. 43:12.

Hickey, J. M., S. Dreisigacker, J. Crossa, S. Hearne, R. Babu, B. M. Prasanna, et al. 2014. Evaluation of genomic selection training population designs and genotyping strategies in plant breeding programs using simulation. Crop Sci. 54:1476–1488.

Huang, Y., J. M. Hickey, M. A. Cleveland, and C. Maltecca. 2012. Assessment of alternative genotyping strategies to maximize imputation accuracy at minimal cost. Genet. Sel. Evol. 44:25.

Jannink, J. L., A. J. Lorenz, and H. Iwata. 2010. Genomic selection in plant breeding: from theory to practice. Brief. Funct. Genomics 9:166–177.

Jonas, E., and D. J. de Koning. 2013. Does genomic selection have a future in plant breeding? Trends Biotechnol. 31:497–504.

Kinghorn, B. P. 1994. Pedigree Viewer-a graphical utility for browsing pedigreed data sets. 5th World Cong. Genet. Appl. Livest. Prod. 22:85–86.

Lee, S. H., M. E. Goddard, P. M. Visscher, and J. H. van der Werf. 2010. Research using the realized relationship matrix to disentangle confounding factors for the estimation of genetic variance components of complex traits. Genet. Sel. Evol. 42:22.

Lin, Z., B. J. Hayes, and H. D. Daetwyler. 2014. Genomic selection in crops, trees and forages: a review. Crop Pasture Sci. 65:1177–1191.

Mackay, I., A. Horwell, J. Garner, J. White, J. McKee, and H. Philpott. 2011. Reanalyses of the historical series of

UK variety trials to quantify the contributions of genetic and environmental factors to trends and variability in yield over time. Theor. Appl. Genet. 122:225–238.

Meuwissen, T. H. E., and M. E. Goddard. 2001. Prediction of identity by descent probabilities from marker-haplotypes. Genet. Sel. Evol. 33:605–634.

Meuwissen, T. H., B. J. Hayes, and M. E. Goddard. 2001. Prediction of total genetic value using genome-wide dense marker maps. Genetics 157:1819–1829.

Moehring, J., E. R. Williams, and H. P. Piepho. 2014. Efficiency of augmented p-rep designs in multi-environmental trials. Theor. Appl. Genet. 127:1049–1060.

Montes, J. M., A. E. Melchinger, and J. C. Reif. 2007. Novel throughput phenotyping platforms in plant genetic studies. Trends Plant Sci. 12:433–436.

Pérez, P., and G. de los Campos. 2014. Genome-wide regression & prediction with the BGLR statistical package. Genetics 198:483–495.doi: 10.1534/genetics.114.164442.

Scutari, M., I. Mackay., and D. J. Balding. 2013. Improving the Efficiency of Genomic Selection. Statistical Applications in Genetics and Molecular Biology. 12:517–527.

Smith, H. F. 1936. A discriminant function for plant selection. Ann. Eugenics 7:240–250.

Tai, G. C. C. 1989. A proposal to improve the efficiency of index selection by "rounding". Theor. Appl. Genet. 78:798–800.

VanRaden, P. M., C. P. Van Tassell, G. R. Wiggans, T. S. Sonstegard, R. D. Schnabel, J. F. Taylor, et al. 2009. Reliability of genomic predictions for North American Holstein bulls. J. Dairy Sci. 92:16–24.

VanRaden, P. M., J. R. O'Connell, G. R. Wiggans, and K. A. Weigel. 2010. Combining different marker densities in genomic evaluation. Interbull Bullet. 42:113.

Wenzl, P. 2013. Seeds of discovery: unlocking the genetic potential of maize and wheat genetic resources. Plant Anim. Genome XXI Conf. P0768.

Whalley, W. R., C. W. Watts, A. S. Gregory, S. J. Mooney, L. J. Clark, and A. P. Whitmore. 2008. The effect of soil strength on the yield of wheat. Plant Soil 306:237–247.

White, J., J. R. Law, I. Mackay, K. J. Chalmers, J. S. C. Smith, A. Kilian, et al. 2008. The genetic diversity of UK, US and Australian cultivars of Triticum aestivum measured by DArT markers and considered by genome. Theor. Appl. Genet. 116:439–453.

Whittaker, J. C., R. Thompson, and M. C. Denham. 2000. Marker-assisted selection using ridge regression. Genet. Res. 75:249–252.

Williams, E., H.-P. Piepho, and D. Whitaker. 2011. Augmented p-rep designs. Biom. J. 53:19–27.

Wishart, J. 1950. Field trials II. The analysis of covariance. Commonwealth Bureau of Plabt Breeding and Genetics, Technical Communication No. 15. School of Agriculture, Cambridge, U.K.

3

A new emphasis on root traits for perennial grass and legume varieties with environmental and ecological benefits

Athole H. Marshall, Rosemary P. Collins, Mike W. Humphreys & John Scullion

Institute of Biological and Environmental Research, Aberystwyth University, Gogerddan, Aberystwyth SY233EE, UK

Keywords
Ecology, environment, grasslands, phenotyping, plant breeding, roots

Correspondence
Athole H. Marshall, Institute of Biological and Environmental Research, Aberystwyth University, Gogerddan, Aberystwyth SY233EE, UK.

E-mail: thm@aber.ac.uk

Funding Information
MH is funded by the Biotechnology and Biological Sciences Research Council. AM, MH, and RC are part funded by the BBSRC-LINK programme SUREROOT which in addition to BBSRC is jointly supported by partners from UK's grass seed, grassland, and livestock industries. MH and JS are part of the Climate-Smart Consortium. Funding to support this work was provided by the Welsh Government and HEFCW through the Sêr Cymru National Research Network for Low Carbon, Energy and Environment.

Abstract

Grasslands cover a significant proportion of the agricultural land within the UK and across the EU, providing a relatively cheap source of feed for ruminants and supporting the production of meat, wool and milk from grazing animals. Delivering efficient animal production from grassland systems has traditionally been the primary focus of grassland-based research. But there is increasing recognition of the ecological and environmental benefits of these grassland systems and the importance of the interaction between their component plants and a host of other biological organisms in the soil and in adjoining habitats. Many of the ecological and environmental benefits provided by grasslands emanate from the interactions between the roots of plant species and the soil in which they grow. We review current knowledge on the role of grassland ecosystems in delivering ecological and environmental benefits. We will consider how improved grassland can deliver these benefits, and the potential opportunities for plant breeding to improve specific traits that will enhance these benefits whilst maintaining forage production for livestock consumption. Opportunities for exploiting new plant breeding approaches, including high throughput phenotyping, and for introducing traits from closely related species are discussed.

Introduction

Grasslands are the main survival resource for about 1 billion people worldwide (Peyraud et al. 2014) and cover nearly 70% of the world's agricultural area (Sousanna and Luscher 2007). They are defined as 'land devoted to the production of forage for harvest by grazing/browsing, cutting or both, or used for other agricultural purposes such as renewable energy production' (Peeters et al. 2014).

In Europe (EU-27), grassland accounts for 39% of the agricultural area (Huyghe et al. 2014). The main reason for the prevalence of grasslands in agriculture worldwide is that they provide a relatively cheap source of feed for ruminants and allow the production of meat, wool, and milk from grazing animals. Furthermore, they are located frequently in marginal land areas deemed otherwise unsuitable for most other agricultural crops. Aside from those natural grasslands considered to be climatically determined, it is grasslands that are anthropogenically generated that have the most significant role in agriculture and are the focus of this review. They are located mainly within temperate climatic regions, where woody vegetation is excluded and herbaceous plant communities are maintained by appropriate human intervention and by livestock agriculture. It is possible to further divide these anthropogenically derived grasslands into long-term naturalized grasslands and those that are cultivated, which differ according to level of intensification. The latter, often termed 'improved grasslands', vary markedly in their coverage across the four countries of the UK, occupying 13.8% of Scotland, 30.3% of England, 42.0% of Wales, and 54.0% of Northern Ireland according to LCM2000 calibrations (Fuller et al. 2002). High levels of grassland improvement in Northern Ireland reflect the importance of livestock farming to the country's economy in comparison to the rest of UK (Cruickshank 1987), whereas economic pressures associated with a smaller average farm size compared to those in England and Scotland are believed to have contributed to the prevalence of the practice in Wales (Eadie 1984).

The capacity of improved grasslands to sustain animal production has been their primary function and consequently the main focus of research for most of the last century. Intensive grassland systems in the UK are currently associated with the widespread use of monocultures (usually perennial ryegrass *Lolium perenne* L.) or binary mixtures that include a legume (usually white clover *Trifolium repens* L.). These swards have a high yield potential and feeding value, can sustain frequent harvesting and/or high stocking rates, and are maintained by moderate-to-high levels of nitrogen (N) input (Wilkins et al. 2002). Peukert et al. (2014) found that rates of soil and phosphorus (P), but not N, losses under intensively managed grassland could be as high as those for other agricultural systems. Such intensive grassland systems are often short-lived and temporary components within a managed crop-rotation and as such have only a limited environmental benefit.

More extensive grassland designs and management practices require the use of mixtures of complementary perennial species in order to achieve more sustainable crop production and greater crop persistency for five-ten years or more before the need for re-sowing. Such "long-lived" pastures can be regarded as multifunctional, and in addition to their provision of forage for livestock use, they also provide a habitat for a vast and diverse ecosystem that can support a multitude of "hidden" attributes in terms of alternative environmental benefits. Grassland ecosystems are dependent on, and affect, a host of other biological organisms in the soil and in adjoining habitats. These organisms provide services such as decomposition, maintenance of soil fertility and provision of clean water.

The grassland acts as a resource for insects, such as bees, required to ensure pollination and future development of populations of wild and cultivated plant species. Grasslands ecosystems also support large populations of invertebrates, many such as earthworms providing key food resources for birds and mammals (Kruuk and Parish 1981; Peach et al. 2004). These functions are often collectively termed "ecosystem services" (Tancoigne et al. 2014) and have wider benefits such as prevention of soil erosion, carbon sequestration, and genetic conservation. Improved soil structure, nutrient retention, and N-fixation comprise some of the additional benefits derived from improved grasslands. These grasslands are now increasingly recognized for their wider contribution to society and their delivery of "public good" (Abberton et al. 2008; Humphreys et al. 2014a).

An important but little acknowledged feature of improved and permanent grassland is the enormous biomass perennial grassland species produce below ground (Newman et al. 1989) – a feature that is considered highly relevant to the provision of ecosystem services (Bardgett et al. 2014). A survey of old grassland in the UK carried out by Dickinson and Polwart (1982), for example, measured a root biomass of around 5 t/ha in the top 15 cm of soil. Plant breeding effort has focused primarily on above-ground traits and has largely neglected the potential to improve root design and function, mainly due to difficulties in root phenotyping and selection (Jahufer et al. 2008). In any case, forage crops grown in highly fertilized monocultures have maximum above ground production and forage quality as the main breeding objectives; here the form and function of the root system are considered to be relatively unimportant. As a result of these factors, many of the varieties of our current forage species are relatively shallow rooting, a trait that will compromise both their long-term persistency and yield potential following onset of stress conditions (Humphreys et al. 2014). Many of the ecological and environmental benefits provided by grasslands emanate from the interactions between the roots of different plant species and the soil in which they grow. Therefore, it is essential that modern plant breeding strategies aiming to promote sustainable crop production take into account the overall effects of growth by the crop on the surrounding biota and ecosystem.

Here, we review current knowledge on the role of grassland ecosystems in delivering ecological and environmental benefits related to soil structure and functioning. The plant breeding focus will be predominantly on root traits, and will also consider how new developments in dynamic root imaging, combined with an increased understanding of the genetic bases of variation in root architecture, could bring about a step-change in our awareness of these root traits. This understanding will allow traits such as root ontogeny, depth, thickness, and distribution to be designed strategically to produce an optimal balance between above-ground biomass productivity and below-ground biotic and abiotic interactions. The review focuses primarily on UK grasslands but many of the concepts and approaches discussed are applicable to the wider temperate grassland areas and to grassland systems in other climatic zones.

Environmental and Ecological Benefits from Grassland

Regulation of water release and flood prevention

Grasslands, particularly in temperate regions, frequently occur in land areas where rainfall is the highest, which for the UK is predominately in western and northern regions. These grasslands are important as catchments for major rivers, where they regulate water acquisition, its quality and its later release from soils. Climate change will inevitably lead to changes in agro-ecosystem functioning. The most recent Intergovernmental Panel on Climate Change (IPCC) report predicts major increases in global mean air temperatures of between 1.8 and 4.0°C by 2100, bringing with it greater uncertainty in weather patterns and also an increased incidence of extreme events (IPCC 2013; WMO 2013). Heavy rainfall impacts on the capacity of agricultural grasslands to effectively retain the rainwater that falls on them and to regulate its rate of loss, thereby leading to significant run-off and flooding of lowland areas. In this regard, the study by Peukert et al. (2014) noted several rainfall events where the runoff coefficient on intensively managed grassland exceeded 100%; these findings point to the need to enhance the capacity of grassland systems to deliver effective water regulation.

The incidence of extreme weather events has become more frequent in the last 20 years (WMO 2013). For example, in Europe, flood and drought events currently estimated to have a frequency of one per 100 years are now predicted to recur every 10–50 years by the 2070s (Lehner et al. 2006; WMO 2013). In temperate European catchments, increased volumes and intensity of winter rainfall are predicted, leading to raised levels of erosive runoff (Sauerborn et al. 1999). The immediate impact of flooding on crop production alone in 2014 in England was estimated at £25 million (ADAS 2014). However, the socio-economic after-effects of these extreme weather events may also persist for many years after the event has occurred, and in some cases have led to destabilization of local communities (Lehner et al. 2006). Despite these risks, our understanding of how extreme events will

impact on crops and soil functioning, and consequently on sustainable livestock farming, remains poor and requires urgent attention.

Hydrologists have long been aware of the pivotal role of vegetation in regulating and buffering the hydrological cycle. Changes in land use have been shown to change the regional climate (Stohlgren et al. 1998). At the single plant scale, biophysical changes in soil hydraulic properties due to root activity have also been demonstrated (Whalley et al. 2005). Thus, root activity tends to increase the number of large pores in soils, and there is a tendency for this property to change their water release characteristics. Detailed studies have illustrated the importance of rooting depth and the vertical variation in root function on soil water uptake. They have also highlighted that soil porosity is not a fixed parameter and that its dynamics are strongly influenced by vegetation (Rodriguez-Iturbe et al. 2001). Rooting depth determines the soil volume from which plants draw water and is influenced by various key soil hydraulic properties; soil texture and rooting depth largely define the plant-available water capacity (MacLeod et al. 2007). Recent unpublished results using X-ray Computed Tomography have revealed very different impacts on soil porosity derived from *Lolium*, *Festuca*, *Festulolium*, and *Trifolium* roots over time and have shown how these effects changed with contrasting soil types (Mooney S.J., pers comm.).

MacLeod et al. (2013) reported how a deep rooting *Festulolium* (a hybrid between a ryegrass and fescue species) variety reduced rainfall run-off compared to both its parental species by 51% compared to perennial ryegrass, and by 43% compared to meadow fescue (*Festuca pratensis*). The *Festulolium* variety was a synthetic version of *Festulolium loliaceum*, a natural grass hybrid involving these ryegrass and fescue species which occurs naturally in waterlogged soils in mature water meadows. Frequently in this habitat its parental species are either absent or present in only low numbers, implying that the hybrid has a selective advantage (Humphreys and Harper 2008). MacLeod et al. (2013) suggest that the deep rooting of the *Festulolium* variety and its subsequent dieback, particularly at depth in the root profile, were enhancing soil structure and through this assisting water retention. Among the temperate legumes, red clover (*Trifolium pratense*) is deeper rooting than white clover (*T. repens*) and while its effects on reducing run-off are currently unknown, its potential benefits in this regard are currently under investigation at IBERS. These investigations follow a dual strategy to produce new grasses and legumes that (i) provide nutritious forage for livestock, and (ii) possess enhanced root traits that can improve soil hydrology, and are carried out as part of a new BBSRC-LINK Programme, SUREROOT (www.sureroot.uk). The novel grasses and legumes are being assessed under diverse livestock management systems at different UK locations. For the first time, two new UK national capabilities facilities are being used in conjunction: IBERS' National Plant Phenomics Centre (NPPC) in Aberystwyth and Rothamsted Research's Farm Platform at North Wyke, Devon. Detailed analysis of changes in root architecture and ontogeny are identified in the NPPC, and their impact on soil hydrology at the field scale is measured on the Farm Platform. Here, the fields are isolated hydrologically to enable rainfall run-off to be accurately determined. The soils at North Wyke are representative of many in western England, are shallow and cover highly impermeable clay-layers, and are thus prone to flooding. Over an 80 h period in November 2012, more than 46×10^6 L of rain fell on the North Wyke Farm Platform, of which 90% was lost as overland flow or in drainage (P. Murray, pers. comm.), illustrating the importance of improving the capacity of grassland to retain water.

Where surface runoff occurs, there is the potential for soil erosion. Grasslands have in general been considered to reduce water erosion in comparison with arable-cropped land, although this view is being increasingly challenged, particularly in relation to impacts on aquatic ecosystems. Protection of the soil surface from raindrop impact, higher surface infiltration rates and enhanced stabilization of otherwise erodible fine soil particles into larger water-stable aggregates are considered mechanisms by which grasslands restrict erosion. It is clear, however, that where these benefits are not achieved grasslands may fail to deliver soil conservation objectives. The physical changes that promote soil conservation arise directly from physical binding by roots, but also through a complex interplay between roots and the soil biological community, driven by root-derived inputs of C. For example, there is abundant evidence that earthworm burrows make a major contribution to surface infiltration, reducing overland flow (e.g., Shipitalo and Butt 1999), and that grasslands support large populations of these burrowing earthworms (e.g., Scullion et al. 2002).

Carbon (C) sequestration

Reducing the "carbon footprint" of food production is fundamental for developing globally sustainable agriculture to support current human population growth in a changing climate. Because of the large amounts of C held as organic matter (SOM) in the world's soils, it is very important to understand how best to conserve or, if possible, increase soil C stocks (Powlson et al. 2014).The cultivation of perennial crops with large and deep rooting systems to increase the input of atmospheric CO_2 to agricultural soils has recently been highlighted as a potential approach for reducing the impact of agricultural production systems on greenhouse gas emissions (Kell 2011).

Whether or not an ecosystem (above- and below-ground) accumulates or loses C is a function of input and output. Sequestration of C occurs when gross primary productivity (GPP) exceeds ecosystem respiration (ER), which in turn is the sum of plant respiration and heterotrophic respiration of nonphotosynthetic organisms. This has been termed net ecosystem productivity (NEP) (Chapin et al. 2006). The final rate of accumulation or loss of C in a particular ecosystem (net ecosystem carbon budget (NECB)), in addition to NEP, will depend on external deposition of C (such as inputs of organic manures and dissolved C in rain water) and also loss through erosion, removal (harvesting), and nonbiological oxidation through fire or UV radiation (Lovett et al. 2006).

We believe that should opportunities for C sequestration in grasslands be enhanced through the widespread use of new varieties with high root biomass turn-over, particularly at depth, it may well provide a low-cost short-term measure to mitigate atmospheric C accumulation until "low" or "zero C" energy sources have opportunities to take effect. The strategy would have advantages over the high economic and environmental costs associated with long-term measures such as engineering techniques of CO_2 capture and injection into geological and oceanic strata.

Plant roots constitute a significant proportion of the C transferred to soils as mediated by processes in the rhizosphere, and perennial plants such as grasses offer significant advantages over annuals as a means of increasing soil C stocks because of their prolonged photosynthetic activity, greater root biomass, and proportionately deeper rooting. Furthermore, grasslands have an advantage over arable crops in not requiring annual ploughing and resowing, which through the soil disturbance generated will inadvertently cause losses of C into the environment.

Kell (2011) highlighted a key focus of research to be root–soil interactions that govern SOM dynamics and the possibility of breeding plants with root traits to deposit more or more-recalcitrant C in soils, especially at depth. In addition to contributing to atmospheric CO_2 sequestration, increased root C deposition may benefit soil quality and function, through better soil physical structural stability, and by favoring beneficial biological communities in the rhizosphere (Peiffer et al. 2013). However, it should be noted that plant breeding activity targeted at improved C sequestration in soils is both highly problematic and complex, and its effects are dependent on many interacting factors. Plant-derived inputs of C to soil include root turnover, root exudation, and mycorrhizal turnover. Each of these has different seasonal and spatial dynamics in the soil, and different dependencies on plant and soil conditions. For example, a plant that maintains roots for longer (i.e., less root turnover) may allocate less C to the production of new roots, but expend more C and energy on maintaining old roots that may be less efficient in nutrient and water capture (Norby and Jackson 2000).

The capacity of a particular soil to accumulate SOM is finite and depends on a complicated interaction between physical, chemical and biological processes (Powlson et al. 2014). It is now acknowledged that new inputs of fresh plant-derived C and other root-induced changes in the soil may have a totally contrary effect to that desired, and may stimulate the turnover of existing SOM and nutrients by soil microbes. Reports of the effects of such "priming" record instances where fresh labile organic matter was found to stimulate the decomposition of older SOM throughout the soil profile (Kuzyakov 2010) and more specifically at depth (Fontaine et al. 2007). What happens to soil C stocks at depth during disturbances such as drying or ploughing, land-use change, etc. is still little understood (Gregory et al. 2011).

Deeper and more highly branched rooting will increase the contact of organic matter, both root exudates and decomposing roots, with soil minerals, and may have an important benefit in terms of soil C sequestration (Jobbagy and Jackson 2000; Kell 2011, 2012). In addition, root C appears to make a higher contribution to soil C than above-ground inputs, especially at depth (Rasse et al. 2005; Mendez-Millan et al. 2010). Deep soil C is an important contributor to overall soil C stocks. Gregory et al. (2011) estimated that 980 Mt of organic C is stored below 30 cm depth in soils in England and Wales, approximately 50% of the total. Evidence from tropical savannah also suggests that the planting of exotic deeper rooting plant varieties leads to significant increases in soil C stocks (Fisher et al. 1994). However, in order for deep rooting plants to have a significant effect, the soil must be deeper than the normal rooting depth and root growth must not be impeded by compaction, nutrient limitation, or waterlogging. On thin soils, or where roots cannot penetrate, deeper rooting plants are not going to provide additional sequestration benefits.

Pasture improvement and the introduction of high yielding forage grasses and legumes have been shown both theoretically (e.g., Soussanna et al. 2004) and practically (e.g., Lal et al. 1998) to substantially increase C stocks as compared to that achieved from some preexisting grasslands or land under arable farming. For example, compared to annuals, the sowing of perennial crops such as *Festuca arundinacea* increased soil C stocks by 17.2% (equivalent to C sequestration of circa 3 Mg C/ha/year over a 6-year period). As far as we are aware, no attempts have been made to breed explicitly for increased soil C sequestration in any economically important temperate forage species,

mainly due to unanswered questions regarding (a) the magnitude, (b) genetic control, and (c) G × E modulation of the various C flux pathways between plant, soil and atmosphere. Consequently, uncertainties about which traits are most likely to confer durable increases in C sequestration remain. The feasibility of evaluating traits likely to be linked with C sequestration and subsequently identifying the associated QTLs (thereby enabling marker-assisted selection) is currently under investigation at IBERS.

Improvement of soil structure

Grassland farming relies largely on long-term production from perennial species, and sustainable grassland production consequently depends on the maintenance of good soil quality. Inherent soil characteristics (e.g., texture and mineralogy) form a significant part of "soil quality", and these are not readily amenable to human manipulation. However, other aspects of soil quality are more dynamic in nature, being strongly affected by recent land use. Such attributes include SOM (discussed in Section 'Carbon (C) sequestration') and soil structural properties (e.g., porosity, permeability, and aggregation) (Carter 2000).

There is an increasing body of evidence that plants differ in their effects on soil structure (Materechera et al. 1994; MacLeod et al. 2007). Soil structure is commonly interpreted through the concept of an aggregate hierarchy (Tisdall and Oades 1982; Miller and Jastrow 1996), and comparative studies have shown that differences in aggregating and stabilizing efficiency in soils vary not only between plant species (Drury et al. 1991; Materechera et al. 1994), but also between varieties within species (Carter et al. 1994), and even potentially between genotypes within varieties (MacLeod et al. 2007). These effects may be direct, resulting from the influence of morphological root traits on binding soil particles, or indirect, for example, through associated biological action on aggregate formation. The latter effects may have a stronger impact on soil structure (Bardgett et al. 2014).

Direct effects

Several investigations have been carried out on forage plant-driven changes in soil structure, and some have compared the effects of different plant species. Mytton et al. (1993) and Holtham et al. (2007) both described visible differences between white clover and perennial ryegrass soil cores in terms of particle aggregation, with much greater aggregation present in the white clover cores. In the latter study, while the soil in the white clover cores was more structured than that in the perennial ryegrass cores, the root biomass present in white clover was substantially lower. Aggregation of soil particles near white clover roots was also observed in glass-fronted rhizotrons by Pugh et al. (1995). The process of aggregation is important for many aspects of soil functioning related to plant growth, and soils with more stable aggregates are also more resistant to surface crusting (Le Bissonnais and Arrouyais 1997), and to compaction (Angers et al. 1987). Such soils consequently favor seedling emergence, root growth, and water infiltration and storage (Angers and Caron 1998). Increases in aggregation also contribute to SOM build-up and thus to nutrient storage. In another component of the study of Mytton et al. (1993), significant improvements in drainage in soil cores containing white clover were observed, founded on differences in soil macropore space. Values of soil macroporosity for pure perennial ryegrass and white clover soil cores were 23.6% and 45.3%, respectively. Holtham et al. (2007) also recorded substantially higher rates of drainage from soil cores under white clover (599 ml/day) than under perennial ryegrass (115 mL/day). They attributed this large difference to the effects of local structuring around the clover roots, creating a more porous soil. The pore structure of the cores was simulated using a void space network model ("Pore-Cor"). This confirmed larger pores beneath white clover, a difference between the species in local soil structuring and a saturated hydraulic conductivity (describing the ease with which water can move through pore spaces) in white clover that was four times greater than under perennial ryegrass. The formation of continuous macropores, key for promoting conductivity, by penetrating roots represents a significant plant-induced change in soil structure. Macropores facilitate aeration, and water movement and temporary storage in the soil, as well as decreasing resistance to further root growth (Angers and Caron 1998). Soil macroporosity and pore size distribution were also measured by Papadopoulos et al. (2006), using high-resolution image analysis in a study comparing soil structuring under five plant-type treatments in a stockless organic rotation. The only inputs to the plots were plant biomass from the treatments themselves. In the spring of the second year of the rotations, the two red clover/perennial ryegrass treatments had by far the highest values of macroporosity and the vetch treatment had the lowest. The high soil porosity values initially observed by Papadopoulos et al. (2006) in the red clover treatments were attributed to the way in which this species rapidly develops a high- density root mat. However, differences between treatments were transient, and by later in the summer had disappeared. Three years later, when all plots had grown winter wheat, it was found that macroporosity and pore size were similar in all plots. Thus, it appears that the measurable and rapid improvements in soil structure and aggregate stability brought about by legume species may not be robust and could be quickly reversed by subsequent arable crops in a rotation

system. In longer term grassland systems, however, the positive effects of legumes may persist.

The roots of growing plants help to aerate the soil by creating channels through which water, soil solutions, microorganisms, and soil invertebrates can move easily. Root system architecture and depth distribution are consequently important contributors to soil quality and therefore merit attention. Broad interspecific differences have been identified between perennial ryegrass and white clover in terms of root dimensions (e.g., mean root diameters of 0.19 and 0.26 mm, respectively, measured by Evans (1977)). Significant intraspecific differences in root system morphology between white clover populations grown as spaced plants in the field have also been described, with some varieties producing large tap-rooted systems with a small proportion of finer, fibrous roots, and others producing no large tap-roots and a high proportion of fibrous roots (Caradus 1977). However, more detailed information on root architecture (e.g., the proportions of root lengths produced in different diameter size classes at different depths in the root profile) is lacking, due to the technical difficulties associated with measuring plant roots *in situ*.

Indirect effects

Soil aggregation processes result from the production of organic-binding agents (e.g., polysaccharides) by microbes, from microorganisms breaking down organic matter, and through the enmeshing effects of plant roots and fungal hyphae (Watson et al. 2002). Due to the temporary nature of plant roots and fungal hyphae (with tissue turnover), their ability to act as soil structuring agents is related to their rates of production and longevity, along with the effects of root products on the activity of the soil biotic community. This suggests that the importance of root system architecture to the formation and stabilization of soil aggregates is related to the kinds of roots produced, as well as to other factors associated with the mycorrhizal condition of the roots (Miller and Jastrow 1990). Specifically, it is thought that coarser-rooted plants are more likely to be dependent on a mycorrhizal association than finer, fibrous-rooted plants (Heterick et al. 1992), and are therefore more likely to have a greater affinity for extra-radicle fungal hyphae in the soil (Miller and Jastrow 1996). Perennial legume species tend to have a high dependence on associations with mycorrhizal fungi, except under high-P conditions (Heterick et al. 1992). It has therefore been proposed that a major part of the influence of legumes on soil structure results from the interaction of their root systems with associated mycorrhizas (Miller and Jastrow 1996). In contrast, perennial ryegrass usually shows no, or only a small positive response to inoculation with mycorrhizas in terms of plant growth and nutrient uptake (Hall et al. 1984). This has been attributed to the fact that perennial ryegrass has an extensive, finely divided root system (Evans 1977), and therefore benefits less from the mycorrhizal symbiosis. In addition to the potential benefits for soil structure arising from physical binding by fungal hyphae, their input of glomalin-like compounds (Rillig and Mummey 2006) has also been implicated in the stabilization of soil aggregates. There is again evidence that mycorrhizas vary in the extent of these inputs and that this variation is affected by root-fungal interrelationships.

Farmland biodiversity

There has been a general deterioration in biodiversity on UK farmland over recent decades with declines seen in plant, insect, and bird populations (Firbank et al. 2013). Grasslands have the potential to deliver biodiversity services but they vary in the extent and nature of the benefits that they deliver. In some aspects (e.g., pollinating insects), the benefits relate to the diversity and management of the plant communities present. Productive grasslands tend to have a limited range of plant species present and any biodiversity benefits relate mainly to their support of large soil faunal communities (e.g., Scullion et al. 2002) based on limited disturbance and large inputs of organic residues; these communities provide a substantial input to above-ground food chains benefitting a range of farmland birds and mammals. Alterations to rooting traits, and associated C inputs to soils, have the potential to affect both soil and above-ground communities. Although there are a number of soil invertebrates contributing to this link, earthworms have been studied most intensively.

A number of the farmland bird species currently in decline forage extensively on grasslands for earthworms and other soil invertebrates. For example, Peach et al. (2004) concluded that recovery of song thrush (*Turdus philomelos*) populations in lowland Britain will require an increase in grassland cover to support these food resources. In addition, maintenance of moist soil surfaces during early summer is necessary to increase prey availability and improve recruitment of fledglings into breeding populations.

Farmland mammals show a similar reliance on prey that are present in greater abundance in grassland compared with other agricultural land uses. Badgers (*Meles meles*) have a strong seasonal dependence on earthworms and insect larvae in their diet (Cleary et al. 2011). These species and others such as moles (*Talpa europaea*) and foxes (*Vulpes vulpes*), have a preference for the earthworm species *Lumbricus terrestris* in their diet, indicating that the composition of prey communities may be as important as their overall biomass (Murchie and Gordon 2013).

Grass breeding initiatives that enhance C input to soils via their root systems are likely to increase grassland soil faunal populations overall. Where changes to root architecture favor a redistribution of C inputs to greater depths, this change may affect the composition of these populations, for example, promoting deeper burrowing earthworm species such as *L. terrestris*, rather than those inhabiting surface layers. Impacts of changes in root architecture on soil profile moisture regimes may also affect prey availability. Greater storage of incident rainfall (see Section 'Regulation of water release and flood prevention') and enhanced water uptake from depth may prolong the period over which surface soils are moist enough to facilitate near surface feeding on soil invertebrates. Therefore, future plant breeding programs may have much wider impacts on agroecosystems than those focussed on soil characteristics alone.

Implications for Plant Breeding Programs

Selection for above versus below ground biomass in forage species

Until recently, relatively little emphasis has been placed in forage species on selection for below-ground biomass, despite recognition of the fact that root system morphology (size and architecture) is crucially important not only for provision of the long-term ecosystem services described above, but also for the basic functions of water and nutrient uptake. Genetic improvement of forage grasses and legumes has traditionally focused on the selection of traits that improve sward yield and animal performance, for example, by increasing production of above-ground biomass and improving forage quality characteristics and plant persistence. The majority of these traits are included in the statutory evaluation of varieties in official trials. However, the major challenges facing forage plant breeders now include selection for traits that deliver environmental benefit (e.g., larger, deeper root systems), and dealing with any resultant trade-offs between environmental and production traits. It is well known that improvements in above-ground yield traits in forage species do not necessarily result in larger root systems, and this may impair the capacity of new varieties to deliver efficient water and nutrient uptake combined with environmental benefits (Crush et al. 2010). The presence of appropriate root system traits in forage species has measurable effects on plant performance under conditions of drought and nutrient stress. In white clover, for example, differences in root system depth were shown to be strongly related to the performance of cultivars and ecotypes under drought stress (Ennos 1985). Also in this species, root traits have been shown to affect the uptake of nutrients such as P (Nichols et al. 2014a,b). In this case, root length and frequency of branching were identified as the key traits for P uptake.

Focusing on white clover, the most widely grown temperate forage legume, some research has been carried out to quantify the potential for genetic improvement of root traits (Ennos 1985; Caradus and Woodfield 1990, 1998; Woodfield and Caradus 1990; Nichols et al. 2007; Jahufer et al. 2008). Significant genetic variation has been observed, and it is thought that a major component of variation in root growth is due to additive gene effects which can be selected for (Woodfield and Caradus 1990). Reported heritability for root system dry weight was higher than for shoot dry weight, and heritabilities for taproot diameter and proportion of the roots with a diameter greater than 1 mm were both high enough to make these traits amenable to selection (Caradus and Woodfield 1990). However, the presence of strong correlations between shoot and root traits within white clover morpho-types (e.g., leaf-size categories) presents a complication when selecting for root characters in this species (Annichiarico et al. 2015). Indirect selection for root traits to enhance drought tolerance has been carried out in white clover, and the results showed limited scope for improvement within the morphotype of germplasm used (Annicchiarico and Piano 2004). The absence of an effect of root traits (in this case overall root biomass) on drought tolerance in the latter study was considered to be due to the existence of concurrent, less efficient methods for controlling water loss. Thus, it appears that the strong correlations in white clover between the shoot and root system morphology may result in a conflict, when considering selection for drought tolerance, between the improved water uptake associated with the deeper and more extensive root systems of larger-leaved, more tap-rooted morphotypes and the superior water conservation associated with small leaf size (Woodfield and Caradus 1987). An alternative approach to improving drought tolerance in white clover involves its hybridization with other, closely related species that exhibit this trait (e.g., Marshall et al. 2001; Nichols et al. 2014a,b). The introgression into white clover of the rhizomatous growth habit from Caucasian clover, *Trifolium ambiguum* (Marshall et al. 2001) has shown promise for improving drought tolerance (Marshall et al. 2015). The latter study concluded that the greater biomass at depth in the root profile in backcross hybrids of the two species contributed to their superior drought tolerance compared with white clover. This change in root system shape was not achieved at the expense of reductions in forage yield or quality in the hybrid germplasm (Marshall et al. 2003, 2004). Nichols et al. (2014a,b) showed that above-ground yield in first generation backcross hybrids between white clover and

Trifolium uniflorum was significantly less affected by long-term drought than the white clover parent, and that biomass allocation to roots increased.

In forage grasses, alternatives to ryegrasses, for example, tall fescue, with large deep root systems and more efficient water use have been used regularly in areas such as in the USA where water supply is suboptimal and droughts occur regularly. *Festulolium*-based research and breeding is gaining increasing acceptance as a way forward that can harness the attributes of rapid growth rates, establishment, and forage quality traits present in ryegrasses with the enhanced stress resistance, root depth, and strength traits found in the fescues (Ghesquiére et al. 2010). The variety "Lueur" developed in France, an amphiploid hybrid combination of *L. multiflorum* and *F. glaucescens*, has been shown to extract water more effectively than ryegrass from depth in soils and therefore to offer greater drought resistance (Durand et al. 2007; Ghesquiére et al. 2010). Novel deep rooting *L. perenne* × *F. glaucescens* and *L.perenne* × *F. mairei* amphiploid hybrids have been developed at IBERS and have demonstrated excellent agronomic traits with the potential for improved efficiency of ruminant nutrition (Humphreys et al. 2014b). An introgression-breeding approach similar to that described above involving Caucasian clover has been used successfully in *Festulolium*. In this case, in two separate breeding programs, *F. glaucescens* and *F. arundinacea* genes were introduced onto different locations on chromosome 3 of *L. multiflorum* and subsequently also transferred into *L. perenne* (Humphreys and Thomas 1993; Humphreys et al. 2005, 2014a). In all instances, the water-use-efficiency was significantly enhanced compared to *Lolium*. Chromosome 3 was a good target for entry of novel allelic variants into *Lolium* for drought resistance as QTL for the trait have been detected throughout the chromosome in *Festuca*, but have never been reported there in *Lolium* species (Turner et al. 2008; Alm et al. 2011). Turner et al. (2008) reported a correlation between increased root growth and enhanced leaf extension under drought when certain *Festuca*-derived genes were transferred onto chromosome 3 of *Lolium*. The forage production and quality of these different combinations of ryegrass and fescue is being studied (Humphreys et al. 2014b) in parallel with analysis of root ontogeny (MacLeod et al. 2013).

Selection for above-ground biomass, in parallel with below-ground biomass, is part of the IBERS forage breeding programme within the "Public Good Plant Breeding Group". The "SUREROOT" project funded through the Biotechnology and Biological Research Council LINK programme includes an element of selection for root architectural traits, especially root depth and thickness. This research has highlighted the challenge facing plant breeders of developing appropriate methodologies to phenotype root traits to quantify the variation in root traits within breeding populations.

The following sections summarize the new technologies being used to quantify variation in root architecture between and within species and in response to specific management strategies.

Phenotyping approaches

Phenotyping of root traits presents technical challenges, particularly in relation to the number of plants that must be screened in a plant breeding program. Typical approaches, such as those described by Chmelikova et al. (2015) for analysis of red clover roots, involve digging up a soil monolith beneath the plant and washing to extract the below-ground organs of the plants from the soil. The root mass of the whole plant is then digitized using a scanner and the number and size of roots quantified. This often leads to underestimation of fine roots through breakage during washing, and the three dimensional spatial distribution is also lost (Mairhofer et al. 2013). Alternative nondestructive approaches have included the analysis of root systems in flowing solution culture (Abberton et al. 2000; Collins et al. 2003), but the relevance of these results to performance in the field is often questioned. A more recent approach has been the application of automated scanning technology to reconstruct a three-dimensional data set. CT X-ray tomography, for example, has been used to analyze root architecture of plants grown in soil (Mairhofer et al. 2013) and to analyze the effect of soil moisture on root architecture (Zappala et al. 2013).

Other phenotyping approaches are also available to study the root architecture of forage grasses and legumes and to quantify the impact of different root architecture on water flow and nutrient dynamics. For example, the NPPC at IBERS provides dynamic (nondestructive) developmental and physiological imaging of shoots and roots of plants automatically moved from controlled environments to imaging systems. This state-of-the art system is designed for automatic, high throughput, nondestructive phenotyping of a wide range of plant material. Near infrared (NIR) thermography provides multiple-sided imaging of roots and soil, to detect root growth and soil water content profile changes for plants grown in root columns. In this system, root columns are transported on carriages identified by RFID tags allowing each plant to be imaged, and provided with precise watering and nutrition (or, if necessary, sprayed). Watering, nutrient, and weighing stations allow precise control on a per-plant basis for controlled drought, nutrient, and water stress experiments. An example of an output of the detailed

monitoring of monthly changes in root distribution and number for a *Festulolium* hybrid-derivative is shown in Fig. 1.

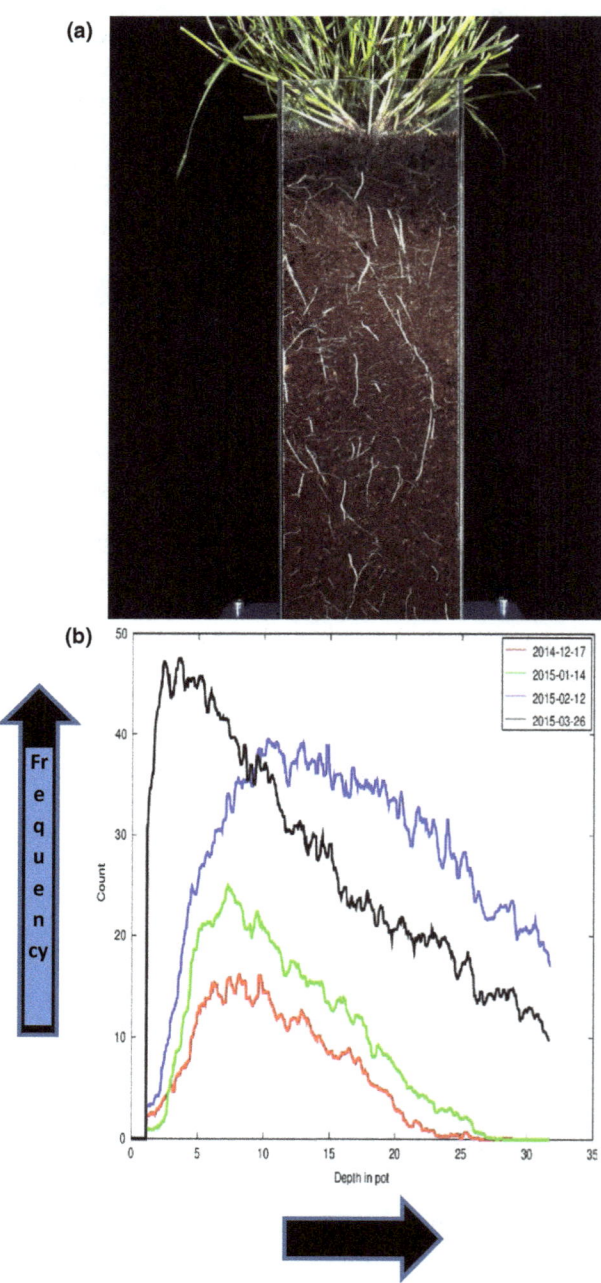

Figure 1. (a) Festulolium grass hybrid growing in potting compost within a 12 × 12 × 50 cm clear column for root analysis over consecutive months in the National Plant Phenomics Centre at IBERS and (b) Root ontogeny measures derived from comparisons in root density scores of a Festulolium grass hybrid taken over 4 consecutive months. Mean root density scores are calculated from 12 merged root images each representing consecutive 5 cm sections along a 50 cm root column. High vis camera images were captured and merged at the National Plant Phenomics Centre at IBERS.

Plant testing systems

The varieties of grasses and legumes used in agricultural and amenity grassland systems within the EU must meet certain standards if they are to be marketed. A series of protocols that determine their DUS (Distinctiveness, Uniformity and Stability) and VCU (Value for Cultivation and Use) are used to identify the best varieties for use (Gilliland and Gensollen 2010). The DUS system is designed to provide protection for the intellectual property rights (IPR) residing in existing novel varieties and is controlled by legislation in all EU member states, conducted in compliance with international guidelines compiled by UPOV (The International Union for the Protection of New Varieties of Plants). Assessing the VCU of forage grasses and legumes is subject to national statutory testing within all EU member states. Such systems seek to ensure that new varieties reach a certain level of performance before seed of those varieties can be sold commercially. New plant varieties are therefore evaluated in agronomic trials with control varieties, and only those that reach a certain level of performance are added to the UK Common Catalogue and can be sold. However, there are no internationally agreed guidelines within the EU and no harmonized testing protocols between official testing authorities. The testing systems used differ between species and countries, and for agricultural species they focus specifically on the evaluation of agronomic traits such as dry matter yield and disease resistance, with relatively limited analysis of forage quality characteristics (Gilliland and Gensollen 2010). Considerable emphasis is now being placed on the measurement of nutritional parameters that can quantify "forage value" to farmers in terms of direct ruminant benefit. Despite these developments, the current systems are not designed to quantify the merits of improved forage varieties in terms of ecosystem services. Development of appropriate tests that analyze the value of forage varieties in terms of delivery of such services is therefore unlikely and probably commercially unsustainable. Testing systems need to be robust with relatively simple protocols, necessitating decisions on which "ecosystem services" are to be targeted, followed by the development of appropriate protocols for measuring the impact of plant traits on these services and ranking them in comparison with agronomic and forage quality traits.

Conclusions

This paper has concentrated on the capacity of grassland ecosystems to deliver environmental benefit, and the challenge of balancing the delivery of these benefits whilst maintaining the production of high-quality forage for livestock production. It has focused on the potential for the genetic improvement of specific root traits within the forage species that are currently being used in UK grassland agriculture. Within the scope of this review, it was not feasible to consider the potential opportunities from inclusion of alternative forage species or to consider the impact that different grassland management systems may have on the ability of grassland ecosystems to deliver these environmental services, but these should be considered important factors when assessing the potential of grassland systems.

Acknowledgments

MH is funded by the Biotechnology and Biological Sciences Research Council. AM, MH, and RC are part funded by the BBSRC-LINK programme SUREROOT which in addition to BBSRC is jointly supported by partners from UK's grass seed, grassland and livestock industries. MH and JS are part of the Climate-Smart Consortium. Funding to support this work was provided by the Welsh Government and HEFCW through the Sêr Cymru National Research Network for Low Carbon, Energy and Environment.

Conflict of Interest

The authors declare that they have no conflict of interest.

References

Abberton, M. T., J. H. MacDuff, S. Vagg, A. H. Marshall, and T. P. T. Michaelson-Yeates. 2000. Nitrogen fixation in hybrids of white clover (*Trifolium repens* L.) and Caucasian clover (*Trifolium ambiguum* M. Bieb). J. Agron. Crop Sci. 185:241–247.

Abberton, M. T., A. H. Marshall, M. W. Humphreys, and J. H. MacDuff. 2008. Genetic improvement of forage species to reduce the environmental impact of temperate livestock grazing systems. Adv. Agron. 98:311–355.

ADAS. 2014. Impact of 2014 winter floods on agriculture in England. Defra, United Kingdom.

Alm, V., C. M. Busso, A. Ergon, H. Rudi, A. Larsen, M. W. Humphreys, et al. 2011. QTL analysis and comparative genetic mapping of frost tolerance, winter survival and drought tolerance in meadow fescue (*Festuca pratensis* Huds.). Theor. Appl. Genet. 123:369–382.

Angers, D. A., and J. Caron. 1998. Plant-induced changes in soil structure: processes and feedbacks. Biogeochem. 42:55–72.

Angers, D. A., B. D. Kay, and P. H. Groenevelt. 1987. Compaction characteristics of a soil cropped to corn and bromegrass. Soil Sci. Soc. Amer. J. 51:779–783.

Annicchiarico, P., and E. Piano. 2004. Indirect selection for root development of white clover and implications for drought tolerance. J. Agron. Crop Sci. 190:28–34.

Annichiarico, P., B. Barrett, E. Charles Brummer, B. Julier, and A. Marshall. 2015. Achievements and challenges in improving temperate perennial forage legumes. Crit. Rev. Pl. Sci. 34(1–3):327–380.

Bardgett, R. D., L. Mommer, and F. T. De Vries. 2014. Going underground: root traits as drivers of ecosystem processes. Trends Ecol. Evol 29:692–699.

Caradus, J. R. 1977. Structural variation of white clover root systems. N. Z. J. Agric. Res. 20:213–219.

Caradus, J. R., and D. R. Woodfield. 1990. Estimates of heritability for, and relationships between, root and shoot characters of white clover. I. Replicated clonal material. Euphytica 46:203–209.

Caradus, J. R., and D. R. Woodfield. 1998. Genetic control of adaptive root characteristics in white clover. Plant Soil 200:63–69.

Carter, M. R. 2000. Organic matter and sustainablity. Pp. 9–22 in R. M. Rees, B. C. Ball, C. D. Campbell and C. A. Watson, eds. Sustainable management of soil organic matter. CAB International, United Kingdom.

Carter, M. R., D. A. Angers, and H. T. Kunelius. 1994. Soil structural form and stability, and organic matter under cool-season perennial grasses. Soil Sci. Soc. Amer. J. 58:1194–1199.

Chapin, F. S., G. M. Woodwell, J. T. Randerson, E. B. Rastetter, G. M. Lovett, D. D. Baldocchi, et al. 2006. Reconciling carbon-cycle concepts, terminology, and methods. Ecosystems 9:1041–1050.

Chmelikova, L., S. Wolfrum, H. Schmid, et al. 2015. Seasonal development of above- and below-ground organs of Trifolium pratense in grass-legume mixture on different soils. J. Plant. Nutrit. Soil Sci. 178:13–24.

Cleary, G. P., L. A. L. Corner, J. O'Keeffe, and N. M. Marples. 2011. Diet of the European badger (Meles meles) in the Republic of Ireland: a comparison of results from an analysis of stomach contents and rectal faeces. Mammalian Biol. 76:470–475.

Collins, R. P., M. Fothergill, J. H. Macduff, and S. Puzio. 2003. Morphological compatibility of white clover and perennial ryegrass cultivars grown under two nitrate levels in flowing solution culture. Ann. Bot. 92:247–258.

Cruickshank, J. G. 1987. Land and land use. Pp. 19–41 in R. H. Buchanan and B. M. Walker, eds. Province, City and People: Belfast and its Region. Greystone Books, Antrim.

Crush, J. R., S. N. Nichols, and L. Ouyang. 2010. Adventitious root mass distribution in progeny of four perennial ryegrass (Lolium perenne L.) groups selected for root shape. N. Z. J. Agric. Res. 53:193–200.

Dickinson, N. M., and A. Polwart. 1982. The effect of mowing regime on an amenity grassland ecosystem: above- and below-ground components. J. App. Ecol. 19:569–577.

Drury, C. F., J. A. Stone, and W. I. Findlay. 1991. Microbial biomass and soil structure associated with corn, grasses and legumes. Soil Sci. Soc. Amer. J. 55:805–811.

Durand, J. L., T. Bariac, M. Ghesquiére, P. Brion, P. Richard, M. W. Humphreys, et al. 2007. Ranking of the depth of water extraction by individual grass plants, using natural ^{18}O isotope abundance. Environ. Exp. Bot. 60:137–144.

Eadie, J. 1984. Agriculture and the environment. ITE Sympos. 13:13–20.

Ennos, R. A. 1985. The significance of genetic variation for root growth within a natural population of white clover (Trifolium repens). J. Ecol. 73:615–624.

Evans, P. S. 1977. Comparative root morphology of some pasture grasses and clover. N. Z. J. Agric. Res. 20:331–335.

Firbank, L., R. B. Bradbury, D. I. McCracken, and C. Stoate. 2013. Delivering multiple ecosystem services from Enclosed Farmland in the UK. Agric. Ecosyst. Environ. 166:65–75.

Fisher, M. J., I. M. Rao, M. A. Ayarza, C. E. Lascano, J. I. Sanz, R. J. Thomas, et al. 1994. Carbon storage by introduced deep-rooted grasses in the South American savannas. Nature 371:236–238.

Fontaine, S., S. Barot, P. Barré, N. Bdioui, B. Mary, and C. Rumpel. 2007. Stability of organic carbon in deep soil layers controlled by fresh carbon supply. Nature 450:277–280.

Fuller, R. M., G. M. Smith, J. M. Sanderson, R. A. Hill, A. G. Thomson, R. Cox, et al. 2002. Countryside Survey 2000 Module 7. Land Cover Map 2000. Final Report. Centre for Ecology and Hydrology, Cambridge.

Ghesquiére, M., M. W. Humphreys, and Z. Zwierzykowski. 2010. Festulolium (Chapter 12). Pp. 293–316 in F. Veronesi, U. Posselt, B. Beart, eds. Handbook on Plant Breeding, Eucarpia Fodder Crops and Amenity Grasses Section. Springer, Dordrecht Heidelberg London New York, ISBN:9781441907592, 5.

Gilliland, T. J., and V. Gensollen. 2010. Review of the protocols used for assessment of DUS and VCU in Europe. Pp. 261–276 in C. Huyghe, ed. Sustainable use of genetic diversity in forage and turf breeding, proceedings of the 27th Conference of the Eucarpia Fodder and Turf Section. Springer, Dordrecht Heidelberg London New York.

Gregory, A. S., A. Whitmore, G. J. D. Kirk, B. Rawlins, K. Ritz, and P. Wallace. 2011. Review of the evidence base for the status and change of soil carbon below 15 cm from the soil surface in England and Wales. DEFRA, United Kingdom.

Hall, I. R., P. D. Johnstone, and R. Dolby. 1984. Interactions between endomycorrhizas and soil nitrogen and phosphorus on the growth of ryegrass. New Phtyol. 97:447–453.

Heterick, B. A. D., G. W. T. Wilson, and T. C. Todd. 1992. Relationships of mycorrhizal symbiosis, rooting strategy and phenology among tallgrass prairie forbs. Can. J. Bot. 70:1521–1528.

Holtham, D. A. L., G. P. Matthews, and D. S. Scholefield. 2007. Measurement and simulation of void structure and hydraulic changes caused by root-induced soil structuring under white clover compared to ryegrass. Geoderma 142:142–151.

Humphreys, M. W., and J. A. Harper. 2008. *Festulolium loliaceum* an understudied natural UK grass hybrid species that may provide benefits to UK grasslands withstanding the onsets of climate change. European Crop Wild Relative Newsletter 6:7–9.

Humphreys, M. W., and H. Thomas. 1993. Improved drought resistance in introgression lines derived from *Lolium multiflorum x Festuca arundinacea* hybrids. Plant Breed. 111:155–161.

Humphreys, J., J. A. Harper, I. P. Armstead, and M. W. Humphreys. 2005. Introgression-mapping of genes for drought resistance transferred from *Festuca arundinacea var. glaucescens* into *Lolium multiflorum*. Theor. Appl. Genet. 110(3):579–587.

Humphreys, M. W., G. O'Donovan, and M. Sheehy-Skeffington. 2014a. Comparing synthetic and natural grasslands for agricultural production and ecosystem service. Grassland Sci. Europe. 19:215–229.

Humphreys, M. W., S. A. O'Donovan, M. S. Farrell, A. Gay, and A. L. Kingston-Smith. 2014b. The potential of novel *Festulolium* (2n=4x=28) hybrids as productive, nutrient-use-efficient fodder for ruminants. J. Food Energy Security 50:1–13.

Huyghe, C., A. De Vliegher, and P. Golinski. 2014. European grasslands overview: temperate region. Grassland Sci. Europe 19:29–40.

IPCC. 2013. Climate change 2013: the physical science basis. Cambridge University Press, United Kingdom.

Jahufer, M. Z. Z., S. N. Nichols, J. R. Crush, L. Ouyang, A. Dunn, J. L. Ford, et al. 2008. Genotypic variation for root trait morphology in a white clover mapping population grown in sand. Crop Sci. 48:487–494.

Jobbagy, E. G., and R. B. Jackson. 2000. The vertical distribution of soil organic carbon and its relation to climate and vegetation. Ecol. Apps. 10:423–436.

Kell, D. B. 2011. Breeding crop plants with deep roots: their role in sustainable carbon, nutrient and water sequestration. Ann. Bot. 108:407–418.

Kell, D. B. 2012. Large-scale sequestration of atmospheric carbon via plant roots in natural and agricultural ecosystems: why and how. Philos. Trans. Royal Soc. Lond B. Biol. Sci. 367:1589–1597.

Kruuk, H., and T. Parish. 1981. Feeding specialization of the European badger *Meles meles* in Scotland. J. Anim. Ecol. 50:773–788.

Kuzyakov, Y. 2010. Priming effects: Interactions between living and dead organic matter. Soil Biol. Biochem. 42:1363–1371.

Le Bissonnais, Y., and D. Arrouais. 1997. Aggregate stability and assessment of soil crustability and erodibility: II. Application to humic loamy soils with various organic carbon contents. Eur. J. Soil Sci. 48:39–48.

Lal, R., J. M. Kimble, R. F Follett, and C. V Cole. 1998. The potential of U.S. cropland to sequester carbon and mitigate the greenhouse effect. Ann Arbor Press, Chelsea, MI.

Lehner, B., P. Döll, J. Alcamo, T. Henrichs, and F. Kaspar. 2006. Estimating the impact of global change on flood and drought risks in Europe: a continental, integrated analysis. Clim. Change 75:273–299.

Lovett, G., J. Cole, and M. Pace. 2006. Is net ecosystem production equal to ecosystem carbon accumulation? Ecosystems 9:152–155.

MacLeod, C. J. A., A. Binley, S. L. Hawkins, M. W. Humphreys, L. B. Turner, W. R. Whalley, et al. 2007. Genetically modified hydrographs: what can grass genetics do for temperate catchment hydrology? Hydrol. Process. 21:2217–2221.

MacLeod, C. J. A., M. W. Humphreys, R. Whalley, L. Turner, A. Binley, C. W. Watts, et al. 2013. A novel grass hybrid to reduce flood generation in temperate regions. Nature Sci. Rep. 3, 1683.

Mairhofer, S., S. Zappala, S. Tracy, S. Sturrock, M. J. Bennett, S. J. Mooney, et al. 2013. Recovering complete plant root system architectures from soil via X-ray μ-Computed Tomography. Plant Methods 9:8.

Marshall, A. H., C. Rascle, M. T. Abberton, T. P. T. Michaelson-Yeates, and I. Rhodes. 2001. Introgression as a route to improved drought tolerance in white clover (*Trifolium repens* L.). J. Agron. Crop Sci. 187:11–18.

Marshall, A. H., T. A. Williams, M. T. Abberton, T. P. T. Michaelson-Yeates, and H. G. Powell. 2003. Dry matter production of white clover (*Trifolium. repens* L.), Caucasian clover (*T. ambiguum* M. Bieb) and their associated hybrids when grown with a grass companion over three harvest years. Grass Forage Sci. 59:91–99.

Marshall, A. H., T. A. Williams, M. T. Abberton, T. P. T. Michaelson-Yeates, P. Olyott, and H. G. Powell. 2004. Forage quality of white clover (*Trifolium. repens* L.) × Caucasian clover (*T. ambiguum* M. Bieb) hybrids when grown with a grass companion over three harvest years. Grass Forage Sci. 59:91–99.

Marshall, A. H., M. Lowe, and R. P. Collins. 2015. Variation in response to moisture stress of young plants of interspecific hybrids between white clover (*T. repens* L.) and Caucasian clover (*T. ambiguum* M. Bieb.). Agriculture 5:353–366.

Materechera, S. A., J. M. Kirby, A. M. Alston, and A. R. Dexter. 1994. Modification of soil aggregation by watering regime and roots growing through beds of large aggregates. Plant Soil 160:57–66.

Mendez-Millan, M., M.-F. Dignac, C. Rumpel, and D. P. Rasse. 2010. Molecular dynamics of shoot vs. root biomarkers in an agricultural soil estimated by natural abundance ^{13}C labelling. Soil Biol. Biochem. 42:169–177.

Miller, R. M., and J. D. Jastrow. 1990. Hierarchy of root and mycorrhizal fungal interactions with soil aggregation. Soil Biol. Biochem. 22:579–584.

Miller, R.M., and J. D. Jastrow, 1996. Contributions of legumes to the formation and maintenance of soil structure. Pp. 105–112 in D. Younie, ed. Legumes in sustainable farming systems. BGS Occasional Symposium No. 30. British Grassland Society, Reading.

Murchie, A. K., and A. W. Gordon. 2013. The impact of the 'New Zealand flatworm', Arthurdendyus triangulatus, on earthworm populations in the field. Biol. Invasions 15:569–586.

Mytton, L. R., A. Cresswell, and P. Colbourn. 1993. Improvement in soil structure associated with white clover. Grass Forage Sci. 48:84–90.

Newman, E. L., K. Ritz, and P. Jupp. 1989. The functioning of roots in the grassland ecosystem. Aspects Appl. Biol. 22:263–269.

Nichols, S. N., J. R. Crush, and D. R. Woodfield. 2007. Effects of inbreeding on nodal root system morphology and architecture of white clover (Trifolium repens L.). Euphytica 156:365–373.

Nichols, S. N., J. R. Crush, and L. Ouyang. 2014a. Phosphate responses of some Trifolium repens x T. uniflorum hybrids grown in soil. Crop Pasture Sci. 65:382–387.

Nichols, S. N., R. W. Hofmann, and W. M. Williams. 2014b. Drought resistance of Trifolium repens x Trifolium uniflorum interspecific hybrids. Crop Pasture Sci. 65:911–921.

Norby, R. G., and R. B. Jackson. 2000. Root dynamics and global change: seeking an ecosystem perspective. New Phytol. 147(1):3–12.

Papadopoulos, A., S. J. Mooney, and N. R. A. Bird. 2006. Quantification of the effects of contrasting crops in the development of soil structure: an organic conversion. Soil Use Manag. 22:172–179.

Peach, W. J., R. A. Robinson, and K. A. Murray. 2004. Demographic and environmental causes of the decline of rural song thrushes Turdus philomelos in lowland Britain. Ibis 146:50–59.

Peeters, A., G. Beaufoy, R. M. Canals, A. De Vliegher, C. Huyghe, J. Isselstein, et al. 2014. Grassland term definitions and classifications adapted to the diversity of European grassland-based systems. Grassland Sci. Europe. 19:743–750.

Peiffer, J. A., A. Spor, O. Koren, Z. Jin, S. G. Tringe, J. L. Dangl, et al. 2013. Diversity and heritability of the maize rhizosphere microbiome under field conditions. Proc. Natl Acad. Sci. USA 110:6548–6553.

Peukert, S., B. A. Griffith, P. J. Murray, C. J. A. Macleod, and R. E. F. Brazier. 2014. Intensive management in grasslands caused diffuse water pollution at the farm scale. J. Envir. Qual. 43:2009–2023.

Peyraud, J. L., A. van den Pol-van Dasselaar, R. P. Collins, O. Huguenin-Elie, P. Dillon, and A. Peter. 2014. Multi-species swards and multi scale strategies for multifunctional grassland-base ruminant production systems: an overview of the FP7-MultiSward project. Grassland Sci. Europe 19:695–715.

Powlson, D. S., P. C. Brookes, P. A. Whitmore, K. W. T. Goulding, and D. W. Hopkins. 2014. Soil organic matters. European J. Soil Sci. 62:1–4.

Pugh, R., J. F. Witty, L. R. Mytton, and F. R. Minchin. 1995. The effect of waterlogging on nitrogen fixation and nodule morphology in soil-grown white clover (Trifolium repens L.). J. Exp. Bot. 46:285–290.

Rasse, D. P., C. Rumpel, and M. F. Dignac. 2005. Is soil carbon mostly root carbon? Mechanisms for a specific stabilisation Plant Soil 269:341–356.

Rillig, M. C., and D. L. Mummey. 2006. Mycorrhizas and soil structure. New Phytol. 171:41–53.

Rodriguez-Iturbe, I., A. Porporato, F. Laio, and L. Ridolfi. 2001. Plants in water-controlled ecosystems: active role in hydrologic processes and response to water stress - I. Scope and general outline. Adv. Water Resour. 24:695–705.

Sauerborn, P., P. Klein, A. Botschek, et al. 1999. Future rainfall erosivity derived from large-scale climate models - methods and scenarios for a humid region. Geoderma 93:269–276.

Scullion, J., S. Neale, and L. Philipps. 2002. Comparisons of earthworm populations and cast properties in conventional and organic arable rotations. Soil Use Manage 18:293–300.

Shipitalo, M. J., and K. R. Butt. 1999. Occupancy and geometrical properties of Lumbricus terrestris L-burrows affecting infiltration. Pedobiologia 43:782–794.

Sousanna, J. F., and A. Luscher. 2007. Temperate grassland and global atmospheric change: a review. Grass Forage Sci. 62:127–134.

Soussanna, J. F., P. Loiseau, N. Viuchard, E. Ceschia, J. Balesdent, T. Chevallier, et al. 2004. Carbon cycling anf sequestration opportunities in temperate grasslands. Soils Use Manage 20:219–230.

Stohlgren, T. J., T. N. Chase, R. A. Pielke, T. G. F. Kittel, and J. S. Barron. 1998. Evidence that local land practices influence regional climate, vegetation, and stream flow patterns in adjacent natural areas. Glob. Change Biol. 4:495–504.

Tancoigne, E., M. Barbier, J. P. Cointet, and G. Richard. 2014. The place of agricultural sciences in the literature on ecosystem services. Ecosystem Serv. 10:35–48.

Tisdall, J. M., and J. M. Oades. 1982. Organic matter and water soluble aggregates in soils. J. Soil Sci. 33:141–163.

Turner, L. B., A. J. Cairns, I. P. Armstead, H. Thomas, M. W. Humphreys, and M. O. Humphreys. 2008. Does fructan have a functional role in physiological traits? Investigation by quantitative trait locus mapping. New Phytol. 179:765–775.

Watson, C. A., D. Atkinson, P. Gosling, L. R. Jackson, and F. W. Rayns. 2002. Managing soil fertility in organic farming systems. Soil Use Manage 18:239–247.

Whalley, W. R., B. Riseley, P. B. Leeds-Harrison, N. R. A. Bird, P. K. Leech, and W. P. Adderley. 2005. Structural differences between bulk and rhizosphere soil. Eur. J. Soil Sci. 56:353–360.

Wilkins, R. J.,J. Bertilsson, C.J. Doyle, J. Noussiainen, C. Paul, and L. Syriala-Qvist 2002. Introduction to the LEGSIL project. Pp. 1–4 in R.J. Wilkins, C. Paul, eds. Legume Silages for Animal Production – LEGSIL. Landbauforschung Völkenrode, FAL Agricultural Research, Sonderheft 234. Braunschweig, Germany.

WMO. 2013. The global climate 2001–2010. A decade of climate extremes. World Meteorological Organization, Geneva.

Woodfield, D. R., and J. R. Caradus. 1987. Adaptation of white clover to moisture stress. Proc. N. Z. Grassld. Assoc. 48:143–149.

Woodfield, D. R., and J. R. Caradus. 1990. Estimates of heritability for, and relationships between, root and shoot characters of white clover II. Regression of progeny on mid-parent. Euphytica 46:211–215.

Zappala, S., S. Mairhofer, S. Tracy, C. J. Sturrock, M. Bennett, T. Pridmore, et al. 2013. Quantifying the effect of souil moisture content on segmenting root system architecture in x-ray computed tomography images. Plant Soil 370:35–45.

Prospects for yield improvement in the Australian wheat industry: a perspective

Michael Robertson[1], John Kirkegaard[1,2], Greg Rebetzke[1,2], Rick Llewellyn[1,3] & Tim Wark[4]

[1]CSIRO Agriculture, Wembley, Western Australia, Australia
[2]CSIRO Agriculture, PO Box 1600, Canberra, Australian Capital Territory 2601, Australia
[3]CSIRO Agriculture, Urrbrae, South Australia, Australia
[4]CSIRO Data 61, Pullenvale, Queensland, Australia

Keywords
Agronomy, breeding, cultivar, food security, productivity, soil, sustainable intensification, wheat, yield.

Correspondence
Michael Robertson, Agriculture, CSIRO, Wembley, Western Australia, Australia.

E-mail: Michael.Robertson@csiro.au

Funding Information
No funding information provided.

Abstract

For the Australian grains industry, recent progress in wheat yield (the dominant crop) due to genetic improvement and advances in agronomy is assessed, and we propose some of the emerging technologies that are likely to contribute to yield gain in the medium (10–20 years) term. Advances in yield will be underpinned by new genetics tailored to agronomic technologies with progress in water-limited yield potential expected to increase from current levels of ca. 0.5% per year. This increase will be achieved by selecting traits with greater water productivity, tolerance to frost and high temperature, and resistance to a range of soil constraints through access to novel genetic diversity, and deployment of targeted biotechnology and tools to improve confidence in phenotyping and environmental characterization. Hybrid cereals should halve the time to cultivar delivery to less than 6 years while allowing for more rapid incorporation and delivery of new traits, and capacity to exploit heterosis. There will be better adoption of recent technologies (such as variable-rate technology, soil testing, soil amelioration, and timely sowing) with the potential to increase yield by 10–80%. Novel technology packages such as earlier sowing systems will be enabled by improvements in integrated weed management, seasonal climate forecasting, plastic mulches, materials science, information and communication technologies, and weather monitoring and soil sensing. With a conservative assumption about maintaining the current rate of genetic progress at 0.5% per year, ongoing adoption of current and new agronomic technologies, continuing investment in R&D, and farm consolidation, at a whole industry level, annual gains in wheat yields of around 20 kg/ha (0.8–1.0%) are feasible over the next 20 years. These gains are likely to be attenuated by only a modest impact (<10% reduction) on crop yield potential due to the negative impacts of climate change.

Introduction

Recent reviews on prospects for meeting future global food demand have highlighted the centrality of achieving ongoing progress in crop yield (e.g., Connor and Mınguez 2012). A number of studies have highlighted the slowing of yield progress in many of the world's staple cereal crops (e.g., Alston et al. 2009) and the closing of the exploitable yield gap, especially in industrialized agriculture (Fischer et al. 2014). While these studies highlight the challenge confronting world agriculture, there is a need for them to be complemented by analyses at the country and subcountry level that examine prospects for yield progress through adoption of new technology.

From 2005 to 2012, Australia produced 3.5% of the world's wheat and 12% of the world's wheat export (ABARES, 2012). Australian grain growing typifies industrialized agriculture, where farmers have taken up new technology and have achieved remarkable gains in the efficient use of inputs (e.g., Turner and Asseng 2005). It is also representative of many industrialized countries, where future increases in food production are likely to come more from intensification rather than bringing more land under agricultural production (Hochman et al. 2013). This paper explores the case for continued yield progress in the Australian grains industry, with lessons that can be extrapolated to other industrialized grain-producing countries where water limitation predominates (e.g., North America, Latin America, Southern Africa, and Mediterranean Basin).

The focus for this review is wheat, the dominant crop in the Australian grains industry, which is almost entirely rainfed and grown in the winter and spring from temperate to subtropical climates. The main climatic factors controlling yield are the timing and amount of rainfall, the incidence of frost events particularly around ear emergence, and the avoidance of high temperatures during grain-filling (Barlow et al. 2015). Historical progress in the yield of wheat in Australia has been well documented (e.g., Kirkegaard and Hunt 2010) together with attribution of various innovations in determining such progress. Aside from an initial phase of system rundown before 1900, yield progress has been characterized by phases of gain interspersed with "plateau periods" where progress slows. The intermittent periods of rapid yield improvement occurred where packages of improved management combined to allow the underlying improvements in genetic yield potential to be realized. Since 2000, there has been a noticeable flattening of this upward progress in average yield and more year-to-year variation; in the last 4 years, wheat yield in Australia has averaged 1.7 t/ha (ABARES 2012).

This paper is part review and part considered prediction. It has four aims. First, recent genetic progress in yield of wheat in Australia is described and predictions of the feasibility of future gains, particularly considering emerging modern breeding technologies. Second, five agronomic management innovations are considered that will contribute the greatest to future gains: timely sowing, fertilizer efficiency, management of subsoil constraints, rotation diversity, and soil surface and residue management. Third, a number of technologies are described, such as integrated weed management (IWM) and seasonal climate forecasting (SCF), which when used in concert with improved genetics and agronomy, will enable gains to occur. Fourth, estimate future yield progress in the medium term (20 years) at a whole-of-industry level is estimated, accounting for increased adoption of current technology and the development and uptake of emerging technology, that will close the yield gap (Hochman et al. 2009b, 2012). A target annual yield gain of 1.0–1.5% is proposed based on goals set by industry (Grains Research and Development Corporation 2012) and for Australian farmers to help meet the global challenge of achieving food security (Keating et al. 2014).

The Role of Genetic Improvement

Agronomic improvements in cereal water-productivity have been paralleled by genetic advances in yield potential and disease resistance as well as significant alterations in grain quality for sale into world markets. In Australia, breeding progress has been reported at around 0.5% per year (Fischer 2009; Sadras and Lawson 2011) although reported gains can vary depending on cultivars and the regions where assessment is undertaken (e.g., Fischer et al. 2014). In this section, those breeding technologies that have contributed to recent rates of genetic improvement are documented. Table 1 lists the key modern breeding technologies that will contribute to the ongoing progress.

Progress in water-limited yield potential to date has largely relied on improvements in harvest index through

Table 1. Key outputs from genetic improvement and breeding underpinning future yield increases in the Australian grains industry and their timing of release to grain growers.

		Years from 2010							
		0	5	10	15	20	25	30	35
Adapted phenology for new early-sown systems	Hunt et al. (2012)								
Alternative dwarfing genes with long coleoptiles	Rebetzke et al. (2007)								
Enhanced TE	Condon et al. (1987)								
Canopy architecture, stomatal conductance	Sadras et al. (2012)								
Vigor for establishment and weed competitiveness	Rebetzke et al. (2007, 2008)								
Herbicide tolerance	Zhou et al. (2003)								
Tolerance to aluminum, salinity and boron	Wasson et al. (2012)								
Disease resistance using "gene cassette" constructs	Zhu et al. (2012)								
Tolerance to frost at flowering	Galiba et al. (1995)								
Enhanced nutrient use efficiency	Barraclough et al. (2010)								
Widely adapted high value grain legume	Vanhercke et al. (2014)								
C4 photosynthesis	Reynolds et al. (2011)								

White shade represents the time to delivery

selection of earlier flowering and reduced plant height (Siddique et al. 1989). These improvements have been undertaken simultaneously with genetic improvements in disease resistance, tolerance to subsoil constraints, and improved grain quality. In the future, selection for traits with greater resistance to root diseases and tolerance to subsoil constraints will maximize water use to increase biomass, while genetic variation exists for traits aimed at maintaining a balance in pre- and post-anthesis water use in terminal droughts (e.g., transpiration efficiency, TE, Rebetzke et al. 2002). The potential for these other traits is still preliminary. However, evidence for trait-based genetic improvement has been reported for TE (Rebetzke et al. 2002), improved establishment (Rebetzke et al. 2007) and rapid leaf area development (Botwright et al. 2002).

The main genetic pathways for improvement in water-use efficiency have been early flowering, improved early vigor to suppress soil evaporation, and improved TE. There has been a steady linear genetic gain in grain yield (Richards et al. 2014), associated with an increase in harvest index and not so much to an intrinsic increase in TE for biomass. There is significant opportunity for genetic increase in intrinsic TE either indirectly through selection for increased biomass and yield or directly in selection for changes in carbon isotope discrimination, a surrogate for TE in C3 species.

Other important traits with potential include small erect upper leaves, higher stomatal conductance, the maintenance of grain fertility through altered carbon metabolism, and capacity to accumulate and remobilize water-soluble carbohydrates (Rebetzke et al. 2008; Richards et al. 2010; Sadras and Lawson 2011). Recent interest has focused on capacity to assess for tolerance to frost and high temperature events previously difficult to phenotype but commonplace to specific regions within the wheat belt (Zheng et al. 2012; Barlow et al. 2015). Studies are currently underway to assess and rank relative trait value using Managed Environment Facilities where carefully structured populations contrasting for target traits are evaluated side-by-side under controlled water deficit (Rebetzke et al. 2013). Importantly, increases in grain yield will be made without compromising grain quality. This could provide some challenge as grain yield is strongly genetically correlated with reductions in grain protein in many wheat breeding programs across the globe (e.g., Bogard et al. 2010). Breeders must release at minimum some resistance in their varieties. This resistance is supplemented agronomically with scheduled fungicide sprays that ensure the maintenance of at least the flag and penultimate flag leaves.

Genetic progress through phenotypic selection of common genetically complex traits including grain yield and biomass is challenging owing to the impact of genotype × environment interaction particularly in water-limited environments, and the limitations of screening large number of lines in field plots (Rebetzke et al. 2014). Further, Australia largely produces hard spring white varieties with over 90% of this grain exported throughout the world. Regional adaptation, tolerance to local subsoil constraints and diseases, and a focus on excellent milling quality has restricted the commercial release of overseas-developed wheat varieties and limited the use of overseas germplasm in breeding programs. Yet, there is capacity to target new and important alleles where useful from overseas programs. Biotechnology and specifically molecular genetics, has provided capacity to screen and enrich populations for important genes (both major and minor) thereby reducing the need for costly phenotyping. Larger numbers of lines can be tested cheaply to ensure only families containing key genes for adaptation and quality are promoted for expensive, multienvironment yield testing. In turn, integration of selection using genetic markers with new genetic diversity for physiological traits in a conventional breeding framework will likely increase genetic gains over the next 50 years. The benefits with genomic selection (Heffner et al. 2009) are yet to be validated for performance of rainfed crops where genotype x environment interaction will reduce confidence when predicting from training environments. That said, the benefits of enriching both small and large-effect alleles in selecting prior to multi-environment testing should allow cost-efficient, rapid accumulation of broadly based, genetic effects specific to individual breeding programs.

Ongoing activity in biotechnology will continue to support conventional breeding and physiology in achieving greater genetic gains for grain yield. Where the effects of genes are large and/or have economic significance, there will be interest in a more detailed understanding of their function, potential for modification in regulation, and identification of new alleles. This effort can lead to gene cloning and the potential for transgenics and genetically modified (GM) wheat. For example, there is strong evidence from several years of field testing in Argentina that the HD Zip HAHB4 transcription factor from sunflower reduces sensitivity to ethylene (Manavella et al. 2008) to increase wheat performance under drought and salinity (G. Watson, pers. comm.). Coupling deployment of transgenes with more efficient selection of genetic background (above) will increase the frequency of "better" adapted lines for advanced yield testing. Increasing the number of "needles in a haystack" should significantly increase the likely identification of lines containing important average genetic effects and also rare combinations of epistatic allelic combinations. The increased confidence and reduced cost with discarding of poor progeny should increase breeding efficiencies and could potentially increase genetic gain per cycle by up to 0.5%.

The significant costs in both discovery and regulation of transgenes will limit their deployment in breeding to those of large phenotypic effect and significant economic value, such as herbicide resistance and tolerance to subsoil constraints, while disease resistance may include multiple genes "stacked" into single "gene cassette" constructs designed to slow breakdown of resistance genes to the pathogen (Zhu et al. 2012). Likely deployment of the first of these will be mid-to-late 2020s and will continue as new genes are identified and regulated with thorough testing. Hybrids provide promise for increased yields through heterosis, rapid incorporation of new traits and critically shorter breeding cycles, yet are expensive to develop (Longin et al. 2013). While reports of midparent heterosis of around 10% for wheat are modest in size (e.g., Jordaan et al. 1999; Longin et al. 2013), the rapid incorporation and fixation of alleles for new traits in the female or male lineage then allows for trait expression in the F1 hybrid without the need for reselection and fixation of many alleles common to development of conventional inbred varieties. This should reduce the duration of the breeding cycle considerably to promote new genes to be deployed while maintaining combinations of favorable alleles for quality and adaptation. Development of hybrid cereals could halve the time for delivery from 12 to 6 years due to the development of heterotic pools appropriate to complementary gene action and the longer term exploiting of heterosis. Apomixis genes introduced using GM techniques from foreign species (Koltunow et al. 2011) will allow for the fixation of favorable heterotic combinations, thereby increasing seed production and the reduced cost of hybrid seed. Finally, the more challenging metabolic and path-dependent traits controlled by many genes will be among the last to be deployed through GM. Frost tolerance (Galiba et al. 1995) and C4 metabolism (Reynolds et al. 2011) relying on multiple genes from alien species will require extensive genetics and physiological understanding, yet will deliver significant potential for improving water-limited productivity. Ongoing deployment of new transgenes will focus on release in the best commercial genetic backgrounds containing superior combinations of the yield genes selected through conventional breeding. The potential increases in genetic gain with hybrids are unclear. However, evidence in rainfed maize in the Midwest US (Duvick et al. 2004) and canola in Canada (Morrison et al. 2016) suggest hybrids contribute to greater yield stability with genetic gains of 1–2% per year.

The bringing together of ongoing conventional breeding for complex genetics with the deployment of novel genes through GM will accelerate genetic progress for water-limited yield potential in four ways. Firstly, water productivity will increase through deployment of new genetics to improve water-use efficiency and capacity to tolerate frost and high temperatures. The effects of high temperature and terminal drought on yield are reduced if anthesis date can be promoted 1–2 weeks. However, frost risk is significantly increased requiring genetic means to reduce the impact of cold temperatures during developmentally sensitive stages from meiosis through to soon after fertilization (Thakur et al. 2010). Secondly, hybrids will permit the incorporation and delivery of new traits more rapidly to increase genetic gain per breeding cycle. Thirdly, traits targeting better resource management (e.g., nutrient-use efficiency and weed competitiveness) will reduce the cost of yield per dollar input to increase grower gross margins. Finally, better integration with management to tailor traits and deployment of cultivars at scales from regions to individual paddocks should lift genetic gains.

In summary, various breeding and genetics technologies will be required in conjunction with conventional breeding to lift genetic gains (Table 1). Technologies such as genomic selection, disease resistance gene cassettes, apomictic hybrids, and C4 photosynthesis vary widely in their likely timing. Within the next five years, cultivars with appropriate phenology for earlier sowing are likely to be available. Cultivars with enhanced TE, early vigor, and long coleoptiles are expected to be available in the next 10 years. Further out, cultivars with tolerance to aluminum, salinity, and boron; frost at flowering; and with enhanced nutrient use efficiency will be likely ready between 10 and 30 years hence.

The following five sections cover five agronomic management innovations that will contribute the greatest to future gains: timely sowing, fertilizer efficiency, management of subsoil constraints, rotation diversity, and soil surface and residue management. The contribution of current technologies to progress and the role of likely future technologies is summarized in Table 2.

Timely Sowing

In many environments in the Australian grains industry, grain-filling is characterized by water deficit and high temperature. In such situations, there is a clear advantage to timely sowing so that flowering and grain-filling occurs after the period of heightened frost risk, but before the effects of late-season water limitation and high temperature can reduce yield. Yield potential will also be increased through an extended duration of the vegetative phase, deeper root systems that can access more soil water during grain-filling, and increased conversion of transpiration to biomass through growth occurring at a time of the year when evaporative demand is lower.

Yield gains from early sowing have been estimated at 10–30% at the level of individual fields and across entire farms (Stephens and Lyons 1998; Sharma et al. 2008;

Table 2. Five key practices underpinning future yield increases and the key current and future technologies to enable these technologies.

The Technology/practice	Key current enabling technologies	Key future practices and enabling technologies
Timely sowing: Timely sowing so that flowering and grain-filling occurs after the period of heightened frost risk, but before the effects of late season water limitation and high temperature can reduce yield Fertilizer management: More soil testing, variable rate application, better matching rates to crop demand to improve efficiency of fertilizer use	• Integrated weed management • Sowing machinery capacity matched to farm size • Matching cultivar and agronomy to yield potential • Decision-support systems using seasonal climate forecasts (e.g., Yield Prophet®) to manage risk of under- or overapplication • Variable rate controllers, mapping software and spatial information • Soil nutrient testing	• Seed coatings to delay imbibition under marginal moisture • Cultivars with appropriate phenology and long coleoptiles to allow deep sowing • Cultivars tolerant to frost at flowering • Further improvements in seasonal climate forecasts • Improved fertilizer formulations to release available nutrients matched to crop demand • On-the-go proximal soil sensing • Real-time sensors for plant available soil water • Cultivars with higher P and N use efficiency
Tillage and residue management: Precision seeding systems under control traffic and with disk openers for minimum disturbance. Soil surface management to minimize soil evaporation and maximize infiltration	• 2 cm precision guidance • Stubble retention • No-tillage	• Sprayable biodegradable plastic mulches to suppress soil evaporation • Further advances in precision seeding systems to sow under marginal moisture, maximize benefits and minimize negatives of on/off previous row sowing
Rotation diversity: Sustainability of high intensity cereal cropping through diverse rotations that minimize yield reducing factors (disease, weeds, pests, N deficiency)	• Improved break crop agronomy • Improved management packages	• Technologies for managing seasonal risk (opportunity cropping, relay cropping or sowing of sacrifice inter-row mulches) • Low-input break options • Novel intercropping facilitated by precision technology • Widely adapted high-value grain legume
Subsoil constraints: Improving the access of crop root systems to water and nutrients stored in the subsoil through amelioration	• Mapping and diagnosing subsoil constraints • Deep placement of ameliorants • Soil profile reconstruction • Where amelioration is not feasible, managing inputs on constrained soils	• Low-cost and easy means to map and diagnose constrained soils • Novel ameliorants and carriers (e.g., organics) • Mechanical and chemical means to increase rate of penetration of ameliorants into sub-soils • Cultivars with tolerance to aluminum, salinity and boron

Fletcher et al. 2015a,b). While earlier sowing has been adopted by many grain growers, there is scope for larger areas of the farm to be sown early and risk-averse farmers to adopt early sowing, enabled by no-till which allows much earlier and deeper planting into more marginal moisture ("moisture seeking") and in response to smaller rainfall events. In regions of the cropping zone where there are livestock and a late summer-early autumn feed gap, an additional driver for early sowing will be dual-purpose grazing of crops. The potential benefits of earlier sowing will not be realized in full without new technologies such as cultivars with appropriate phenology (Hunt et al. 2012) and alternative dwarfing genes with long coleoptiles to allow safe establishment of deep-sown crops, seed coatings that delay imbibition and germination of seed under marginal soil water content (Johnson et al. 2004), and weed control technologies to avoid having to delay sowing after the first rains in order to reduce weed burdens. More skilful seasonal climate forecasts (see below) and sensing of stored soil water status (Viscarra Rossel et al. 2010) will guide the decision on which seasonal circumstances to sow early.

Soil Surface and Residue Management

Improvements in soil surface and residue management are an important component of creating more frequent and safer opportunities for timely sowing through great capture and storage of rainfall in the soil, and less soil evaporation. Studies on the water balance of dryland cropping systems have highlighted the large component of rainfall that is currently "lost" as evaporation from soil, deep drainage, and runoff, and which represents foregone yield if this water could be diverted toward transpiration (Passioura and Angus 2010; Sadras and McDonald 2012). Several management strategies have been developed and adopted with efficient water capture and use as a primary focus (Kirkegaard and Hunt 2010; Kirkegaard et al. 2014) and include management for improved capture and storage, and improvements in crop vigor to reduce evaporative losses. Importantly, such direct and indirect management often occurs in the years and months prior to sowing the crop, although much attention is given to management at sowing and during the crop growth phase. No-till weed-free seeding systems with stubble retention have a direct short-term impact on water capture and storage by improving water infiltration, reducing run-off, and minimizing evaporative loss.

In the future, new precision seeding systems under control traffic and with disk openers for minimum disturbance may further reduce unnecessary water loss related to soil disturbance (Rainbow and Derpsch 2011). In the longer term, soil structural improvements on degraded soils may improve under no-till systems, further improving water capture and storage. Further innovations in seeding technology to improve precision and timeliness are likely (Rainbow and Derpsch 2011). New herbicide chemistry delivered using "weed seeking" color sensitive on-the-move spray rigs are improving the economics of weed control during the fallow, while pre-emergent herbicides with efficacy in undisturbed no-till systems are providing a greater capacity for timely sowing without crop damage or significant weed competition (Kleemann et al. 2014).

In environments with frequent rains and fine soil texture, unproductive evaporative loss of water during the crop establishment stage can often represent 30–50% of the total evapotranspiration during the crop phase (Sadras and McDonald 2012). A range of novel strategies are being considered to reduce losses including early sowing, vigorous crop cultivars, and cultivars with longer coleoptiles using alternative dwarfing genes.

The potential use of biodegradable plastic mulches to reduce unproductive water loss has received recent attention. Even though Australian grains production has relied on crop and pasture residues to provide natural mulches, they are often not available in adequate amounts, quality is inconsistent, and they do not always provide anticipated benefits. The use of plastic mulch in world agriculture has increased dramatically in the last 10 years, particularly in vegetable production. Kasirajan and Ngouajio (2012) have reviewed the use of polyethylene and biodegradable mulches in agriculture and pointed out that despite multiple benefits, the removal and disposal of conventional polyethylene mulches remains a major agronomic, economic, and environmental constraint to adoption. At present, cost and uncertainty about benefits constrains the use of such products in broadacre grains production in Australia. As cheaper biodegradable materials become available, opportunity for using polymers as barriers to soil evaporation may be viable. Such materials made out of starch have been available in horticulture for the last few years and degrade at desired rates to generate yield benefits. However, high cost is the main disincentive for wider use. More work to develop sprayable, biodegradable mulches for use in broadacre agriculture could provide significant improvements in dryland crop productivity. Early investigations with biodegradable films in Australian cotton production systems (Braunack and Price 2012) give cause for hope that such technology may have a role in grains production.

In summary, a range of studies have quantified the yield gains from improved surface and residue management. Taking advantage of increased availability of soil water through the improved technologies and practices listed above will be facilitated by more frequent, timely and cheap methods for monitoring stored soil water (Corke et al. 2010), and decision aids to assist with feasible management responses (e.g., increase inputs) (Hochman et al. 2009a).

Fertilizer Management

The key pathway by which fertilizer management will contribute to yield progress will be via maintaining the level of input in the soil at the minimum level required to meet plant demand throughout the production cycle and maintain soil organic matter levels, through improved risk management, greater and more accurate soil testing, and improved fertilizer formulations. A side benefit of improvements in yield through genetic gain and the removal of other production constraints will be improvements in nutrient use efficiency. Dobermann and Cassman (2004) attributed significant improvements in nitrogen-use efficiency of cereals to higher yields of modern cultivars and the improved management of production factors other than N.

Due to risk management objectives, farmers in variable rainfall environments, are known to use rates of nitrogen (N) well below the profit-maximizing rate, and thereby are likely to miss out on greater returns from more favorable production years (Monjardino et al. 2013). Tools that integrate all factors pertaining to fertilizer decisions in the context of seasonal risk, such as Yield Prophet® (Hochman et al. 2009a), currently enable farmers to make best use of soil test information to optimize fertilizer inputs. Advances in the skill of SCF will be an important enabler (see below). More accurate, cheaper and easier soil testing to monitor soil fertility will increase confidence in fertilizer use. At present, it is estimated that only 50% of fields are tested each year (Edwards et al. 2013). Improved product formulations and methods of fertilizer application will also enhance crop uptake. For phosphorus (P), McLaughlin et al. (2011) have suggested that fluids, slow release products, better placement, even into subsoils, and modification of soil chemistry around the fertilizer granule could improve P use efficiency, but few have provided solid field evidence for efficacy. For N, management practices alone will not prevent all losses (e.g., by denitrification), and it may be necessary to use enhanced efficiency fertilizers, such as controlled release products, and urease and nitrification inhibitors (Chen et al. 2008).

There are prospects of increased adoption of variable-rate application fertilizer from the current 20–30% of grain growers to over 80% due to the widespread use of various enabling technologies that are becoming cheaper and more accessible, and the stimulus of rising costs of fertilizer (Robertson et al. 2012). Yield gains and cost savings will come through improvement in uptake of nutrients and less under- and overfertilization.

Rotation Diversity

Despite the diversity of pulse and oilseed crops available in Australia, and the demonstrated benefits to cereal crop yields in the rotation (Angus et al. 2015), cropping systems today remain dominated by intensive cereal production where wheat and barley make up 80% of the cropped area. Cereals are attractive for a number of reasons including wide adaptation, the ease of management and marketing, and lower risk due to reduced up-front costs and more reliable performance in difficult seasons. However, broadleaf oilseed and legume break crops can be profitable, especially when whole-of-sequence impacts are included. Yet their inclusion in the sequence at levels lower than would seem optimal in many areas belies these perceived benefits (Robertson et al. 2010).

Currently, improved cultivars with traits to facilitate simpler management of pests and diseases, along with associated management packages should lead to further increases in profitable break crop areas. Beyond this, there are three novel approaches to increase crop diversity and productivity (Table 2). The first is low-input multiple benefit break crops. Intensive cereal systems that have developed multiple-resistant herbicide-tolerant weeds may become unprofitable without a non-cereal break. On crop-intensive farms with no livestock, an option gaining attention is a low-input break crop where input costs are minimized and weed control or N contribution are maximized (Peoples et al. 2013). These approaches include grazing, cutting the crops for hay or using the crops as "green" (incorporated) or "brown" (desiccated) manures

that are not harvested, thereby preserving soil water and N for following crops. Early removal of a low-input break crop could create an opportunity for a short-duration spring-sown warm season crop (e.g., sunflower, safflower) in some agro-ecological zones (e.g., Robertson et al. 2005).

The second option is the use of novel intercropping. The new precision agriculture (PA) (Robertson et al. 2012) and controlled traffic (CT) systems (Rainbow and Derpsch 2011) may in future facilitate novel options to more readily manage mixtures of crops (or crop cultivars) in the same year as intercrops or relay crops to capture the rotational or synergistic benefits that can arise from such diversity. Australian studies reporting the benefits of such systems are rare (Fletcher et al. 2015a,b) and more work is needed. The logistical difficulties previously associated with these options are more manageable with new PA technologies and a range of new herbicide-resistant crop types.

The third option is the creation of a new widely adapted high-value legume crop. The rapid and widespread adoption of canola in Australia during the 1990s demonstrated the willingness of Australia farmers to adopt new technologies that are readily adoptable and profitable and complement their existing cropping system. However, in the longer term, as pasture area declines and the cost of fertilizer N increases, there will be an increasing need for a profitable legume break-crop similarly adapted to the larger areas of acid soils where lupin was formerly grown to complement profitable oilseed break crops such as canola. One option is to utilize new genetic oil modification strategies (Lu et al. 2011; Vanhercke et al. 2014) to create a high-oil lupin crop that could provide the same opportunities that soybean provides in the northern hemisphere as a high-value oil crop for human consumption.

Managing Subsoil Constraints

Physiochemical subsoil constraints (salinity, sodicity, acidity, alkalinity, boron, high-soil strength) affect an estimated 60% of cropped land across south-eastern (Adcock et al. 2007), western (DAFWA 2013), and northern (Dang et al. 2010) Australian cropping regions. Management strategies include (i) various amelioration techniques, (ii) plant breeding for tolerance, and (iii) avoidance through agronomic or agro-engineering solutions to manage yield potential when the constraints are intractable (Adcock et al. 2007). Improving the access of crop root systems to water and nutrients stored in the subsoil through amelioration can produce spectacular yield increases of −0.6 t/ha from as little as 10 mm of extra water accessed from an extra 100 mm of soil depth during grain-filling (Kirkegaard et al. 2007). Current innovations such as placement of gypsum, lime, nutrients and organic matter at depth, and soil profile reconstruction through mechanical means (spading, moldboard ploughing) have demonstrated sustained yield increases of 20–80% in a range of recent studies (Hamza and Anderson 2002, McBeath et al. 2010; Gill et al. 2012; Kirkegaard et al. 2014). There appears to be significant scope to increase the levels of current adoption of existing technologies such as lime application to acid soils, which at 10% is well below that required to recover soil pH to target levels (Edwards et al. 2013).

In the future, ongoing development of more novel amelioration approaches, such as deep placement of ameliorants (Gill et al. 2012) will provide new opportunities for significant yield improvement. Efforts to identify where economic gains will be greatest, and to develop more cost-effective application technologies are ongoing, but upfront investment costs are likely to remain a constraint. In the meantime, significant yield and WUE improvements can be made on constrained or variable soils by identifying and managing inputs according to yield potential (Monjardino et al. 2012; Oliver and Robertson 2013). For example, McBeath et al. (2012) demonstrated that yield increased of up to 90% by reallocating resources such as N fertilizer from nonresponsive to responsive soils using variable rate techniques on constrained Mallee soils. In terms of genetic technologies, the discovery of genes (and associated markers) for tolerance to aluminum, salinity, and boron offers promise, but the need to have tolerance to multiple constraints on many constrained Australian soils (McDonald et al. 2012) means that progress from breeding for tolerance is unlikely in the short to medium term (Fischer 2011).

Enabling Technologies

A number of enabling technologies will be instrumental in facilitating the future realization of the full benefits of agronomic and genetic innovations. Some enabling technologies, such as IWM and SCF, are in use by grain growers now and will be subject to ongoing refinement and improvement over the next 30 years. Other enabling technologies such as those enabled by the digital revolution are only just being developed for use by grain growers.

Integrated weed management

Maintaining the weed-free status of most crops, while not directly contributing to yield progress, will enable the expression of improved water-limited yield potential and improved agronomic practices described above, such as early sowing, rotation diversity, and new soil surface and residue management systems.

Weed management and herbicide resistance is often a primary determinant of crop choice, crop sequence, and time of sowing. Adoption of IWM and a small, but important supply of alternative herbicide options has allowed Australian grain growers to successfully manage increasingly herbicide-resistant weed populations and maintain low weed densities (Llewellyn et al. 2009). This has also enabled the shift to earlier sowing times with less opportunity for presowing weed control. IWM employs a range of tactics for the effective, long-term management of weed populations (e.g., pre- and postemergent herbicides, sowing competitive crops, destruction of weed seeds after harvest, spray-topping to prevent weed seed set). When faced with major resistance to herbicides, Australian grain growers have successfully responded by increasing their emphasis on controlling return of viable weed seed to the soil (Llewellyn et al. 2007). This has included introduction of practices such as crop topping (treating weeds late in the cropping season with a herbicide that prevents weed seed set) and harvest weed seed control methods such as technology to place weed seeds from the harvester into windrows for later burning and other forms of seed capture and destruction (Walsh et al. 2013). There is still the opportunity for greater use of these practices as weed management challenges increase in many regions. In the long term, the advent of a wider range of herbicide options and herbicide-tolerant crops will lead to advances in the convenience and attractiveness of weed management options for Australian farmers, although other genetic approaches such as cultivars with vigorous canopies and root systems will also contribute. Innovation in methods to maintain low-weed seed bank densities by killing weed seeds in crop, at harvest and in, or at the soil surface will remain important.

Seasonal climate forecasting

Managing seasonal risks, while not directly contributing to yield progress, will enable the expression of improved water-limited yield potential and all of the improved agronomic practices described above.

An important development in the last 20 years has been the advent of SCF to aid risk management by cropping farmers. SCF will have an important role to play in the future in maximizing the benefits of improved fertilizer management practices, weed management practices, decisions about timely sowing, and effective use of break crops. Hayman et al. (2007) estimates adoption of SCF, principally based on the El Nino Southern Oscillation Index, by Australian farmers is 30–50% with the benefits, based on a perfect forecast, estimated in a number of studies to be AUS$12–60/ha. The benefits likely to accrue to grain growers through the use of SCF will remain limited while forecasting skill is modest. Progress in seasonal forecast skill is likely to parallel the improvement of short-term weather forecast skill over the last few decades (McIntosh et al. 2007). Since 1980, weather forecasts have increased their lead time at a defined level of skill by about 1 day per decade for the northern hemisphere, and 1 day per 3 years for the southern hemisphere (Simmons and Hollingsworth 2002). Increase in physical understanding of climate together with improvements in observations, modeling techniques and computer speed will all lead to an increase in seasonal forecast skill (McIntosh et al. 2007).

Digital technologies

Just as the agricultural revolution of the 19th century built on the industrial and scientific revolutions which were taking place around the same time, the rapid growth of information and communication technologies (ICT) over the past decades is expected to also have a similar effect in driving new directions for agriculture.

Automation is already relatively commonplace among agricultural systems, such as in automated guidance systems and automated weed-spraying (Slaughter et al. 2008). While full automation of large vehicles is occurring in mining, this is unlikely to happen in agriculture for cost and safety reasons. However, relatively cheap, lightweight robotic platforms for both ground and air are becoming commercially available – with functions such as navigation, path-planning, and obstacle avoidance and potential for undertaking tasks such as planting, weed control, and pest management.

The rapid growth in use of 2D mapping services such as Google Maps and widespread availability of satellite data such as SPOT or Landsat means that there is an increasing availability of spatial data to improve estimates of plant quality and biomass over large areas, as well as navigation or tracking location of assets. Emerging technologies such as low-cost and portable laser-ranging units and cheap stereo cameras now mean that there is the possibility to rapidly form 3D maps of the quality and quantity of grains from devices mounted on vehicles or hand-held units (Weiss and Biber 2011; Bosse et al. 2012). In addition to this is the use of various geophysical methods to map and monitor soils (electromagnetic induction, gamma radiometrics). The only use of in situ sensing for tactical management decisions is some limited use of soil water sensing and NDVI for nitrogen fertilizer topdressing. A variety of data acquisition and access systems are used, usually association with a particular proprietary sensing system.

From the late 1990s, as Moore's Law saw the exponentially decreasing size and cost of computer chips, many started to predict a future of "Smart Dust" – networks of tiny devices which could sense, store, and communicate information about the environment into which they were distributed. While limitations around energy storage and communications hardware has prevented the smart-dust vision becoming reality to date, there has been significant progress in cheap, low-power, wireless data loggers (Wark et al. 2007). Within the next decade, it is possible that tiny, disposable devices which could be buried to soil to monitor moisture levels or scattered among crops to monitor for pests and diseases will become available.

The rise of the Internet over the past three decades was largely driven by the desire to reduce transaction costs in communication, storage, and analysis of information. Developments in mobile devices, data stored on remote cloud servers, and high-speed broadband networks will allow emerging infrastructure to develop new services which can integrate both local data from the farm and integrate with external information such as weather or price forecasts (Taylor et al. 2013).

Synthesis: Prospects for Yield Improvement

As a summary of the evidence above, Figure 1 presents a best estimate of the current and future potential levels of adoption of current management technologies by Australian grain growers, together with average benefits in yield and/or cost savings. Current levels of adoption were based on national or regional surveys of grain growers and are sourced from Edwards et al. (2013), Robertson et al. (2010, 2012), McNee et al. (2015), Hayman et al. (2007), and Llewellyn et al. (2007, 2009, 2011). Potential future levels of adoption were estimated by us based on the framework described by Kuehne et al. (2015) and applied by Hayman et al. (2007). Briefly, the framework accounts for characteristics of the technology, characteristics of the population of potential adopters, the actual advantage of using the technology, and the ease of learning of the actual advantage of the innovation. Rapid and high levels of adoption are achieved with technologies that have clear benefits to adopters, are easy to learn, adopt and disadopt, and are applicable to a broad proportion of the population. Conversely, technologies that are complex and difficult to learn or identify clear benefits from, or target a small proportion of growers will have a slower rate and lower final level of adoption. When applying the framework here, technologies that would be adopted rapidly and widely would have an eventual adoption rate of 80–90%, whereas technologies that would have a lower rate of take-up might have an eventual adoption rate of 40–60% of grain growers.

Benefits to yield and growing costs were based on review studies where the increases in crop yield or decreases in growing costs were quantified and summarized, and were sourced from Robertson et al. (2010, 2012), Hayman et al. (2007), Llewellyn et al. (2007, 2009), McIntosh et al. (2007), Hochman et al. (2009b), Dang et al. (2010), Fletcher et al. (2015a,b), Hamza and Anderson (2005), Kirkegaard and Hunt (2010), Kirkegaard et al. (2014), McBeath et al. (2010), Simpfendorfer (2012), Blackwell et al. (2004), and McCallum (2005).

Figure 1 indicates that levels of adoption by grain growers of current technologies span the full spectrum from around 10% (e.g., use of decision support systems for risk management) to 90% (autosteer and guidance on farm vehicles). There are a significant cluster of technologies that are currently adopted by 30% or less of grain growers and which could potentially be adopted by 70% or more. The size of the "bubbles" in Figure 1 represents the average percentage benefit to the grain grower of adopting this technology. Many technologies generate around a 10% benefit, however, there are around seven technologies that can generate 20% or more average benefit when adopted: timely sowing, amelioration of subsoil constraints, clean fallow management, seasonal climate forecasts, broader adoption of break crops, IWM, and more frequent and widespread soil nutrient testing. Technologies that have a large capacity for increased adoption and large benefits are obvious candidates for promotion and extension with grain growers. In our analysis, these include soil amelioration, timely sowing, soil nutrient testing, and SCF.

Realizing yield gains will require integration of innovation in genetics, agronomy, materials, and information

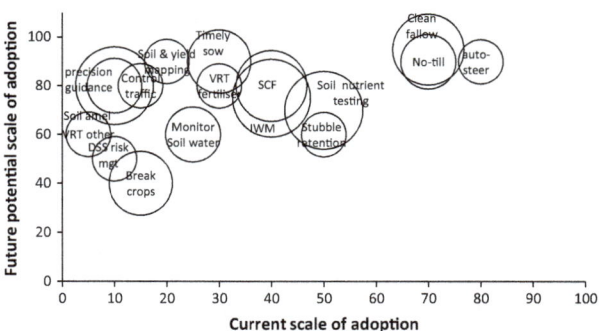

Figure 1. Gains in yield or cost saving expressed for key current technologies or practices. Points are plotted as the current level of adoption (% of Australian grain growers) versus the future potential level of adoption, with the diameter of the bubble representing the average percentage gain due to the technology or practice. SCF = seasonal climate forecasting, IWM=integrated weed management, VRT= variable rate technology, DSS=decision support system. See text for evidence of current and future levels of adoption and productivity gains..

technology to deliver gains that neither could in isolation. Figure 2 presents an example of this idea that combine elements of the focus areas discussed above. In this example, soil water sensors and seasonal climate forecasts will reduce riskiness around the decision to sow earlier than currently possible. Seed technology (e.g., coatings that

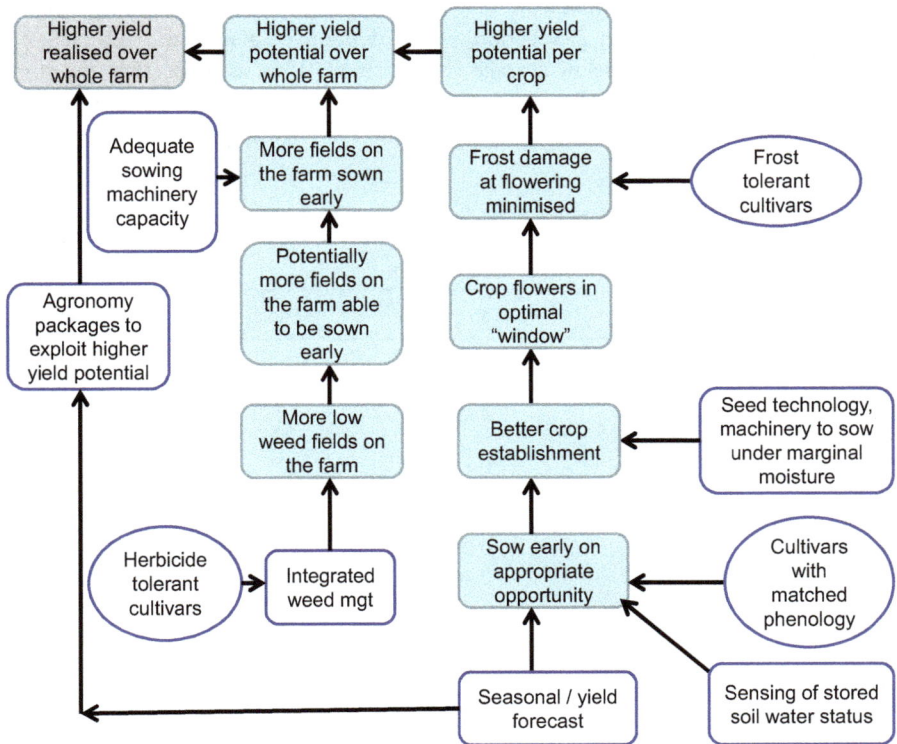

Figure 2. Schematic of the impact of various current and future genetic (hollow circles) and management (hollow boxes) technologies on the intermediate variables (shaded boxes) leading to the potential whole-farm yield gains to be made with timelier sowing.

delay imbibition) and seeding equipment of adequate sowing capacity along with cultivars with superior early growth characteristics will ensure more successful establishment across the entire farm. Appropriate phenology will allow the crop to flower within the optimal window that minimizes damage to the crop by temperature extremes, aided by improvements in tolerance to these stresses. Crops will have higher yield potential and so appropriate agronomic packages (fertilizer, pest, weed, and disease control) will be required to exploit this.

This section estimates the prospects for future yield progress comprising of improvements in underlying genetic potential, resource supply needed to realize that potential (most often attributed to improved agronomy, but also including the agronomy x genetic interactions), and accounting for the rate of adoption by farmers of improved cultivars and agronomic practices, which translates potential progress into realized progress on-farm. The prospects for yield improvement over 20 years for whole-of-industry mean wheat yield (t/ha) (Table 3) are calculated, with six assumptions.

1. The population of grain growers is considered as three groups comprised as top performers and early adopters of new technology (25% of land), a middle group (50% of land) who perform to the industry average and will adopt technology more slowly than the top group, and the bottom group (25% of land) who perform below the industry average and will not adopt new technology.
2. Farm consolidation contributes to gains in production through transfer of land ownership from below-average farmers to above-average farmers at a rate of 0.5% per year.

3. There is a background genetic gain in yield potential at 0.5% per year (Fischer and Edmeades 2010). To maintain the current rate of progress at 0.5% per year will require an increase in the absolute rate (Fischer 2015), so a modest target of maintaining progress in percentage terms is used. The realization of improved yield potential will vary with farmer group due to adoption of high yielding cultivars. It is assumed that increased carbon dioxide concentration in the atmosphere could make an additional 0.2% contribution to increase in yield potential in a C3 crop like wheat (Fischer 2009), but that this is offset by the negative effects of a warming and drying climate.
4. Adoption of existing practices varies by farmer group. It is assumed that the top group has already fully adopted a practice (such as variable rate fertilizer) that generates a benefit of 5% (or 0.1 t/ha), and the bottom performing group will never adopt it. The middle performing group moves from the current 20% to 100% adoption between years 1 and 10.
5. Adoption of a new and relatively simple practice that increases available soil water used by crops by 10 mm (equivalent 0.15 t/ha), due to amelioration of a soil constraint or use of plastic mulch to reduce soil evaporation. Because these are relatively simple, the new practices are fully adopted by the top performing group between years 1 and 10. The middle and bottom groups will not adopt this within the 20 year period as it is a comparatively new technology and relatively untested.
6. Adoption of a new and relatively complex practice by farmers. It is assumed that the top performing group adopts this complex new practice/technology, but takes the full 20-year period to reach 100% adoption because

Table 3. Calculations of projected yield increase in the Australian wheat industry over 20 years.

	Farmer group			
	Top	Middle	Bottom	Weighted all
Situation at year 1				
% of farmer population in each group	25	50	25	100
Yield (t/ha)	2.2	1.7	1.2	1.7
Between year 1 and 20 the separate contributions of				
Genetic improvement (t/ha)	0.22	0.10	0.02	0.22
Adoption of existing practices (t/ha)	0.00	0.10	0.00	
Adoption of a new and simple practice (t/ha)	0.15	0.00	0.00	
Adoption of a new and complex practice (t/ha)	0.07	0.00	0.00	
Total gain (t/ha)	0.44	0.20	0.02	
Situation at year 20				
Yield (t/ha)	2.6	1.9	1.2	2.1
% of farmer population in each group	35	50	15	100
Overall net improvement				
Average annual yield gain (kg/ha per year)	19	10	1	17
Average annual yield gain (% of year 1 yield)	0.8	0.6	0.1	1.0
Average annual yield gain (% of year 20 yield)	0.7	0.5	0.1	0.8

Groups refer to segments of the farmer population varying in their production performance and adoption of existing and new technology.

it takes longer to develop and be adapted by farmers to their particular circumstances (Llewellyn et al. 2011). An example would be a package of new agronomy and suitable cultivars for sowing 10 days earlier than currently practiced, worth on average 0.1 t/ha. The middle and bottom groups will not adopt this as it is comparatively new and complex technology and relatively untested.

The results in Table 3 show that over a 20-year period that with the above assumptions, yield increases from 1.7 to 2.1 t/ha, which is an average annual yield gain of 17 kg/ha per year or 1% and expressed relative to the year 1 yield and 0.8% when expressed relative to the year 20 yield. Because of farm consolidation, the weighted yield in year 20 is halfway between that of the middle and top performing group because after 20 years, the top group now comprises 35% of the total land area. The poor yield gain by the bottom group (3 kg/ha per year) does not influence the overall result significantly because there are only 16% of farmers in this category by year 20.

The assumptions made in our analysis are conservative. For instance, it is assumed that there is no improvement in genetic gain in percentage terms, despite the optimism expressed above for the impact of modern breeding technologies to lift this rate. The impacts of the three categories of new agronomic technologies of 5–10% are modest when compared to published responses reviewed above. A sensitivity analysis of the key parameters in this modeling exercise (data not shown) shows the annual yield gain varying from 10 to 50 kg/ha per year in response to variation in parameters, which is equivalent to 0.5–1.7% per year when expressed relative to year 20 yields. The rates of yield increase modeled here are consistent with recent historical yield gains of 13 kg/ha per year in good seasons (Richards et al. 2014), and 30 kg/ha per year in South Australia and 40 kg/ha per year in Western Australia between 1980 and 2000 when technology uptake was rapid (Turner and Asseng 2005; Richards et al. 2014). Our conclusion is that an annual rate of yield gain of 1% is a minimum expectation and 1.5% is achievable.

One of the largest contributors to future uncertainty about whether these yield gains are achievable is the impact of changes in the climate. Richards et al. (2014) show that yields in the Australian wheat industry are still increasing linearly between 1890 and 2010 both for the better (at 13 kg/ha per year) and drier (9 kg/ha per year) seasons, giving some confidence that improved genetics and practices are contributing to yield gains under both good and poor conditions. The most recent crop model estimates using climate projections from downscaled GCMs, assuming moderate to high emission scenarios, incorporating CO_2 fertilization effects, and allowing for some adaptation of sowing date, suggest that over the next 20–30 years there will be modest (<10% reduction) impacts on wheat yield potential (O'Leary et al. 2011; Potgieter et al. 2013; Yang et al. 2014). Other impact studies predict median wheat yield to decrease by up to 30% late in the 21st century (e.g., Anwar et al. 2007) under the most likely climate change scenarios. These impacts are significant, but 70 years hence, and are preceded by gradual changes over the next 20 years.

One of the likely consequences of climate change will be greater interannual variability in seasonal conditions, and this is already obvious in greater fluctuations in yield and revenue for wheat farmers in the last 10 years (Kingwell 2011). Most of the technologies listed above ought to reduce this variability as they are aimed at raising yields in water-limited situations, rather than increasing the water-nonlimited yield.

Conclusion

Our review has canvassed the prospects for 1–1.5% annual increases in cereal grain yields in Australia over the next 20 years, consistent with industry targets and the global food security imperative. It is clear that genetic gain, better adoption of existing practices/technologies, or development and adoption of new technologies and practices will each not deliver the desired gain alone. Rather, it will be an additive combination of all three that could result in gains of ca. 20 kg/ha per year in industry-level wheat yields. Gains could be even higher if G × M synergies result in multiplicative effects. Research, development, and extension will all be required to deliver these gains. The degree to which climate change is likely to constrain these gains is unclear beyond 20 or so years, however, for now, the impacts are likely to be of secondary importance relative to the impact of improved adoption, advances in technology, and practices and farm consolidation.

Conflict of Interest

None declared.

References

ABARES. 2012. Agricultural commodity statistics 2012. Australian Bureau of Agricultural and Resource Economics and Sciences, Canberra.

Adcock, D., A. M. McNeill, G. K. McDonald, and R. D. Armstrong. 2007. Subsoil constraints to crop production on neutral and alkaline soils in south-eastern Australia: a review of current knowledge and management strategies. Aust. J. Exp. Agric. 47:1245–1261.

Alston, J. M., J. M. Beddow, and P. G. Pardey. 2009. Agricultural research, productivity, and food prices in the long run. Science 325:1209–1210.

Angus, J. F., J. A. Kirkegaard, J. R. Hunt, M. H. Ryan, L. Ohlander, and M. B. Peoples. 2015. Break crops and rotations for wheat. Crop Pasture Sci. 66:523–552.

Anwar, M. R., G. O'Leary, D. McNeil, H. Hossain, and R. Nelson. 2007. Climate change impact on rainfed wheat in south-eastern Australia. Field. Crop. Res. 104:139–147.

Barlow, K. M., B. P. Christy, G. J. O'Leary, P. A. Riffkin, and J. G. Nuttall. 2015. Simulating the impact of extreme heat and frost events on wheat crop production: a review. Field. Crop. Res. 171:109–119.

Barraclough, P. B., J. R. Howarth, J. Jones, R. Lopez-Bellido, S. Parmar, C. E. Shepherd, et al. 2010. Nitrogen efficiency of wheat: genotypic and environmental variation and prospects for improvement. Eur. J. Agron. 3:1–11.

Blackwell, P. S., B. Webb, J. Lemon, and G. Reithmuller. 2004. Tramline farming systems technical manual. Dept. Agric. WA Bull. 4607:11.

Bogard, M., V. Allard, M. Brancourt-Hulmel, E. Heumez, J. M. Machet, M. H. Jeuffroy, et al. 2010. Deviation from the grain protein concentration–grain yield negative. J. Exp. Bot. 61: 4303–4312.

Bosse, M., R. Zlot, and P. Flick. 2012. Zebedee: design of a spring-mounted 3-d range sensor with application to mobile mapping. IEEE Trans. Rob. 28:1104–1119.

Botwright, T. L., A. G. Condon, G. J. Rebetzke, and R. A. Richards. 2022. Field evaluation of early vigour for genetic improvement of grain yield in wheat. Aust. J. Agric. Res. 53, 1137–1145.

Braunack, M., and J. Price. 2012. The potential for thin biodegradable film in the Australian cotton industry. in 'Capturing Opportunities and Overcoming Obstacles in Australian Agronomy: 16th Australian Society of Agronomy Conference', 14–18 October 2012, Armidale, Australia. Available at http://www.regional.org.au/au/asa/2012/crop-production/7956_braunack.htm (accessed 15 January 2015).

Chen, D., H. Suter, A. Islam, R. Edis, J. R. Freney, and C. N. Walker. 2008. Prospects of improving efficiency of fertiliser nitrogen in Australian agriculture: a review of enhanced efficiency fertilisers. Aust. J. Soil Res. 46:289–301.

Condon, A. G., R. A. Richards, and G. D. Farquhar. 1987. Carbon isotope discrimination is positively correlated with grain-yield and dry-matter production in field-grown wheat. Crop Sci. 27:996–1001.

Connor, D. J., and M. I. Mínguez. 2012. Evolution not revolution of farming systems will best feed and green the world. Global Food Secur. 1:106–113.

Corke, P., T. Wark, R. Jurdak, W. Hu, P. Valencia, and D. Moore. 2010. Environmental wireless sensor networks. Proc. IEEE 98:1903–1917.

DAFWA. 2013. Report Card on sustainable natural resource use in agriculture. Available at https://www.agric.wa.gov.au/soil-constraints/report-card-south-west-western-australia (accessed 1 July 2015).

Dang, Y. P., R. C. Dalal, S. R. Buck, B. Harms, R. Kelly, Z. Hochman, et al. 2010. Diagnosis, extent, impacts, and management of subsoil constraints in the northern grains cropping region of Australia. Aust. J. Soil Res. 48:105–119.

Dobermann, A., K. G. Cassman. 2004. Environmental dimensions of fertiliser nitrogen: What can be done to increase nitrogen use efficiency and ensure global food security?. Pp. 261–278 in A. R. Mosier, et al, eds. Agriculture and the nitrogen cycle: assessing the impacts of fertiliser use on food production and the environment. SCOPE 65. Island Press, Washington, DC.

Duvick, D. N., J. S. C. Smith, and M. Cooper. 2004. Long-term selection in a commercial hybrid maize breeding program. Plant Breeding Rev. 24:109–152.

Edwards, J., A. Umbers, and S. Wentworth. 2013. GRDC Farm Practices Survey Report 2012. Grains Research and Development Corporation. 97 pp.

Fischer, R. A. 2009. Farming systems of Australia: exploiting the synergy between genetic improvement and agronomy. Pp. 23–54 in V. O. Sadras and D. Calderini, eds. Crop physiology: applications for genetic improvement and agronomy. Academic Press, Burlington, MA.

Fischer, R. A. 2011. Wheat physiology: a review of recent developments. Crop Pasture Sci. 62:95–114.

Fischer, R. A. 2015. Definitions and determination of crop yield, yield gaps, and of rates of change. Field. Crop. Res. 182:9–18.

Fischer, R. A., and G. O. Edmeades. 2010. Breeding and cereal yield progress. Crop Sci. 50:S-85–S-98.

Fischer, R. A., D. Byerlee, and G. O. Edmeades. 2014. Crop yields and global food security—will yield increase continue to feed the world? ACIAR Monograph No. 158. Australian Centre for International Agricultural Research: Canberra.

Fletcher, A., M. Robertson, D. Abrecht, D. Sharma, and D. Holzworth. 2015a. Dry sowing increases farm level wheat yields but not production risks in a Mediterranean environment. Agric. Syst. 136:114–124.

Fletcher, A., M. Peoples, J. Kirkegaard, M. Robertson, J. Whish, and T. Swan. 2015b. A review of annual intercrops in rainfed farming systems of southern Australia. "Building Productive, Diverse and Sustainable Landscapes". Proceedings of the 17th ASA Conference, 20–24 September 2015, Hobart, Australia. Available at www.agronomy2015.com.au (accessed 1 July 2015).

Galiba, G., S. A. Quarrie, J. Sutka, A. Morgounov, and J. W. Snape. 1995. RFLP mapping of the vernalization (Vrn1) and frost resistance (Fr1) genes on chromosome 5A of wheat. Theor. and Appl. Genet. 90:1174–1179.

Gill, J. S., G. J. Clark, P. W. Sale, R. R. Peries, and C. Tang. 2012. Deep placement of organic amendments in dense sodic subsoil increases summer fallow efficiency and the use of deep soil water by crops. Plant Soil 359:57–69.

Grains Research and Development Corporation. 2012. Strategic Research & Development Plan 2012-17. Available at http://strategicplan2012.grdc.com.au/ investment_themes_and_outcomes/part3_theme_2_ improving_crop_yield.html (accessed 1 July 2015).

Hamza, M. A., and Anderson W. K. 2002. Improving soil physical fertility and crop yield on a clay soil in Western Australia. Aust. J. Agric. Res. 53:615–620.

Hamza, M. A., and W. K. Anderson. 2005. Soil compaction in cropping systems—a review of the nature, causes and possible solutions. Soil Till. Res. 82:121–145.

Hayman, P., J. Crean, J. Mullen, and K. Parton. 2007. How do probabilistic seasonal climate forecasts compare with other innovations that Australian farmers are encouraged to adopt? Aust. J. Agric. Res. 58:975–984.

Heffner, E. L., M. E. Sorrells, and J. L. Jannink. 2009. Genomic selection for crop improvement. Crop Sci. 49:1–12.

Hochman, Z., H. Van Rees, P. S. Carberry, et al. 2009a. Re-inventing model-based decision support with Australian dryland farmers: 4. Yield Prophet® helps farmers monitor and manage crops in a variable climate. Crop Pasture Sci. 60:1057–1070.

Hochman, Z., D. Holzworth, and J. R. Hunt. 2009b. Potential to improve on-farm wheat yield and WUE in Australia. Crop Pasture Sci. 60:708–716.

Hochman, Z., D. Gobbett, D. P. Holzworth, T. McClelland, H. van Rees, O. Marinoni, et al. 2012. Quantifying yield gaps in rainfed cropping systems: a case study of wheat in Australia. Field. Crop. Res. 136:85–96.

Hochman, Z., P. S. Carberry, M. J. Robertson, D. Gaydon, and L. W. Bell. 2013. Prospects for ecological agricultural intensification in Australia. Eur. J. Agron. 44:109–123.

Hunt, J. R., N. Fettell, J. Midwood, P. Breust, R. Peries, J. S. Gill, et al. 2012. Optimising flowering time, phase duration, HI and yield of milling wheat in different rainfall zones of southern Australia. Capturing Opportunities and Overcoming Obstacles in Australian Agronomy. Proceedings of 16th Agronomy Conference 2012, University of New England in Armidale, NSW, 14–18 October 2012. Available at http://regional.org.au/au/asa/2012/crop-development/8186_ huntjr.htm (accessed 1 July 2015).

Johnson, E. N., P. R. Miller, R. E. Blackshaw, Y. Gan, K. N. Harker, G. W. Clayton, et al. 2004. Seeding date and polymer seed coating effects on plant establishment and yield of fall-seeded canola in the Northern Great Plains. Can. J. Plant Sci. 84:955–963.

Jordaan, J. P., S. A. Engelbrecht, J. H. Malan, and H. A. Knobel. 1999. Wheat and heterosis. The genetics and exploitation of heterosis in crops. ASA, CSSA, and SSSA, Madison, WI, 411–421.

Kasirajan, S., and M. Ngouajio. 2012. Polyethylene and biodegradable mulches for agricultural applications: a review. Agron. Sustain. Dev. 32:501–529.

Keating, B. A., M. Herrero, P. S. Carberry, J. Gardner, and M. Cole. 2014. Food wedges: farming the global food demand and supply challenge towards 2050. Global Food Secur. 3:125–132.

Kingwell, R. S. 2011. Revenue volatility faced by Australian wheat farmers. In: Australian agricultural and resource economics society in its series 2011 conference (55th), Melbourne, 8–11 February 2011

Kirkegaard, A. J., and J. R. Hunt. 2010. Increasing productivity by matching farming system management and genotype in water-limited environments. J. Exp. Bot. 61:4129–4143.

Kirkegaard, J. A., J. M. Lilley, G. N. Howe, and J. M. Graham. 2007. Impact of subsoil water use on wheat yield. Aust. J. Agric. Res. 58:303–315.

Kirkegaard, J. A., M. K. Conyers, J. R. Hunt, C. A. Kirkby, M. Watt, and G. J. Rebetzke. 2014. Sense and nonsense in conservation agriculture: principles, pragmatism and productivity in Australian mixed farming systems. Agric. Ecosyst. Environ. 187:133–145.

Kleemann, S. G. L., C. Preston, and G. S. Gill. 2014. Influence of seeding system disturbance on preplant incorporated herbicide control of rigid ryegrass (Lolium rigidum) in wheat in Southern Australia. Weed Technol. 28:323–331.

Koltunow, A. M., S. D. Johnson, J. Rodrigues, T. Okada, Y. Hu, T. Tsuchiya, et al. 2011. Sexual reproduction is the default mode in apomictic Hieracium subgenus Pilosella, in which two dominant loci function to enable apomixis. Plant J. 66:890–902.

Kuehne, G., R. Llewellyn, D. J. Pannell, R. Wilkinson, P. Dolling, J. Ouzman, et al. 2015. A tool for predicting adoption and diffusion of agricultural innovations: ADOPT. Technological forecasting and social change (in review).

Llewellyn, R. S., R. K. Lindner, D. J. Pannell, and S. B. Powles. 2007. Herbicide resistance and the adoption of integrated weed management by Western Australian grain growers. Agric. Econ. 36:123–130.

Llewellyn, R. S., F. H. D'Emden, M. J. Owen, and S. B. Powles. 2009. Herbicide resistance in rigid ryegrass (Lolium rigidum) has not led to higher weed densities in Western Australian cropping fields. Weed Sci. 57:61–65.

Llewellyn, R. S., F. H. D'Emden, and G. Kuehne. 2011. Extensive use of no-tillage in grain growing regions of Australia. Field. Crop. Res. 132:204–212.

Longin, C. F. H., J. Mühleisen, H. P. Maurer, H. Zhang, M. Gowda, and J. C. Reif. 2013. Hybrid breeding in autogamous cereals. Theoret. Appl. Genet. 125:1087–1096.

Lu, C., J. A. Napier, T. E. Clemente, and E. B. Cahoon. 2011. New frontiers in oilseed biotechnology: meeting the global demand for vegetable oils for food, feed, biofuel, and industrial applications. Curr. Opin. Biotechnol. 22:252–259.

Manavella, P. A., C. A. Dezar, G. Bonaventure, I. T. Baldwin, and R. L. Chan. 2008. HAHB4, a sunflower HD-Zip protein, integrates signals from the jasmonic acid and ethylene pathways during wounding and biotic stress responses. Plant J. 56:376–388.

McBeath, T. M., C. D. Grant, R. S. Murray, and D. J. Chittleborough. 2010. Effects of subsoil amendments on soil physical properties, crop response, and soil water quality in a dry year. Soil Res. 48:140–149.

McBeath, T. M., M. Monjardino, R. Llewellyn, B. Jones, J. Ouzman, B. Davoren, et al. 2012. Can soil-specific inputs of N fertilizer be a risk-reducing strategy? In 'Capturing Opportunities and Overcoming Obstacles in Australian Agronomy: 16th Australian Society of Agronomy Conference', 14–18 October 2012, Armidale, Australia. Available at http://www.regional.org.au/au/asa/2012/nutrition/8113_mcbeath.htm (accessed 28 November 2013).

McCallum, M. 2005. Inter-row sowing shows yield benefits. No-till J. 2:10.

McDonald, G. L., J. D. Taylor, A. Verbyla, and H. Kuchel. 2012. Assessing the importance of subsoil constraints to yield of wheat and its implications for yield improvement. Crop Pasture Sci. 63:1043–1065.

McIntosh, P. C., M. J. Pook, J. S. Risbey, S. N. Lisson, and M. Rebbeck. 2007. Seasonal climate forecasts for agriculture: towards better understanding and value. Field. Crop. Res. 104:130–138.

McLaughlin, M. J., T. M. McBeath, R. Smernik, S. P. Stacey, A. Babasola, and C. Guppy. 2011. The chemical nature of P accumulation in agricultural soils-implications for fertiliser management and design: an Australian perspective. Plant Soil 349:69–87.

McNee, M., A. Fletcher, D. Minkey, and L. Celenza. 2015. Extent and attitudes of growers to dry seeding in the agroecological zones of Western Australia. "Building Productive, Diverse and Sustainable Landscapes" Proceedings of the 17th ASA Conference, 20 – 24 September 2015, Hobart, Australia. Available at www.agronomy2015.com.au (accessed 1 July 2015).

Monjardino, M., T. M. McBeath, L. E. Brennan, and R. S. Llewellyn. 2013. Are farmers in low-rainfall cropping regions under-fertilising with nitrogen? A risk analysis. Agric. Syst. 116:37–51.

Morrison, M., N. Harker, R. Blackshaw, C. Holzapfel, and J. Odonovan. 2016. Canola yield improvement on the Canadian Prairies from 2000 to 2013. Crop Pasture Sci. (in press)

O'Leary, G. J., B. Christy, A. Weeks, J. Nutall, P. Riffken, C. Beverly, et al. 2011. Downscaling global climatic predictions to the regional level: a case study of regional effects of climate change on wheat crop production in Victoria, Australia. Pp. 12–26 in S. S. Yadav, R. J. Redden, J. L. Hatfield, H. Lotze-Campen and A. E. Hall, eds. Crop adaptation to climate change. Wiley, Chichester.

Oliver, Y. M., and M. J. Robertson. 2013. Quantifying the spatial pattern of the yield gap within a farm in a low rainfall Mediterranean climate. Field. Crop. Res. 150:29–41.

Passioura, J. B., and J. F. Angus. 2010. Improving productivity of crops in water-limited environments. In Advances in Agronomy, Vol 106. D. L. Sparks, editor 37–75.

Peoples, M. B., J. Brockwell, J. R. Hunt, A. D. Swan, L. Watson, R. C. Hayes, et al. 2013. Factors affecting the potential contributions of N2 fixation by legumes in Australian pasture systems. Crop Pasture Sci. 63: 759–786.

Potgieter, A., H. Meinke, A. Doherty, V. O. Sadras, G. Hammer, S. Crimp, et al. 2013. Spatial impact of projected changes in rainfall and temperature on wheat yields in Australia. Clim. Change. 117:163–179.

Rainbow, R., and R. Derpsch. 2011. Advances in no-till farming technologies and soil compaction management in rainfed farming systems. Pp. 991–1014 in P. Tow, I. Cooper, I. Partridge and C. Birch, eds. Rainfed Farming Systems. Springer, Dordrecht, TheNetherlands (Chapter 39).

Rebetzke, G. J., A. G. Condon, R. A. Richards, and G. D. Farquhar. 2002. Selection for reduced carbon isotope discrimination increases aerial biomass and grain yield of rainfed bread wheat. Crop Sci. 42:739–745.

Rebetzke, G. J., R. A. Richards, N. A. Fettell, M. Long, A. G. Condon, R. I. Forrester, et al. 2007. Genotypic increases in coleoptile length improves stand establishment, vigour and grain yield of deep-sown wheat. Field. Crop. Res. 100:10–23.

Rebetzke, G. J., A. F. van Herwaarden, C. Jenkins, M. Weiss, D. Lewis, S. Ruuska, et al. 2008. Quantitative trait loci for water-soluble carbohydrates and associations with agronomic traits in wheat. Aust. J. Agric. Res. 59:891–905.

Rebetzke, G. J., B. Biddulph, K. Chenu, D. Deery, J. Mayer, C. Moeller, et al. 2013. Development of a multisite, managed environment facility for targeted trait and germplasm evaluation. Funct. Plant Biol. 40:1–13.

Rebetzke, G. J., R. A. Fischer, A. F. van Herwaarden, D. G. Bonnett, K. Chenu, A. R. Rattey, et al. 2014. Plot size matters: interference from intergenotypic competition in plant phenotyping studies. Funct. Plant Biol. 41:107–118.

Reynolds, M., D. Bonnett, S. C. Chapman, R. T. Furbank, Y. Manès, D. E. Mather, et al. 2011. Raising yield potential of wheat. I. Overview of a consortium approach and breeding strategies. J. Exp. Bot. 62:439–452.

Richards, R. A., G. J. Rebetzke, M. Watt, A. G. Condon, W. Spielmeyer, and R. Dolferus. 2010. Breeding for improved water–productivity in temperate cereals: phenotyping, quantitative trait loci, markers and the selection environment. Func. Plant Biol. 37:1–13.

Richards, R. A., J. R. Hunt, J. A. Kirkegaard, and J. B. Passioura. 2014. Yield improvement and adaptation of wheat to water-limited environments in Australia – a case study. Crop Pasture Sci. 65:676–689.

Robertson, M. J., D. Gaydon, D. J. M. Hall, A. Hills, and S. Penny. 2005. Production risks and water use benefits of summer crop production on the south coast of Western Australia. Aust. J. Agric. Res. 56:597–612.

Robertson, M. J., R. A. Lawes, A. Bathgate, F. Byrne, P. White, and R. Sands. 2010. Determinants of the proportion of break crops on Western Australian broadacre farms. Crop Pasture Sci. 61:203–213.

Robertson, M. J., R. S. Llewellyn, R. Mandel, R. Lawes, R. G. V. Bramley, L. Swift, et al. 2012. Adoption of variable rate fertiliser application in the Australian grains industry: status, issues and prospects. Precision Agric. 13:181–199.

Sadras, V. O., and C. Lawson. 2011. Genetic gain in yield and associated changes in phenotype, trait plasticity and competitive ability of South Australian wheat varieties released between 1958 and 2007. Crop Pasture Sci. 62:533–549.

Sadras, V. O., and G. McDonald. 2012. Water use efficiency of grain crops in Australia: principles, benchmarks and management. Grains Research and Development Corporation, South Australian Research and Development Institute and University of Adelaide.

Sadras, V. O., C. Lawson, and A. Montoro. 2012. Photosynthetic traits in Australian wheat varieties released between 1958 and 2007. Field. Crop. Res. 134:19–29.

Sharma, D. L., M. F. D'Antuono, W. K. Anderson, B. J. Shackley, C. M. Zaicou-Kunesch, and M. Amjad. 2008. Variability of optimum sowing time for wheat yield in Western Australia. Aust. J. Agric. Res. 59:958–970.

Siddique, K. H. M., R. K. Belford, M. W. Perry, and D. Tennant. 1989. Growth, development and light interception of old and modern wheat cultivars in a Mediterranean-type environment. Aust. J. Agric. Res. 40:473–487.

Simmons, A. J., and A. Hollingsworth. 2002. Some aspects of the improvement in skill of numerical weather prediction. Q. J. R. Meteorol. Soc. 128:647–677.

Simpfendorfer, S. 2012. Inter-row sowing reduces crown rot in winter cereals. in 'Proceedings of the First International Crown Rot Workshop for Wheat Improvement'. 22–23 October. (Eds. RIS Brettell, JM Nicol) (Organising Committee of the 1st International Crown Rot Workshop: Narrabri).

Slaughter, D. C., D. K. Giles, and D. Downey. 2008. Autonomous robotic weed control systems: a review. Comput. Electron. Agric. 61:63–78.

Stephens, D. J., and T. J. Lyons. 1998. Variability and trends in sowing dates across the Australian wheatbelt. Aust. J. Agric. Res. 49:1111–1118.

Taylor, K., D. Lamb, C. Griffith, G. Falzon, L. Laurent, R. Gaire, et al. 2013. Farming the web of things. IEEE Intell. Syst. 28:12–19.

Thakur, P., S. Kumar, J. A. Malik, J. D. Berger, and H. Nayyar. 2010. Cold stress effects on reproductive development in grain crops: an overview. Environ. Exp. Bot. 67:429–443.

Turner, N. C., and S. Asseng. 2005. Productivity, sustainability, and rainfall-use efficiency in Australian rainfed Mediterranean agricultural systems. Aust. J. Agric. Res. 56:1123–1136.

Vanhercke, T., J. R. Petrie, and S. P. Singh. 2014. Energy densification in vegetative biomass through metabolic engineering. Biocatal. Agric. Biotechnol. 3:75–80.

Viscarra Rossel, R. A., A. B. McBratney, and B. Minasny. 2010. P. 472 in Proximal soil sensing. Springer Science & Business Media, Sydney.

Walsh, M., P. Newman, and S. Powles. 2013. Targeting weed seeds in-crop: a new weed control paradigm for global agriculture. Weed Technol. 27:431–436.

Wark, T., P. Corke, P. Sikka, L. Klingbeil, Y. Guo, C. Crossman, et al. 2007. Transforming agriculture through pervasive wireless sensor networks. IEEE Pervasive Comput. 6:50–57.

Wasson, A. P., R. A. Richards, R. Chatrath, S. C. Misra, S. S. Prasad, G. J. Rebetzke, et al. 2012. Traits and selection strategies to improve root systems and water uptake in water-limited wheat crops. J. Exp. Bot. 63:3485–3498.

Weiss, U., and P. Biber. 2011. Plant detection and mapping for agricultural robots using a 3D LIDAR sensor. Robot. Auton. Syst. 59:265–273.

Yang, Y., D. L. Liu, M. R. Anwar, H. Zuo, and Y. Yang. 2014. Impact of future climate change on wheat production in relation to plant-available water capacity in a semiarid environment. Theoret. Appl. Climatol. 115:391–410.

Zheng, B., K. Chenu, F. M. Dreccer, and S. C. Chapman. 2012. Breeding for the future: what are the potential impacts of future frost and heat events on sowing and flowering time requirements for Australian bread wheat (Triticum aestivum) varieties? Glob. Change Biol. 18:2899–2914.

Zhou, H., J. D. Berg, S. E. Blank, C. A. Chay, G. Chen, S. R. Eskelsen, et al. 2003. Field efficacy assessment of transgenic Roundup Ready wheat. Crop Sci. 43:1072–1075.

Zhu, S., Y. Li, J. H. Vossen, R. G. Visser, and E. Jacobsen. 2012. Functional stacking of three resistance genes against Phytophthora infestans in potato. Transgenic Res. 21:89–99.

5

Selection of soybean lines exhibiting resistance to stink bug complex in distinct environments

Fabiani da Rocha, Caio Canella Vieira, Mônica Christina Ferreira, Kênia Carvalho de Oliveira, Fabiana Freitas Moreira & José Baldin Pinheiro

Departamento de Genética, Universidade de São Paulo, Escola Superior de Agricultura "Luiz de Queiroz", Avenida Pádua Dias, 11, 13.418-900 Piracicaba, SP, Brazil

Keywords

Euschistus heros, genetic gain, *Glycine max*, heritability, *Nezara viridula*, *Piezodorus guildinii.*

Correspondence

José Baldin Pinheiro, Universidade de São Paulo, Escola Superior de Agricultura "Luiz de Queiroz", Departamento de Genética, Avenida Pádua Dias, 11, 13.418-900, Piracicaba, SP, Brazil.
E-mail: jbaldin@usp.br

Funding Information

The authors thank the National Council for Scientific and Technological Development (CNPq), Brazilian Federal Agency for Support and Evaluation of Graduate Education (CAPES), and São Paulo Research Foundation (FAPESP) for granting a scholarship and funding this study.

Abstract

In soybean, stink bugs are considered the most important pest insect as they feed directly from the grain, causing significant losses in seed yield and quality. The use of resistant genotypes is a promising strategy to control these insects. Focusing on selection of soybean lines with resistance and high yield potential, 251 recombinant inbred lines (RILs), derived from a cross between IAC-100 (resistant) and CD-215 (susceptible), were evaluated in two experiments, designed as alpha-lattice, with three replicates in Piracicaba, during the growing seasons of 2012/13 and 2013/14. The evaluated traits were as follows: number of days to maturity (NDM), plant height at maturity (PHM), grain filling period (GFP), lodging (L), agronomic value (AV), grain yield (GY), weight of a hundred seeds (WHS), leaf retention (LR), and healthy seeds weight (HSW). Variance components were estimated by the Restricted Maximum Likelihood method (REML). Heritability and selection gain (SG) parameters were also calculated. Selection was carried out based on 2012/13 season, considering the genotypes that exhibited a minimum HSW of 2908.26 kg ha^{-1} (acceptable losses of 20% from the average GY). Insect population was monitored by cloth beating. An increase in stink bug population was observed during the grain filling period, with the highest population density occurring in the season 2012/13. Estimates of the variance components demonstrated the elevate influence of the interaction genotype x environment on GY and HSW, which exhibited the lowest estimates of heritability (23 and 34%, respectively). The estimate selection gain, calculated from the predicted means of GY and HSW, was of 665.4 and 482.4 kg ha^{-1} season 2012/13. Therefore, the applied selection allowed the identification of the genotypes exhibiting higher yields and resistance to the stink bug complex. From the RIL population, lines or genotypes potentially useful to generate novel cultivars were identified.

Introduction

Will the available agricultural production be enough to supply the demands of the world growing population? Currently, it is a debatable topic in sustainability discussions (Odegard and Voet 2014). Estimates have demonstrated that the required increase in agricultural production has to be from 100 to 110% to prevent the failure of food supply (Tilman et al. 2011). According to Ray et al. (2013), the production increase must come from yield improvement, instead of extending the cultivated area.

Soybean is among the four crops responsible for providing 2/3 of calories derived from agriculture worldwide (Ray et al. 2013). Therefore, a consistent production of this legume crop is necessary to guarantee food security. Plant breeding is an important tool to these goals, as it generates superior, high yielding, and adapted genotypes to a wide range of adverse conditions.

Several challenges need to be overcome for soybean production to sustain the predicted global population growth rate. The current yield increase rate is of 1.3%, however, it is required an increase around 2.4% to supply the food demands by 2050 (Ray et al. 2013). An important challenge is the increasing number of pests attacking the crop and causing yield losses (Belorte et al. 2003). Stink bugs are considered the most important pest insect to soybean. In Brazil, three species, consisting in the stink bug complex, are predominant: small green stink bug (*Piezodorus guildinii*), green stink bug (*Nezara viridula*), and neotropical brown stink bug (*Euschistus heros*) (Guedes et al. 2012). The brown stink bug is the predominant specie in soybean growing areas (Corrêa-Ferreira et al. 2009). *Piezodorus guildinii* is a neotropical specie, found from the South of the United States down to Argentina. This specie has secondary importance, as it occurs at low densities, although it is responsible for more severe damages to soybean due to the larger area of the insect feeding apparatus (Depieri and Panizzi 2010). *Nezara virudula* occurs mainly in the states of Santa Catarina and Rio Grande do Sul (Hoffmann-Campo et al. 2000), but has also expanded toward the Central-West regions of the country (Wiest and Barreto 2012).

Larger nymphs, from the 3rd to 5th instars, and adults cause direct and indirect damages, irreversible to seed development (Panizzi and Slansky 1985; Prado et al. 2010). The insects feed directly from the pods, by inserting its sucking mouth apparatus, reaching the grains (Corrêa-Ferreira 2000). The most important injuries are associated to the injection of digestive enzymes, leading to deformation, abortion, loss in germination, and seed vigor (Oliveira 2010). Moreover, the attack also allows the transmission of pathogens, such as the yeast *Eremothecium coryli*, delay in the physiological maturity or leaf retention, impairing mechanical harvesting of the crop (Gazzoni and Moscardi 1998; Silva et al. 2013). The losses caused by the stink bug complex may reach up to 125 kg ha^{-1}, considering the presence of a single stink bug per square meter (Guedes et al. 2012).

Chemical control has been widely employed to control the insects in soybean (Musser and Catchot 2008). However, besides the costs, the excessive use of insecticides has induced the appearance of insect populations exhibiting resistance to certain molecules (Corrêa-Ferreira et al. 2013); thus, requiring higher numbers of applications and the use of broad spectrum chemicals (Temple et al. 2009). The current scenario of chemical control consists in the prohibition of some chemicals and shortage of innovation (Guedes et al. 2012). Moreover, environmental pollution consequences are drastic. According to Hart and Pimentel (2002), only 0.1% of the pesticides applied reach the target insect and 99.9% impact the surrounding environment. Therefore, the use of resistant genotypes is an interesting alternative or substitute to insect chemical control (Smith 2005), offering a series of benefits from the environmental and economical standpoints.

Thus, the current work aimed to identify high yielding soybean lines exhibiting resistance to the stink bug complex.

Materials and Methods

Development of the population

The F$_8$ population (recombinant inbred lines, RIL) used in this study was developed from the cross between IAC-100 (resistant to the stink bug complex) and CD-215 (susceptible). Seeds of the F$_1$ generation were harvested in 2007/2008, sown in a greenhouse and gave rise to a total of 251 progenies that were advanced up to F$_8$ under these conditions using the single-pod descendent (SPD) method.

Experiments

The experiments were carried out in the field, at the Experimental Station Anhumas, in Piracicaba, from the Genetics Department at Luiz de Queiroz College of Agriculture (ESALQ/USP). In 2012/2013, the used experimental design was alpha-lattice 10 × 25 with three replicates, totalizing 248 RILs and the respective parents. The experimental plot consisted of two lines of 4 m in length. In 2013/2014, 251 RILs and the respective parents were analyzed, along with three checks (BMX Potência, Vmax, and BMX Apolo). The 256 treatments were arranged as 32 × 8 alpha-lattice and each experimental plot consisted of four lines of 5 m in length.

Line spacing for the experiments was of 0.5 m, with 18 seeds per linear meter. Natural stink bug infestation was evaluated and the insect population density was recorded weekly according to the cloth beating method (Stürmer et al. 2012).

Plant phenotyping

The agronomical performance of the lines was evaluated for the following traits:
1. NDM: number of days to maturity, counted from seeding up to the date when 95% of the pods were ripe;
2. PHM: plant height (cm) at maturity, measured from the base of the plant (on the ground) up to the apex of the main stem;
3. L: lodging evaluated at maturity by a scale of visual grades ranging from 1 to 5, where 1 corresponds to the erect plant and 5, to lodged plants;
4. AV: agronomic value, evaluated at maturity according to visual grading system ranging from 1 to 5, where 1 corresponds to plants with no agronomic value and

5, to plants with excellent agronomic features (large number of pods, height superior to 60 cm, vigorous, erect plants, absence of green stems and leaf retention, absence of pod shattering, and absence of disease symptoms);

5. GY: grain yield, presented as kg ha^{-1}.

Moreover, four traits associated with insect resistance were evaluated:

1. GP: grain filling period (days) was obtained by the difference between the reproductive stages R7 and R5;
2. WHS: weight of a hundred seeds (g), obtained from a randomly selected sample after standardization of the moist contents;
3. LR: leaf retention, determined according a grading system ranging from 0 for plants with normal senescence to 5 for plants with several green stems and leaves (mechanical harvest impracticable);
4. HSW: healthy seeds weight (kg ha^{-1}), weight of seeds with no damage from stink bug attack, evaluated after grain harvest and processing. Seeds were processed by spiraling in order to remove empty, green, and ill-formed seeds by centrifugal and gravitational forces.

Data analyses

The data analyses considered all parameters of the model as random. The variance components (by the restricted maximum likelihood method - REML), minimum significant difference, coefficient of variation and heritability were estimated by combining the two environments (seasons). Predicted genetic means were estimated by the sum of the grand mean and the random estimate of the effect of each genotype (Best Linear Unbiased Prediction – BLUP).

The data were analyzed using the PROC MIXED, via META suite (Vargas et al. 2013) at SAS, according to the following statistical model (eq. 1):

$$Y_{ijkl} = \mu + G_i + A_j + R_k(A_j) + B_l(A_jR_k) + G_i * A_j + \varepsilon_{ijkl} \quad (1)$$

where:

Y is the value of the observation corresponding to the genotype i at replicate k in environment j;

μ is the mean of the principal effect;

G_i is the effect of genotype i;

A_j is the effect of the environment (season) j;

R_k is the effect of the replicate k within environment j;

B_l is the effect of the block l within environment j within replicate k;

$G_i * A_j$ is the effect of the interaction genotype x environment; and

ε_{ijkl} is the error or random residue.

Heritability was estimated based on the mean of the genotypes (eq. 2):

$$h^2 = \frac{\sigma_g^2}{\sigma_g^2 + \frac{\sigma_{ga}^2}{n\text{Env}} + \frac{\sigma_e^2}{n\text{Env} \times n\text{Rep}}} \quad (2)$$

Where:

σ_g^2, σ_{ga}^2 and σ_e^2 are variances: genotypic, from the interaction Genotype × Environment and error, respectively;

nEnv is the number of environments where the experiments were conducted; and

nRep is the number of replicates.

The verification of genetic progress was performed from the selection based on the target environment – presence of stink bugs. Then, the verification of the progress for the season 2013/2014 was carried out based on the selection of individuals from the previous season. This approach was chosen based on the higher population density of stink bugs in 2012/13, in comparison to 2013/14. Moreover, the climate conditions of the later growing season were atypical, consisting of high temperatures and low precipitation.

The selection target trait was HSW, using a classification based on the ideotype: assuming 2908.26 kg ha^{-1} as minimum HSW value. The value was estimate considering the maximum acceptable loss of 20% in comparison to the general yield average of the genotypes used in the experiment. The consideration of a single trait for selection, namely HSW, was decided after the demonstration by Rocha et al. (2014) that it is a useful trait for the simultaneous selection of genotypes exhibiting resistance and high yield.

Selection gain (SG) was estimated based on the formula (eq. 3):

$$GS = \overline{X}_S - \overline{X}_O \quad (3)$$

Where:

$\overline{X}_S$ represents the means predicted by BLUP for the selected lines; and $\overline{X}_O$ represents the means predicted by BLUP for the original population.

Results and Discussion

Stink bug population

The expressive increase in the stink bug population at the end of the crop cycle was observed during both seasons (Figs. 1, 2), which is due to the presence of pods in the plants that are directly correlated to the presence of stink bugs migrating from harvest neighboring areas (Panizzi et al. 2000).

In 2012/13, the maximum number of stink bugs was as high as six (Fig. 1), whereas in 2013/14, it reached 3.5 (Fig. 2). In the later season, the infestation was below

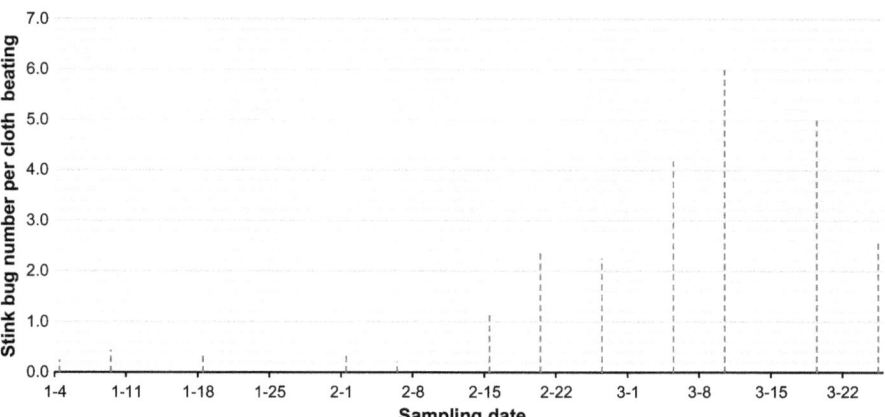

Figure 1. Fluctuation of stink bug population in the experimental area in the growing season 2012/2013 along soybean phenological stages R3–R8, evaluated by the cloth beating method (2 m line), for soybean genotypes.

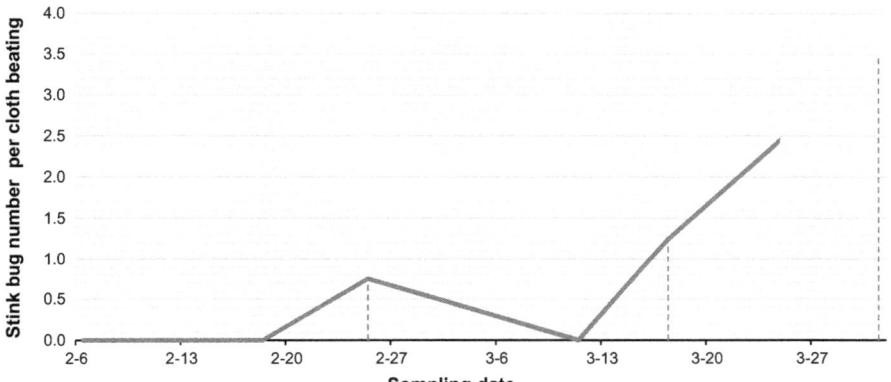

Figure 2. Fluctuation of stink bugpopulation in the experimental area in the growingseason 2013/2014 along soybean phenological stages R3–R8, evaluated by the cloth beating method (2 m per line), for soybean genotypes.

the economic injury level. The action threshold for all stink bug species is 4 stink bugs per 2 m row on a drop cloth, considering grain production and 2 if we consider seed production (Corrêa-Ferreira and Panizzi 1999).

In 2012/13, an elevate increase in the number of insects per cloth beating was observed when the majority of the genotypes was entering stage R5, when soybean is most susceptible to attack, the called critical period (Corrêa-Ferreira and Panizzi 1999). Although stink bugs can be detected at vegetative growth, their negative effects on grain filling and seed quality are noteworthy during pod formation (Corrêa-Ferreira et al., 2013). The population of stink bugs reached its maximum growth in the beginning of March, when the majority of the genotypes were mature. At harvest, the stink bugs are dispersed to alternative host plants, and, considering the brown stink bug, to diapause niches (Corrêa-Ferreira and Panizzi 1999).

During the harvesting season 2013/14, insects were absent from the first two evaluations (Fig. 2). The experiment was installed in the last days of the seeding window, with the expectations that the number of insects per cloth beating would be high. However, the

inexistence of soybean crops in the neighboring areas, which could favor stink bug migration, the occurrence of drought and high temperatures during the period may have interfered with the growth of the insect population.

Variance components and Heritability

Variance components, determined by the restricted maximum likelihood (REML), minimum significant difference, variation coefficient, and heritability combined for two environments (seasons) for the traits PHM, NDM, GFP, LR, AV, L, GY, WHS, and HSW are shown in Table 1. For the majority of the investigated traits, variation coefficients were below 20%, indicating experimental precision (Pimentel Gomes 2000).

The use of the random model focus the results on the estimation of the variance components (Table 1), from which it is possible to estimate the genetic parameters, such as heritability and gains obtained from selection. The estimate components of the phenotypic variance demonstrate the strong influence of the environment on the expression of the majority of the traits.

Table 1. Variance components, mean, LSD (least significant difference), coefficient of variation (CV), and heritability (h^2) for the traits plant height at maturity (PHM in cm), number of days to maturity (NDM), grain filling period (GFP in days), leaf retention (LR in grading scale from 1 to 5), agronomic value (AV in grading scale ranging from 1 to 5), lodging (L in grading scale ranging from 1 to 5), grain yield (GY in kg ha^{-1}), healthy seeds weight (HSW in kg ha^{-1}), and weight of a hundred seeds (WHS in g), evaluated in soybean recombinant inbred lines with resistance alleles to stink bug complex in two environments.

Statistics	PHM	NDM	GFP	LR	AV	L	GY	HSW	WHS
σ_a^2	10.187	43.759	19.798	0.092	0.009	0.564	311513.111	227615.949	2.336
σ_g^2	41.132	20.440	8.613	0.273	0.045	0.069	30662.445	39439.256	2.117
σ_{ga}^2	30.875	8.700	3.203	0.103	0.077	0.043	111196.233	65697.742	0.126
σ_e^2	57.358	15.062	27.754	0.489	0.381	0.337	262668.872	249105.875	0.975
Mean	61.685	117.290	35.899	1.826	3.364	1.765	2771.313	2516.049	12.913
LSD	14.347	7.439	7.080	1.046	0.918	0.807	903.984	788.923	1.380
CV	11.809	3.220	10.014	29.104	13.859	23.215	16.562	15.920	5.425
h^2	0.622	0.749	0.580	0.672	0.306	0.471	0.236	0.347	0.904

σ_a^2: environmental variance; σ_g^2: genotypic variance; σ_{ga}^2: genotype × environment interaction variance; and σ_e^2: residual variance.

PHM and LR were the only traits displaying higher genetic variance (30 and 28% of the total variation, respectively) than environmental and genotype × environment interaction effect (G × E). These results indicate that for these traits the variability existent among the genotypes was more expressive than the remaining variances present in the model. While for the trait AV, genetic variance had higher values than the environment, but was the component of G × E interaction that had the most determinant influence on the mean of the genotypes.

Genetic variation for the trait leaf retention is highly important, as it is associated to soybean resistance to the stink bug complex. Genotypes exhibiting lower scores for the trait are desirable. The evaluation of leaf retention takes into account the amount of green stems and leaves in the plant. Green stem corresponds to the maintenance of green primary and secondary stems, and leaf retention is the absence of leaf loss, even after physiological seed maturity (Silva et al. 2013). Stink bug attacked soybean plants at the reproductive phase may lose grains and pods and the stress causes the retention of green leaves and stems. Leaf retention impairs mechanical harvesting and seed storage and processing, thus reducing the quality. Fernandes et al. (1994) have demonstrated the lower leaf retention indices of IAC-100, in comparison to the genotypes IAC-8, IAC-12, IAC-17, Dourados, Emgopa 304, and Emgopa 309, suggesting that the cultivar is less prone to leaf retention even under stink bug attack.

For the traits NDM, GFP, L, GY, HSW, and WHS, environmental variance was of higher magnitude in comparison to genotypic and G × E interaction variances. For the trait AV, the component of G × E interaction had the most determinant influence on the mean of the genotypes. Moreover, for GFP and HSW, the effect of G × E interaction displayed higher estimates than that of genotypic variance.

Heritability is one of the genetic parameters underlying the success and delimitation of selection strategies (Laviola et al. 2010). Higher values for the estimates of the parameter correspond to higher possible gains by selection (Gravois and Bernhardt 2000). Considering the investigated genotypes, the estimates of heritability coefficients, based on the means for the genotypes, ranged from 0.23 to 0.90. Values at the lower end correspond to the traits HSW and GY. In general, grain yield exhibits medium to low values due to the quantitative inheritance of the trait (Bueno et al. 2013). Traits controlled by several genes are highly influence by the environment. The complex nature of the inheritance prevents the selection of genetically superior genotypes (Li et al. 2003), as the correlation between the phenotype and the genotype is reduced (Coimbra et al. 2009).

Heritability values for GY and HSW were of 0.24 and 0.35, respectively. The HSW is directly correlated to GY, as it is measured by the weight of seeds from a given plot, after the removal of the grains injured by stink bug attack. Bueno et al. (2013) have found a heritability value superior to 83.50% for grain yield, when four investigated environments were considered as a group. According to the authors, the high heritability values are explained by the elevated genetic variance. In contrast, for Lopes et al. (1997), the heritability was of 57.92% in an experiment with natural stink bug infestation. The heritability is a parameter that could help the breeder choosing the breeding strategies for the interested traits. When the trait has low heritability or high dominance effect, it is not recommended to make selections in initial generations of selfing. Thus, according the hereditability values found in this work for GY and HSW, the selection should be made after F_5 generations, when it is possible to consider that the individuals have enough level of homozigose. GY and HSW are related. So, according Rocha et at. (2014) if the selections are based on HSW in an environment with stink bug stress, it is possible identify that genotypes with resistance and high yield potential.

For GFP, the estimated heritability value was of 0.58, higher than the number found by Godoi and Pinheiro (2009). The heritability for GFP in the previous work was of 36.06%, in narrow sense. However, the estimates were based on plots. Estimates based on genotypes mean tend to result in higher values of heritability, as the effect of the residual variance is divided by the number of blocks and by the number of locations. For LR, heritability value was of 67%, also superior to 20% found by Godoi and Pinheiro (2009) and 20.6%, by Santos (2012).

The heritability values for PHM, NDM, and AV were 0.622, 0.749 and 0.306, respectively. These values were inferior to 90, 81, and 63%, respectively, found by Santos (2012). For L, Lopes et al. (1997) have found heritability estimates of 82.97%, whereas in our current work, the value was 0,471 or 47,1%.

Selection gain

The reaction of the genotypes to stink bug infestation could only be effectively considered during the 2012/13 season, thus allowing their classification according to resistance to the insect attack. This work focus on the identification of resistance and high yield lines, we have selected the genotypes exhibiting high performance for the traits associated with resistance and elevated yield for the 2012/13 season and estimate their gain for the 2013/14 season.

Based on HSW, genotypes producing more than 2908.26 kg ha^{-1} were selected. In order to reduce the number of lines and increase selection gain, a second criterion was adopted: minimum GY of 3635.32 kg ha^{-1} (20% superior to the mean).

Selection gains estimated based on the season 2012/13 (Table 2) were of (1) PHM: +8.8 cm – the mean height for the selected genotypes was of 72.6 cm, thus, the gain is considered satisfactory. According to Garcia et al. (2007), values higher than 60 cm are ideal to minimize grain losses during harvest; (2) NDM: +5.1 days – the increase in the number of days to maturity is not positive for insect resistance, as it causes higher exposition periods of the plants to the attacking insects; (3) GFP: +1.5 – the period between R5 and R7 is the most susceptible phase of the crop. Thus, genotypes exhibiting shorter grain filling periods tend to suffer less damage by stink bugs, it is a pseudo resistance mechanism of the host evasion type, when the time during the most susceptible phase is shortened. The use of the mechanism has been suggested to reduce damages by stink bug attack; (4) LR: +0.2 – mechanical harvest, seed storage, and processing are difficult by leaf retention. The highest

note in the grading system is 5 and the mean of the selected lines correspond to 2.1, a reasonable performance and a nonexpressive increment; (5) AV: +0.0 - the selection gain was null for the trait. The parameter considers a general plot pattern for the number of pods, plant height, vigor, lodging, green stems and leaf retention, pod shedding, and disease symptoms. Although no selection gain was detected for the isolated parameter, it was positive for its components; (6) L: +0.0 – the selection gain was null for lodging, which is a positive aspect as higher grades represent lodge of plants in the plot. Moreover, the parents are cultivars, which means that they have undergone selection for the trait; (7) GY: +665.4 kg ha^{-1} – an excellent gain, corresponding to more than 11 sacs per ha; (8) HSW: +482.4 kg ha^{-1} – elevated and simultaneous with yield gain. Our results demonstrate that it is possible to concomitantly select for stink bug resistance and grain yield; (9) WHS: +0.2 g – the weight of a hundred seeds is another pseudo resistance mechanism, defined as damage dilution type. It can be used to reduce losses due to insect attack. Lower WHS correspond to a higher number of seeds per plant and smaller proportional number of damaged seeds. Therefore, the increase in the mean for the trait is not considered positive for stink bug resistance.

The selection of superior progenies based only on traits that are strongly influenced by the environment and often correlated could provoke effects in other. As the results showed, the selection based on GY and HSW increased all the other traits evaluated, expect for AV. No changes in AV could be justified because both parents involved in the cross were commercial cultivars, i.e. have good agronomic performance. PHM and GY have near QTLs reported in the literature, and shared the same direction of additive effects (Liu et al. 2013), so when the selection focus the increase of one between the two traits cited, it is expect higher values for both. In the same way, Lee et al. (1996) identified same QTLs for plant height, lodging, and maturity. But, correlations between L and GY (Panthee et al. 2007) and LR and GY (Lopes et al., 2007) are negative. Therefore, we did not expect increase in these traits making selection based in GY. However, the changes were just 0.1 and 0.2, values that we believe that did not have affect the harvest processes, as well GY. The increase in the GFP is supported by it positive correlation with GY, in order of 0.42 (Panthee et al. 2007). Yield-component traits, as WHS, are responsible in determining GY, thus the increase in WHS was expect, even it is not a positive aspect when the breeding program is focusing in the reduction of stink bug damage.

During the season 2013/14 (Table 3), selection gains were estimated to be in the same direction, however,

Table 2. Performance of 29 soybean lines selected based on the genetic mean estimated by BLUP (Best linear unbiased prediction) for plant height at maturity (PHM in cm), number of days to maturity (NDM), grain filling period (GFP in days), leaf retention (LR in grading scale from 1 to 5), agronomic value (AV in grading scale ranging from 1 to 5), Lodging (L in grading scale ranging from 1 to 5), grain yield (GY in kg ha^{-1}), healthy seeds weight (HSW in kg ha^{-1}), and weight of a hundred seeds (WHS in g) during the growing season 2012/13.

Genotype	PHM	NDM	GFP	LR	AV	L	GY	HSW	WHS
17	74.6	124.5	39.4	2.4	3.2	1.4	3703.5	3271.0	13.6
18	77.2	128.7	37.4	2.3	3.3	1.6	3669.4	3227.0	13.1
22	71.9	127.6	41.6	2.5	3.1	1.3	4021.6	3540.9	14.3
32	82.8	124.2	38.0	1.3	3.7	1.3	3842.7	3432.8	11.7
40	77.1	124.2	40.8	2.0	3.3	1.2	3747.8	3316.1	15.7
62	82.2	115.8	34.9	2.0	3.5	1.2	3648.4	3235.8	16.8
69	76.1	141.0	43.0	4.1	3.2	1.1	3648.6	3181.9	14.9
71	71.8	139.1	42.6	1.8	3.2	1.3	3752.6	3104.3	15.0
85	69.9	129.4	39.4	2.7	3.2	1.6	3890.5	3081.7	15.8
89	63.3	125.3	40.8	2.5	3.1	1.4	3798.8	3297.1	16.6
93	72.9	131.5	43.8	1.8	3.4	1.2	3936.7	3456.0	14.3
101	65.3	124.1	40.8	2.2	3.3	1.2	3669.0	3137.2	14.0
108	63.8	118.3	40.2	2.4	3.3	1.3	3693.4	3295.8	14.8
117	78.8	124.8	40.8	2.2	3.3	1.6	4089.5	3416.5	16.5
126	68.3	123.2	41.8	1.8	3.4	1.1	3858.4	3528.2	13.3
127	72.3	124.9	40.8	3.2	2.9	1.2	4137.3	3700.1	12.8
132	79.4	134.6	39.6	2.3	3.0	1.3	3833.2	3246.9	12.7
142	71.0	137.1	43.0	3.6	3.2	1.2	3758.7	3129.6	16.2
155	66.0	129.8	42.8	2.4	3.3	1.2	3812.4	3329.8	12.8
167	78.5	132.2	36.9	2.0	3.4	1.3	4044.9	3443.2	14.1
178	71.5	132.6	42.2	2.0	3.2	1.3	3656.2	2974.9	13.4
195	68.2	120.3	39.0	1.8	3.6	1.1	3783.6	3453.8	13.5
198	73.0	119.5	43.5	1.5	3.6	1.3	3796.7	3440.5	13.0
202	65.5	131.0	42.8	3.7	3.1	1.1	3978.8	3277.1	16.4
215	71.7	127.3	38.0	1.8	3.3	1.2	3950.5	3450.2	15.0
219	71.5	125.1	40.8	1.8	3.3	1.6	3776.6	3329.3	13.2
224	75.0	124.5	41.0	2.5	3.4	1.1	3916.4	3603.9	13.9
245	62.6	123.4	40.6	1.8	3.3	1.2	3707.1	3313.1	12.9
251	83.8	120.6	39.0	1.6	3.1	1.9	3925.1	3455.0	12.1
CD-215	71.3	115.7	40.2	1.3	3.8	1.1	2973.7	2777.7	15.6
IAC-200	64.8	135.9	44.1	1.4	3.1	1.1	2676.2	2372.3	11.4
μ_L	63.8	121.9	39.0	2.1	3.3	1.2	3163.9	2851.0	14.0
μ_S	72.6	127.1	40.5	2.3	3.3	1.3	3829.3	3333.4	14.2
SG	8.8	5.1	1.5	0.2	0.0	0.1	665.4	482.4	0.2

μ_L: general mean from the lines included in the experiment; μ_S: mean from the selected lines; and SG: selection gain.

at smaller proportion. Selection based on data from the season 2013/14 would have the minimum selectable parameters at 2723.32 kg ha^{-1} for GY (15% superior to the mean) and 2178.65 kg ha^{-1} for HSW (acceptable loss of 20% of the mean GY). The selectable values are inferior to those considered for the previous harvesting season and only three genotypes would have been selected, without correspondence to those selected in 2012/13. In the light of the conditions during both growing seasons and the problems that occurred in the later one, the most plausible decision was to perform indirect selection, based on the results from 2012/13.

The lack of consistency between the selected genotypes for each growing season, along with the variance components, shows the magnitude of the G × E interaction component for GY and HSW. The G × E interaction may occur in two forms: simple, when there are differences in the genotypes variation for distinct environments, but their relative ranking is not altered; or complex, when the genotypes responses are different depending on the environment (Cruz and Carneiro 2006). In the later situation, breeders' decision making is difficult (Coelho et al. 2010).

The effects and the genotypic values for the 29 selected soybean lines based on yield and resistance is presented in Table 4. In order to facilitate data interpretation, the general mean was added to each deviation. Positive values indicate that the given genotype has contributed to increase the general mean of the assay, whereas, negative

Table 3. Performance of 29 soybean lines selected based on the genetic mean estimated by BLUP (Best linear unbiased prediction) for plant height at maturity (PHM in cm), number of days to maturity (NDM), grain filling period (GFP in days), leaf retention (LR in grading scale from 1 to 5), agronomic value (AV in grading scale ranging from 1 to 5), Lodging (L in grading scale ranging from 1 to 5), grain yield (GY in kg ha^{-1}), healthy seeds weight (HSW in kg ha^{-1}), and weight of a hundred seeds (WHS in g) during the growing season 2013/14.

Genotype	PHM	NDM	GP	LR	AV	L	GY	HSW	WHS
17	67.4	113.2	33.5	2.3	3.1	2.5	2505.7	2284.8	11.0
18	67.2	117.1	37.9	1.5	3.5	2.5	2590.7	2382.5	10.9
22	64.6	114.6	37.1	1.7	3.4	2.6	2431.3	2233.5	13.0
32	80.6	115.0	32.6	1.3	3.4	2.5	2588.4	2423.0	9.8
40	68.2	112.8	33.4	1.6	3.2	2.5	2334.8	2166.7	13.9
62	71.0	112.0	32.3	1.4	3.5	2.4	2415.3	2232.3	13.0
69	68.9	118.1	34.5	1.9	3.7	2.3	2472.8	2180.0	12.1
71	60.9	116.6	35.0	1.8	3.4	2.4	2577.7	2323.4	12.5
85	53.8	115.5	35.8	1.2	3.6	2.4	2189.3	2019.6	12.4
89	57.8	114.5	36.2	1.2	3.6	2.3	2472.2	2303.5	12.4
93	58.9	117.3	35.1	1.2	3.7	2.1	2464.3	2310.1	12.5
101	48.2	113.1	34.2	1.5	3.0	2.6	2362.8	2178.4	11.5
108	58.2	113.6	34.0	1.4	3.6	2.3	2478.0	2296.3	12.9
117	56.5	114.1	34.1	2.0	3.6	2.5	2424.8	2212.2	13.3
126	66.8	111.0	34.7	1.5	3.4	2.4	2640.6	2472.4	11.4
127	56.4	114.2	35.1	2.0	3.3	2.4	2282.0	2107.9	10.4
132	78.2	116.8	33.9	1.2	3.4	2.6	2354.5	2159.9	10.3
142	61.3	115.3	35.4	1.9	3.5	2.3	2432.2	2084.7	15.3
155	68.3	117.0	35.5	2.1	3.0	2.5	2360.6	2187.3	11.1
167	66.2	114.7	35.9	1.8	3.2	2.3	2304.3	2108.8	12.3
178	60.9	115.4	32.6	1.1	3.6	2.4	2243.5	2022.6	10.7
195	63.4	109.5	30.2	1.1	3.8	2.3	2725.9	2516.3	10.9
198	66.6	110.5	42.9	1.4	3.2	2.4	2304.4	2170.3	11.4
202	49.5	118.0	35.7	2.1	3.2	2.3	2076.9	1878.4	15.4
215	62.5	116.3	34.9	1.5	3.4	2.5	2399.0	2225.0	12.2
219	55.2	116.0	31.0	1.5	3.2	2.4	2044.3	1863.2	12.6
224	57.2	115.8	35.2	2.4	3.4	2.3	2468.9	2267.1	11.9
245	59.9	104.8	31.5	1.4	3.0	2.6	2497.8	2291.5	11.3
251	66.2	114.1	29.2	1.9	3.3	2.3	2408.1	2220.8	10.6
CD-215	68.4	110.9	30.6	1.2	3.7	2.4	2633.4	2450.3	13.0
IAC-100	55.2	116.3	36.1	1.7	3.3	2.4	2252.2	2076.5	9.6
BMX Potência	71.4	115.6	37.3	2.3	3.7	2.3	2697.2	2457.8	13.7
Vmax	60.8	118.7	37.1	2.1	3.2	2.3	2453.8	2108.2	14.0
BMX Apolo	53.8	104.7	28.1	1.3	3.9	2.1	2214.6	1963.4	12.1
μ_L	59.3	112.6	32.7	1.6	3.4	2.3	2366.5	2173.1	11.8
μ_S	62.8	114.4	34.5	1.6	3.4	2.4	2408.7	2211.1	12.0
GS	3.5	1.8	1.7	0.0	−0.1	0.1	42.2	38.1	0.2

μ_L: general mean from the lines included in the experiment; μ_S: mean from the selected lines; and SG: selection gain.

values indicate the opposite effect. Thus, RIL 62 contributed to reduce the general mean in 2.36 days for GFP, whereas RIL 198 contributes to increase this trait in 7.52 days. For LR, RIL 32 reduced the general mean up to 0.43 points, whereas RIL 69 increased the parameter up to 1.10 points of the grading system. For GY, the totality of the investigated genotypes contribute to increase the general mean, as expected, as they were selected based on the desired ideotype. For HSW, only RIL 178 contributes to reduction of the general mean in 38 kg, whereas, the effect of the remaining lines was positive. For WHS,

RIL 32 contributed to a 2.20 g reduction, whereas RIL 202 promoted an increase of 3.10 g.

Although the predicted selection gains for GFP, LR, and WHS, traits associated with resistance to stink bug complex, were positive; the results indicate that among the selected 29 lines those contributing negatively to the mean of the traits can still be identified. The most noteworthy RILs contributing to those traits are 32, 195, 219, 245, and 251, which contribute to the reduction of GP, LR, and WHS means and to the increase in GY and HSW, among the selected lines.

Table 4. Genotypic effect (GE) and genotypic value (GV) for 29 selected lines, considering the traits involved in soybean resistance to stink bug complex: leaf retention (LR grading scale from 1 to 5), grain filling period (GFP in days), grain yield (GY in kg ha^{-1}), healthy seeds weight (HSW in kg ha^{-1}), and weight of a hundred seeds (WHS in g), for two assays conducted in the growing seasons 2012/13 and 2013/14.

Genotype	GFP		LR		GY		HSW		WHS	
	GE	GV	GE	GV	GE	GV	GE	GV	GE	GV
17	0.71	36.60	0.55	2.37	109.14	2875.62	154.03	2667.11	−0.55	12.36
18	2.11	38.00	0.12	1.94	120.08	2886.56	165.38	2678.46	−0.91	12.00
22	3.60	39.49	0.23	2.05	140.43	2906.91	206.99	2720.07	0.76	13.67
32	−0.41	35.48	−0.43	1.39	155.60	2922.08	249.76	2762.84	−2.20	10.71
40	1.18	37.07	0.01	1.83	62.10	2828.58	97.23	2610.31	1.98	14.89
62	−2.36	33.53	−0.10	1.72	66.22	2832.70	100.42	2613.50	2.04	14.95
69	2.95	38.84	1.10	2.92	86.26	2852.74	81.30	2594.38	0.50	13.41
71	2.76	38.65	0.01	1.83	108.37	2874.85	86.82	2599.90	0.85	13.76
85	1.64	37.53	0.12	1.94	68.79	2835.27	16.97	2530.05	1.32	14.23
89	2.67	38.56	0.01	1.83	94.31	2860.79	127.83	2640.91	1.70	14.61
93	3.32	39.21	−0.32	1.50	121.70	2888.18	182.02	2695.10	0.52	13.43
101	1.55	37.44	0.12	1.94	76.84	2843.32	82.08	2595.16	−0.15	12.76
108	1.46	37.35	0.12	1.94	118.69	2885.17	185.36	2698.44	0.86	13.77
117	1.36	37.25	0.23	2.05	139.19	2905.67	154.46	2667.54	1.93	14.84
126	2.30	38.19	−0.21	1.61	152.93	2919.41	266.03	2779.11	−0.69	12.22
127	2.11	38.00	0.66	2.48	128.83	2895.31	211.11	2724.19	−1.38	11.53
132	1.01	36.90	−0.10	1.72	101.69	2868.17	110.86	2623.94	−1.45	11.46
142	3.60	39.49	0.99	2.81	102.36	2868.84	42.75	2555.83	2.85	15.76
155	3.14	39.03	0.44	2.26	92.12	2858.60	123.81	2636.89	−0.97	11.94
167	0.81	36.70	0.12	1.94	117.62	2884.10	146.28	2659.36	0.20	13.11
178	1.36	37.25	−0.21	1.61	29.28	2795.76	−38.55	2474.53	−0.90	12.01
195	−1.34	34.55	−0.32	1.50	150.30	2916.78	248.53	2761.61	−0.77	12.14
198	7.52	43.41	−0.32	1.50	78.28	2844.76	146.93	2660.01	−0.74	12.17
202	3.23	39.12	0.99	2.81	62.05	2828.53	23.32	2536.40	3.10	16.01
215	0.71	36.60	−0.21	1.61	120.30	2886.78	170.83	2683.91	0.58	13.49
219	−0.03	35.86	−0.10	1.72	20.29	2786.77	32.23	2545.31	−0.03	12.88
224	2.30	38.19	0.66	2.48	134.78	2901.26	234.38	2747.46	0.03	12.94
245	0.34	36.23	−0.21	1.61	111.84	2878.32	169.39	2682.47	−0.88	12.03
251	−1.71	34.18	−0.10	1.72	136.29	2902.77	201.82	2714.90	−1.58	11.33
CD-215	−0.50	35.39	−0.54	1.28	31.13	2797.61	67.54	2580.62	1.38	14.29
IAC-100	4.25	40.14	−0.27	1.55	−112.82	2653.66	−188.34	2324.74	−2.53	10.38
BMX Potência	3.46	39.35	0.52	2.34	76.98	2843.46	112.25	2625.33	1.92	14.83
Vmax	3.34	39.23	0.38	2.20	13.79	2780.27	−34.55	2478.53	1.93	14.84
BMX APolo	−3.52	32.37	−0.29	1.53	−50.81	2715.67	−103.10	2409.98	0.27	13.18

Conclusions

The generated population exhibits variation for the traits of interest, thus, allowing the identification of lines exhibiting adequate agronomical performance and resistance to the stink bug complex.

Heritability estimates were higher for the traits: plant height at maturity (PHM), number of days to maturity (NDM), grain filling period (GFP), leaf retention (LR), lodging (L), and weight of a hundred seeds (WHS). For agronomical value (AV), grain yield (GY), and healthy seeds weight (HSW), the values were intermediate to low.

From the investigated 251 genotypes, 29 were selected with a minimum GY of 3635.32 kg ha^{-1} (15% higher than the general mean) and a minimum HSW of 2908.26 kg ha^{-1} (acceptable loss of 20% of the mean GY), thus allowing the identification of lines that simultaneously exhibit high yield and resistance to the stink bug complex.

The predicted selection gain for GY and HSW was of 665.4 and 482.4 kg ha^{-1} for the growing season of 2013/13, and for 2013/14, it had the same direction, but distinct magnitude.

It is possible to select cultivars from the current RIL (recombinant inbred line) population.

RIL 32 stands out due to its negative genotypic effects on GFP, LR, and WHS HSW and positive effects on GY and HSW.

Acknowledgments

The authors thank the National Council for Scientific and Technological Development (CNPq), Brazilian Federal Agency for Support and Evaluation of Graduate Education (CAPES), and São Paulo Research Foundation (FAPESP) for granting a scholarship and funding this study.

Conflict of Interest

None declared.

References

Belorte, L. C. C., Z. A. Ramiro, and A. M. Faria, 2003. Levantamento de percevejos pentatomídeos em cinco cultivares de soja [*Glycine max* (L.) Merrill, 1917] na região de Araçatuba, SP. Arquivo Instituto de Biologia, São Paulo, v. 70, n. 4, pp. 447–451.

Bueno, R. D., L. L. Borges, K. M. A. Arruda, L. L. Bhering, E. G. Barros, and M. A. Moreira. 2013. Genetic parameters and genotype x environment interaction for productivity, oil and protein content in soybean. Afr. J. Agric. Res., 8:4853–4859.

Coelho, M. A. O., A. B. T. Condé, C. H. Yamanaka, and H. R. Corte. 2010. Avaliação da produtividade de trigo (*Triticum aestivum* l.) de sequeiro em Minas Gerais. Biosci. J., 26:717–723.

Coimbra, J. L. M., J. G. Bertoldo, T. E. Haroldo, S. Hemp, N. M. Vale, D. Toaldo, et al. 2009. Mineração da interação genótipo x ambiente em *Phaseolus vulgaris* L. para o Estado de Santa Catarina. Cienc. Rural, 39:355–363.

Corrêa-Ferreira, B. S. 2000. Controle biológico do complexo de percevejos da soja. In: Soja: Tecnologia da produção II, Piracicaba, P: ESALQ/USP – Departamento de produção vegetal, Pp. 01-17.

Corrêa-Ferreira, B. S., and A. R. Panizzi, 1999. Percevejos da soja e seu manejo. Londrina: EMBRAPA-CNPSO, 45 p.

Corrêa-Ferreira, B. S., L. C. Castro, S. Roggia, N. Cesconetto, B. S. Corrêa-Ferreira, F. C. Krzyzanowski, et al., 2009. Percevejos e a qualidade da semente de soja – Série Sementes. Londrina: Embrapa Soja. 15 p. (Embrapa Soja: Circular Técnica, 67).

Corrêa-Ferreira, B. S., de Castro L. C., S. Roggia, N. L. Cesconetto, J. M. da Costa, and de Oliveira M. C. N. 2013 MIP-Soja: resultados de uma tecnologia eficiente e sustentável no manejo de percevejos no atual sistema produtivo da soja. Disponível em: <http://ainfo.cnptia.embrapa.br/digital/bitstream/item/87596/1/Doc-341.pdf>. Acesso em: 03 dez. 2014.

Depieri, R. A., and A. R. Panizzi. 2010. Rostrum length, mandible serration, and food and salivary canals areas of selected species of stink bugs (Heteroptera, Pentatomidae). Rev. Bras. Entomol., 54:584–587.

Fernandes, M. F., F. L. M. Athayde, and M. F. Lara. 1994. Comportamento de cultivares de soja no campo em relação ao ataque de percevejos. Pesqui. Agropecu. Bras., 29:363–367.

Garcia, A., A. E. Pípolo, I. O. N. Lopes, and F. A. F. Portugal, 2007. Instalação da lavoura de soja: época, cultivares, espaçamento e população de plantas. Pp. 11. Embrapa Soja, Londrina

Gazzoni, D. L., and F. Moscardi. 1998. Effect of defoliation levels on recovery of leaf area, on yield and agronomic traits of soybeans. Pesqui. Agropecu. Bras., 33:411–424.

Godoi, C. R. C., and J. B. Pinheiro. 2009. Genetic parameters and selection strategies for soybean genotypes resistant to the stink bug-complex. Gen. Mol. Biol., 32:328–336.

Gravois, K. A., and J. L. Bernhardt. 2000. Heritability x environment interactions for discoloured rice kernels. Crop Sci., 40:314–318.

Guedes, J. V. C., J. A. Arnemann, G. R. Stürmer, A. A. Melo, M. Bigolin, C. R. Perini, et al., 2012 Percevejos da soja: novos cenários, novo manejo. Revista plantio Direto, Passo Fundo, v.janeiro/fevereiro, p.28-34.

Hart, K. A., and D. Pimentel,2002 Environmental and economic costs of pesticide uses. Pp. 237–239 in D. Pimentel, ed. Encyclopedia of pest management. Marcel Dekker, Inc., New York, 927.

Hoffmann-Campo, C. B., F. Moscardi, B. S. Corrêa-Ferreira, L. J. Oliveira, D. R. Sosa-Gomez, A. R. Panizzi, et al. 2000 Pragas da soja no Brasil e seu manejo integrado. Pp. 70. Embrapa Soja, Londrina.

Laviola, B. G., T. B. Rosado, A. K. K. Bhering, and M. D. V. Resende. 2010. Genetic parameters and variability in physic nut accessions during early developmental stages. Pesqui. Agropecu. Bras., 45:1117–1123.

Li, Z. K., S. B. Yu, H. R. Lafitte, N. Huang, B. Courtois, S. Hittalmani, et al. 2003. Qtl x environment interactions in rice I. Heading date and plant height. Theor. Appl. Gen., 108:141–153.

Liu, Y., Y. Li, J. C. Reif, M. F. Mette, Z. Liu, B. Liu, S. Zhang, L. Yan, R. Chang, L. Qiu, L. Identification of quantitative trait loci underlying plant height and seed weight in soybean. The Plant Genome, Madison, 6(3):1–11 2013.

Musser, F. R., and A. L. Catchot. 2008. Mississippi soybean insect losses. Midsouth Entomol., 1:29–36.

Odegard, Y. R., and E. van der Voet. 2014. The future of food — Scenarios and the effect on natural resource use in agriculture in 2050. Ecol. Econ., 97:51–59.

Oliveira, J. R.. 2010. Ovos de *Euschistus heros* parasitados por *Telenomus podisi*. Informativo do Manejo Ecológico de Pragas, 63:738–739.

Panizzi, A. R., and F. Slansky. 1985. Review of phytophagous pentatomids (Hemiptera: Pentatomidae) associated with soybean in the Americas. Fla. Entomol., 68:184–214.

Panizzi, A. R., J. E. McPherson, D. G. James, M. Javahery, and R. M. McPherson 2000 Stink bugs (Pentatomidae). Pp. 432–434 in C. W. Schaefer, A. R. Panizzi, eds. Heteroptera of economic importance. Vol. 13. CRC, Boca Raton, FL.

Panthee, D. R., V. R. Pantalone, A. N. Saxton, D. R. West, and C. E. Sams Quantitative trait loci for agronomic traits in soybean. Plant Breeding, Berlin, 126:51–57, 2007.

Pimentel Gomes, F. 2000 Curso de estatística experimental. 14 ed. Pp. 477. Degaspari, Piracicaba.

Prado, E. P., C. G. Raetano, H. O. Aguiar-Júnior, R. S. Christovam, M. H. F. A. D. Pogetto, and M. J. Gimenes. 2010. Velocidade do fluxo de ar em barra de pulverização no controle químico de *Anticarsia gemmatalis*, Hübner e percevejos na cultura da soja. Bragantia, 69:995–1004.

Ray, D. K., N. D. Mueller, P. C. West, and J. A. Foley. 2013. Yield trends are insufficient to double global crop production by 2050. Proc. Natl Acad. Sci. 8:e.66428.

Rocha, F., F. Bermudez, M. C. Ferreira, K. C. Oliveira, and J. B. Pinheiro. 2014. Effective selection criteria for assessing the resistance of stink bugs complex in soybean. Crop Breed. Appl. Biotechnol., 14:174–179.

Santos, M. F. 2012 Mapeamento de QTL e expressão gênica associados à resistência da soja ao complexo de percevejos.

2012. 119 p. Tese (Doutorado em Genética e Melhoramento de Plantas) – Escola Superior de Agricultura "Luiz de Queiroz", Universidade de São Paulo, Piracicaba.

Silva, A. J., M. G. Canteri, and A. L. Silva. 2013. Haste verde e retenção foliar na cultura da soja. Summa Phytopathol., 39:151–156.

Smith, C. M.. 2005. Plant resistance to arthropods molecular and conventional approaches. Pp. 423. Springer Netherlands, Dordrecht.

Stürmer, G. R., A. Cargnelutti-Filho, L. S. Stefanelo, and J. V. C. Guedes. 2012. Eficiência de métodos de amostragem de lagartas e de percevejos na cultura de soja. Ciênc. Rural, 42:2105–2111.

Temple, J. H., B. R. Leonard, J. A. Davis, and K. Fontenot. 2009. Insecticide efficacy against red banded stink bug, *Piezodorus guildinii* (Westwood), a new stink bug pest of Louisiana soybean. Midsouth Entomol., 2:68–69.

Tilman, D., C. Balzer, J. Hill, and B. L. Befort. 2011. Global food demand and the sustainable intensification of agriculture. Proc. Natl Acad. Sci., 108:20260–20264.

Vargas, M., E. Combs, G. Alvarado, G. Atlin, K. Mathews, and J. Crossa. 2013. META: A suite of SAS programs to analyze multienvironment breeding trials. Agron. J., 105:11–19.

Wiest, A., and M. R. Barreto. 2012. Evolução dos insetos-praga na cultura da soja no Mato Grosso. EntomoBrasilis, 5:84–87.

6

The contribution of wheat to human diet and health

Peter R. Shewry[1,2] & Sandra J. Hey[1]

[1]Rothamsted Research, Harpenden, Hertfordshire AL5 2JQ, UK
[2]University of Reading, Whiteknights, Reading Berkshire RG6 6AH, UK

Keywords
Diet and health, dietary fiber, grain composition, phytochemicals, wheatwheat

Correspondence
Peter R. Shewry, Rothamsted Research, Harpenden, Hertfordshire AL5 2JQ, UK.

E-mail: peter.shewry@rothamsted.ac.uk

Funding Information
We are grateful to CGIAR and the CGIAR WHEAT program for financial support to prepare a review (Project No. A403 1.09.47) on which this article is based. Rothamsted Research receives strategic funding from the Biotechnological and Biological Sciences Research Council (BBSRC) of the UK.

Abstract

Wheat is the most important staple crop in temperate zones and is in increasing demand in countries undergoing urbanization and industrialization. In addition to being a major source of starch and energy, wheat also provides substantial amounts of a number of components which are essential or beneficial for health, notably protein, vitamins (notably B vitamins), dietary fiber, and phytochemicals. Of these, wheat is a particularly important source of dietary fiber, with bread alone providing 20% of the daily intake in the UK, and well-established relationships between the consumption of cereal dietary fiber and reduced risk of cardio-vascular disease, type 2 diabetes, and forms of cancer (notably colo-rectal cancer). Wheat shows high variability in the contents and compositions of beneficial components, with some (including dietary fiber) showing high heritability. Hence, plant breeders should be able to select for enhanced health benefits in addition to increased crop yield.

Introduction

The economic importance of wheat and its contribution to the diets of humans and livestock cannot be disputed. Currently available figures show an average annual global production of about 680 million tonnes (mt) over the 5-year period from 2008 to 2012, with almost 700 mt being produced in 2011 (FAOStat http://faostat.fao.org/site/291/default.aspx). This makes it the third most important crop in terms of global production, the comparative values for the production of the two other major cereals over the same period being 704 mt for rice and 874 mt for maize. However, wheat is unrivaled in its range of cultivation, from 67°N in Scandinavia and Russia to 45°S in Argentina, including elevated regions in the tropics and subtropics (Feldman 1995). Furthermore, there is an increasing demand for wheat in new markets beyond its region of climatic adaptation.

Increasing global demand for wheat is based on the ability to make unique food products and the increasing consumption of these with industrialization and westernization. In particular, the unique properties of the gluten protein fraction allows the processing of wheat to produce bread, other baked goods, noodles and pasta, and a range of functional ingredients. These products may be more convenient to produce or consume than traditional foods, and form part of a "western lifestyle".

Wheat species

The major wheat species grown throughout the world is *Triticum aestivum*, a hexaploid species usually called "common" or "bread" wheat. However, the total world production includes about 35–40 mt of *T. turgidum* var. *durum*, a tetraploid species which is adapted to the hot dry conditions surrounding the Mediterranean Sea and similar climates in other regions. This is used for making pasta and is often referred to either as "pasta wheat" or "durum wheat". Other wheat species are only cultivated on small areas, either for cultural reasons or for the expanding market in health foods. These are einkorn (diploid *T. monococcum* var. *monococcum)*, emmer (tetraploid *T. turgidum* var. *dicoccum*), and spelt (*T. aestivum* var. *spelta*), the latter being a cultivated form of hexaploid wheat. Spelt, emmer, and most forms of einkorn differ from bread and durum wheats in being hulled (i.e., the glumes remain tightly closed over the grain and are not removed by threshing).

Changes in the consumption pattern of wheat

Wheat is a global commodity, with about 150 mt being traded annually (World Agricultural Outlook Board, 2014). Hence, the pattern of production does not necessarily reflect the pattern of consumption. This is particularly significant for wheat, with increased consumption being associated with the adoption of a "western lifestyle" in countries undergoing urbanization and industrialization. These include countries in which wheat is not readily grown. It is not possible to provide a global view of wheat production and consumption in the present article and we have therefore selected nine countries to illustrate trends. Data on actual consumption are not available for most countries so we use data from the FAO Food Balance Sheets (http://faostat3.fao.org/faostat-gateway/go/to/download/FB/FBS/E) on food availability *per capita*. Hence the values reflect food wasted in the home as well as food consumption.

The nine countries were selected to represent traditional wheat-consuming areas of Western Europe (UK, Finland) and West Asia and North Africa (Turkey, Egypt), rapidly urbanizing and industrializing countries in Sub-Saharan Africa (Nigeria, South Africa) and Asia (China, India), and Mexico in which maize and wheat are staple foods.

Figure 1 shows the changes in the availability of wheat (kcal/day) in these countries between 1961 and 2011. It is notable that the contribution of wheat to total kcal increased significantly in Nigeria (from less than 1% to 6.64%), India (11.85% to 20.41%), and China (12.20% to 17.83%), although the percentage contributions of all cereals declined in these three countries. Hence, in these three countries increased wheat consumption has occurred at the expense of other cereals, particularly minor cereals (millets and sorghum). It is also of interest to compare the figures for consumption with changes in wheat production and imports over the same period, summarized

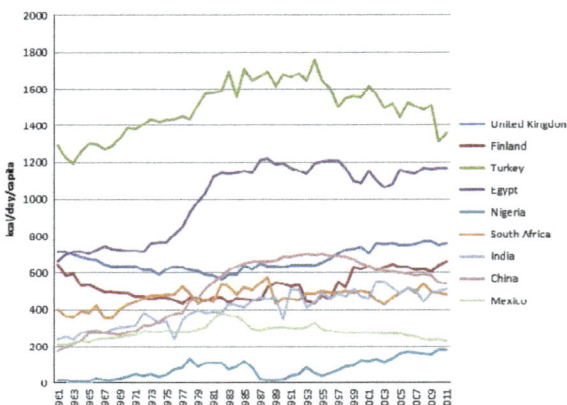

Figure 1. Changes in the availability of wheat (kcal/day) in nine countries between 1961 and 2011. Data from the FAO Food Balance Sheets (http://faostat3.fao.org/faostat-gateway/go/to/download/FB/FBS/E).

in Table 1. Whereas increased wheat production has been accompanied by decreased imports in the UK, China, and India, imports have risen dramatically in African countries, Turkey and Mexico, despite increased production.

Contribution of wheat to dietary intake of nutrients

Wheat is often considered primarily as a source of energy (carbohydrate) and it is certainly important in this respect. However, it also contains significant amounts of other important nutrients including proteins, fiber, and minor components including lipids, vitamins, minerals, and phytochemicals which may contribute to a healthy diet.

The UK National Diet and Nutrition Survey (NDNS) (Bates et al. 2014a,b) is an annual survey of the food consumption and nutritional status of a UK representative sample of 1000 people per year (500 children and 500 adults) aged 18 months and above. It is probably the most complete national survey that is available with the most recent release covering the 4 years 2008/9 to 2011/12. Data for adults (aged 19–65) are shown in Table 2.

Cereals and breads were the main source of energy for all age groups, contributing 31% for adults, and of non-starch polysaccharides (dietary fiber, DF), with bread alone contributing about a fifth of the average daily intake. In addition cereals, including wheat, contribute significantly to the daily intake of protein, B vitamins, and iron. These high contributions of wheat to essential nutrients in the UK, a comparatively prosperous country with a varied diet, underline its importance to nutrition globally (not just in less developed countries).

Table 1. Changes in wheat production and imports in nine representative countries.

	Production		Imports	
	1961–1965	2009–2013	1961–1965	2007–2011
United Kingdom	3519	13,878	4045	1178
Finland	448	872	155	24
Turkey	8584	20,844	552	3311
Egypt	1458	8472	907	9217
Nigeria	17.6	107	35	3580
South Africa	834	1813	127	1391
India	11,191	87,349	4521	606
China	19,119	118,094	5292	1912
Mexico	1783	3610	2	3358

Data are in 1000 tonnes and are 5 year averages 1961–1965 and 2009–2013 (FAOStat http://faostat.fao.org/site/291/default.aspx).

Table 2. Percentage contributions of cereals and cereal products to average daily intake of some essential nutrients of adults (aged 19–64) in the UK (Bates et al. 2014a,b) (NDNS Data Released 14/05/2014) https://www.gov.uk/government/publications/national-diet-and-nutrition-survey-results-from-years-1-to-4-combined-of-the-rolling-programme-for-2008-and-2009-to-2011-and-2012).

	Total		Breads	
	Men	Women	Men	Women
Energy	31	31	12	10
Protein	23	23	11	11
Carbohydrates	45	45	21	18
NSP	40	37	21	18
Thiamin (B1)	36	35	16	15
Riboflavin (B2)	22	22	5	4
Niacin (eq) (B3)	25	26	11	10
Vitamin B6	16	18	5	4
Folates (B9)	27	27	12	12
Iron	40	38	16	15
Calcium	32	30	19	15
Magnesium	28	28	13	13
Sodium	31	31	19	18
Zinc	25	25	12	12
Copper	34	32	15	14
Selenium	28	27	13	12

NSP, nonstarch polysaccharides (dietary fiber).

Limited data are available for less developed countries but some comments can be made. In India, the diets of the rural poor are based predominantly on cereals, which provide 80% of energy and other nutrients except vitamins A and C. Cereals therefore require supplementation with other food groups such as pulses, vegetables, fruits, or animal products to make the diet more balanced and adequate, particularly with respect to vitamin A, iron, and riboflavin. However, Gopalan et al. (2012) suggest that the diets of the poor can also be improved to reduce the incidence of major nutrient deficiencies (such as vitamin A deficiency and iron deficiency anemia) by replacing a cereal diet (such as rice) with mixed cereals, including millets.

Changes in the Chinese diet over the past 50 years are well documented with the consumption of cereals increasing in rural communities and decreasing in urban communities between 1952 and 1992 (Du et al. 2014).

Wheat grain composition in relation to diet and health

A vast volume of literature exists on wheat grain composition, much of which is collated in the monograph "Wheat: Chemistry and Technology", particularly the third (Pomeranz 1988) and fourth (Khan and Shewry 2009) editions. We will therefore focus on components of direct relevance to human nutrition and health and compare data for two fractions, wholegrain and white flour. We will also consider data on the locations of components within the different tissues of the grain, as this is relevant to their recovery in milling fractions, and briefly discuss the roles of grain components in diet and health. Some grain components, such as protein and B vitamins, have clearly established roles in the growth and health of humans and these roles will not be discussed in detail here. In other cases, such as DF and phytochemicals, the benefits are less well established and will be discussed more fully. They are also briefly summarized in Table 3.

Protein

Protein content

Grain protein content is determined by genetic and environmental factors, notably the availability of nitrogen fertilization. The protein content of 12,600 lines in the USDA World Wheat Collection has been reported to range from 7% to 22% of the dry weight (Vogel et al. 1976), but generally varies from about 10–15% of the dry weight

Table 3. Summary of proposed and established health benefits of components present in wheat grain. A number of components have been described as having "antioxidant" properties, but the in vivo significance of this broad but readily measured activity is debated.

Component	Proposed health benefit (for cereals or other foods)	Supported by approved* health claim?
Dietary fiber	Reduces postprandial glycemic response (and risk of type 2 diabetes)	Yes
	Reduces intestinal transit time	
	Increases fecal bulk	
	Reduces cholesterol and risk of coronary heart disease	
	Reduces risk of colo-rectal cancer	No
	Reduces risk of breast cancer	
	Reduces risk of stroke	
	Prebiotic effects	
	Stimulate immune responses	
Resistant starch	Reduces postprandial glycaemic response	Yes
	Other benefits as part of dietary fiber above	
Fructans	Prebiotic effects	No
	Promote calcium (and iron?) absorption	
Betaine	Normal homocysteine metabolism (reduced risk of coronary heart disease)	Yes (not for cereals)
Choline	Normal homocysteine metabolism (reduced risk of coronary heart disease)	No
Phenolic acids	Improve vascular function	No
	Antitumor properties	
	"Antioxidant"	
Alkylresorcinols	Antimicrobial	No
	"Antioxidant"	
Lignans	Phytoestrogen	No
	Antitumor	
	Antimicrobial	
Sterols, stanols, and derivatives	Reduce serum cholesterol and risk of coronary heart disease	Yes
	Anticancer effects	No
Tocols	Vitamin E activity	Yes
	Prevention of neurodegeneration	No
	Induction of immune responses	
	Anticancer	
	Cholesterol lowering	
	"Antioxidant"	

*Approved by EFSA (EU) or FDA (USA).

for wheat cultivars grown under field conditions. Where separate cultivars are bred for livestock and food, these may differ in protein content by about 2% protein (dw basis) when grown under the same conditions (Snape et al. 1993).

Protein is unevenly distributed in the grain, with values of 5.1% reported for the pericarp, 5.7% for the testa, 22.8% for the aleurone, and 34.1% for the germ (Jensen and Martens 1983). Broadly similar values have been reported by other authors (discussed by Pomeranz 1988b). The protein content of the starchy endosperm (white flour) is generally about 2% dry weight lower than wholegrain protein content and varies with the environment (particularly nitrogen availability), in line with the variation in wholegrain protein content.

Genetic sources of high grain protein content which have been exploited in breeding programs include Atlas 50 and Atlas 66 which are derived from the South American cultivar Frondoso (Johnson et al. 1985) and Nap Hal from India (www.indiaresource. org). Other sources of "high protein genes" have come from related wild species. For example, the Kansan variety Plainsman V contains a gene(s) from *Aegilops* which is thought to increase grain protein by 2–3% (Finney 1978). However, the most widely exploited sources of high protein genes in wheat are wild emmer (*Triticum turgidum* var *dicoccoides*) lines from Israel and in particular the accession FA15-3 which is able to accumulate over 40% protein when given adequate nitrogen (Avivi 1978). The locus controlling this trait has been mapped to chromosome 6B and designated *Gpc-B1* (Joppa and Cantrell 1990; Olmos et al. 2003; Distelfeld et al. 2004, 2006). Transfer of this gene into hard red spring wheats results in up to 3% more protein than in the parental lines (Khan et al.1989) and this effect has been exploited in the high protein hard red spring cultivar Glupro (quoted in Khan et al. 2000; Mesfin et al. 2000).

The *Gpc-B1* gene is now known to encode a transcription factor that accelerates senescence resulting in increased mobilization and transfer of nitrogen and minerals (zinc, iron) to the developing grain (Uauy et al. 2006). Hence, lines expressing this allele contain higher amounts of iron and zinc in their grain as well as higher protein (Distelfeld et al. 2006; Uauy et al. 2006).

QTLs for grain protein have also been mapped on chromosomes 5A, 5D, 2D, 2B, 6A, 6B, and 7A of bread wheat (Snape et al. 1993; Worland and Snape 2001; Blanco et al. 2002; Groos et al. 2003; Turner et al. 2004) and on chromosome 5B of emmer wheat (Gonzalez-Hernandez et al. 2004).

Protein quality

Protein nutritional quality is determined by the proportions of essential amino acids, as these cannot be synthesized by animals and hence must be provided in the diet. If only one essential amino acid is limiting, the others will be broken down and excreted resulting in restricted growth in humans and loss of nitrogen present in the diet. Ten amino acids are strictly essential: lysine,

isoleucine, leucine, phenylalanine, tyrosine, threonine, tryptophan, valine, histidine, and methionine. However, cysteine is often also included as it can only be synthesized from methionine, with combined proportions of cysteine and methionine often being presented. The requirements for essential amino acids are lower for adults where amino acids are required only for maintenance, than for children where they are also required for growth.

Essential amino acids in wheat grain

Typical contents of essential amino acids reported for wholemeal wheat and white flour are compared with the minimum physiological requirements for adults in Table 4.

The data in Table 4 support the widely accepted view that the first limiting amino acid in wheat grain is lysine with other essential amino acids being present in adequate amounts, at least for adults. The lower contents of essential amino acids in white flour compared with wholegrain relate to the high content of lysine-poor prolamin storage proteins (gluten proteins) in the starchy endosperm. These proteins are restricted to the starchy endosperm cells, where they account for about 80% of the total

Table 4. Minimum physiological requirements (g/100 g protein) for essential amino acid for adults (g/100 g protein) [WHO/FAO/UNU Expert Consultation (2002, Geneva, Switzerland)] and ranges of % total protein (%N × 5.7) (as is basis) and essential amino acid compositions (g/100 g protein) for wholemeal and white wheat flour.

	WHO adult intake	Sample	No. Samples	Range Min	Max	Mean	References
Total Protein		Wholemeal	22	7.70	17.20	14.35	2, 4
		White flour	29	7.50	15.80	13.09	1, 2, 3, 4
Tryptophan	0.6	Wholemeal	–	–	–	–	–
		White flour	7	0.68	1.01	0.85	1, 3
Threonine	2.3	Wholemeal	22	1.74	3.10	2.54	2, 4
		White flour	29	1.60	2.80	2.24	1, 2, 3, 4
Isoleucine	3.0	Wholemeal	22	2.05	4.00	3.14	2, 4
		White flour	29	2.13	4.30	3.09	1, 2, 3, 4
Leucine	5.9	Wholemeal	22	3.79	7.10	5.94	2, 4
		White flour	29	3.93	7.00	5.65	1, 2, 3, 4
Lysine	4.5	Wholemeal	22	2.50	3.82	2.88	2, 4
		White flour	29	1.70	2.90	2.22	1, 2, 3, 4
Methionine	1.6 ⎤	Wholemeal	16	1.00	1.40	1.20	4
	2.2	White flour	29	0.83	1.50	1.13	1, 2, 3, 4
Cysteine	0.6 ⎦	Wholemeal	16	2.10	2.80	2.43	4
		White flour	29	1.40	3.30	2.17	1, 2, 3, 4
Phenylalanine	⎤	Wholemeal	22	1.79	4.90	3.90	2, 4
	3.8	White flour	29	1.87	5.00	3.75	1, 2, 3, 4
Tyrosine	⎦	Wholemeal	22	1.10	1.90	1.54	2, 4
		White flour	29	1.02	1.80	1.39	1, 2, 3, 4
Valine	3.9	Wholemeal	22	2.56	4.80	3.88	2, 4
		White flour	29	2.34	4.40	3.54	1, 2, 3, 4
Histidine	1.5	Wholemeal	22	2.20	3.66	2.66	2, 4
		White flour	29	1.90	3.71	2.69	1, 2, 3, 4

1. Tkachuk 1966; 2. Simmonds 1962; 3. McDermott and Pace 1957; 4. Shoup et al. 1966.

proteins, and have unusual amino acid compositions with high contents of glutamine and proline and low contents of lysine (reviewed by Shewry and Halford 2002; Shewry 2007; Shewry et al. 2009a). This contrasts with the proteins present in the other grain tissues which are more lysine-rich.

Free amino acids

The pools of free amino acids in cereal grain are small (generally regarded as 5% or less of total grain nitrogen) and subjected to strict feedback regulation. Hence, analyses are rarely included when balance sheets of wheat grain composition are compiled. However, they are of interest as targets for increasing grain lysine content by transgenesis (see section below).

High lysine wheat

High lysine cereals have been a target for over 40 years, since Mertz et al. (1964) described the high lysine *opaque-2* mutant of maize (reviewed by Coleman and Larkins 1999; Shewry 2007). Many other high lysine mutants were subsequently reported in maize and in other diploid cereals (sorghum and barley). However, all high lysine genes are associated with detrimental pleiotropic effects on yield and grain structure and/or composition, which have proved difficult to separate in plant breeding programs. Hence, despite a vast effort only the *opaque-2* mutation of maize has been successfully incorporated into commercial cultivars. The molecular mechanisms responsible for *opaque-2* and several other high lysine mutants of maize have also been determined (Schmidt et al. 1992; Coleman et al. 1995; Hartings et al. 2011). High lysine mutants have not been identified in wheat, probably due to its hexaploid constitution (dominant mutations being very rare). However, the identification of an opaque-2-like factor (called storage protein activator, SPA) (Albani et al. 1997) indicates that opaque-2 like lines could be developed in transgenic wheat.

A genetic engineering approach could also be taken in wheat, as in maize where metabolic engineering has been used to increase the content of free lysine (Huang et al. 2005). However, this is unlikely to find consumer acceptance in the near future, particularly in countries where wheat is used mainly for human food rather than livestock feed.

Carbohydrates

At maturity, the wheat grain consists of 85% (w/w) carbohydrate, 80% of which is starch (present only in the starchy endosperm); approximately 7% low molecular mass mono-, di-, and oligosaccharides (present in the aleurone, starchy endosperm and tissues of the embryonic axis) and fructans (present in the starchy endosperm and bran); and about 12% cell wall polysaccharides (present in all tissues) (see Stone and Morell 2009). Starch from wheat and other cereal grains is the predominant source of human dietary carbohydrate, whereas the cell wall polysaccharides are the major components of DF which is important for human health as well as having impacts on grain utilization and end-use quality (Stone and Morell 2009).

Monosaccharides and oligosaccharides

The low molecular mass carbohydrate fraction includes the reducing aldohexose monosaccharide D-glucose (0.03–0.09% of dw) and its ketohexose isomer, D-fructose (0.06–0.08% of dw) (Lineback and Rasper 1988) and minor amounts of their phosphorylated forms, which are intermediates in carbohydrate metabolism (Stone and Morell 2009).

The most abundant oligosaccharides in wheat grain are polymers of fructose: fructo-oligosaccharides and fructans. The distinction between these two groups is not clear so we will consider them as a single group of carbohydrates which comprise three or more fructose units, with some forms also having a single glucose unit. Cereals contain graminan-type (branched) fructans with both β-(2→1) and β-(2→6) linkages (Ritsema and Smeekens 2003; Roberfroid 2005). Wholegrain fructans have been reported to have an average degree of polymerization (DP) of 5–7, with 45–50% of those in flour (Nilsson et al. 1986) and 60% of those in wholegrain (Henry and Saini 1989) having DP below 6. Studies of wheat flour have reported DP of up to 7–8 (van Loo et al. 1995) and greater than 16 (Nilsson et al. 1986).

The fructan content of whole wheat grain has been reported to vary from 0.84% to 1.85% (mean 1.28%) in 129 winter wheat varieties grown on a single site (Andersson et al. 2013), from 0.9% to 1.8% in five cultivars from five sites (Fretzdorff and Welge 2003), from 1.5% to 2.3% in 19 genotypes grown in the field (from 0.7% to 1.6% for the same lines grown in the glasshouse) (Huynh et al. 2008).

High levels of fructans are present in "sweet wheat", a high sugar line produced by combining mutations in the granule bound starch synthase 1 (GBBS1) and starch synthase IIa (SSIIa) genes (Nakamura et al. 2006; Shimbata et al. 2011).

Analysis of milling fractions from three wheat cultivars showed the highest fructan contents in bran (3.4, 3.7, 4.0%) and shorts (3.2, 3.5, 4.1%) and less in the germ (1.7, 2.5%) and flour (1.4, 1.5, 1.7%) (Haska et al. 2008).

Wheat also contains significant amounts of the disaccharides sucrose (comprising glucose and fructose units) (0.54–1.55% of dw) and maltose (two glucose units) (0.05–0.18% of dw) and the trisaccharide raffinose (galactose, glucose, and fructose units) (0.19–0.68% of dw) (Lineback and Rasper 1988). Huynh et al. (2008) reported from 0.39% to 0.49% raffinose in 19 genotypes grown in the field, and from 0.15% to 0.30% in the same lines grown in the glasshouse.

Health benefits of fructans

Fructans form part of the DF fraction and are highly fermentable in the colon, with inulin (a form of fructan derived from Jerusalem artichoke and other plants) being widely used as a standard prebiotic (discussed below under dietary fiber).

However, in wheat, fructans and raffinose are the major components of a group of small fermentable carbohydrates which have been termed FODMAPs (Fermentable, Oligo-, Di-, Mono-saccharides, and Polyols). It has been suggested that a low FODMAP diet improves the management of irritable bowel syndrome (IBS) and inflammatory bowel disease (Crohn's disease and ulcerative colitis), by reducing fermentation in the colon (Gibson and Shepherd 2010; Muir and Gibson 2013). Hence, low fructan wheats could be of interest for developing low FODMAP food products.

Fructans have received particular attention in relation to human health because of their reported ability to promote the absorption of calcium (Lopez et al. 2000; Scholz-Ahrens et al. 2001; Abrams et al. 2007a,b). They have also been suggested to increase iron absorption, but the evidence for this is less conclusive (see, for example, Ohta et al. 1998; Patterson et al. 2009). Of particular interest is a recent report that wheat lines differing in arabinoxylan and fructan content had no effect on iron status when fed to iron deficient broiler chickens (Tako et al. 2014).

Starch

Starch constitutes about 60–70% of the mass of wheat grain, and about 20% more of the total mass (i.e., about 70–85%) in white flour (Toepfer et al. 1972). It influences aspects of end-product quality and is crucial for human nutrition, being the main source of dietary carbohydrate.

Starch is a mixture of two glucose polymers: amylose, which comprises single unbranched (1→4) α-linked chains of up to several thousand glucose units and amylopectin which is highly branched (with (1→6) α-linkages as well as (1→4) α-linkages) and may comprise over 100,000 glucose unit residues. In most species, including wheat, amylose and amylopectin occur in a ratio of 1:3 amylose: amylopectin.

Resistant starch

Most starch is digested in the small intestine but a proportion may escape digestion and is termed "resistant starch" (RS) (Englyst and Cummings 1985; Asp 1992). RS is classified into five types, with starch entrapped in the food matrix and therefore physically inaccessible being termed RS1, native (uncooked) starch granules RS2, retrograded starch formed after starch gelatinization RS3, chemically modified starch RS4 and starch capable of forming complexes between amylose and long branch chains of amylopectin with lipids RS5 (Thompson 2000; Sharma et al. 2008; Hasjim et al. 2010; Birt et al. 2013). In fact, the resistance of starch to digestion is influenced by many properties of the starch granule including size, shape, and crystallinity, and the contents of amylose, lipids, proteins, and phosphate (Themeier et al. 2005; Tester et al. 2006) and also depends on the processing conditions (Alsaffar 2011).

The RS content of food is highly variable, with cereals generally containing about 3% RS and green bananas about 75%. RS forms part of DF and has similar properties to nonstarch polysaccharides (NSP) in that it is fermented by colonic microorganisms into short chain fatty acids (SCFAs) (acetate, butyrate, and propionate). However, it is associated with the production of higher levels of butyrate compared to other types of DF (Topping and Clifton 2001; Topping et al. 2008) and hence may have greater health benefits. Similarly to NSP, RS increases fecal bulk, by increasing the volume of bacteria, and may also have beneficial effects on insulin sensitivity (Robertson et al. 2003, 2005; Lobley et al. 2013) and reduce the risk of colo-rectal cancer (Keenan et al. 2012; Humphreys et al. 2014). However, RS is a minor component compared with NSPs in western diets (Lobley et al. 2013). For example, the daily intake of RS in European countries is estimated as 4.1 g/day RS compared with 15–20 g/day of fermentable fiber (Cummings 1983). Hence, the beneficial effects of RS in western diets are minor in comparison with those of NSPs in bran, fruit, and vegetables (Cummings et al. 2004).

The demonstration that the amount of RS2 and the ability to form RS3 are influenced by the amylose content of cereal starches (Morell et al. 2004; Rahman et al. 2007) has resulted in interest in identifying or developing lines with increased amylose content.

Mutations affecting starch composition

Naturally occurring and induced mutations affecting the ratio of amylose: amylopectin were identified many years ago in diploid cereals (maize, barley, rice) but the polyploidy nature of wheat limited their identification and exploitation in wheat until the last 25 years.

Waxy wheat

Low amylose (waxy) cereals have essentially 100% amylopectin and result from mutations in granule bound starch synthase (GBSS), the single enzyme that catalyses amylose synthesis. Waxy cereal starches have unusual properties (notably high viscosity and water-retention) (Kim et al. 2003) that are exploited in the food industry, particularly for the production of refrigerated and frozen foods and as fat replacers (Jobling 2004), whereas waxy (glutinous/sticky) rice is widely consumed in Asia. In polyploid bread wheat, the waxy phenotype requires mutations in all three GBSS genes and has not been reported to occur naturally. However, the inactivation of one or two of these genes results in the production of partial waxy lines which occur quite commonly. For example, Yamamori et al. (1994) screened 1960 wheat cultivars and identified null mutations at all three loci, although these differed in frequency and geographical distribution: Wx-1A nulls occurred in 16.2% of Japanese, 10.8% of Korean, and 51.9% of Turkish wheats but were less common in Chinese, South Asian, Australian, North American, Western European, and Russian wheats, whereas Wx-1B nulls were commonest in Australian and Indian wheats. Only one Wx-1D null was identified (in a Chinese cultivar), whereas nine Japanese cultivars were double Wx-1A/Wx-1B nulls. Demeke et al. (1997) similarly reported high incidences of Wx-1A and Wx1-B nulls in Japanese and US wheats, respectively; while Graybosch et al. (1998) reported 10% of a sample of 200 US wheats had null Wx alleles: six Wx-1A nulls, 13 Wx-1B nulls and one double Wx-1-A/Wx-1B null. These partial waxy wheats have lower amylose contents than normal wheats, with the Wx-1B null having a greater effect than the Wx-1A null, and the double null a greater effect than the single nulls (Yamamori et al. 1994; Graybosch 1998; Graybosch et al. 1998).

Partial null lines can be readily crossed to obtain complete waxy types of bread and durum wheat with very low amylose contents (about 0–2%) (Nakamura et al. 1995; Kiribuchi-Otobe et al. 1997; Graybosch 1998; Lafiandra et al. 2010). Polymorphism at the waxy loci has also been shown to affect the amount of amylose (increased or decreased) in durum and bread wheats (Yamamori 2009).

In contrast to high amylose starch (below), waxy starch is highly digestible and hence has a high glycemic index.

High amylose wheat

Analysis of diploid cereals showed that the high amylose phenotype could arise from reduced activity of starch synthase (SSII) or starch branching enzymes (SBEIIa or SBEIIb) (reviewed by Lafiandra et al. 2014). As with waxy wheat, mutations in all three genomes are required to significantly affect the amylose content and naturally occurring high amylose lines have not been reported. Yamamori and Endo (1996) therefore used biochemical screening to identify mutants lacking each of the three SSIIa proteins. Combining these three mutations resulted in lines with about 20% less starch but amylose contents of 37% (Yamamori et al. 2000). Similarly, Rakszegi et al. (2014) recently reported amylose contents of about 40% when the three mutations were introgressed into three commercial bread wheat cultivars, whereas Lafiandra et al. (2010) have used the mutations on the A and B genomes to produce a SSIIa *null* line of the durum wheat cultivar Svevo with an amylose content of 43.6% (compared to 23% in the control).

Hogg et al. (2013) identified two A genome SSIIa nulls but no B genome nulls from a screen of 255 *T. durum* accessions. These were crossed into a commercial cultivar and mutated to generate double SSIIa nulls with 44.3% and 42.8% amylose compared to 28.7% in the control.

The development of TILLING technology for wheat (Slade et al. 2005) has facilitated the identification of further mutations in genes of starch synthesis (Uauy et al. 2009; Sestili et al. 2010). Thus, Botticella et al. (2011) identified knock-out mutants in the three *SBEIIa* homeologues of bread wheat and showed that combining two of these resulted in increases in amylose content from 33.2% in the control to between 38.6% and 39.9%, whereas Hazard et al. (2012) showed that double *SBEIIa* mutants in durum wheat had an increase of 22% in amylose content and 115% in resistant starch content. Similarly, Slade et al. (2012) combined mutations in *SBEIIa* genes to produce durum and bread wheat lines containing 47–55% amylose and with elevated levels of resistant starch compared with wild-type wheat.

However, high amylose wheats differ in their processing properties from conventional wheats, particularly in starch swelling and viscosity (Van Hung et al. 2005; Yamamori et al. 2006; Schirmer et al. 2013), which means that processes and products will need to be modified.

Yields of starch mutants

Because commercial waxy and high amylose cultivars of wheat have not been developed data on comparable yields are not readily available. However, it is expected that both phenotypes would be associated with yield deficits, as reported for commercial high amylose and waxy maize hybrids. This is currently about 3–4% for waxy maize but greater for high amylose lines. Similarly, transgenic high amylose barley lines were shown to have a yield deficit of over 20% (Carciofi et al. 2012). Effects on yield are not unexpected, as starch is the major determinant of yield and any changes in composition are likely to affect the highly organized packaging of the amylose and amylopectin polymers in the starch granules.

Nevertheless, it should be possible to produce lines with improved health benefits, and acceptable yields and processing properties, by increasing the amylose content by a modest amount (e.g., by using mutations in only one or two of the genomes).

Cell wall polysaccharides

The major components of wheat grain DF are cell wall polysaccharides, lignin, fructan, and resistant starch (see above and Table 5).

Lignin is a complex polymer of aromatic alcohols and is characteristic of the secondary cell walls of woody tissues. In wheat and other cereals it is only present in the pericarp/seed coat (Stone and Morell 2009) and hence is enriched in the bran and absent from white flour (Table 6). The detailed structure of cereal grain lignin has been described by Bunzel et al. (2004).

The major cell wall polysaccharides of wheat grain are arabinoxylan (AX) and $(1\rightarrow3, 1\rightarrow4)$-$\beta$-D-glucan ($\beta$-glucan), with smaller amounts of cellulose ($(1\rightarrow4)$-β-D-glucan) and glucomannan (Table 5). Other minor polysaccharides, including callose ($(1\rightarrow3)$-β-D-glucan), xyloglucan, and pectins, can be detected by immunocytochemistry (Pellny et al. 2012; Chateigner-Boutin et al. 2014; Palmer et al, 2015) or sugar analysis.

AX comprises a backbone of β-D-xylopyranosyl (xylose) residues linked through $(1\rightarrow4)$ glycosidic linkages with some residues being substituted with α-L-arabinofuranosyl (arabinose) residues at either one or two positions. Some arabinose residues present as single substitutions on xylose

may also be substituted with ferulic acid at the five position, allowing the oxidation of ferulate present on adjacent AX chains to give dehydrodimers (diferulate cross-links). The extent of diferulate cross-linking is important as it affects the physio-chemical properties (notably solubility and viscosity) of AX and hence the behavior in food processing and also probably the health benefits. AX is therefore often divided into two classes, depending on whether it is extractable (WE-AX) or unextractable (WU-AX) with water.

β-glucan comprises glucose residues joined by $(1\rightarrow3)$ and $(1\rightarrow4)$ linkages. Single $(1\rightarrow3)$ linkages are usually separated by two or three $(1\rightarrow4)$ linkages, but longer stretches of up to 14 $(1\rightarrow4)$ linked glucan units (sometimes referred to as "cellulose-like" regions) have been reported for wheat bran β-glucan (Li et al. 2006). Unlike β-glucan in oats and barley, wheat β-glucan shows low solubility, with about 10–15% of the total in wholemeal samples being soluble in hot water (Nemeth et al. 2010).

The amounts and proportions of cell wall polysaccharides vary between tissues, as summarized in Table 6.

The cell walls of the starchy endosperm (i.e., white flour) account for about 2–3% of the dry weight and comprise about 70% AX and 20% β-glucan, with 2% cellulose and 7% glucomannan (Mares and Stone 1973). In addition, immunolabeling of developing tissues has shown the presence of callose ($(1\rightarrow3)$-β-D-glucan), xyloglucan and pectin (Pellny et al. 2012; Chateigner-Boutin et al. 2014; Palmer et al. 2015).

Starchy endosperm AX contains only low levels of ferulic acid: 0.2–0.4% (w/w) of WE-AX and 0.6–0.9% (w/w) of WU-AX (Bonnin et al. 1998).

The aleurone cells have thick cell walls accounting for about 35–40% of the dry weight (Barron et al. 2007). These comprise 29% β-glucan, 65% arabinoxylan and 2% each of cellulose and glucomannan (Bacic and Stone 1981). The aleurone AX are highly esterified and cross-linked with about 3.2% of the AX dw being ferulic acid and 0.45% being diferulic acid (Antoine et al. 2003; Parker et al. 2005). Additional esterification with *p*-coumaric acid and acetyl groups also occurs (Rhodes and Stone 2002; Antoine et al. 2004).

The outer layers comprise about 45–50% cell wall material (Barron et al. 2007). The major pericarp tissue comprises about 30% cellulose, 60% arabinoxylan, and 12% lignin (reviewed by Stone and Morell 2009). The pericarp AX also has a complex highly branched structure, with galactose and glucuronic acid residues, and is often termed glucuronoarabinoxylan (GAX). It also has high contents of ferulic acid and diferulic acid (Saulnier and Thibault 1999; Antoine et al. 2003; Parker et al. 2005) and acetylation (Mandalari et al. 2005) with significant amounts of ferulic acid trimer (Barron et al. 2007). Immunolabeling shows that the aleurone and pericarp

Table 5. Contents of total dietary fiber and dietary fiber components in 129 winter wheat varieties (taken from data in Andersson et al. 2013).

	Range	Mean
Total dietary fiber (%)	11.5–15.5	13.4
Klason lignin (%)	0.74–2.03	1.33
Arabinoxylan (%)	5.53–7.42	6.49
Cellulose (%)	1.67–3.05	2.11
β-Glucan (%)	0.51–0.96	0.73
Fructan (%)	0.84–1.85	1.28

Table 6. Contents and compositions of cell wall in wheat grain tissues (% dw) (from Shewry et al. 2010).

Tissue	Cell walls (% dw)	Components				
		Cellulose	Lignin	Xylan	β-glucan	Glucomannan
Starchy endosperm	2–3	2	0	70	20	7
Bran		29	8	64	6	–
Aleurone	40	2–4	0	62–65	29–34	–
Outer pericarp (beeswing)		30	12	60	–	–

cell walls also contain pectic polysaccharides (Chateigner-Boutin et al. 2014; Palmer et al. 2015).

Barron et al. (2007) reported that the scutellum and embryonic axis of the germ contained about 12% and 25% of neutral carbohydrate, respectively, with arabinose and xylose (presumably derived from AX) accounting for about 65% of the total. Other sugars released were glucose (presumably from β-glucan), galactose and, for the embryonic axis only, mannose (possibly from mannans or glucomannans).

Variation in cell wall polysaccharides in wholegrain and white flour

Table 7 summarizes analyses of total dietary fiber (TDF) and the major components (AX and β-glucan) in wholemeal and white flour, with the AX being determined as total, water-extractable (i.e., soluble) and water-unextractable (insoluble). Substantial variation occurs in all the total contents of all fractions, with wholegrain being richer in TDF and individual DF components than white flour.

Of particular interest are the datasets from the EU HEALTHGRAIN study which analyzed lines grown together on a single site (Gebruers et al. 2008, 2010; Andersson et al. 2013). Hence the data are directly comparable. In this study TOT-AX in white flour varied by over twofold (from 1.35 to 2.75% dw), and WE-AX by over fourfold (0.30–1.40% dw). The proportion of soluble AX ranged from 20% to 50% of TOT-AX.

Health benefits of dietary fiber

There is a massive volume of literature supporting the health benefits of wheat fiber, although a wide range of fractions have been studied. The UK Scientific Advisory Committee on Nutrition (SACN) (2015) has recently reviewed the evidence for health benefits of cereal fiber, as part of a wider review of dietary carbohydrates. Table 8 therefore summarize its conclusions on the health benefits of wheat fiber, including data for "total fiber", "soluble fiber" and "insoluble fiber" fractions which include fiber from other sources (fruit, vegetables, legumes). These tables also include data for widely studied β-glucan fractions from oats and barley. However, it should be noted that oat and barley β-glucans differ from wheat β-glucan in their structures and properties, being much more soluble and giving more highly viscous solutions (Li et al. 2006; Lazaridou and Biliaderis 2007).

Table 7. Ranges of cell wall polysaccharides in wholegrain wheat and white flour (summarized from Shewry 2013).

Components (g/100 g dw)	Sample	No. samples	Range		Mean	References
			Min	Max		
Total Dietary Fiber	Wholemeal	138	10.26	15.5	13.39	1, 2, 3, 4, 5
	White flour	10	1.94	6.27	3.52	2, 4, 5, 6
Total AX	Wholemeal	173	5.53	8.88	6.60	1, 7, 8, 9
	White flour	110	1.88	3.58	2.64	9
WE-AX	Wholemeal	166	0.29	1.62	0.57	10
	White flour	110	0.30	0.91	0.58	9
WU-AX	Wholemeal	20	5.87	8.16	6.61	8
	White flour	90	1.52	2.93	2.10	9
β-glucan	Wholemeal	166	0.29	1.10	0.81	10
	White flour	–	–	–	–	–

1. Andersson et al. 2013; 2. USDA National Nutrient Database R26; 3. UK Cofids database; 4. Finland Finelli database; 5. South African Food Composition database; 6. Turkey Türkomp Food Composition database; 7. Barron et al. 2007; 8. Saulnier et al. 1995; 9. Saulnier et al. 2007; 10. Gebruers et al. 2008.

Table 8. Summary of the conclusions of SACN 2015 on associations and effects (in italics) of cereal and other dietary fiber fractions on improved cardio-metabolic and colo-rectal health.

Health outcome	Fraction
Cardio-metabolic health	
CVD	TDF, insoluble fiber, soluble fiber, total cereals, wholegrains
Coronary events	TDF, insoluble fiber, cereal fiber, high fiber breakfast cereals
Stroke	TDF, wholegrains
Hypertension	Wholegrains
Blood pressure	*Oat bran/oat or barley β-glucans*
Fasting blood total cholesterol, LDL cholesterol, triacylglycerol	*Oat bran/oat or barley β-glucans*
Type 2 diabetes	TDF, insoluble fiber, soluble fiber, cereal fiber, high fiber breakfast cereals, whole grain bread, wholegrains
Colo-rectal health	
Fecal weight	*TDF, wheat fiber, non-wheat cereal fiber*
Intestinal transit time	*TDF, wheat fiber, non-wheat cereal fiber*
Intestinal transit time in patients with constipation	*Wheat fiber*
Constipation	*Cereal fiber*
Colo-rectal cancer	TDF, cereal fiber
Colon cancer	TDF, cereal fiber, wholegrains
Rectal cancer	TDF, wholegrains

SACN 2015 is the most comprehensive and critical review so far published but even this did not cover the full literature: the authors noted that: "due to the wealth of data available and because of the concerns around their limitations, case-control, cross-sectional and ecological studies were not considered" with "only prospective cohort studies and randomized controlled trials" being considered.

It is therefore important to mention other studies which did not meet the SACN criteria but are nevertheless relevant to this review. In particular, several studies have reported decreased risk of other types of cancer. For example, cereal fiber and breast cancer in premenopausal women (Cade et al. 2007), fiber/wholegrain intake and small intestinal cancer in men and women (Schatzkin et al. 2008) and wholegrains/high fiber foods and pancreatic cancer (Chan et al. 2007).

In most countries claims for health benefits of foods and food ingredients must be approved by regulatory authorities, which are the FDA in the USA and EFSA in the EU. These bodies require high standards of evidence with a relatively small proportion of applications being approved. Nevertheless, 11 out of about 250 health claims approved by EFSA relate to wheat and related cereals

(barley, oats, rye) (http://ec.europa.eu/nuhclaims/), and in particular to fiber components and their effects on GI tract function (reduced transit time, increased fecal bulk), reduction in postprandial glucose responses, or maintenance of blood cholesterol concentrations. These claims can be regarded as generally accepted.

Mechanism of action of dietary fiber

A full discussion of the mechanisms of action of DF is outside the scope of this review, except to note that a number of mechanisms probably contribute, including physical properties (fecal bulk, viscosity of soluble fiber fractions) and fermentation in the colon to produce SCFAs which have physiological effects on the colon and other tissues. A number of reviews of this topic are available (Topping 2007; Buttriss and Stokes 2008; Theuwissen and Mensink 2008; Anderson et al. 2009; Brownlee 2011; Lafiandra et al. 2014).

The concept of stimulating the growth of beneficial colonic bacteria by manipulating the content and composition of nondigestible carbohydrates in the diet (prebiotics) was introduced by Gibson and Roberfroid (1995), following the introduction of microbial (probiotic) supplements. Although these two approaches have the same aim and are essentially complementary, it is considered that the effects of prebiotics may be less transitory than those of probiotics. The effects of prebiotics may be determined experimentally in culture, using either sophisticated model colon systems or simple cultures of fecal bacteria, with effects on bacterial populations (in particular increases in *Bifidobacterium* and *Lactobacillus* spp.) and the production of SCFAs (particularly butyrate) being measured.

The most widely studied prebiotic (which is usually used as a standard) is inulin, a β-(2→1) linked fructan from tubers of Jerusalem artichoke (*Helianthus tuberosus*). Fructans are components of the DF fraction in wheat (see above) and other DF components (resistant starch and all nonstarch polysaccharides) are similarly fermented, although the extent depends on their solubility and other factors such as lignification, particle size and effects of food processing. In an excellent review, Cummings and Macfarlane (1991) estimate that at least 50% of the cellulose and 80% of the noncellulosic polysaccharides in the human diet are fermented in the colon.

There are also differences in the proportions of individual SCFAs produced by fermentation of different oligosaccharides and polysaccharide substrates. Although the precise ratios of SCFAs produced by fermentation in vitro vary between reports, two parallel studies using in vitro fecal cultures showed that wheat arabinoxylan fractions gave broadly similar proportions of SCFAs to inulin whereas five β-glucan fractions from oats and barley gave substantially higher proportions of butyrate than inulin (Hughes et al. 2007, 2008). Hence, fructans and other prebiotic oligosaccharides (including the arabinoxylan oligosaccharides (AXOS) which have approved health benefits) should be considered as highly fermentable DF components.

Dietary fiber and immune function

There is increasing evidence that mixed-linkage β-glucans are able to regulate the immune responses that are involved in fighting infection, attacking tumors and various inflammatory conditions. Most studies have been carried out on $(1→3)$ $(1→6)$ β-glucans from yeast, mushrooms, other fungi and seaweed. Cereal β -glucans differ for these species in their linkage pattern (with $(1→3)$ $(1→4)$ as opposed to $(1→3)$ $(1→6)$ linkages) but may also have immune modulatory properties (Brown and Gordon 2001; Rice et al. 2005). Immune stimulatory effects have also been proposed for arabinoxylan (Capek and Matulova 2013; Li et al. 2015).

These effects remain to be conclusively demonstrated in vivo and the mechanisms at the organism, as opposed to cellular, level remain unclear. However, they could prove to be an important facet of the contribution of wheat to human health.

Phytochemicals: phenolics and terpenoids

Wheat grain contains two major groups of phytochemicals derived from different biosynthetic pathways: phenolics and terpenoids. A range of health benefits have been proposed for these components but few have been established with sufficient scientific rigor to result in health claims approved by FDA or EFSA, the main exceptions being tocols (vitamin E) which have a number of established functions and plant sterols and stanol esters which have accepted benefits in reducing blood cholesterol and therefore the risk of cardiovascular disease.

Phenolic compounds

These contain at least one aromatic ring bearing at least one hydroxyl group. They exhibit immense diversity in structure, forming the largest and most complex group of secondary products present in cereal grain.

Phenolic acids (PAs) are the major group of phytochemicals in wheat grain. They contain a phenolic ring and an organic carboxylic function and fall into two groups, which are derived either from cinnamic acid or benzoic acid. They also occur in three forms, either as free compounds, as soluble conjugates bound to low molecular weight compounds such as sugars, and as bound forms which are linked to cell wall polysaccharides (particularly arabinoxylan in cereal grain) by ester bonds.

Alkylresorcinols are phenolic lipids comprising a 1,3-dihydroxylated benzene ring with an alkane chain at position 5.

Flavonoids are a large and structurally highly diverse group of phenolic compounds which are based on a 15 carbon ring structure (Jende-Strid 1993). Over 5000 have been characterized from plants and classified into a number of groups. Of particular importance and interest in wheat are the flavanols (flavan-3-ols), which are based on a 2-phenyl-3,4-dihydro-2H-chromen-3-ol skeleton and include the proanthcyanidin pigments which are present in the testa of red wheats, and the anthocyanidins which are glycosylated to form anthocyanins.

Lignans are polyphenols derived from phenylalanine via dimerization of substituted cinnamic alcohols, to a dibenzylbutane skeleton.

Terpenoids

Terpenoids are derived from five carbon isoprene units.

Plant sterols are steroid alcohols, comprising a tetracyclic cyclopenta[α]phenanthrene ring with a hydroxyl group at the C4 position and a flexible side chain at the C17 carbon position. They are divided into three types, the 4-desmethyl sterols which are the major components in plant tissues and the minor 4α-monomethyl sterols and 4,4-dimethyl sterols which are precursors of the 4-desmethyl sterols. The major plant 4-desmethyl sterols have a Δ^5 double bond in the B ring and modifications at the C24 position in the side chain (Piironen et al. 2000). Cereals also contain significant amounts of saturated sterols, which are called stanols. A substantial proportion of the sterols and stanols present in wheat are modified, with the 3OH group on ring A being esterified to a fatty acid or phenolic acid to form sterol esters, or β-linked to a carbohydrate to form a sterol glycoside, with the latter also sometimes being acylated. β-sitosterol (the major wheat sterol) is present in approximately equal amounts in all four forms (free, esterified, glycosides, and acylated glycosides) (Chung et al. 2009). However, the modified forms are usually hydrolyzed during preparation to release the free sterols. Unless otherwise noted all data presented here are for total sterols (after hydrolysis).

Tocols comprise a chromanol ring with a C16 phytol side chain, which can be either saturated (tocopherols, T) or contain three double bonds at carbons 3, 7, and 11 (tocotrienols, T-3). Each type also exists in four forms, which differ in the positions of methyl groups on the chromanol ring and are called α (5,7,8-trimethyl), β (5,8-dimethyl), γ (7,8-dimethyl), and δ (8-methyl).

Carotenoids are isoprenoids derived from long polyene chains of 35–40 carbons. They occur in two major forms, the oxygen-containing xanthophylls (which include lutein and zeaxanthin) and the unoxygenated carotenes (which include α-carotene and β-carotene). Some carotenoids, including the carotenes, are converted to vitamin A (retinol) in mammals, and hence are also referred to as pro-vitamin A. The levels of carotenoids in cereal grain are generally low, but are of particular interest in durum wheat where lutein in the major determinant of the yellow color which is a quality trait for breeders and consumers (Borrelli et al. 2000).

Content and composition of phytochemicals in wholegrain and flour

Although many studies of phytochemicals in wheat have been published, very few provide comparative analyses of multiple genotypes which have been grown under the same conditions and analyzed using the same methods. Hence,

much of the discussion below is based on the outcomes of the HEALTHGRAIN study which compared wholemeal samples of 150 wheat genotypes grown together on a site in Hungary (Ward et al. 2008), with 26 lines then being grown in five additional environments (Shewry et al. 2010). This material was analyzed for phytochemicals and other putative "bioactive" components and provides the most complete database on wheat grain composition so far reported. Most other studies have also been carried out on wholemeal with limited data on white flour being available.

Table 9 therefore summarizes wholegrain data from the HEALTHGRAIN study and from the most extensive reported studies for carotenoids (Moore et al. 2005) and lignans (Smeds et al. 2009), with data on total phenolic acids in two samples of white flour. The reader is referred to Piironen et al. (2009) and Chung et al. (2009) for detailed recent reviews of phenolics and terpenoids in wheat.

Table 9. Variation in the contents of phytochemicals (phenolics and terpenoids), in wholegrain (WG) of wheat and white flour.

Component	Fraction	Lines	Range	Fold variation	Mean	References
Phenolics						
Total phenolic acids	WG	150	326–1171 µg/g dm	3.4	657 µg/g dm	1
Total phenolic acids	White flour	2	171–190 µg/g dm	1.1	180 µg/g dm	2
Free phenolic acids	WG	150	3–30 µg/g dm	10	10.6 µg/g dm	1
Conjugated phenolic acids	WG	150	76–297 µg/g dm	3.9	162.5 µg/g dm	1
Bound phenolic acids	WG	150	208–878 µg/g dm	4.2	484.9 µg/g dm	1
Bound ferulic acid	WG	150	162–721 µg/g dm	4.5	367.4 µg/g dm	1
Alkylresorcinols	WG	150	241–677 µg/g dm	2.8	432 µg/g dm	1
Lignans	WG	73	3.4–22.70 µg/g db	6.7	10.5 µg/g db	3
Terpenoids						
Total tocols	WG	150	27.6–79.7 µg/g dm	2.9	49.9 µg/g dm	4
Tocopherols	WG	150	12.3–33.2 µg/g dm	2.7	19.9 µg/g dm	4
α-Tocopherol (vitamin E)	WG	150	9.1–19.9 µg/g dm	2.2	13.6 µg/g dm	4
Tocotrienols	WG	150	12.5–52.0 µg/g dm	4.2	30.02 µg/g dm	4
Total sterols (inc stanols)	WG	150	670–959 µg/g dm	1.43	844 µg/g dm	5
% stanols	WG	150	11–29%	2.6	23.9%	5
β-carotene	WG	8	0.11–0.24 µg/g db	2.1	0.19 µg/g db	6
Lutein	WG	8	0.93–1.3 µg/g db	1.4	1.14 µg/g db	6
Zeaxanthin	WG	8	0.23–0.44 µg/g db	1.9	0.33 µg/g db	6

Where original data were presented on an "as is" basis, a moisture content of 14% was assumed to convert to dry matter basis (db). References: 1. Li et al. 2008; 2. Mattila et al. 2005; 3. Smeds et al. 2009; 4. Lampi et al. 2008; 5. Nurmi et al. 2008; 6. Moore et al. 2005. Stanols are expressed as % total sterols+stanols.

PAs are the major group, ranging in amount from about 300 to 1500 μg/g dry weight. Bound PAs account for about 70–80% of the total PAs, with ferulic acid being the major bound component (and hence the major component of total PAs). Ferulic acid is known to be bound to arabinoxylan by ester linkages to the arabinose side chains. This allows the formation of cross-links between arabinoxylan chains, by oxidation to give diferulates (dehydrodimers) or more rarely triferulate. The amounts and compositions of PAs vary immensely between cultivars, with free PAs being particularly variable (Li et al. 2008). Alkylresorcinols and lignans also show wide variation, from 241 to 677 μg/g dry weight and from 3.4 to 22.7 μg/g dry weight, respectively (Table 9).

Wide variation also occurs in the contents of terpenoids, with data for the major fractions and α-tocopherol determined in large scale studies being given in Table 9. It is notable that the content of total sterols (including stanols) varied less than those of other phytochemicals (by only ×1.43), possibly because sterols play essential functions as components of plant cell membranes. There was also variation in the percentage of stanols, by 2.6-fold (from 11 to 29% total sterols).

Carotenoids are minor components in bread wheat with few detailed studies. The "major" component in wholegrain is lutein, with lower amounts of β-carotene and zeaxanthin. These vary by up to twofold between cultivars (Table 9).

Distribution of phytochemicals in mature grain

All phytochemicals are concentrated in the aleurone and bran (embryo and outer layer), with data for white flour fractions rarely being reported. For example, the two white flour samples analyzed for PAs by Mattila et al. (2005) contained 171 and 190 μg/g dm compared with 326–1171 μg/g dm in the wholemeals of the 150 HEALTHGRAIN lines (Li et al. 2008) (Table 9).

More detailed information on the distribution of phytochemicals within the grain comes from analyses of individual tissues isolated by hand dissection and of bran, germ and flour fractions from milling. However, direct comparison of these datasets is difficult as different tissues/components have been analyzed in individual reports. Antoine et al. (2003, 2004) and Barron et al. (2007) reported the distributions of major PAs (p-coumaric, sinapic and monomeric, dimeric and trimeric forms of ferulate) in hand-dissected tissues. The latter study, which compared two cultivars, showed a mean total ferulate content of 1040 μg/g dm in whole grain, 50 μg/g dm in white flour, 500 μg/g dm in the embryonic axis, 547 μg/g dm in the intermediate layer, 3630 μg/g dm in the scutellum, 6345 μg/g dm in the outer pericarp, 8415 μg/g dm in the aleurone and 10975 μg/g dm in the hyaline layer. p-Coumaric acid was not detected in white flour but was concentrated in the aleurone and, to a lesser extent, the outer grain layers with sinapic acid also being concentrated in the aleurone layer with only traces in white flour.

Chung and Ohm (2000) determined the contents and compositions of tocols in dissected wheat grain tissues, whereas Piironen et al. (1986) and Holasova (1999) reported analyses of milling fractions, with generally good agreement between the two approaches. The content of total tocols in wholegrain ranged from 40 to 58 μg/g, with low contents in the purest white flour/starchy endosperm fractions (17–20 μg/g) and higher contents in the bran layers (76–95 μg/g) and germ (from hand dissection only) (181–320 μg/g).

Chen and Geddes (1945) (quoted in Chung et al. 2009) reported that carotenoids are concentrated in the germ with a total concentration of 4.13–11.04 μg/g compared with 0.88–2.22 μg/g in bran and 1.57–2.18 μg/g in the endosperm. Zhou and Yu (2005) have also recently reported the presence of β-cryptoxanthin in wheat bran as well as β-carotene, lutein, and zeaxanthin.

Health benefits of phytochemicals

Phenolic acids exhibit strong antioxidant activity and the total phenolic content is strongly correlated with total antioxidant activity (Adom et al. 2003; Beta et al. 2005). The relevance of antioxidant properties for human health is widely debated, with insufficient evidence being available to convince many health professionals or support health claims. However, there is increasing evidence that phenolic compounds, including ferulic acid (the major phenolic acid in wheat), improve vascular function in humans (Katz et al. 2001, 2004; Vauzour et al. 2010; Rodriguez-Mateos et al. 2013) and animal models (Alam et al. 2013; Badawy et al. 2013; Suzuki et al. 2007). Whereas free (and possibly also conjugated) phenolic acids are absorbed in the small intestine, a high proportion of the total phenolic acids in wheat is ferulic acid, and to a lesser extent p-coumaric acid, which are bound to arabinoxylan fiber. Bound ferulate, p-coumarate and diferulate are all released by fermentation in the colon (Buchanan et al. 1996; Kroon et al. 1997; Andreasen et al. 2001) making them available for uptake (reviewed by Vitaglione et al. 2008). Ferulic and p-coumaric acids have also been reported to have antiproliferative effects on human Caco-2 colon cancer cells, but this effect has not been confirmed in vivo.

Studies have also shown activities of cereal alkylresorcinols (phenolic lipids) but only in cell and animal models (reviewed by Piironen et al. 2009).

Phytosterols and tocols both have established health benefits which do not apply specifically to wheat. Phytosterols are integral components of plant cell membranes and have well-documented cholesterol-lowering effects in humans with accepted health claims in Europe (EU Register on Nutrition and Health Claims, http://ec.europa.eu/nuhclaims/).

Although the name "vitamin E" is commonly applied to all tocols they differ in their biological activity with α-tocopherol being the most active form. It is therefore usual to define their relative activities in International Units of α-tocopherol equivalents, which are 1 mg per mg for α-tocopherol and 0.5, 0.1, and 0.33 mg α-tocopherol equivalents per mg for β, γ, and δ-tocopherols, respectively. Of the tocotrienols only the α form has significant vitamin E activity, with an α-tocopherol equivalent of 0.3 mg per mg. The biological activity of tocols is reviewed by Bramley et al. (2000) and Piironen et al. (2009).

B vitamins

The B vitamin complex comprises eight water-soluble components which often occur together in the same foods and were initially considered to be a single compound. Cereals are dietary sources of several B vitamins, particularly thiamine (B1), riboflavin (B2), niacin (B3), pyridoxine (B6), and folates (B9). There has been considerable debate on whether to introduce fortification of flour with folates (B9), on a voluntary or compulsory basis, with no internationally accepted policy (Lawrence et al. 2009; WHO, 2009).

All B vitamins are concentrated in the bran and/or germ (Piironen et al. 2009), with white flour containing significantly lower contents than wholemeal (Table 10). Variation in the contents of B vitamins has also been reported, particularly in the contents of thiamine (B1), riboflavin (B2), and niacin (B6) in white flour (Table 10).

Niacin (B3) is of particular concern as only a proportion of the total present in cereals is bioavailable. Niacin deficiency leads to pellagra, which was historically associated with consumption of diets based largely on maize flour. Carter and Carpenter (1982) reported that wheat bran contained only traces of free niacin, and that only 24% of the bound niacin was bioavailable. The bioavailability increased to 62% when the bound niacin was treated with alkali to convert it to free nicotinic acid. The bioavailable forms of niacin (the sum of nicotinamide and nicotinic acid after acid hydrolysis) were also determined by Shewry et al. (2011) on wholemeal flours of the HEALTHGRAIN winter wheat lines on four sites (UK, Poland, France and Hungary) in 2007. The bioavailable niacin content of these samples (0.16–1.74 μg/g dw) was about 10–20% of the total niacin content reported in previous studies.

Table 10. Variation in the content of B vitamins in wheat.

Component	Number of lines	Range of content (μg/g dw)	Fold variation	Mean content (μg/g dw)	References
Wholegrain					
Folates (B9)	150	0.32–0.77	2.4	0.56	1
Thiamine (B1)	26	6.90–11.7	1.7	8.55	2
Thiamine (B1)	49	2.59–6.13	2.4	3.74	3
Riboflavin (B2)	26	0.86–1.07	1.2	0.96	2
Riboflavin (B2)	49	0.48–1.07	2.2	0.71	3
Niacin (B3)	26	0.69–1.18	1.7	0.87	2
Pyridoxine (B6)	26	1.6–2.2	1.4	1.91	2
Pyridoxine (B6)	49	1.44–3.16	2.2	2.2	3
White flour					
Folates (B9)	3	0.16–0.204	1.3	0.185	4
Thiamine (B1)	9	1.25–2.2	1.8	1.68	3
Thiamine (B1)	95	0.57–15.2	26.7	2.72	5
Riboflavin (B2)	9	0.43–0.58	1.3	0.49	3
Riboflavin (B2)	95	0.11–8.4	76.4	0.84	5
Niacin (B3)	95	6.84–100.3	14.7	21.51	5
Pyridoxine (B6)	9	0.27–0.52	1.93	0.43	3
Pyridoxine (B6)	16	0.074–0.349	4.7	0.209	6

References: 1. Piironen et al. 2008; 2. Shewry et al. 2011; 3. Batifoulier et al. 2006; 4. Gujska and Kuncewicz 2005; 5. Ranum et al. 1980; 6. Sampson et al. 1996.

Methyl donors (betaine, choline)

Although wheat is rarely discussed as a source of methyl donors (betaine and choline), it is one of the richest known dietary sources of betaine. Data from the HEALTHGRAIN project showed a threefold range in the content of betaine in wholemeal and a 1.55-fold range in the content of choline (Corol et al. 2012) (Table 11).

Both compounds are concentrated in the bran fractions with low levels in white flour. Bruce et al. (2010) reported ranges from 165.9 to 325.8 μg/g dw betaine and 53.6 to 64.6 μg/g dw choline in six white wheat flours, compared with 747.0 to 1502.7 μg/g dw betaine and 75.9 to 134.9 μg/g dw choline in five wholegrain samples (Table 11).

Similarly, analyses of fractions from a single wheat line showed 8670 μg/g dw betaine and 1020 μg/g dw choline in bran, 15530 μg/g dw betaine and 2090 μg/g dw choline in aleurone and 230 μg/g dw betaine and 280 μg/g dw choline in white flour (Graham et al. 2009). Likes et al. (2007) showed that the contents of choline and betaine were 14.4 and 291.2 mg/100 g in wholemeal, 47.3 and 1293.3 mg/100 g in bran, 114.9 and 1163.5 mg/100 g in germ, and 3.6 and 71.8 mg/100 g in the purest white flour fraction.

Health benefits of methyl donors

High plasma homocysteine (hyperhomocysteinemia) is a major risk factor in CVD, with homocysteine produced by demethylation of methionine being removed either by remethylation to methionine, metabolism to give cysteine or conversion to S-adenosylhomocysteine. The remethylation of homocysteine requires a methyl donor, either folate (vitamin B9) or betaine ((N,N,N,-trimethyl) glycine) or choline (which can be converted to betaine in animals). Betaine and choline can also substitute for folate in other methylation reactions including the methylation of DNA (Zeisel and Blusztajn 1994; Niculescu and Zeisel 2002; Ueland et al. 2005).

Humans obtain betaine almost solely from their diet, but it can also be produced in humans by the irreversible

conversion of choline. Wheat contains the highest reported levels of betaine of all plant foods, 12.9 and 15 mg/g in bran and 2.91 mg/g in wholegrain, with lower levels of free choline (about 0.5 mg/g in bran and 0.14 mg/g in wholegrain) (Zeisel et al. 2003; Likes et al. 2007). However, the content of betaine in 150 wheat lines varied by threefold (0.97–2.94 mg/g) and of choline from 0.18 to 0.28 mg/g (Corol et al. 2012).

A cross-sectional analysis in 1477 women (part of the Nurses' Health Study) showed that the total dietary intake of choline+betaine was inversely associated with total homocysteine in plasma (Chiuve et al. 2007), indicating the importance of adequate dietary intake. Intervention studies with aleurone-rich foods (Price et al. 2010) and with wholegrain foods (Ross and Bruce 2011) have also shown that increased total intakes and plasma concentrations of betaine were associated with improvements in a number of biomarkers of health, including decreased total plasma homocysteine and LDL cholesterol. A more recent study has shown that the consumption of minimally processed wheat bran, and particularly the aleurone fraction, resulted in substantial postprandial increases in plasma betaine concentrations, including when aleurone fractions were incorporated into bread (Keaveney et al. 2015). EFSA have accepted a health claim relating to betaine, but this can only be applied to products containing at least 500 mg/portion (contributing to a daily intake of 1.5 g) (http://ec.europa.eu/nuhclaims/). This claim has not been applied to wheat products but it is likely even the lower betaine levels present in wheat products will contribute to reduced risk of CVD.

Heritability of variation in wheat grain composition

It is clear that there is substantial variation in wheat grain composition, including the content and composition of DF and putative "bioactive" components. However, this variation can result from three effects: genetic differences between lines (genotype), environmental conditions

Table 11. Variation in the contents of methyl donors (betaine and choline) in wholemeal of wheat.

Component	Number of lines	Range of content	Fold variation	Mean content	References
Wholemeal					
Betaine	150	970–2940 μg/g dm	3.03	1596 μg/g dm	1, 2
Betaine	5	747–1503 μg/g dw	2.01	1020 μg/g dm	3
Choline	150	180–280 μg/g dm	1.55	221 μg/g dm	1, 2
Choline	5	76–135 μg/g dw	1.78	110 μg/g dm	3
White flour					
Betaine	6	166–326 μg/g dw	1.96	239 μg/g dm	3
Choline	6	54–65 μg/g dw	1.21	58 μg/g dm	3

References: 1. Shewry et al. 2013; 2. Corol et al. 2012; 3. Bruce et al. 2010.

(including the weather, soil conditions and agronomy) and interactions between the genotype and environment. Comparison of the compositions of sets of wheat lines allows the variation to be apportioned between these three effects, with a high contribution of the genotype meaning that the trait is highly heritable and hence the variation is available to breeders. We will therefore review our current knowledge of the genetic control and heritability of wheat grain composition.

The "broad sense heritability" of grain components can be calculated by comparing the compositions of samples of multiple genotypes grown in multiple environments (sites and or years). Relatively little information is available on the heritability of most grain components, with the most complete series of studies being carried out under the HEALTHGRAIN program. This included multisite trials, in which 23–26 cultivars were grown in either four environments (for B vitamins) or six environments (for other components) (Shewry et al. 2010, 2011; Corol et al. 2012). The results of these studies are summarized in Fig. 2. Alkylresorcinols, sterols and tocols all show high heritability (above 50%), whereas B vitamins, methyl donors (choline, betaine), and phenolic acids all have low heritability with high effects of environment (or G × E interactions).

DF components show high heritability, in both white flour and bran fractions (data for wholemeal are not available). In particular, TOT-AX and WE-AX in white flour show heritabilities of about 70% and 60%, respectively.

Similar high heritabilities for AX have been shown in several other reported studies. Hong et al. (1989) analyzed 18 wheat lines (seven hard red winter, seven hard white winter and four club wheats) grown on two sites in Washington State, USA, and calculated that the genotypic variance for water-soluble pentosans (i.e., WE-AX) in wholemeal was 1.6 times the environmental variance and for total pentosans (i.e., TOT-AX) 2.4 times. Martinant et al. (1999) reported broad sense heritabilities of 0.75 for WE-AX of flour and 0.80 for the viscosity of aqueous extracts of flour (which is largely determined by WE-AX) from 19 cultivars grown on three locations in France, whereas Dornez et al. (2008) reported broad sense heritabilities of 0.53 for TOT-AX and 0.96 for WE-AX in wholemeals of 14 cultivars grown in Belgium for 3 years. Similar studies of wholemeal samples of five durum wheat cultivars grown under four agronomic regimes gave genotype/environment ratios of 4.5 for TOT-AX and 4.9 for WE-AX (Lempereur et al. 1997). High heritability of WE-AX and TOT-AX in flour was also reported by Finnie et al. (2006) who analyzed seven spring wheat

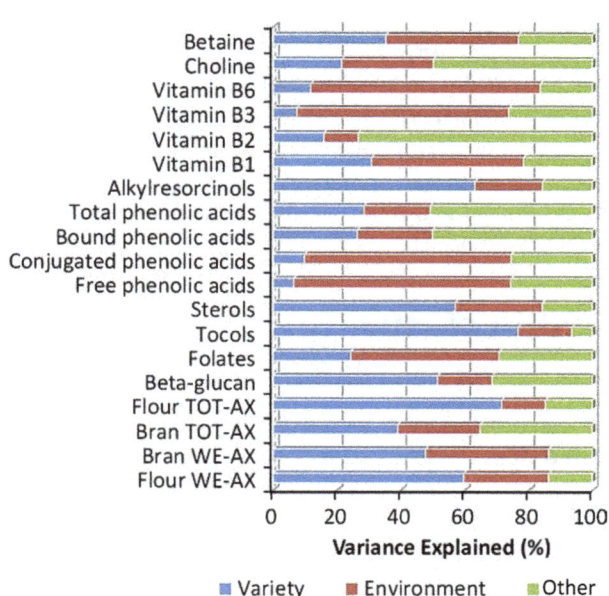

Figure 2. Summary of the heritability of dietary fiber and other components in wheat grain, based on the HEALTHGRAIN study. "Other" includes variance ascribed to Genotype × Environment interactions and/or error.

lines grown in 10 environments and 20 winter wheat lines grown in 12 environments.

However, Li et al. (2009) reported contrasting results for wholemeals of 25 hard winter wheats and 25 hard spring wheats grown at three locations. They showed that environment had a much greater effect than genotype on WE-AX and TOT-AX in the winter lines, by more than an order of magnitude, and a greater impact than genotype on WE-AX (but not TOT-AX) in spring wheats. Hence, the relative effects of G and E depended on the genotypes and environments.

Conclusions

The consumption of wheat is increasing globally, including in countries with climates that are not suitable for wheat production. Wheat-based foods provide a range of essential and beneficial components to the human diet, including protein, B vitamins, DF, and phytochemicals. These components may also vary widely in amount and composition due to effects of genotype and environment. DF is particularly important as consumption is associated with reduced risk of CVD, type 2 diabetes, and certain forms of cancer. DF components also have high heritability and their amount should therefore be amenable to manipulation by breeding, particularly if molecular markers can be established to reduce the need for expensive chemical analyses during screening.

Acknowledgments

We are grateful to CGIAR and the CGIAR WHEAT programme for financial support to prepare a review (Project No. A403 1.09.47) on which this article is based. Rothamsted Research receives strategic funding from the Biotechnological and Biological Sciences Research Council (BBSRC) of the UK.

Conflict of Interest

None declared.

References

Abrams, S. A., I. J. Griffin, and K. M. Hawthorne. 2007a. Young adolescents who respond to an inulin-type fructan substantially increase total absorbed calcium and daily calcium accretion to the skeleton. J. Nutr. 137:25245–25265.

Abrams, S. A., K. M. Hawthorne, O. Aliu, P. D. Hicks, Z. Chen, and I. J. Griffin. 2007b. An inulin-type fructan enhances calcium absorption primarily via an effect on colonic absorption in humans. J. Nutr. 137:2208–2212.

Adom, K. K., M. E. Sorrells, and R. H. Liu. 2003. Phytochemical profiles and antioxidant activity of wheat varieties. J. Agric. Food Chem. 51:7825–7834.

Alam, M. A., C. Sernia, and L. Brown. 2013. Ferulic acid improves cardiovascular and kidney structure and function in hypertensive rats. J. Cardiovasc. Pharmacol. 61:240–249.

Albani, D., M. C. U. Hammond-Kosack, C. Smith, S. Conlan, V. Colot, M. Holdsworth, et al. 1997. The wheat transcriptional activator SPA: a seed-specific bZIP protein that recognizes the GCN4-like motif in the bifactorial endosperm box of prolamin genes. Plant Cell 9:171–184.

Alsaffar, A. A. 2011. Effect of food processing on the resistant starch content of cereals and cereal products – a review. Int. J. Food Sci. Technol. 46:455–462.

Anderson, J. W., P. Baird, R. H. Jr Davis, S. Ferreri, M. Knudtson, A. Koraym, et al. 2009. Health benefits of dietary fiber. Nutr. Rev. 67:188–205.

Andersson, A. A. M., R. Andersson, V. Piironen, A.-M. Lampi, L. Nystrom, D. Boros, et al. 2013. Contents of dietary fibre components and their relation to associated bioactive components in whole grain wheat samples from the HEALTHGRAIN diversity screen. Food Chem. 136:1243–1248.

Andreasen, M. F., P. A. Kroon, G. Williamson, and M.-T. Garcia-Conesa. 2001. Intestinal release and uptake of phenolic antioxidant diferulic acids. Free Radic. Biol. Med. 31:304–314.

Antoine, C., S. Peyron, F. Mabille, C. Lapierre, B. Bouchet, J. Abecassis, et al. 2003. Individual contribution of grain outer layers and their cell wall structure to the mechanical properties of wheat bran. J. Agric. Food Chem. 51:2026–2033.

Antoine, C., S. Peyron, V. Lullien-Pellerin, J. Abecassis, and X. Rouau. 2004. Wheat bran tissue fractionation using biochemical markers. J. Cereal Sci. 39:387–393.

Asp, N. G.. 1992. Resistant starch. Proceedings from the second plenary meeting of EURESTA: european FLAIR-Concerted Action No 11 on the physiological implications of the consumption of resistant starch in man. Eur. J. Clin. Nutr. 46: Supplement 2 S1.

Avivi, L. (1978) High grain protein content in wild tetraploid wheat *Triticum dicoccoides* Korn. In: Fifth International Wheat Genetics Symposium, New Delhi, India. 23–28 February 1978, Pp. 372–380.

Bacic, A., and B. A. Stone. 1981. Chemistry and organisation of aleurone cell wall components from wheat and barley. Aust. J. Plant Physiol. 8:475–495.

Badawy, D., H. M. El-Bassossy, A. Fahmy, and A. Azhar. 2013. Aldose reductase inhibitors zopolrestat and ferulic acid alleviate hypertension associated with diabetes: effect on vascular reactivity. Can. J. Physiol. Pharmacol. 91:101–107.

Barron, C., A. Surget, and X. Rouau. 2007. Relative amounts of tissues in mature wheat (*Triticum aestivum* L.) grain and their carbohydrate and phenolic acid composition. J. Cereal Sci. 45:88–96.

Bates, B., A. Lennox, A. Prentice, C. Bates, P. Page, S. Nicholson, et al. (Eds) (2014a) National diet and nutrition survey: results from years 1-4 (combined) of the rolling programme (2008/2009 – 2011/2012). Executive Summary, Public Health England, London, UK.

Bates, B., A. Lennox, A. Prentice, C. Bates, P. Page, S. Nicholson, et al. (Eds) (2014b) National diet and nutrition survey: results from years 1-4 (combined) of the rolling programme (2008/2009–2011/2012). Public Health England, London, UK.

Batifoulier, F., M.-A. Verny, E. Chanliard, C. Remesy, and C. Demigne. 2006. Variability of B vitamin concentrations in wheat grain, milling fractions and bread products. Eur. J. Agron. 25:163–169.

Beta, T., S. Nam, J. E. Dexter, and H. D. Sapirstein. 2005. Phenolic content and antioxidant activity of pearled wheat and roller-milled fractions. Cereal Chem. 82:390–393.

Birt, D. F., T. Boylston, S. Hendrich, J. L. Jane, J. Hollis, L. Li, et al. 2013. Resistant starch: promise for improving human health. Adv. Nutr. 4:587–601.

Blanco, A., A. Pasqualone, A. Troccoli, N. Di Fonzo, and R. Simeone. 2002. Detection of grain protein content QTLs across environments in tetraploid wheats. Plant Mol. Biol. 48:615–623.

Bonnin, E., A. Le Goff, L. Saulnier, M. Charand, and J. F. Thibault. 1998. Preliminary characterisation of endogenous wheat arabinoxylan-degrading enzymic extracts. J. Cereal Sci. 28:53–62.

Borrelli, G. M., A. Troccoli, N. Di Fonzo, and C. Fares. 2000. Durum wheat lipoxygenase activity and other quality parameters that affect pasta color. Cereal Chem. 76:335–340.

Botticella, E., F. Sestili, A. Hernandez-Lopez, A. Phillips, and D. Lafiandra. 2011. High resolution melting analysis for the detection of EMS induced mutations in wheat *SbeIIa* genes. BMC Plant Biol. 11:156.

Bramley, P. M., I. Elmadfa, E. A. Kafatos, F. J. Kelly, Y. Manios, H. E. Roxborough, et al. 2000. Vitamin E. J. Sci. Food Agric. 80:913–938.

Brown, G., and S. Gordon. 2001. A new receptor for β-glucans. Nature 413:36–37.

Brownlee, I. A. 2011. The physiological roles of dietary fiber. Food Hydrocolloids 25:238–250.

Bruce, S. J., P. A. Guy, S. Rezzi, and A. B. Ross. 2010. Quantitative measurement of betaine and free choline in plasma, cereals and cereal products by isotope dilution LC-MS/MS. J. Agric. Food Chem. 58:2055–2061.

Buchanan, C. J., G. Wallace, and S. C. Fry. 1996. In vivo release of ^{14}C-labelled phenolic groups from intact dietary spinach cell walls during passage through the rat intestine. J. Sci. Food Agric. 71:459–469.

Bunzel, M., J. Ralph, F. Lu, R. D. Hatfield, and H. Steinhart. 2004. Lignins and ferulate-coniferyl alcohol cross-coupling products in cereal grains. J. Agric. Food Chem. 52:6496–6502.

Buttriss, J. L., and C. S. Stokes. 2008. Dietary fibre and health, an overview. Nutr. Bull. 33:186–200.

Cade, J. E., V. J. Burley, D. C. Greenwood, and the UK Women's Cohort Study Steering Group. 2007. Dietary fibre and risk of breast cancer in the UK Women's Cohort Study. Int. J. Epidemiol. 36:431–438.

Capek, P., and M. Matulova. 2013. An arabino(glucurono) xylan isolated from immunomodulatory active hemicellulose fraction of *Salvia officinalis* L. Int. J. Biol. Macromol. 59:396–401.

Carciofi, M., A. Blennow, S. L. Jensen, S. S. Shaik, A. Henriksen, A. Buléon, et al. 2012. Concerted suppression of all starch branching enzyme genes in barley produces amylose-only starch granules. BMC Plant Biol. 12:223.

Carter, E. G. A., and K. J. Carpenter. 1982. The bioavailability for humans of bound niacin from wheat bran. Am. J. Clin. Nutr. 36:855–861.

Chan, J. M., F. Wang, and E. A. Holly. 2007. Whole grains and risk of pancreatic cancer in a large population-based case-control study in the San Francisco Bay area, California. Am. J. Epidemiol. 166:1174–1185.

Chateigner-Boutin, A. L., B. Bouchet, C. Alvarado, B. Bakan, and F. Guillon 2014. The wheat grain contains pectic domains exhibiting specific spatial and development-associated distribution. PLoS ONE PMID:24586916.

Chen, T. K., and W. F. Geddes 1945. Studies on the wheat pigment. MS thesis, University of Minnesota, Minnesota, MN.

Chiuve, S. E., E. L. Giovannucci, S. E. Hankinson, S. H. Zeisel, L. W. Dougherty, W. C. Willett, et al. 2007. The association between betaine and choline intakes and the plasma concentrations of homocysteine in women. Am. J. Clin. Nutr. 86:1073–1081.

Chung, O. K., and J. B. Ohm (2000) Cereal Lipids. Pp. 417–477. In: K. Kulp, J. G. Ponte Jr., eds. Handbook of cereal science and technology, 2nd edn. Marcel Dekker, New York, NY.

Chung, O. K., J.-B. Ohm, M. S. Ram, S.-H. Park, and C. A. Howitt 2009. Wheat Lipids. pp. 363–399. *in* K. Khan, P. R. Shewry, eds. Wheat: chemistry and technology, 4th edn. AACC, St Paul, MN.

Coleman, C. E., and B. A. Larkins 1999. The prolamins of maize. Pp. 109–139. *in* P. R. Shewry, R. Casey, eds. Seed proteins. Kluwer Academic Publishers, Dordrecht.

Coleman, C. E., M. A. Lopes, J. W. Gillikin, R. S. Boston, and B. A. Larkins. 1995. A defective signal peptide in the maize high-lysine mutant floury2. Proc. Natl Acad. Sci. USA 92:6828–6831.

Corol, D. I., C. Ravel, M. Raksegi, Z. Bedo, G. Charmet, M. H. Beale, et al. 2012. Effects of genotype and environment on the contents of betaine, choline, and trigonelline in cereal grains. J. Agric. Food Chem. 60:5471–5481.

Cummings, J. H. 1983. Fermentation in the human large intestine: evidence and implications for health. Lancet 321:1206–1209.

Cummings, J. H., and G. T. Macfarlane. 1991. The control and consequences of bacterial fermentation in the human colon. J. Appl. Bacteriol. 70:443–459.

Cummings, J. H., L. M. Edmond, and E. A. Magee. 2004. Dietary carbohydrates and health: do we still need the fiber concept? Clin. Nutr. Suppl. 1:5–17.

Demeke, T., P. Hucl, R. B. Nair, T. Nakamura, and R. N. Chibbar. 1997. Evaluation of Canadian and other wheats for waxy proteins. Cereal Chem. 74:442–444.

Distelfeld, A., C. Uauy, S. Olmos, A. R. Schlatter, J. Dubcovsky, and T. Fahima. 2004. Microlinearity between a 2-cM region encompassing the grain protein content locus Gpc-6B1 on wheat chromosome 6B and a 350-kb region on rice chromosome 2. Funct. Integr. Genomics 4:59–66.

Distelfeld, A., C. Uauy, T. Fahima, and J. Dubcovsky. 2006. Physical map of the wheat high-grain protein content gene Gpc-B1 and development of a high-throughput molecular marker. New Phytol. 169:753–763.

Dornez, E., K. Gebruers, I. J. Joye, B. de Ketelaere, J. Lenartz, and C. Masseax. 2008. Effects of genotype, harvest year and genotype-by-harvest year interactions on arabinoxylan endoxylanase activity and endoxylanase inhibitor levels in wheat kernels. J. Cereal Sci. 47:180–189.

Du, S. F., H. J. Wang, B. Zhang, F. Y. Zhai, and B. M. Popkin. 2014. China in the period of transition from scarcity and extensive undernutrition to emerging nutrition-related non-communicable diseases, 1949-1992. Obes. Rev. 15:8–15.

Englyst, H. N., and J. H. Cummings. 1985. Digestion of the polysaccharides of some cereal foods in the human small intestine. Am. J. Clin. Nutr. 42:778–787.

Feldman, M. 1995. Wheats. Pp. 185–192. in J. Smartt, N. W. Simmonds, eds. Evolution of crop plants. Longman Scientific and Technical, Harlow, UK.

Finland Finelli database http://www.fineli.fi/index. php?lang=en

Finney, K. F. 1978. Genetically high protein hard winter wheat. Bakers Digest, 32–35.

Finnie, S. M., A. D. Bettge, and C. F. Morris. 2006. Influence of cultivar and environment on water-soluble and water-insoluble arabinoxylans in soft wheat. Cereal Chem. 83:617–623.

Fretzdorff, B., and N. Welge. 2003. Fructan and raffinose contents in cereals and pseudo-cereal grain. Getreide Mehl und Brot 57:3–8.

Gebruers, K., E. Dornez, D. Boros, A. Fras, W. Dynkowska, Z. Bedo, et al. 2008. Variation in the content of dietary fibre and components thereof in wheats in the HEALTHGRAIN Diversity Screen. J. Agric. Food Chem. 56:9740–9749.

Gebruers, K., E. Dornez, Z. Bedo, M. Rakszegi, C. M. Courtin, and J. A. Delcour. 2010. Variability in xylanase and xylanase inhibitor activities in different cereals in the HEALTHGRAIN diversity screen and contribution of environment and genotype to this variability in common wheat. J. Agric. Food Chem. 58:9362–9371.

Gibson, G. R., and M. B. Roberfroid. 1995. Dietary modulation of the human colonic microbiota: introducing the concept of prebiotics. J. Nutr. 125:1401–1412.

Gibson, P. R., and S. J. Shepherd. 2010. Evidence-based dietary management of functional gastrointestinal symptoms: the FODMAP approach. J. Gastroenterol. Hepatol. 25:252–258.

Gonzalez-Hernandez, J. L., E. M. Elias, and S. F. Kianian. 2004. Mapping genes for grain protein concentration and grain yield on chromosome 5B of Triticum turgidum (L.) var. dicoccoides. Euphytica 139:217–225.

Gopalan, C., B. V. Rama Sastri, and S. C. Balasubramanian. 2012. Nutritive value of Indian foods. National Institute of Nutrition, Hyderabad, India.

Graham, S. F., J. H. Hollis, M. Migaud, and R. A. Browne. 2009. Analysis of betaine and choline contents of aleurone, bran, and flour fractions of wheat (Triticum aestivum L.) using ^{1}H nuclear magnetic resonance (NMR) spectroscopy. J. Agric. Food Chem. 57:1948–1951.

Graybosch, R. A. 1998. Waxy wheats: origin, properties, and prospects. Trends Food Sci. Technol. 9:135–142.

Graybosch, R. A., C. J. Peterson, L. E. Hansen, S. Rahman, A. Hill, and J. H. Skerritt. 1998. Identification and characterization of US wheats carrying null alleles at the wx loci. Cereal Chem. 75:162–165.

Groos, C., N. Robert, E. Bervas, and G. Charmet. 2003. Genetic analysis of grain protein content, grain yield and thousand kernel weight in bread wheat. Theor. Appl. Genet. 106:1032–1040.

Gujska, E., and A. Kuncewicz. 2005. Determination of folate in some cereals and commercial cereal-grain products consumed in Poland using trienzyme extraction and high-performance liquid chromatography methods. Eur. Food Res. Technol. 221:208–213.

Hartings, H., M. Lauria, N. Lazzaroni, R. Pirona, and M. Motto. 2011. The Zea mays mutants opaque-2 and opaque-7 disclose extensive changes in endosperm metabolism as revealed by protein, amino acid, and transcriptome-wide analyses. BMC Genom. 12:41.

Hasjim, J., S.-O. Lee, S. Hendrich, S. Setiawan, Y. Ai, and J. Jane. 2010. Characterization of novel resistant-starch and its effects on postprandial plasma-glucose and insulin responses. Cereal Chem. 87:257–262.

Haska, L., M. Nyman, and R. Andersson. 2008. Distribution and characterisation of fructan in wheat milling fractions. J. Cereal Sci. 48:768–774.

Hazard, B., X. Zhang, P. Colasuonno, C. Uauy, D. M. Beckles, and J. Dubcovsky. 2012. Induced mutations in the Starch Branching Enzyme II (SBEII) genes increase amylose and resistant starch content in durum wheat. Crop Sci. 52:1754–1766.

Henry, R. J., and H. S. Saini. 1989. Characterization of cereal sugars and oligosaccharides. Cereal Chem. 66:362–365.

Hogg, A. C., K. Gause, P. Hofer, J. M. Martin, R. A. Graybosch, L. E. Hansen, et al. 2013. Creation of a high-amylose durum wheat through mutagenesis of starch synthase II (SSIIa). J. Cereal Sci. 57:377–383.

Holasova, M. 1999. Distribution of tocopherols and tocotrienols in the main products of wheat and rye milling. Potravin. Vedy 15:343–350.

Hong, B. H., G. L. Rubenthaler, and R. E. Allen. 1989. Wheat pentosans. I. Cultivar variation and relationship to kernel hardness. Cereal Chem. 66:369–373.

Huang, S., D. E. Kruger, A. Frizzi, R. L. D'Ordine, C. A. Florida, W. R. Adams, et al. 2005. High-lysine corn produced by the combination of enhanced lysine biosynthesis and reduced zein accumulation. Plant Biotechnol. J. 3:1467–7652.

Hughes, S. A., P. R. Shewry, L. Li, G. R. Gibson, M. L. Sanz, and R. A. Rastall. 2007. In vitro fermentation by human fecal microflora of wheat arabinoxylans. J. Agric. Food Chem. 55:4589–4595.

Hughes, S. A., P. R. Shewry, G. R. Gibson, B. V. McCleary, and R. A. Rastall. 2008. In vitro fermentation of oat and barley derived β-glucans by human faecal microbiota. FEMS Microbiol. Ecol. 64:482–493.

Humphreys, K. J., M. A. Conion, G. P. Young, D. L. Topping, H. Ying, J. M. Winter, et al. 2014. Dietary manipulation of oncogenic microRNA expression in human rectal mucosa: a randomized trial. Cancer Prev. Res. 7:786–795.

Huynh, B.-L., L. Palmer, D. E. Mather, H. Wallwork, R. D. Graham, R. M. Welch, et al. 2008. Genotypic variation in wheat grain fructan content revealed by a simplified HPLC method. J. Cereal Sci. 48:369–378.

Jende-Strid, B. 1993. Genetic control of flavonoid biosynthesis in barley. Hereditas 119:187–204.

Jensen, S. A., and H. Martens. 1983. The botanical constituents of wheat and wheat milling fractions. II. Quantification by amino acids. Cereal Chem. 60:170–177.

Jobling, S. 2004. Improving starch for food and industrial applications. Curr. Opin. Plant Biol. 7:210–218.

Johnson, V. A., P. J. Mattern, C. J. Peterson, and S. L. Kuhr. 1985. Improvement of wheat protein by traditional breeding and genetic techniques. Cereal Chem. 62:350–355.

Joppa, L. R., and R. G. Cantrell. 1990. Chromosomal location of genes for grain protein content of wild tetraploid wheat. Crop Sci. 30:1059–1064.

Katz, D. L., H. Nawaz, J. Boukhalil, W. Chan, R. Ahmadi, V. Giannamore, et al. 2001. Effect of oat and wheat cereals on endothelial responses. Prev. Med. 33:476–484.

Katz, D. L., M. A. Evans, W. Chan, H. Nawaz, B. Patton Comerford, M. S. Hoxley, et al. 2004. Oats, antioxidants and endothelial function in overweight, dyslipidemic adults. J. Am. Coll. Nutr. 23:397–403.

Keaveney, E. M., R. K. Price, L. L. Hamill, J. M. W. Wallace, H. McNulty, M. Ward, et al. 2015. Postprandial plasma betaine and other methyl donor-related responses after consumption of minimally processed wheat bran or wheat aleurone, or wheat aleurone incorporated into bread. Br. J. Nutr. 113:445–553

Keenan, M. J., R. J. Martin, A. M. Raggio, K. L. McCutcheon, I. L. Brown, A. Birkett, et al. 2012. High-amylose resistant starch increases hormones and improves structure and function of the gastrointestinal tract: a microarray study. J. Nutrigenet. Nutrigenomics 5:26–44.

Khan, K., and P. R. Shewry, eds. 2009. Wheat chemistry and technology. 4th edn 467 pp. American Association of Cereal Chemists, St Paul, MN.

Khan, K., R. Frohberg, T. Olsen, and L. Huckle. 1989. Inheritance of gluten protein components of high protein hard red spring wheat lines derived from *Triticum turgidum* var. *dicoccoides*. Cereal Chem. 166:397–401.

Khan, I. A., J. D. Procunier, D. G. Humphreys, G. Tranquilli, A. R. Schlatter, S. Marcucci-Poltri, et al. 2000. Development of PCR- based markers for a high grain protein content gene from Triticum turgidum spp. Dicoccoides transferred to bread wheat. Crop Sci. 40:518–524.

Kim, W., J. W. Johnson, R. A. Graybosch, and C. S. Gaines. 2003. Physicochemical properties and end-use quality of wheat starch as a function of waxy protein alleles. J. Cereal Sci. 37:195–204.

Kiribuchi-Otobe, C., T. Nagamine, T. Yanagisawa, M. Ohnishi, and I. Yamaguchi. 1997. Production of hexaploid wheats with waxy endosperm character. Cereal Chem. 74:72–74.

Kroon, P. A., C. B. Faulds, P. Ryden, J. A. Robertson, and G. Williamson. 1997. Release of covalently bound ferulic acid from fiber in the human colon. J. Agric. Food Chem. 45:661–667.

Lafiandra, D., F. Sestili, R. D'Ovidio, M. Janni, E. Botticella, G. Ferrazzano, et al. 2010. Approaches for modification of starch composition in durum wheat. Cereal Chem. 87:28–34.

Lafiandra, D., G. Riccardi, and P. R. Shewry. 2014. Improving cereal grain carbohydrates for diet and health. J. Cereal Sci. 59:312–326.

Lampi, A.-M., T. Nurmi, V. Ollilainen, and V. Piironen. 2008. Tocopherols and tocotrienols in wheat genotypes in the HEALTHGRAIN diversity screen. J. Agric. Food Chem. 56:9716–9721.

Lawrence, M. A., W. Chai, R. Kara, I. H. Rosenberg, J. Scott, and A. Tedstone. 2009. Examination of selected national policies towards mandatory folic acid fortification. Nutr. Rev. 67:S73–S78.

Lazaridou, A., and C. G. Biliaderis. 2007. Molecular aspects of cereal β-glucan functionality: physical properties, technological applications and physiological effects. J. Cereal Sci. 46:101–118.

Lempereur, I., X. Rouau, and J. Abercassis. 1997. Genetic and agronomic variation in arabinoxylan and ferulic acid contents of durum wheat (*Triticum durum* L.) grain and its milling fractions. J. Cereal Sci. 25:103–110.

Li, W., S. W. Cui, and Y. Kakuda. 2006. Extraction, fractionation, structural and physical characterization of wheat β-D-glucans. Carbohydr. Polym. 63:404–408.

Li, L., P. R. Shewry, and J. L. Ward. 2008. Phenolic acids in wheat varieties in the HEALTHGRAIN Diversity Screen. J. Agric. Food Chem. 56:9732–9739.

Li, S., C. F. Morris, and A. D. Bettge. 2009. Genotype and environment variation for arabinoxylans in hard winter and spring wheats in the U.S. Pacific Northwest. Cereal Chem. 86:88–95.

Li, W., S. Zhang, and C. Smith. 2015. The molecular structure features-immune stimulatory activity of arabinoxylans derived from the pentosan faction of wheat flour. J. Cereal Sci. 62:81–86.

Likes, R., R. L. Madl, S. H. Zeisel, and S. A. S. Craig. 2007. The betaine and choline content of a whole wheat flour compared to other mill streams. J. Cereal Sci. 46L:93–95.

Lineback, D. R., and V. F. Rasper 1988. Wheat carbohydrates. Pp. 277–372. *in* Y. Pomeranz, ed. Wheat chemistry and technology, 3rd edn, Vol 1. American Association of Cereal Chemists, St Paul, MN.

Lobley, G. E., G. Holtop, D. M. Bremner, A. G. Calder, E. Milne, and A. M. Johstone. 2013. Impact of short term consumption of diets high in either non-starch polysaccharides or resistant starch in comparison with moderate weight loss on indices of insulin sensitivity in subjects with metabolic syndrome. Nutrients 5:2144–2172.

van Loo, J., P. Coussement, L. De Leenheer, H. Hoebregs, and C. Smits. 1995. On the presence of inulin and oligofructose as natural ingredients in the Western diet. Crit. Rev. Food Sci. Nutr. 35:525–552.

Lopez, H. W., C. Coudray, M.-A. Levrat-Verny, C. Feillet-Coudray, C. Demigne, and C. Remesy. 2000. Fructooligocsaccharides enhance mineral apparante absorption and counteract the deleterious effects of phytic acid on mineral homeostasis in rats. J. Nutr. Biochem. 11:500–508.

Mandalari, G., C. B. Faulds, A. I. Sancho, G. Bisignano, R. LoCurto, and K. W. Waldron. 2005. Fractionation and characterisation of arabinoxylans from brewers' spent grain and wheat bran. J. Cereal Sci. 42:205–212.

Mares, D. J., and B. A. Stone. 1973. Studies on wheat endosperm. I. Chemical composition and ultrastructure of the cell walls. Aust. J. Biol. Sci. 26:793–812.

Martinant, J.-P., A. Billot, A. Bouguennec, G. Charmet, L. Saulnier, and G. Branlard. 1999. Genetic and environmental variations in water-extractable arabinolylans content and flour extract viscosity. J. Cereal Sci. 30:45–48.

Mattila, P., J.-M. Pihlava, and J. Hellström. 2005. Contents of phenolic acids, alkyl- and alkenylresorcinols, and avenanthramides in commercial grain products. J. Agric. Food Chem. 53:8290–8295.

McDermott, E. E., and J. Pace. 1957. The content of amino-acids in white flour and bread. Br. J. Nutr. 11:446–452.

Mertz, E. T., L. S. Bates, and O. E. Nelson. 1964. Mutant gene that changes protein composition and increases lysine content of maize endosperm. Science 145:279–280.

Mesfin, A., R. C. Frohberg, K. Khan, and T. C. Olson. 2000. Increased grain protein content and its association with agronomic and end-use quality in two hard red spring wheat populations derived from *Triticum turgidum* L. var. dicoccoides. Euphytica 116:237–242.

Moore, J., Z. Hao, K. Zhou, M. Luther, J. Costa, and L. Yu. 2005. Carotenoid, tocopherol, phenolic acid and antioxidant properties of Maryland-grown soft wheat. J. Agric. Food Chem. 53:6649–6657.

Morell, M. K., C. Konik-Rose, R. Ahmed, Z. Li, and S. Rahman. 2004. Synthesis of resistant starches in plants. J. AOAC Int. 87:740–748.

Muir, J. G., and P. R. Gibson. 2013. The low FODMAP diet for treatment of irritable bowel syndrome and other gastrointestinal disorders. J. Gastroenterol. Hepatol. 9:450–452.

Nakamura, T., M. Yamamori, H. Hirano, S. Hidaka, and T. Nagamine. 1995. Production of waxy (amylose-free) wheats. Mol. Gen. Genet. 248:243–259.

Nakamura, T., T. Shimbata, P. Vrinten, M. Saito, J. Yonemaru, Y. Seto, et al. 2006. Sweet wheat. Genes Genet. Syst. 81:361–365.

Nemeth, C., J. Freeman, H. D. Jones, C. Sparks, T. K. Pellny, M. D. Wilkinson, et al. 2010. Down-regulation of the CSLF6 gene results in decreased (1,3;1,4)-β-D-glucan in endosperm of wheat. Plant Physiol. 152:1209–1218.

Niculescu, M. D., and S. H. Zeisel. 2002. Diet, methyl donors and DNA methylation: interactions between dietary folate, methionine and choline. J. Nutr. 132:2333S–2335S.

Nilsson, U., A. Dahlquist, and B. Nilsson. 1986. Cereal fructosans: part 3. Characterization and structure of wheat fructosans. Food Chem. 22:95–106.

Nurmi, T., L. Nyström, M. Edelmann, A.-M. Lampi, and V. Piironen. 2008. Phytosterols in wheat genotypes in the HEALTHGRAIN Diversity Screen. J. Agric. Food Chem. 56:9710–9715.

Ohta, A., M. Ohtsuki, M. Uehara, A. Hosono, M. Hirayama, T. Adachi, et al. 1998. Dietary fructooligosaccharides prevent postgastrectomy anaemia and osteopenia in rats. J. Nutr. 128:485–490.

Olmos, S., A. Distelfeld, O. Chicaiza, A. R. Schlatter, T. Fahima, V. Echenique, et al. 2003. Precise mapping of a locus affecting grain protein content in durum wheat. Theor. Appl. Genet. 107:1243–1251.

Palmer, R., V. Cornuault, S. E. Marcus, J. P. Knox, P. R. Shewry, and P. Tosi. 2015. Comparative in situ analyses of cell wall matrix polysaccharide dynamics in developing rice and wheat grain. Planta 241:669–685

Parker, M. L., A. Ng, and K. W. Waldron. 2005. The phenolic acid and polysaccharide composition of cell walls of bran layers of mature wheat (*Triticum aestivum* L. cv. Avalon) grains. J. Sci. Agric. Food Chem. 85:2539–2547.

Patterson, J. K., M. A. Rutzke, S. L. Fubini, R. P. Glahn, R. M. Welch, X. Lei, et al. 2009. Dietary inulin supplementation does not promote colonic iron absorption in a porcine model. J. Agric. Food Chem. 57:5250–5256.

Pellny, T. K., A. Lovegrove, J. Freeman, P. Tosi, C. G. Love, P. Knox, et al. 2012. Cell walls of developing wheat starchy endosperm: comparison of composition and RNA-Seq transcriptome. Plant Physiol. 158:612–627.

Piironen, V., E. Syväoja, P. Varo, K. Salminen, and P. Koivistoinen. 1986. Tocopherols and tocotrienols in cereal products from Finland. Cereal Chem. 63:78–81.

Piironen, V., D. G. Lindsay, T. A. Miettinen, J. Toivo, and A.-M. Lampi. 2000. Plant sterols: biosynthesis, biological function and their importance to human nutrition. J. Sci. Food Agric. 80:939–966.

Piironen, V., M. Edelmann, S. Kariluoto, and Z. Bedo. 2008. J. Agric. Food Chem. 56:9726–9731.

Piironen, V., A.-M. Lampi, P. Ekholm, M. Salmenkallio-Marttila, and K.-H. Liukkonen. 2009. Micronutrients and phytochemicals in wheat grain. Pp. 179–222 in K. Khan and P. R. Shewry, eds. Wheat: chemistry and technology, 4th edn. AACC, St Paul, MN.

Pomeranz, Y., ed. 1988. Wheat chemistry and technology, 3rd edn. 2 Vols. (514 and 562 pp.). American Association of Cereal Chemists, St Paul, MN.

Price, R. K., E. M. Keaveney, L. L. Hamill, J. M. W. Wallace, M. Ward, P. M. Ueland, et al. 2010. Consumption of wheat aleurone-rich foods increases fasting plasma betaine and modestly decreases fasting homocysteine and LDL-cholesterol in adults. J. Nutr. 140:2153–2157.

Rahman, S., A. Bird, A. Regina, Z. Li, J. Philippe Ral, S. McMaugh, et al. 2007. Resistant starch in cereals: exploiting genetic engineering and genetic variation. J. Cereal Sci. 46:251–260.

Rakszegi, M., B. N. Kisgyörgy, T. Kiss, F. Sestili, L. Láng, D. Lafiandra, et al. 2014. Development and characterization of high-amylose wheat lines. Starch/Stärke 66:1–8.

Ranum, P. M., F. F. Barrett, R. J. Loewe, and K. Kulp. 1980. Nutrient levels in internationally milled wheat flours. Cereal Chem. 57:361–366.

Rhodes, D. I., and B. A. Stone. 2002. Proteins in walls of wheat aleurone cells. J. Cereal Sci. 36:83–101.

Rice, P. J., E. A. Adams, T. Ozment-Skelton, A. J. Gonzalez, M. P. Goldman, B. E. Lockhart, et al. 2005. Oral delivery and gastrointestinal absorption of soluble glucans stimulate increased resistance to infectious challenge. J. Pharmacol. Exp. Ther. 314:1079–1086.

Ritsema, T., and S. Smeekens. 2003. Fructans: beneficial for plants and humans. Curr. Opin. Plant Biol. 6:223–230.

Roberfroid, M. B. 2005. Inulin-type fructans: functional foods ingredients. CRC Press, Boca Raton, FL.

Robertson, M. D., J. M. Currie, L. M. Morgan, D. P. Jewell, and K. N. Frayn. 2003. Prior short-term consumption of resistant starch enhances postprandial insulin sensitivity in healthy subjects. Diabetologia 46:659–665.

Robertson, M. D., A. S. Bickerton, A. L. Dennis, H. Vidal, and K. N. Frayn. 2005. Insulin-sensitizing effects of dietary resistant starch and effects on skeletal muscle and adipose tissue metabolism. Am. J. Clin. Nutr. 82:559–567.

Rodriguez-Mateos, A., C. Rendeiro, B. Bergillos-Meca, S. Tabatabaee, T. W. George, C. Heiss, et al. 2013. Intake and time dependence of blueberry flavonoid-induced improvements in vascular function: a randomized, controlled, double-blind, crossover intervention study with mechanistic insights into biological activity. Am. J. Clin. Nutr. 98:1179–1191.

Ross, A. B., and S. J. Bruce. 2011. A whole-grain cereal-rich diet increases plasma betaine, and tends to decrease total and LDL-cholesterol compared with a refined-grain diet in healthy subjects. Br. J. Nutr. 105:1492–1502.

Sampson, D. A., Q.-B. Wen, and K. Lorenz. 1996. Vitamin B6 and pyridoxine glucoside content of wheat and wheat flours. Cereal Chem. 73:770–774.

Saulnier, L., and J. F. Thibault. 1999. Ferulic acid and diferulic acids as components of sugar beet pectins and maize bran heteroxylans. J. Sci. Food Agric. 79:396–402.

Saulnier, L., N. Peneau, and J.-F. Thibault. 1995. Variability in grain extract viscosity and water-soluble arabinoxylan content in wheat. J. Cereal Sci. 22:259–264.

Saulnier, L., P.-E. Sado, G. Branlard, G. Charmet, and F. Guillon. 2007. Wheat arabinoxylans: exploiting variation in amount and composition to develop enhanced varieties. J. Cereal Sci. 46:261–281.

Schatzkin, A., Y. Park, M. F. Leitzmann, A. R. Hollenbeck, and A. J. Cross. 2008. Prospective study of dietary fiber, whole grain foods, and small intestinal cancer. Gastroenterology 135:1163–1167.

Schirmer, M., A. Höchstötter, M. Jekle, E. Arendt, and T. Becker. 2013. Physicochemical and morphological characterization of different starches with variable amylose/amylopectin ratio. Food Hydrocolloids 32:52–63.

Schmidt, R. J., M. Ketudat, M. Aukerman, and G. Hoschek. 1992. OPAQUE-2 is a transcriptional activator that recognizes a specific target site in 22kD-zein genes. Plant Cell 4:689–700.

Scholz-Ahrens, K. E., G. Schaafsma, E. G. H. M. van den Heuvel, and J. Schrezenmeir. 2001. Effects of prebiotics on mineral metabolism. Am. J. Clin. Nutr. 73:459S–4564S.

Scientific Advisory Committee on Nutrition (SACN) (2015) Carbohydrates and Health Report. https://www.gov.uk/government/publications/sacn-carbohydrates-and-health-report

Sestili, F., E. Botticella, Z. Bedo, A. Phillips, and D. Lafiandra. 2010. Production of novel allelic variation for genes involved in starch biosynthesis through mutagenesis. Mol. Breeding 25:145–154.

Sharma, A., B. S. Yadav, and Ritika. 2008. Resistant starch: physiological roles and food applications. Food Rev. Int. 24:193–234.

Shewry, P. R. 2007. Improving the protein content and composition of cereal grain. J. Cereal Sci. 46:239–250.

Shewry, P. R. 2013. Improving the content and composition of dietary fibre in wheat. Pp. 154–169 in J. A. Delcour and K. Poutanen, eds. Fibre-rich and wholegrain foods – improving quality. Woodhead Publishing, Ltd., Sawston, Cambridge.

Shewry, P. R., and N. G. Halford. 2002. Cereal seed storage proteins: structures, properties and role in grain utilization. J. Exp. Bot. 53:947–958.

Shewry, P. R., R. D'Ovidio, D. Lafiandra, J. A. Jenkins, E. N. C. Mills, and F. Bekes. 2009a. Wheat grain proteins. Pp. 223–298 in K. Khan and P. R. Shewry, eds. Wheat: chemistry and technology, 4th edn. AACC, St Paul, MN, USA.

Shewry, P. R., V. Piironen, A.-M. Lampi, M. Edelmann, S. Kariluoto, T. Nurmi, et al. 2010. The HEALTHGRAIN wheat diversity screen: effects of genotype and environment on phytochemicals and dietary fiber components. J. Agric. Food Chem. 58:921–928.

Shewry, P. R., F. Van Schaik, C. Ravel, G. Charmet, M. Rakszegi, Z. Bedo, et al. 2011. Genotype and environment effects on the contents of vitamins B1, B2, B3 and B6 in wheat grain. J. Agric. Food Chem. 59:10564–10571.

Shewry, P. R., M. J. Hawkesford, V. Piironen, A.-M. Lampi, K. Gebruers, D. Boros, et al. 2013. Natural variation in grain composition of wheat and related cereals. J. Agric. Food Chem. 61:8295–8303.

Shimbata, T., T. Inokuma, P. Vrinten, M. Saito, T. Takiya, and T. Nakamura. 2011. High levels of sugars and fructan in mature seed of sweet wheat lacking GBSSI and SSIIa enzymes. J. Agric. Food Chem. 59:4794–4800.

Shoup, F. K., Y. Pomeranz, and C. W. Deyoe. 1966. Amino acid composition of wheat varieties and flours varying widely in bread-making potentialities. J. Food Sci. 31:94–93.

Simmonds, D. H. 1962. Variations in the amino acid composition of Australian wheats and flours. Cereal Chem. 39:445–454.

Slade, A. J., S. I. Fuerstenberg, D. Loeffler, M. N. Steine, and D. Facciotti. 2005. A reverse genetic, nontransgenic approach to wheat crop improvement by TILLING. Nat. Biotechnol. 23:75–81.

Slade, A. J., C. McGuire, D. Loeffler, J. Mullenberg, W. Skinner, G. Fazio, et al. 2012. Development of high amylose wheat through TILLING. BMC Plant Biol. 12:69.

Smeds, A. I., L. Jauhiainen, E. Tuomola, and P. Peltonen-Sainio. 2009. Characterisation of variation in the lignin content and composition of winter rye, spring wheat and spring oat. J. Agric. Food Chem. 57:5837–5842.

Snape, J. W., V. Hyne, and K. Aitken 1993. Targeting genes in wheat using marker mediated approaches. Pp. 749–759. in Proceedings of eighth international wheat genetics symposium, Beijing.

South African Food Composition database http://safoods.mrc.ac.za/

Stone, B., and M. K. Morell 2009. Carbohydrates. Pp. 299–362. in K. Khan, P. R. Shewry, eds. Wheat chemistry and technology, 4th edn. American Association of Cereal Chemists, St Paul, MN.

Suzuki, A., M. Yamamoto, H. Jokura, A. Fujii, I. Tokimitsu, T. Hase, et al. 2007. Ferulic acid restores endothelium-dependent vasodilation in aortas of spontaneously hypertensive rats. Am. J. Hypertens. 20:508–513.

Tako, E., R. P. Glahn, M. Knez, and J. C. R. Stangoulis. 2014. The effect of wheat prebiotics on the gut bacterial population and iron status of iron deficient broiler chickens. Nutr. J. 13:58.

Tester, R. F., X. Qi, and J. Karkalas. 2006. Hydrolysis of native starches with amylases. Anim. Feed Sci. Technol. 130:39–54.

Themeier, H., J. Hollmann, U. Neese, and M. G. Lindhauer. 2005. Structural and morphological factors influencing the quantification of resistant starch II in starches of different botanical origin. Carbohydr. Polym. 61:72–79.

Theuwissen, E., and R. P. Mensink. 2008. Water-soluble dietary fibers and cardiovascular disease. Physiol. Behav. 94:285–292.

Thompson, D. B. 2000. Strategies for the manufacture of resistant starch. Trends Food Sci. Technol. 11:245–253.

Tkachuk, R. 1966. Amino acid composition of flours. Cereal Chem. 43:207–223.

Toepfer, E. W., M. M. Polansky, J. F. Eheart, H. T. Slover, F. N. Hepburn, and F. W. Quackenbush. 1972. Nutrient composition of selected wheats and wheat products. XI Summary. Cereal Chem. 49:173–186.

Topping, D. 2007. Cereal complex carbohydrates and their contribution to human health. J. Cereal Sci. 46:220–229.

Topping, D. L., and P. M. Clifton. 2001. Short-chain fatty acids and human colonic function: roles of resistant starch and nonstarch polysaccharides. Physiol. Rev. 81:1031–1064.

Topping, D. L., B. H. Bajka, A. R. Bird, J. M. Clarke, L. Cobiac, M. A. Conlon, et al. 2008. Resistant starches as a vehicle for delivering health benefits to the human large bowel. Microb. Ecol. Health Dis. 20:103–108.

Turkey Türkomp database http://www.turkomp.gov.tr/ (Turkish Food Composition database)

Turner, A. S., R. P. Bradburne, L. Fish, and J. W. Snape. 2004. New quantitative trait loci influencing grain texture and protein content in bread wheat. J. Cereal Sci. 40:51–60.

Uauy, C., A. Distelfeld, T. Fahima, A. Blechl, and J. Dubcovsky. 2006. A NAC gene regulating senescence improves grain protein, zinc and iron content in wheat. Science 314:1298–1301.

Uauy, C., F. Paraiso, P. Colasuonno, R. K. Tran, H. Tsai, S. Berardi, et al. 2009. A modified TILLING approach to detect induced mutations in tetraploid and hexaploid wheats. BMC Plant Biol. 9:115.

Ueland, P. M., P. I. Holm, and S. Hustad. 2005. Betaine: a key modulator of one-carbon metabolism and homocysteine status. Clin. Chem. Lab. Med. 43:1069–1075.

UK Cofids database http://tna.europarchive. org/20110116113217/http://www.food.gov.uk/science/ dietarysurveys/dietsurveys/

USDA National Nutrient Database R26 2014. http://www.ars. usda.gov/Services/docs.htm?docid=8964

Van Hung, P., M. Yamamori, and N. Morita. 2005. Formation of enzyme-resistant starch in bread as affected by high-amylose wheat flour substitutions. Cereal Chem. 82:690–694.

Vauzour, D., E. J. Houseman, T. W. George, G. Corona, R. Garnotel, K. G. Jackson, et al. 2010. Moderate Champagne consumption promotes and acute improvement in acute endothelial-independent vascular function in health human volunteers. Br. J. Nutr. 103:1168–1178.

Vitaglione, P., A. Napolitano, and V. Fogliano. 2008. Cereal dietary fibre: a natural functional ingredient to deliver phenolic compounds into the gut. Trends Food Sci. Technol. 19:451–463.

Vogel, K. P., V. A. Johnson, and P. J. Mattern. 1976. Protein and lysine content of grain, endosperm, and bran of wheats from USDA World Wheat collection. Crop Sci. 16:655–660.

Ward, J. L., K. Poutanen, K. Gebruers, V. Piironen, A.-M. Lampi, L. Nyström, et al. 2008. The HEALTHGRAIN cereal diversity screen: concept, results and prospects. J. Agric. Food Chem. 56:9699–9709.

WHO. 2002. Technical Report Series 935 Protein and amino acid requirements in human nutition. Report of a Joint WHO/FAO/UNU Expert Consultation, Geneva, Switzerland.

WHO. 2009. WHO/NMH/NHD/MNM/09.1. World Health Organisation, Geneva.

Worland, T., and J. W. Snape. 2001. Genetic basis of worldwide wheat varietal improvement. Pp. 59–100 in A. P. Bonjean and W. J. Angus, eds. The world wheat book – a history of wheat breeding. Groupe Limagrain, Paris.

World Agricultural Outlook Board (2014) World Agricultural Supply and Demand Estimates. WASDE-530. USDA-ERS, FAS (online). http://www.usda.gov/oce/ commodity/wasde/latest.pdf

Yamamori, M. 2009. Amylose content and starch properties generated by five variant Wx alleles for granule-bound starch synthase in common wheat (Triticum aestivum L.). Euphytica 165:607–614.

Yamamori, M., and T. R. Endo. 1996. Variation of starch granule proteins and chromosome mapping of their coding genes in common wheat. Theor. Appl. Genet. 93:275–281.

Yamamori, M., T. Nakamura, T. R. Endo, and T. Nagamine. 1994. Waxy protein deficiency and chromosomal locations of coding genes in common wheat. Theor. Appl. Genet. 89:179–184.

Yamamori, M., S. Fujita, K. Hayakawa, J. Matsuki, and T. Yasui. 2000. Genetic elimination of a starch granule protein, SGP-1, of wheat generates an altered starch with apparent high amylose. Theor. Appl. Genet. 101:21–29.

Yamamori, M., M. Kato, M. Yui, and M. Kawasaki. 2006. Resistant starch and starch pasting properties of a starch synthase IIa – deficient wheat apparent high amylose. Aust. J. Agric. Res. 57:531–535.

Zeisel, S. H., and J. K. Blusztajn. 1994. Choline and human nutrition. Annu. Rev. Nutr. 14:269–296.

Zeisel, S. H., M.-H. Mar, R. C. Howe, and J. M. Holden. 2003. Concentrations of choline-containing compounds and betaine in common foods. J. Nutr. 133:1302–1307. and erratum 133: 2918–2919.

Zhou, K., and L. Yu. 2005. Phenolic acid, tocopherol and carotenoid compositions, and antioxidant functions of hard red winter wheat bran. J. Agric. Food Chem. 53:3916–3922.

Zhou, K., L. Su, and L. Yu. 2004. Phytochemicals and antioxidant properties in wheat bran. J. Agric. Food Chem. 52:6108–6114.

7

Genetics-based dynamic systems model of canopy photosynthesis: the key to improve light and resource use efficiencies for crops

Qingfeng Song[1], Chengcai Chu[2], Martin A. J. Parry[3] & Xin-Guang Zhu[1]

[1]CAS Key Laboratory for Computational Biology and State Key Laboratory of Hybrid Rice, Partner Institute for Computational Biology, Chinese Academy of Sciences, Shanghai 200031, China
[2]The State Key Laboratory of Plant Genomics and National Center of Plant Gene Research (Beijing), Institute of Genetics and Developmental Biology, CAS, Beijing 100101, China
[3]Lancaster Environment Centre, Lancaster University, Lancaster LA1 4YQ, UK

Keywords
Canopy photosynthesis, design crop systems, genetics-based model of canopy photosynthesis, heterogeneity, microclimates.

Correspondence
Xin-Guang Zhu, CAS-MPG Partner Institute for Computational Biology, CAS, Shanghai 200031, China.

E-mail: zhuxinguang@picb.ac.cn

Funding Information
The authors X-GZ, QS and CC thank CAS strategic leading project "Modular Designer Crop Breeding" (grant # XDA08020301). MAJP acknowledges financial support from the BBSRC grants BB/I017372/1, BB/I024488/1, BB/I002545/1, BB/J/00426X/1 20:20 Wheat®.

Abstract

Improving canopy photosynthetic light use efficiency instead of leaf photosynthesis holds great potential to catalyze the next "green revolution". However, leaves in a canopy experience different biochemical limitations due to the heterogeneities of microclimates and also physiological parameters. Mechanistic dynamic systems models of canopy photosynthesis are now available which can be used to design the optimal canopy architectural and physiological parameters to maximize CO_2 uptake. Rapid development of modern crop genetics research now makes it possible to link such canopy models with genetic variations of crops to develop genetics-based dynamic systems models of canopy photosynthesis. Such models can guide marker-assisted breeding or genomic selection or engineering of crops to enhance light and nitrogen use efficiencies for different regions under future climate change scenarios.

Introduction

Photosynthesis is the primary determinant of plant biomass. Canopy photosynthesis is the sum of photosynthetic rates for all photosynthetic tissues (e.g., leaves, stems ears) within the canopy. Since the room to further increase crop harvest index is small, improving canopy photosynthesis and hence biomass production are now widely recognized as a major avenue to increase crop yields. A number of options to increase canopy photosynthesis has been proposed, such as optimizing Rubisco kinetic parameters, increasing the speed of recovery from photoprotective states, decreasing the antenna size of photosystems, etc. (Zhu et al. 2010; Parry et al. 2013; Carmo-Silva et al. 2015; Lin et al. 2014). There is existing substantial heritable variation in photosynthetic traits (see for example Driever et al. 2014) and biotechnological targets (Parry et al. 2011; Ort et al. 2015) that are being exploited to increase yield.

These options are now under exploration to realize their potential in improving crop yields in many international teams though a number of major international projects, including the Bill and Melinda Gates Foundation funded C4 Rice (http://c4rice.irri.org/) and Realizing Improved Photosynthetic Efficiency (http://www.ripe.uiuc.edu) projects, the Center of Excellence for Photosynthesis Research (http://photosynthesis.org.au/). Though it is recognized that there are several targets and approaches to improve canopy photosynthesis, most of the current research focuses on photosynthetic efficiency at the leaf level, more efforts should be taken to explore the impacts of different engineering options on the canopy photosynthesis. In fact, there has been many reports regarding the lack of correlation between instantaneous measurements of leaf photosynthesis and crop yields (Evans and Dunstone 1970). Understanding limitations of canopy photosynthesis and identify leaf features that can confer higher canopy photosynthetic CO_2 uptake rates are required to facilitate the application of photosynthesis in current crop breeding for higher yield potential (Zhu et al. 2010).

Canopy photosynthesis is inherently complex. It is influenced by a large number of factors, including physiological and architectural parameters, and external environmental conditions (Peng et al. 2000; Zhu et al. 2012). As a result, leaves inside a canopy each experience different biochemical limitations and these change over time. Dependent on the crop architecture, the growing location, planting density, planting direction, and leaf physiological parameters at different layers of a canopy, dramatic difference exists in the proportion of leaves limited by light absorption, electron transfer, RuBP regeneration, or Rubisco (Farquhar et al. 1980). As a result of this, the impacts of manipulating leaf photosynthetic properties on canopy photosynthesis are highly non-linear and even Counter-intuitive (Zhu et al. 2014b). In recent years, there has been good progress in both the theoretical approach to modeling canopy photosynthesis and experimental methods to measure canopy photosynthesis, which promise to rapidly advance our ability to pinpoint the most effective approach to engineer canopy photosynthesis in a crop- and in a region-specific manner. The purpose of this perspective paper is to discuss the current status on canopy photosynthesis from the perspective defining options to engineer improved canopy photosynthesis to support a region-specific breeding. Specifically, we emphasize the potential factors that will influence the total canopy photosynthetic CO_2 uptake rates, models and experimental approaches to quantify canopy photosynthesis and propose a new concept of model-guided design of an ideal canopy for future crops. Readers can refer to (Zhu et al. 2012) about the heterogeneity of physiology inside a canopy and the overall structure and rationale for developing a new mechanistic dynamic systems model of canopy photosynthesis.

Quantification of Canopy Photosynthesis

Quantification of canopy photosynthesis requires accurate estimation of photosynthetic rates of all leaves inside a canopy. This is inherently challenging because the microclimate inside a canopy, such as light, CO_2, temperature, and humidity, is highly heterogeneous both spatially and temporally (Pearcy 1990; Zhu et al. 2004a, 2012; Song et al. 2013). As a result, leaves at different layers of a canopy normally experience different biochemical limitations. Leaves at the top of a canopy are usually light saturated; while leaves at the bottom layers of a canopy are usually light limited. The heterogeneity of light is also reflected in the appearance of sunflecks and shade-flecks inside a canopy, which in turn is influenced by the canopy architectures and wind inside a canopy (Pearcy 1990). The physiological status of leaves at different layers of the canopy also vary dramatically (Evans and Poorter 2001; Niinemets 2007). On a broad scale, leaves at the top layers of a canopy are usually thicker, having higher chlorophyll $a{:}b$ ratio, higher Rubisco concentration, etc. compared to the leaves at the lower layers (Terashima and Evans 1988; Hikosaka and Terashima 1995; Evans and Poorter 2001). Furthermore, plants adjust their metabolism to cope with different sunflecks and shade-fleck patterns within the lower layers of a canopy. For example, understory plants usually have leaves that have much higher assimilatory charge, which enable plants to rapidly utilize the incoming sunflecks etc. (Pearcy 1990).

Even though being inherently challenging, canopy photosynthesis has been modelled since the 1950s (Monsi and Saeki 1953). Different models, each with different degree of simplifications regarding the heterogeneity of microclimates inside the canopy, have been constructed in the past, see review by Zhu et al. (2012). Among these, the big leaf model made the assumption that the total canopy photosynthetic CO_2 uptake rate can be effectively represented by a single leaf. Due to its simplicity, it has been used as a basic model in large-scale general circulation model (GCM) (Sellers et al. 1996); however, the connection between this model with canopy architecture and physiological parameters are largely missing. The sunlit-shaded model assumes that leaves in the canopy can be effectively classified as being either shaded leaf or sunlit leaf, each with an associated leaf area index. The leaf physiological parameters, such as maximal rate of RuBP and CO_2 saturated rate of RuBP carboxylation (V_{cmax}) and maximal rate of photosynthetic electron transfer (J_{max}), can be effectively represented in such a model. Furthermore, the sunlit-shaded model is relatively simple and easy to use (dePury and Farquhar 1997). As a result, it is used widely in the research community of photosynthesis physiology, ecology and agronomy. Recently, Song et al. (2013) developed a more mechanistic canopy photosynthesis model. This model can predict the detailed light environments inside a canopy by using realistic 3D reconstruction of a canopy with defined architecture combined with a forward ray tracing algorithm. The different physiological parameters for individual leaves can be incorporated into this model. As a result, the Song et al. (2013) enables evaluation of different architectural and physiological properties on canopy photosynthetic CO_2 uptake rates. With this model, even the impacts of varying growth regions and planting density, planting direction on the total canopy photosynthetic CO_2 uptake rates can be evaluated as well. Figure 1 shows that the growth location and planting direction greatly influence canopy photosynthetic rates. At a particular latitude of a particular growth region, the ideal canopy architecture for optimal canopy photosynthetic CO_2 uptake henceforth should be defined. Different methods to directly measure canopy photosynthesis have been developed. The Bowen ratio/energy balance method is appropriate to quantify the total canopy gas exchange for a large area (Cellier and Olioso 1993). Canopy chamber approach, including both the open system chamber and the closed system chamber, has been developed to evaluate canopy photosynthesis at a plot scale (Dugas 1993; Dugas et al. 1997; Johnson et al. 2003; Song et al. 2016). These measurement systems hold great potential in model development and evaluation of germplasm to select lines with enhanced photosynthetic efficiencies.

Canopy Photosynthesis and Crop Engineering and Breeding

In wheat, screening for improved leaf photosynthetic efficiency did not directly lead to enhanced canopy photosynthesis and crop yield and in fact the light saturated rate of leaf photosynthesis in wheat is negatively correlated with the leaf area index (Evans and Dunstone 1970). Increased leaf photosynthetic efficiency was gained by increasing leaf thickness, which unfortunately is correlated to decreased leaf area index (Evans and Dunstone 1970). However, the total canopy photosynthesis, rather than just leaf photosynthesis, is positively correlated with the biomass accumulation (Zelitch 1982). There are complex nonlinear interactions among crop architectural parameters and leaf physiological parameters, which jointly determine the optimal parameters to gain increased canopy photosynthetic CO_2 uptake rate. This is clearly demonstrated by the impacts of different leaf area index on the potential gain of manipulation Rubisco kinetic properties on canopy CO_2 uptake rates: at a higher leaf area index, there is increased benefit of engineering a Rubisco with higher specificity into a canopy (Zhu et al. 2004). Similarly, it is expected that decreasing leaf chlorophyll concentration will have different consequences for canopies with different architecture (Ort et al. 2011). The increased leaf photosynthetic properties is usually associated with increased leaf chlorophyll concentrations, which can lead to altered light environments inside a canopy and hence altered canopy photosynthetic CO_2 uptake rate as well. All these complex interactions necessitate application of detailed canopy systems models to design optimal parameters for enhanced canopy photosynthesis.

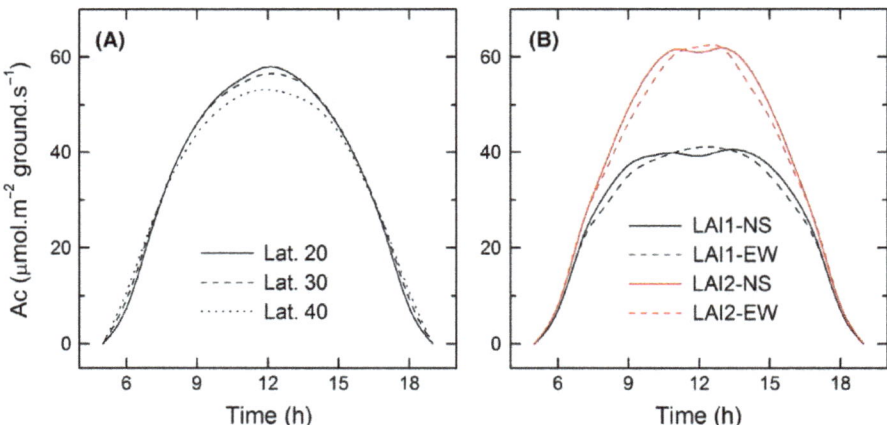

Figure 1. The impacts of varying growth location and growth direction on canopy photosynthetic rates. The simulation was conducted for a rice canopy based on methods for 3D canopy reconstruction (Song et al. 2013). (A) The impacts of varying the growth latitudes on the diurnal canopy photosynthetic rates; (B) The impacts of varying the leaf area index (LAI) and also the growth direction on the diurnal canopy photosynthesis. NS (North-south direction); EW (East-west direction).

Over the last decade, a number of options to manipulate photosynthesis for enhanced photosynthetic efficiency have been identified. These different options were designed to overcome the limitation of photosynthesis at different biophysical or biochemical steps (Zhu et al. 2010; Long et al. 2015). For example, manipulation of leaf chlorophyll content mainly dealing with the excess energy at top layers and increase light availability at lower layers of a canopy; the potential impacts of this manipulation rely on not only leaf area index, canopy architecture, but also growth locations (Song 2004). Manipulation of the recovery from the photoprotective state aims to overcome the loss of quantum yield after plants are shifted from high light to low light (Zhu et al. 2004). Engineering C_4 photosynthesis into a C_3 leaf aims to overcome the limitation of the Rubisco specificity factor on leaf photosynthetic rate under the current atmospheric CO_2 conditions (Hibberd et al. 2008).

Considering that leaves at different locations of a canopy experience different microclimates and hence different biochemical limitations, it is important to design different crop ideal types with potentially different leaf photosynthetic properties for leaves at different layers of a canopy. A "smart canopy" concept has been proposed where the upper leaves should be more vertical, equipped with Rubisco with higher specificity, and smaller antenna size, as compared to lower layers of a canopy; furthermore, leaves at the lower layers of the canopy can be engineered to have enhanced chlorophyll d concentration to better fit the local light environments (Ort et al. 2015).

Canopy Photosynthesis Underlines the Ideal-type Breeding

What are the major components for the ideal-type design from a canopy photosynthesis perspective? They include (1) the canopy architectural parameters, for example leaf length, leaf angle, leaf curvature, shape, leaf number, tiller number, planting density etc.; (2) physiological and biochemical parameters, which include leaf chlorophyll content, nitrogen content, content and activation state of key enzymes related to photosynthesis and parameters related to the recovery from the photoprotective state, parameters related to engagement of cyclic electron transfer, parameters related to stomatal responses, etc. The ideo-type design is to identify optimal combination of these different parameters for a crop grown under a defined geological location (Fig. 2A). Though not guided by a mechanistic canopy photosynthesis model, ideotype-guided breeding has been practiced for a long time. In fact, the concept of ideotype breeding was first proposed by Donald (1968). Since then, it has been widely practiced by breeders from both private and public sectors. Rice breeders in China and also in the International

Rice Research Institute(IRRI) proposed various ideotypes for Japonica and indica, see review by Peng et al. (2008). For example, IRRI proposed the features of ideotype (or a new plant type) include low tillering number, few unproductive tillers, 200–250 grains per panicle, leaves that are thick, dark green, and erect, a plant height of 90–100 cm, thick and sturdy stems, vigorous root system, a growth duration of 100–130 days, and an increased harvest index (Peng et al. 1994). At IRRI, breeding based largely on these features with slight modifications has led to generation of many breeding lines and also the release of cultivars with increased yield potential (Peng et al. 2008). It is worth mentioning that most of the ideotype characteristics were determined based on computer simulations based on models with a simplified canopy architecture description. In China, Prof Longping Yuan proposed the features of rice ideotype, which include: moderate tillering capacity, heavy and drooping panicle at maturity, a plant height of 100 cm, panicle length of 60 cm at maturity, and specific features for the top three leaves, and a harvest index being above 0.55 (Yuan 2001). These ideotypes describe the morphology of the top three leaves, including their length, width, thickness, erectness, leaf angle, and also define the leaf area index (Yuan 2001). Most of these ideotype-related features are related to canopy photosynthesis. Theoretically, different ideotypes should be developed for different growth regions or environments, as reflected in the difference in the ideotypes defined by IRRI and Prof. Yuan. From this perspective, the mechanistic model of canopy photosynthesis, which incorporates the detailed three-dimensional canopy architectural parameters and physiological parameters of leaves at different locations, is needed to provide an objective and systematic method to tailor the features of ideotype to gain the optimal canopy photosynthesis and productivity for crops that are grown at defined locations.

Toward a Genetics-based Crop Systems Models to Guide Ideal-type Design

As discussed earlier, the Song et al. (2013) model can be used to explore the optional canopy architecture, metabolic features, planting density, etc. for a crop at a particular geological location (Fig. 2A). However, to enable such a model to be used effectively in crop breeding, the direct linkage of the model parameters to its genetic basis needs to be established. In other words, the alleles or genes or molecular markers controlling each feature in the mechanistic crop systems model need to be defined. It is important to point out here that for a particular trait, such as leaf angle, there are often multiple different alleles controlling it (Wang and Li 2008). Depending on the various

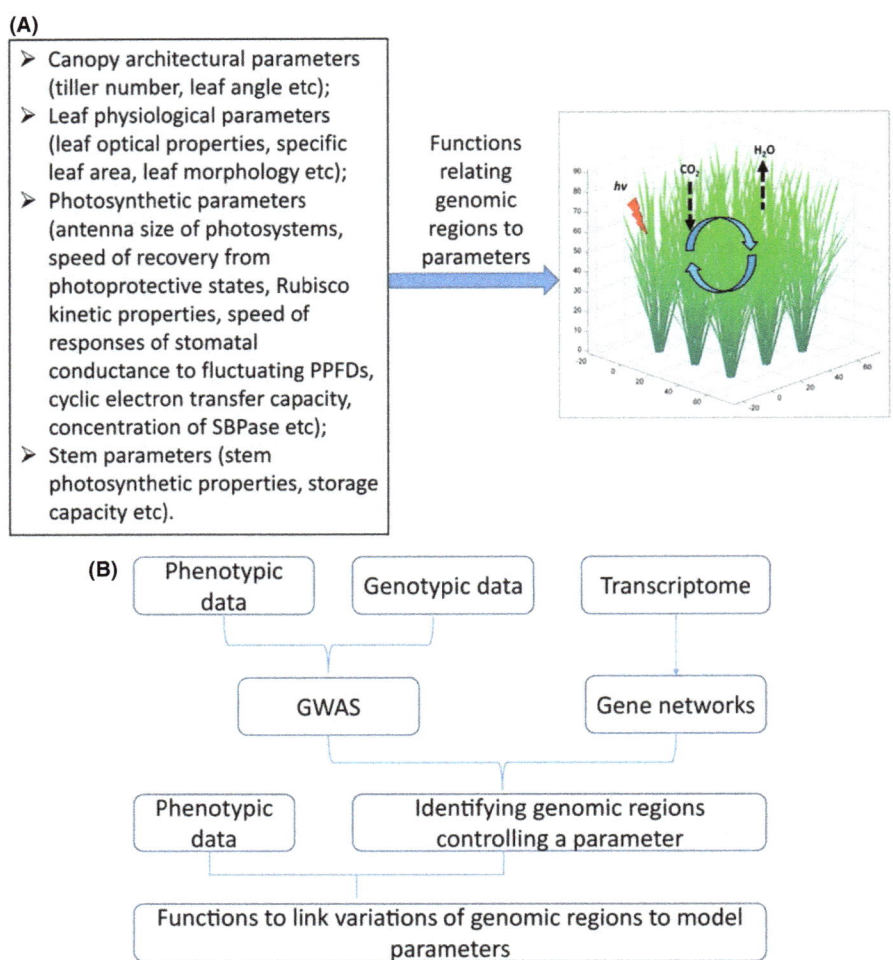

Figure 2. The parameters required for developing an idea type of a particular crop. (A) The general structure of the model. The model will incorporate both the detailed description of the canopy architectural parameters and also the detailed description of the photosynthetic processes. Functions relating genomic variations to variations of parameters will be used in the model so that the model can predict the consequences of different genetic variations on photosynthetic properties. (B) The procedure to establish the function to linking genetic variations to parameters used in the genetics-based dynamic systems model of canopy photosynthesis.

alleles in a particular rice accession, rice shows different leaf angles. Similarly, other parameters required for defining ideo-type are also controlled by a number of alleles. Methods to link such genetic variations with parameters in systems model have been developed and used to predict a number of critical physiological and developmental parameters, such as flowering time (Reymond et al. 2003; van Eeuwijk et al. 2005; Yin et al. 2005). If a model with direct linkage between genetic variations and model parameters can be built, such a model can be immediately used to optimize allele combinations to gain maximal canopy photosynthetic rates at a particular location for a defined crop.

How far are we from realizing such a genetics-based canopy photosynthesis model for one plant species, such as rice? Rice possibly represents the best studied crop

species so far. Many functional relationships between allele variations and canopy architectural parameters have been established already for rice (Zuo and Li 2013). For other crops, such functional studies are much less established comparatively. For photosynthesis-related parameters, so far, little is known about their association with allelic variations. Coordinated efforts are needed to establish new relations which can be used to predict photosynthetic parameters based on allelic contents and environmental conditions. The rapid advances in the modern phenomics facility and NGS technology are now offering an unprecedented opportunity to realize this. To do this, for each particular cultivar, using a large-scale phenomics facility to measure the photosynthetic parameters under a diverse set of environmental conditions for a panel of genetically diverse accessions will

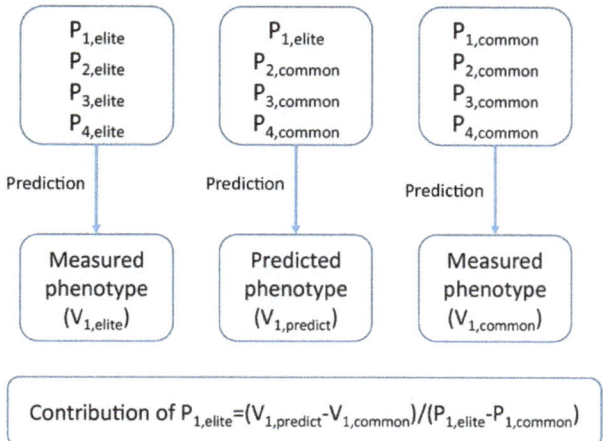

Figure 3. The routine for identifying the key parameters controlling the canopy photosynthetic light or nitrogen use efficiencies of a crop. $P_{1,elite}$, $P_{2,elite}$, ··· ··· $P_{1,common}$, $P_{2,common}$ are parameters for the elite or common cultivars. The $V_{1,elite}$ or $V_{1,common}$ are the predicted value of a particular phenotype. Synthetic cultivar is a hypothetical cultivar in which the value of parameter 1 from common cultivar ($P_{1,common}$) is used to replace the parameter value of the elite cultivar ($P_{1,elite}$).

be the first step (Lawson et al. 2012). This information, coupled with the genome wide association studies, QTL analysis, traditional genetics, and network inference approaches, can be used to identify the major alleles controlling photosynthetic efficiency under different environmental conditions (Fig. 2B). Once a genetics-based systems model for rice is established, the same systems approach can be extended for all major food and energy crops to guide breeding and engineering for enhancing yields (Chu 2015; Long et al. 2015, Zhu et al. 2011).

There are two potential applications of using a genomics-based model to guide the ideo-type design. On one side, the genetics-based model can be used to identify the most limiting step or parameter for light or nitrogen use efficiency for a particular crop grown at a particular region (Fig. 3). It can be used to identify optimal allele combinations to gain maximal CO_2 uptake for a particular crop species. This optimal allelic combination can then be used to guide parental line selection and marker-assisted breeding of new cultivars. On the other side, the genetics-based model can be used to explore the best breeding or engineering strategy for a particular rice accession. In other words, by parameterizing such a genetics-based model for a particular accession, we can use the model to identify the step exerting the highest control over canopy photosynthesis and further define the optimal allele to use to improve canopy photosynthetic efficiency in this particular rice line.

Summary

The heterogeneity of microclimate inside a canopy requires using a mechanistic model of canopy photosynthesis to identify the optimal architectural and physiological parameters to support modern crop breeding or breeding. A mechanistic model of canopy photosynthesis is now available which enables one to evaluate impacts of manipulating canopy architectural and physiological parameters on canopy photosynthesis. The model can be used to define region-specific optimal crop parameters and management practices. The challenge now is to develop a genetics-based model of canopy photosynthesis by incorporating functional relationship between allelic variations and canopy parameters. Such a genetics-based model holds great potential in guiding marker-assisted breeding or genomic selection in the post-genomics era.

Acknowledgments

The authors X-GZ, QS and CC thank the CAS strategic leading project "Modular Designer Crop Breeding" (grant # XDA08020301). MAJP acknowledges financial support from the BBSRC grants BB/I017372/1, BB/1024488/1, BB/I002545/1, BB/J/00426X/1 20:20 Wheat®.

Conflict of Interest

None declared.

References

Carmo-Silva, E., J. C. Scales, P. J. Madgwick, and M. A. J. Parry. 2015. Optimizing Rubisco and its regulation for greater resource use efficiency. Plant Cell Environ. 38:1817–32.

Cellier, P., and A. Olioso. 1993. A simple system for automated long-term Bowen ratio measurement. Agric. For. Meteorol. 66:81–92.

Chu, C. 2015. A new era for crop improvement – From model-guided rationale design to practical engineering. Mol. Plant. 8:1299–1301.

Donald, C. M. 1968. The breeding of crop ideotypes. Euphytica 17:385–403.

Driever, S. M., T. Lawson, P. J. Andralojc, C. A. Raines, and M. A. J. Parry. 2014. Natural variation in photosynthetic capacity, growth and yield in 64 field grown wheat genotypes. J. Exp. Bot. doi:10.1093/jxb/eru253.

Dugas, W. A. 1993. Micrometeorological and chamber measurements of CO_2 flux from bare soil. Agric. For. Meteorol. 67:115–128.

Dugas, W. A., D. C. Reicosky, and J. R. Kiniry. 1997. Chamber and micrometeorological measurements of CO_2

and H$_2$O fluxes for three C$_4$ grasses. Agric. For. Meteorol. 83:113–133.

van Eeuwijk, F. A., M. Malosetti, X. Y. Yin, P. C. Struik, and P. Stam. 2005. Statistical models for genotype by environment data: from conventional ANOVA models to eco-physiological QTL models. Aust. J. Agric. Res. 56:883–894.

Evans, L. T., and R. L. Dunstone. 1970. Some physiological aspects of evolution in wheat. Aust. J. Biol. Sci. 23:725–741.

Evans, J. R., and H. Poorter. 2001. Photosynthetic acclimation of plants to growth irradiance: the relative importance of specific leaf area and nitrogen partitioning in maximizing carbon gain. Plant, Cell Environ. 24:755–767.

Farquhar, G. D., S. von Caemmerer, and J. A. Berry. 1980. A biochemical model of photosynthetic CO$_2$ assimilation in leaves of C$_3$ species. Planta 149:78–90.

Hibberd, J. M., J. E. Sheehy, and J. A. Langdale. 2008. Using C$_4$ photosynthesis to increase the yield of rice– rationale and feasibility. Curr. Opin. Plant Biol. 11:228–231.

Hikosaka, K., and I. Terashima. 1995. A model of the acclimation of photosynthesis in the leaves of C$_3$ plants to sun and shade with respect to nitrogen use. Plant, Cell Environ. 18:605–618.

Johnson, D. A., N. Z. Saliendra, J. W. Walker, and J. R. Hendrickson. 2003. Bowen ratio versus canopy chamber CO$_2$ fluxes on sagebrush rangeland rangeland. J. Range Manag. 56:517–523.

Lawson, T., D. M. Kramer, and C. A. Raines. 2012. Improving yield by exploiting mechanisms underlying natural variation of photosynthesis. Curr. Opin. Biotechnol. 23:215–220.

Lin, M. T., A. Occhialini, J. P. Andralojc, M. A. J. Parry, and M. R. Hanson. 2014. A faster Rubisco with potential to increase photosynthesis in crops. Nature 513:547–550.

Long, S. P., A. M. Marshall, and X. G. Zhu. 2015. Engineering crop photosynthesis and yield potential to meet global food demand of 2050. Cell 161:56–66.

Monsi, M., and T. Saeki. 1953. Uber den Lichtfaktor in den Pflanzengesellschaf- u ur die Stoffproduktion. Jpn. J. Bot. 14:22–52.

Niinemets, U. 2007. Photosynthesis and resource distribution through plant canopies. Plant, Cell Environ. 30:1052–1071.

Ort, D. R., X. G. Zhu, and A. Melis. 2011. Optimizing antenna size to maximize photosynthetic efficiency. Plant Physiol. 155:79–85.

Ort, D. R., S. S. Merchant, J. Alric, A. Barkan, R. E. Blankenship, R. Bock, et al. 2015. Redesigning photosynthesis to sustainably meet global food and bioenergy demand. Proc. Natl Acad. Sci. USA 112:8529–8536.

Parry, M. A. J., M. Reynolds, M. E. Salvucci, C. Raines, P. J. Andralojc, X.-G. Zhu, et al. 2011. Raising Yield Potential of Wheat: (II) Increasing photosynthetic capacity and efficiency. J. Exp. Bot. 62:453–468.

Parry, M. A. J., P. J. Andralojc, J. C. Scales, M. E. Salvucci, H. Alonso, and S. M. Whitney. 2013. Rubisco Activity and regulation as targets for crop improvement. J. Exp. Bot. 64:709–715.

Pearcy, R. W. 1990. Sunflecks and photosynthesis in plant canopies. Ann. Rev. Plant Physiol. Plant Mol. Biol. 41:421–453.

Peng, S., G. S. Khush, and K. G. Cassman. 1994. Evaluation of a new plant ideotype for increased yield potential. Pp. 5–20 in K. G. Cassman, ed. Breaking the yield barrier: proceedings of a workshop on rice yield potential in favourable environments. International Rice Research Institute, Los Ban¯os, Philippines.

Peng, S. 2000. Single-leaf and canopy photosynthesis of rice. Studies Plant Sci. 7:213–228.

Peng, S., G. S. Khush, P. Virk, Q. Y. Tang, and Y. Zou. 2008. Progress in ideotype breeding to increase rice yield potential. Field. Crop. Res. 108:32–38.

dePury, D. G. G., and G. D. Farquhar. 1997. Simple scaling of photosynthesis from leaves to canopies without the errors of big-leaf models. Plant, Cell Environ. 20:537–557.

Reymond, M., B. Muller, A. Leonardi, A. Charcosset, and F. Tardieu. 2003. Combining quantitative trait loci analysis and an ecophysiological model to analyze the genetic variability of the responses of maize leaf growth to temperature and water deficit. Plant Physiol. 131:664–675.

Sellers, P. J., D. A. Randall, G. J. Collatz, J. A. Berry, C. B. Field, D. A. Dazlich, et al. 1996. A revised land surface parameterization (SiB2) for atmospheric GCMs.1. Model formulation. J. Clim. 9:676–705.

Song, Q.-F. 2004. Development, validation and application of integrated C$_3$ canopy photosynthesis models. PhD thesis. The Chinese Academy of Sciences.

Song, Q., H. Xiao, X. Xiao and X.-G. Zhu. 2016. A new canopy photosynthesis and transpiration measurement system (CAPTS) for canopy gas exchange research. Agric. For. Meteorol. (Accepted)

Song, Q., G. Zhang, and X.-G. Zhu. 2013. Optimal crop canopy architecture to maximise canopy photosynthetic CO$_2$ uptake under elevated CO$_2$ – a theoretical study using a mechanistic model of canopy photosynthesis. Funct. Plant Biol. 40:108–124.

Terashima, I., and J. R. Evans. 1988. Effects of light and nitrogen nutrition on the organization of the photosynthetic apparatus in spinach. Plant Cell Physiol. 29:143–155.

Wang, Y., and J. Li. 2008. Molecular basis of plant architecture. Annu. Rev. Plant Biol. 59:253–279.

Yin, X. Y., P. C. Struik, F. A. van Eeuwijk, P. Stam, and J. J. Tang. 2005. QTL analysis and QTL-based prediction of flowering phenology in recombinant inbred lines of barley. J. Exp. Bot. 56:967–976.

Yuan, L. 2001. Breeding of super hybrid rice. Pp. 143–149 *in* S. Peng and B. Hardy, eds. Rice research for food security and poverty alleviation. International Rice Research Institute, Los Banos, Philippines.

Zelitch, I. 1982. The close relationship between net photosynthesis and crop yield. Bioscience 32:796–802.

Zhu, X.-G., D. R. Ort, J. Whitmarsh, and S. P. Long. 2004a. The slow reversibility of photosystem II thermal energy dissipation on transfer from high to low light may cause large losses in carbon gain by crop canopies. A theoretical analysis. J. Exp. Bot. 55:1167–1175.

Zhu, X.-G., Jr. A. R. Portis, S. P. Long. 2004b. Would transformation of C_3 crop plants with foreign Rubisco increase productivity? A computational analysis extrapolating from kinetic properties to canopy photosynthesis. Plant Cell Environ. 27:155–165.

Zhu, X.-G., S. P. Long, and D. R. Ort. 2010. Improving photosynthetic efficiency for greater yield. Ann. Rev. Plant Biol. 61:235–261.

Zhu, X., G. Zhang, D. Tholen, Y. Wang, C. Xin, and Q. Song. 2011. The next generation models for crops and agro-ecosystems. Sci. China Inf. Sci. 54:589–597.

Zhu, X.-G., Q.-F. Song, and D. R. Ort. 2012. Elements of a dynamic systems model of canopy photosynthesis. Curr. Opin. Plant Biol. 15:237–244.

Zuo, J., and J. Li. 2013. Molecular dissection of complex agronomic traits of rice: a team effort by Chinese scientists in recent years. Nat. Sci. Rev. 1:253–276.

8

The potential of novel *Festulolium* (2n=4x=28) hybrids as productive, nutrient-use-efficient fodder for ruminants

Mike W. Humphreys, Sally A. O'Donovan, Markku S. Farrell, Alan P. Gay & Alison H. Kingston-Smith

Institute of Biological, Environmental and Rural Sciences, Aberystwyth University, Aberystwyth, Wales, SY23 3EE, United Kingdom

Keywords

Festuca arundinacea var. *glaucescens*, *Festuca mairei*, *Festulolium*, field performance, plant-mediated proteolysisplant-mediated proteolysis

Correspondence

Mike W. Humphreys, Institute of Biological, Environmental and Rural Sciences, Aberystwyth University, Aberystwyth, Ceredigion, Wales, SY23 3EE United Kingdom.

E-mail: mkh@aber.ac.uk

Funding Information

We thank HCC and EBLEX for funding Sally O'Donovan's PhD program at IBERS some of whose research outcomes on use of *Festulolium* hybrids for improved ruminant nutrition are presented here. We also acknowledge BBSRC for funding the research inputs from Alison Kingston-Smith, Mike Humphreys and Alan Gay. The IBERS field trials and seed multiplication were funded through an industrial DefraLINK Consortium (DefraLink Programme LK0688) led by Mike Humphreys and undertaken by Markku Farrell.

Abstract

The field performance and potential future use of F_1 *Lolium multiflorum* and *Lolium perenne* × *Festuca arundinacea* var. *glaucescens* and *Festuca mairei* hybrids (2n = 4x = 28) are described. Foliar trait expression in the hybrids was largely determined by the *Lolium* rather than their *Festuca* parent ensuring maintenance of high-forage quality. All four *Festulolium* populations comprised high-yielding genotypes, but the *L. multiflorum* populations were particularly erect and tall, while the *L. perenne* populations had significantly higher numbers of tillers and were prostrate. Forage yields of the *Festulolium* populations assessed in field plot trials were either not significantly different from, or were superior to leading *L. multiflorum* and *L. perenne* cultivars used as controls. Endogenous plant proteases contribute to excessive proteolysis in the rumen which causes environmental N pollution. Protein degradation due to plant-mediated proteolysis was assessed by in vitro exposure of leaves to the environmental conditions of the rumen (39°C, anaerobic) and calculated based on the time taken for protein levels to be reduced to half their original levels ($t^{1/2}$). Leaf proteins were significantly more stable in *L. multiflorum* × *F. arundinacea* var. *glaucescens* and *L. perenne* × *F. arundinacea* var. *glaucescens* F_1 hybrids ($t^{1/2}$ 18–21 h) than in their respective *Lolium* parental genotypes ($t^{1/2}$ 4–5 h), and there was a highly significant genome interaction. The $t^{1/2}$ in the majority of the *L. multiflorum* × *F. arundinacea* var. *glaucescens* F_1 hybrids studied often exceeded 24 h, whereas $t^{1/2}$ of their *Lolium* and *Festuca* parents was consistently <14 h. Although inferior to the F_1, *F. arundinacea* var. *glaucescens* genotypes tested had significantly greater $t^{1/2}$ than *L. perenne* under rumen-simulated conditions. Significant variation in protein stability was apparent within the F_1 and their respective parent species' groups. The initial protein content of the F_1 hybrids was lower than their respective parents, but following 24-h exposure to anoxia at 39°C, the protein content of both parent and hybrid genotypes was similar. The differences in protein stability between parental and hybrid genotypes was due to the greater rate of protein decline observed in the *Lolium* genotypes. Hence, uptake of these *Festulolium* hybrids as forage crops has potential to directly mitigate environmental impact of livestock farming without affecting production capacity.

Introduction

In mild temperate climates such as that found in the UK ryegrass (*Lolium* spp.) is often the forage grass of choice due to its high yield and nutritious value. However,

recently *Festulolium* varieties are increasingly gaining interest as sources of reliable, productive, and nutritive fodder for use in livestock agriculture and for their potential for ecosystem services (MacLeod et al. 2013). Importantly, *Festulolium* also has a higher tolerance to stresses such as drought or cold than perennial ryegrass (Ghesquière et al. 2010). *Festulolium* is the result of conventional hybridization of either *Lolium perenne* (perennial ryegrass) or *Lolium multiflorum* (Italian ryegrass) with any related *Festuca* (fescue) species, and may as such be marketed under its own grass category throughout Europe. *Festulolium* varieties may be either amphiploids with combined genome sets of ryegrass and fescue chromosomes, or they may be introgressive forms. In the latter, a limited number of donor gene sequences, most frequently derived from a fescue species, are incorporated into the recipient (ryegrass) genome through a backcross breeding program (Humphreys et al. 2003; Ghesquière et al. 2010). The IBERS-bred variety AberNiche, the first *Festulolium* to gain entry onto the UK National Recommended List is an example of an introgression form and is around 75% Italian ryegrass (*L. multiflorum*) and 25% meadow fescue (*Festuca pratensis*) (Cernoch and Kopecky, pers. comm.) while the French variety Lueur is an example of the amphiploid type with a more balanced ryegrass: fescue genome complement (Ghesquière et al. 2010). The variety Lueur derives from the hybridization of *L. multiflorum* (2n = 4x = 28) with *F. arundinacea* var. *glaucescens* (2n = 4x = 28), a species gaining interest due in particular to its drought and heat tolerance derived from its Mediterranean origin. The introgression breeding approach (Humphreys et al. 2005) has also been used to combine the complementary attributes of high-yielding ryegrass with the drought-tolerance of *F. arundinacea* var. *glaucescens*. This has been achieved through targeted marker-assisted transfers of a single small genome sequence of *F. arundinacea* var. *glaucescens* onto a terminal location of chromosome 3 (Humphreys et al. 2005). This fescue sequence has subsequently been transferred into breeders' lines of both Italian and perennial ryegrass (Humphreys et al. 2012).

Lolium multiflorum × *F. arundinacea* var. *glaucescens* (2n = 4x = 28) hybrid populations have been reported to be highly palatable with voluntary intake, in vivo digestibility of organic matter (DOM), and net energy expressed in fodder units for milk, similar to some of most palatable Italian ryegrass available at that time (Ghesquière et al. 1996). The drought resistance of the variety Lueur derives, at least in part, from its large deep root system and its ability to extract water from depth in the soil profile (Durand et al. 2007). *Festuca mairei* (Atlas fescue), which is related closely to *F. arundinacea* var. *glaucescens* and is indigenous to North Africa, is an alternative source of fescue genes for drought and heat tolerance and has been used (Wang and Bughrara 2005) as a source of novel genome variation for ryegrass through an introgression breeding approach similar to that described by Humphreys et al. (2005).

Farmers are increasingly seeking new grass varieties that provide resilience to climatic stresses, especially if these are accompanied by enhanced opportunities for environmentally sustainable livestock management. Described herein, for the first time is a comparison of the agronomic potential of four amphiploid *Festulolium* populations: *L. multiflorum* × *F. arundinacea* var. *glaucescens*, *L. perenne* × *F. arundinacea* var. *glaucescens* (2n = 4x = 28), and *L. multiflorum* × *F. mairei* and *L. perenne* × *F. mairei* hybrids (all 2n = 4x = 28). They were assessed initially over a 1-year field trial as spaced plants from which plants were selected to provide seed for a subsequent small replicated 1-year field plot trial where their performance was compared with that of elite ryegrass varieties.

In addition to measures of their field performance, the *L. multiflorum* × *F. arundinacea* var. *glaucescens*, *L. perenne* × *F. arundinacea* var. *glaucescens* (2n = 4x = 28), hybrid combinations were investigated for their potential, compared to ryegrass, to improve the efficiency of ruminant nutrition. Ruminant feeds are notorious for their inefficient use due to the poor conversion of ingested protein to milk and meat product and as a consequence, contribute to the high greenhouse gas emissions of nitrous oxide and environmental pollution with ammonia (Ripple et al. 2014). In fresh forage feeding situations, rapid postingestion rates of endogenous (plant) proteolysis can contribute to rates of protein breakdown in excess of that used by microbes (Zhu et al. 1999; Wallace et al. 2001; Kingston-Smith et al. 2010). This occurs during autolysis and cell wall breakdown (Edwards et al. 2008), and through an imbalance between protein supply and energy availability for microbial growth (Johnson 1976). Various plant breeding initiatives have attempted to reduce gaseous emissions by livestock either by encouraging increased rates of microbial N conversion using high sugar ryegrasses (Wilkins and Humphreys 2003), or by delaying the degradation of plant protein in ingested feed (e.g., by stimulating polyphenol oxidase (PPO) expression, Lee et al. 2004), to allow more time for N assimilation by the animal. Shaw (2006) reported that *F. arundinacea* var. *glaucescens* protein was more slowly degraded under rumen conditions and had far greater protein retention than either *L. perenne* or *L. multiflorum*. Shaw (2006) speculated that this may be due to the presence of protein protective mechanisms such as heat-shock proteins that had evolved in the fescue species sufficient to safeguard its adaptation to stresses encountered in Mediterranean conditions. Shaw (2006) also demonstrated slower breakdown of protein than its ryegrass parent in a F_1 *L. multiflorum* × *F. arundinacea* var. *glaucescens* (2n = 4x = 28) hybrid in rumen-simulated conditions. In order to verify and expand the findings of the pilot experiments undertaken by Shaw (2006), both the same and alternative

Lolium spp. × *F. arundinacea* var. *glaucescens* parental genotypes and their respective hybrids were grown under summer-like conditions in a controlled environment (CE), with plant-mediated proteolysis under the temperature and low oxygen conditions of the rumen assessed in vitro as described previously (Kingston-Smith et al., 2010). Shaw (2006) reported the protein retention of *F. mairei* to be inferior to that of *F. arundinacea* var. *glaucescens* when exposed to these conditions. For this reason, the *Lolium* spp. × *F. mairei* F1 hybrids were excluded from the protein stability assessment.

Materials and Methods

Production of *Festulolium* populations

The *L. multiflorum* × *Festuca arundinacea* var. *glaucescens* (2n = 4x = 28) hybrids were produced by hybridizing either autotetraploid *L. multiflorum* cvs Danergo, Gemini, and Roberta, and a Breeders' Line Bb2534 (all 4n = 4x = 28) with genotypes of *F. arundinacea* var. *glaucescens* (2n = 4x = 28) selected from a natural accession Bn354. The fescue derived from an original INRA, Plantes Fourrageres, Rouen, France collection extracted from an Alpine field location 700 m a.s.l. in the Hautes-Alpes, France.

The *L. perenne* × *F. arundinacea* var. *glaucescens* (2n = 4x = 28) hybrids were produced by hybridizing autotetraploid *L. perenne* cvs AberDell and Dunluce (both 4n = 4x = 28) with genotypes of *F. arundinacea* var. *glaucescens* (2n = 4x = 28) also selected from the natural accession Bn354.

The *L. multiflorum* × *F. mairei* (2n = 4x = 28) hybrids were produced by hybridizing either autotetraploid *L. multiflorum* cv Gemini or Bb2534 (an IBERS Breeders' line derived from a chromosome-doubled variety AberEpic) (both 4n = 4x = 28) with genotypes of *F. mairei* (2n = 4x = 28) derived from an IBERS Accession Bs3065 donated by CSIRO Canberra, Australia and collected from the Atlas Mountains in Morocco.

The *L. perenne* × *F. mairei* (2n = 4x = 28) hybrids were produced by hybridizing autotetraploid *L. perenne* cvs AberDell and Dunluce (both 4n = 4x = 28) with genotypes of *F. mairei* (2n = 4x = 28) also selected from the natural accession Bs3065.

For brevity species names and species' hybrids will be described throughout by the following nomenclature: *L. multiflorum* (4x) = Lm; *L. perenne* (4x) = Lp; *F. arundinacea* var. *glaucescens* (4x) = Fg; *L. multiflorum* × *F. arundinacea* var. *glaucescens* (F_1)(4x) = LmFg; *L. perenne* × *F. arundinacea* var. *glaucescens* (F_1)(4x) = LpFg. *F. mairei* (4x) = Fm; *L. multiflorum* × *F. mairei* (F_1) (4x) = LmFm; *L. perenne* × *F. mairei* (F_1)(4x) = LpFm.

Embryo rescue incorporating use of modified Gamborg and Miller B5 Medium (as described in Humphreys et al. 2005) was employed throughout to generate all the LmFg, LpFg, LmFm, and LpFm F_1 hybrids. The LmFg populations used in the field study derived from seven parental Lm genotypes of variety Danergo, six genotypes of both variety Gemini and Bb2534, and one genotype of Roberta. The Lm genotypes were hybridized onto six alternative Bn354 Fg genotypes to generate the LmFg populations. Four of these Fg genotypes were also used as parents with three genotypes of Lp variety Dunluce, and one genotype of Ba14076 and of AberDell to generate all the LpFg populations. The LmFm populations derived from three parental Lm genotypes of Bb2534, and two of variety Gemini which were hybridized onto four alternative genotypes of Fm Accession Bs3065. One of these Fm plants together with an alternative Fm genotype from Accession Bs3065 were used with two Lp genotypes of variety AberDell and two Lp cv Dunluce genotypes (also used to produce LpFg) to generate all the LpFm populations.

Genotypes of LmFg, LpFg, LmFm, and LpFm following embryo rescue were transferred to pots filled with potting compost and established to maturity in a frost-free glasshouse. Following growing conditions sufficient for vernalization and inflorescence induction (>10 weeks of short days and temperatures of circa 4–10°C) and prior to ear emergence 50 F_1 LmFg genotypes, 34 F_1 LpFg genotypes, 23 F_1 LmFm genotypes, and 27 LpFm F_1 genotypes were transferred in their four groups to separate pollen-proof glasshouses to interpollinate. Seed from each polycross was germinated to produce populations for field assessment as spaced plants.

Field assessments

Spaced-plant field trial

For the field study, four populations each of 300 genotypes of LmFg, LpFg, LmFm, and LpFm hybrids were established as individual spaced plants in a field at IBERS, Aberystwyth University in mid-Wales, UK. Heading date was recorded in accordance with IBERS field assessment protocols as number of days between 1st April and ear emergence. Aftermath heading was scored as 1 = no additional inflorescence to 9 many secondary inflorescences. Other phenotype measures were growth habit (based on a scale of 1 = erect; 5 = prostrate), plant height (cm) at ear emergence, leaf width (based on scale 1 = narrow; 9 = wide), plant width at ear emergence (cm), tiller density (based on a scale of low tiller density = 1; high tiller density = 9), disease score (based on the presence of any rust [*Puccinia* spp.]) infection (score 1 = highly infected; 9 = no infection), and on their comparative plant size recorded at 5 time points throughout the growing season (score 1 = small; 9 = large). The first cut was taken on 27 July 2011 with the second on the 1 September 2011.

Fifty high-yielding plants were selected from each population based on average fresh weight over the two harvests, and forage quality of this subset of plants was determined. Forage quality was measured as %water soluble carbohydrate (WSC), %dry organic matter digestibility (DOMD), and %nitrogen and total protein content.

Small-plot field trial

The subsets of 50 plants were then extracted from the field, repotted in potting compost and transferred to pollen-proof isolation houses for seed multiplication. Seed from each LmFg, LpFg, LmFm, and LpFm polycross was sown in a second field trial as three replicate field plots 1 × 3 m at IBERS. Three replicate plots of control varieties Lm cv Danergo, and Gemini (4n = 4x = 28) and Lp cv AberGlyn and AstonEnergy (4n = 4x = 28) were also sown with all plots randomized. Seed of all tetraploid varieties and populations was sown at 3.3 g/m² (in accordance with standardized NIAB and IBERS field protocols). The harvests of dry matter forage yield (DMY) were compared over six cuts taken in 2013. Forage quality: (%WSC, %dry matter digestibility (DMD) and %N and crude protein were assessed at Cuts 1, 2, and 4 using NIR technologies and in complete accordance with the standard IBERS protocols. The percentage ground cover at the end of the growing season for all grasses was compared. Detailed meteorological records for IBERS field trials in 2011 and 2013 are provided in: https://share.aber.ac.uk/dept/ibers/intranet/research/weather/default.aspx.

All field data were analyzed according to standard procedures with the menu-driven options within Genstat 13.2 for Windows (VSN International Ltd., http://www.vsni.co.uk) software. For the spaced-plant field trial, least significant differences (LSD) between LpFg, LpFm, LmFg, and LmFm population means (P < 0.05) were calculated for heading date, growth habit, plant height, leaf width, plant width, tiller density, disease score, and for plant size. For the small-plot field trial, LSD (P < 0.05) was calculated for forage yield during each of 6 cuts, and also for forage quality (%WSC, %DMD, %N) at cuts 1, 2, and 4 between the LpFg, LpFm, LmFg, and LmFm populations and their respective tetraploid Lolium control cultivars Lm cv Danergo and Gemini and Lp cv AstonEnergy and AberGlyn.

In vitro determination of plant-mediated proteolysis

The plants selected for protein analysis comprised clonal replicates of the same genotypes of Fg (4x), Lm (4x), and

Lp (4x) and their respective LmFg and LpFg F₁ hybrids (all 4x) used as parents for progeny assessed in the field study. They are listed in Table 1. All plants were maintained in 6″ pots in Levington's multipurpose compost under identical conditions in a frost-free glasshouse at IBERS under natural illumination and watered, and when required, fertilized, cut, and repotted to encourage active and consistent plant growth throughout. Plants for protein analysis were maintained to achieve an equivalent ontogeny and with no indication of inflorescence induction to minimize potential interactions due to age difference. To further enhance consistency, all plants were transferred into a CE facility (Gallenkamp PLC Monarch Way, Belton Park, Loughborough, UK and Skye Instruments Lighting, Ddole Enterprise Park, Llandrindod Wells, Powys, UK) and acclimated to constant UK summer conditions for a minimum of 6 weeks under 18 h light at 600 μmol.m^{-2}.sec^{-1} at 22°C, and with 6 h darkness at 14°C, all at 72% humidity.

Mature nonsenescent leaves at equivalent developmental stage were selected from each plant genotype, removed and incubated over three concurrent days producing three replicate results. Replicate groups of parent plants and their F₁ were incubated on separate weeks due to the large number of samples. The leaves of each plant genotype were cut 5 cm above the soil into 1 cm lengths and weighed to provide equal 0.1 g fresh-weight samples for 4 incubation time points; 0, 2, 6, and 24 h. These time points were selected based on previous results (as in Shaw 2006). Leaf sections were transferred into Hungate tubes and filled with 5 mL of anaerobic buffer warmed to 39°C (Van Soest 1967) and, except for those at 0 h, capped quickly under a stream of CO_2, and placed into a water bath heated to 39°C. The tubes containing incubated plant

Table 1. Genotypes of Lolium multiflorum (Lm), and L. perenne (Lp) varieties hybridized onto Festulolium glaucescens var. arundinacea (Fg) (all 4x = 28) to produce allotetraploid F1 LmFg and F1 LpFg hybrids (2n = 4x = 28).

Lolium spp. parent variety/plant no.	Festuca glaucescens parent accession/ plant no.	LmFg and LpFg (2n = 4x = 28)
Lm cv Roberta/1[1]	X Bn354/4[1]	= LmFg1[1]
Lm cv Gemini/2	X Bn354/8	= LmFg2
Lm cv Gemini/2	X Bn354/8	= LmFg3
Lm cv Gemini/5	X Bn354/8	= LmFg4
Lm cv Danergo/6	X Bn354/35	= LmFg5
Lm cv Danergo/9	X Bn354/35	= LmFg6
Lp cv AberDell/5	X Bn354/15	= LpFg1
Lp cv AberDell/5	X Bn354/8	= LpFg2
Lp cv Dunluce/5	X Bn354/17	= LpFg3
Lp cv Dunluce/5	X Bn354/17	= LpFg4
Lp cv Dunluce/6	X Bn354/8	= LpFg5

[1] Also used previously by Shaw (2006).

material were removed from the water bath after 2, 6, and 24 h, respectively. Plant material was recovered under vacuum filtration and rinsed with deionized water. Samples were then placed into microcentrifuge tubes, flash frozen in liquid N, and stored at −80°C.

Procedures for protein extraction and measurement

Batches of circa 30 incubated samples in microcentrifuge tubes were removed from the −80°C freezer, placed on ice, and transferred into racks in a freeze dryer for 24 h. The samples were then stored in the dark prior to milling. For milling, freeze-dried samples with two tungsten beads added were placed into 2×24 sample milling boxes and milled in a Retsch (GmbH Haan Germany) MM 300 mill at a frequency of 30 sec for 1 min on each side following rotation of the boxes. Where necessary supplementary hand grinding was applied to ensure that the samples were completely homogeneous and ground fully and uniformly. Processed samples were subsequently stored in a cool, dark, and dry place in preparation for protein extraction.

Protein in the ground residue was extracted by grinding in a mortar and pestle with prechilled (4°C) extraction buffer (0.1 mol/L HEPES, pH 7.5 containing 1 mmol/L EDTA, 0.1% (v/v) Triton-X 100 and 0.5% protease inhibitor cocktail (Sigma UK Ltd, Gillingham, UK) and 2 mmol/L dithiothreitol) added to the samples at a ratio of 40 μL/g dry weight. The sample homogenate was transferred into individual 2 mL microcentrifuge tubes, flash frozen in liquid N, and stored at −80°C, until protein analysis.

Homogenate samples were thawed on ice and samples centrifuged at 13,000g at 4°C for 10 min and protein content of supernatant was determined with the Bio-Rad Protein Assay kit (Bio-Rad UK Ltd., Hemel Hempstead, UK) against a BSA calibration curve (working range 0–5 μg). Sample volumes between 1 and 10 μL were used to ensure results remained within range of the standard curve and water and reagent controls in a total volume of 200 μL were included to determine background absorbance. The absorbance was read at 595 nm on a BioTek ElX 808 microtiter plate reader (Bio Tek (Bedfordshire, UK), Fisher Scientific) after an incubation time of 15–20 min at room temperature.

The protein degradation time courses were fitted with exponential decay curves of the form

$$c = br^t \qquad (1)$$

where c is protein content (mg g^{-1} dry weight), t is time from start of incubation, (hours), b is a fitted parameter describing protein content at time zero, and r is a fitted parameter describing the rapidity of the decay of protein.

This equation was rearranged to allow calculation of the time taken in hours for protein to decay to half its value at time 0 ($t^{1/2}$).

$$t^{1/2} = \frac{log_e 0.5}{log_e r} \qquad (2)$$

This parameter was chosen as the greater the $t^{1/2}$, the greater the resistance of the plant proteins to degradation. Nonlinear curve fitting was performed using a Maximum Likelihood Program (Ross 1987) and parallel curve analysis (Ross 1990) was used to determine significant differences between fitted curves, and to estimate standard errors of fitted parameters and $t^{1/2}$.

Results

Fertility in the LmFg, LpFg, LmFm, and LpFm F_1 genotypes employed as parents in their respective polycross combinations was high and more than sufficient to provide seed for the four plant populations used in the initial field study. The second seed production program all used 50 plants/population selected from the first field trial and as such provided an accurate comparison of population seed-set. Overall differences in seed production between the LpFm, LmFm, and LmFg populations were insignificant (total seed produced 277 g (LpFm), 279 g (LmFm), 271 g (LmFg), but total seed production amongst the LpFg was significantly lower (seed produced = 200 g; χ^2 = 17.29, $P < 0.001$).

Spaced-plant field trial

The mean values for the plant traits scored in the spaced-plant field trial and any significant difference ($P < 0.05$) found between the four populations are shown in Table 2. While variation for heading date was evident within each amphiploid hybrid combination, overall no significant difference was observed between the early-heading populations LpFm and LmFg (both population mean date for heading: day 37). However, populations LmFm and LpFg were both significantly later heading ($P < 0.05$, population mean dates for heading: day 38 and day 43, respectively). Aftermath heading was low and was not significantly different in either of the Lp-based populations (LpFg and LpFm), but was significantly greater ($P < 0.05$) in both the Lm-based populations, especially in LmFg.

Although there were significant differences ($P < 0.05$) between all four populations, overall growth habit was distinct and different between the very erect Lm-based and the far more prostrate Lp-based *Festulolium* populations. LmFg genotypes were the most erect and were significantly more erect than LmFm ($P < 0.05$). Conversely, LpFg was more prostrate than LpFm ($P < 0.05$) indicating that growth habit was determined more by the *Lolium*

Table 2. Mean plant traits scored in a 1 year spaced-plant field trial of 300 plant *Festulolium* populations LmFg, LpFg, LmFm, and LpFm; all 2n = 4x = 28.

Grass hybrid	Mean heading date (day no.)	Aftermath heading score = 1–9	Growth habit score = 1–5	Leaf width score = 1–9	Plant height (cm)	Tiller number score = 1–9	Plant width (cm)	Disease resistance score = 1–9
LmFg (4x)	37 a	6.62 c	1.57 a	7.23 c	62.20 c	6.41 b	43.58 b	5.65 a
LpFg (4x)	43 c	1.85 a	4.21 d	3.18 a	22.00 a	6.10 ab	56.81 d	6.12 b
LmFm (4x)	38 b	5.92 b	1.77 b	6.54 b	51.30 b	5.46 a	38.56 a	6.37 b
LpFm (4x)	37 a	1.80 a	3.87 c	3.28 a	22.20 a	5.76 a	53.27 c	6.22 b
LSD (P < 0.05)	0.89	0.27	0.14	0.22	1.80	0.30	2.12	0.31
SED	0.45	0.14	0.07	0.11	0.91	0.15	1.07	0.16

For each value within a column, populations having $P < 0.05$ difference are indicated by an alternative letter.

than the *Festuca* parent. LmFg was taller than LmFm ($P < 0.05$). The Lm-based populations were taller than the more prostrate LpFg and LpFm populations.

There was considerable within population variation in plant size but species' effects were evident. While the Lm-based *Festulolium* populations were taller, the overall plant width of the Lp-based populations was greater. The mean plant width of LpFg was greater than that found in LpFm ($P < 0.05$), while width of the Lm-based populations LmFg was larger than LmFm ($P < 0.05$). While differences in plant habit, height, and breadth related to their *Lolium* parent, differences in plant mean tiller number in the *Festulolium* populations corresponded more to their *Festuca* species parent. Significantly higher tiller numbers ($P < 0.05$) were observed in the Fg-based populations LmFg and the LpFg, which were themselves not significantly different. The Fm-based populations LmFm and LpFm with a lower tiller frequency were not significantly different.

Leaf width was determined more by the *Lolium* parent than by the *Festuca* parent, with Lm-based hybrid leaves being significantly broader than those involving Lp. The mean leaf width for LmFg was significantly greater than for LmFm ($P < 0.05$), but both Lm-based populations had significantly wider leaves ($P < 0.01$) than LpFg and LpFm, which were not significantly different.

Rust (predominantly *Puccinia coronata*) infection was compared amongst the four amphiploid hybrid populations. Variation in disease susceptibility was evident throughout, but overall, infection was low with 90% of all plants in the field trial either with low or no infection (scores 6–9). However, LmFg was more susceptible than the other three *Festulolium* populations and had significantly higher frequencies of rust infection ($P < 0.05$).

Plants in all four plant populations continued to grow and increased in plant size throughout the growing season with the 50 largest genotypes from each population selected for seed multiplication (Table 3). The *Festulolium* populations containing Fg (LmFg and LpFg) comprised

larger plants than the corresponding populations with Fm (LmFm and LpFm, respectively), but the influence of their *Lolium* parent was also evident. The LmFg commenced growth and developed more extensively during the spring as compared to the other three populations, but it was the LpFg population that demonstrated most growth later in the growing season.

The dry matter yields and forage quality measures for the 50 plant LmFg, LmFm, LpFg, and LpFm selections are presented in Table 4. The mean dry weight for cut 1 (27 July 2011) for the 50 selected plants from LmFg was significantly higher ($P < 0.05$) than for LmFm which in turn was significantly higher ($P < 0.05$) than for LpFm and LpFg. The mean dry weights from each population at cut 2 (1 September 2011) were more comparable, with LmFg and LmFm not significantly different, but both were superior to LpFm and LpFg ($P < 0.05$).

The %WSC of the Lp-based populations, LpFg and LpFm was not significantly different at cut 1 while LmFg and LmFm were significantly ($P < 0.05$) lower. The superior %WSC of the Lp-based populations was maintained at cut 2 ($P < 0.05$), but between the Lm-based populations, LmFm was significantly higher in %WSC than LmFg. The %DMD of the Lp-based populations were not significantly different and also were significantly higher ($P < 0.05$) than those involving Lm both at cut 1 and cut 2. The %N and total protein were also higher in the Lp-based populations over both cut 1 and cut 2

Table 3. Mean ranking of plant size (1 = small – 9 = large) in diverse *Festulolium* populations during the course of a growing season.

Score date	LmFg	LmFm	LpFg	LpFm	LSD P < 0.05 (SED)
14 march	5.62 a	4.82 b	4.21 c	3.78 d	0.338 (0.17)
20 April	6.52 a	5.49 c	5.97 b	5.83 b	0.325 (0.16)
14 July	6.62 a	5.95 b	6.67 a	6.06 b	0.274 (0.14)
23 August	6.45 b	5.86 c	6.9 a	6.38 b	0.291 (0.15)

Significant differences ($P < 0.05$) in plant size between populations within each row on each date are indicated by alternative letters.

Table 4. Mean dry matter yield (DMY), %water soluble carbohydrate (WSC), %dry matter digestibility (DMD), and %nitrogen (N) [and total protein*] of 50 genotypes of *Festulolium* populations LmFg, LpFg, LmFm, and LpFm over two consecutive harvests during summer and autumn 2011; (SE).

Harvest date	Population	Yield DMY(g) (SE)	%WSC (SE)	%DMD (SE)	%N [+total protein*] (SE)
27 July 2011 Cut 1	LmFg (4x)	112.8 a (25.6)	19 b (2.5)	65 b (2.0)	1.5 c (0.3) [9.38]
	LpFg (4x)	48.3 g c (14.0)	23 a (3.4)	79 a (2.0)	2.4 a (0.3) [15]
	LmFm (4x)	93.9 g b (28.1)	19 b (2.2)	66 b (2.4)	1.7 b (0.2) [10.63]
	LpFm (4x)	53.3 g c (12.2)	23 a (4.1)	80 a (2.3)	2.5 a (0.3) [15.63]
LSD (*P* < 0.05)	Cut 1	8.15	1.24	0.86	0.10
1 September 2011 Cut 2	LmFg (4x)	73.2 g a (18.8)	22 c (2.5)	69 b (2.7)	1.6 b (0.2) [10]
	LpFg (4x)	58 g b (13.3)	25 a (3.4)	77 a (2.1)	2.1 a (0.3) [13.13]
	LmFm (4x)	68.1 g a (18.4)	24 b (2.6)	70 b (3.3)	1.6 b (0.3) [10]
	LpFm (4x)	60.9 g b (11.4)	26 a (3.7)	77 a (2.6)	2.0 a (0.2) [12.5]
LSD (*P* < 0.05)	Cut 2	5.95	1.24	1.10	0.10

For each harvest, within column values with the same letter were not significantly different at *P* < 0.05.

(P < 0.05). The %N and total protein of LmFg and LmFm at cut 2 was the same but at cut 1 was lower in LmFg (P < 0.05).

Small plot field trial

Spring 2013 had prolonged low temperatures delaying growth with the consequence that the first harvest was delayed until 24th May. Five further cuts were made with the final cut taken on 24th October. Throughout the growing season, the performance of all four *Festulolium* populations compared well with that of the control varieties (Fig 1). DMY for LmFg and LmFm and the Lm control varieties early in the year (cuts 1 and 2) did not differ significantly. Likewise LpFg and LpFm although lower in yield than their Lm-based counterpart populations (P < 0.05) did not differ for DMY from their Lp control varieties. By cut 3 LmFg had a higher yield than LmFm (P < 0.05) but neither was significantly different from the Lm control varieties. The DMY of LpFm and LpFg at cut 3 did not differ significantly. The *Festulolium* populations did not differ in yield from the highest yielding Lp control variety (Aston Energy). At cut 4 the DMY of LmFg and LmFm and the highest yielding Lm control (variety Gemini) did not differ significantly. Likewise there was no significant difference in DMY between either LpFg and LpFm and the Lp control varieties. By cut 5 and 6, the forage yield of all varieties (except the inferior yielding control Lp cv AberGlyn in cut 5) irrespective of whether they were Lm- or Lp-based did not differ significantly.

The forage quality of the *Festulolium* populations harvested on Cut 1 (24th May), Cut 2 (25th June), and Cut 4 (3rd August) was consistently equivalent to or better than their respective Lm and Lp (4x) controls. The %DMD of LpFg although the highest in the field trial was not significantly superior to LpFm and Lp cv Aston Energy. All were superior to the Lp control AberGlyn

(P < 0.05). LmFg and LmFm had similar %DMD to the tetraploid control varieties Lm cv Danergo and cv Gemini. The %WSC of LmFm and LmFg was equivalent to Lm control cvs Danergo and Gemini. LpFm and LpFg had the same %WSC content of Lp control variety Aston Energy with all superior (P < 0.05) to Lp cv AberGlyn. There was no significant difference (P > 0.05) in %N content and total crude protein between LpFg, LpFm, and the two Lp tetraploid cultivars Aston Energy and AberGlyn. Similarly, there was no significant difference (P > 0.05) in %N and total crude protein content between LmFg, LmFm, and the Lm controls Danergo and Gemini.

The %ground cover of the Fg-based populations LpFg and LmFg was superior to all grasses used in the field trial and was significantly greater than their respective controls Lp cv Aston Energy and AberGlyn and Lm cv Gemini (P < 0.05).

Protein losses through plant-mediated proteolysis

Figure 2 shows that the initial protein content of Lm and Fg was higher than in their F_1 hybrid progeny and differed significantly (P < 0.001). Similarly, the initial protein content of Lp and Fg was higher (P < 0.001 for Lp, P < 0.01 for Fg), than their respective F_1 progeny. Within both the *Lolium* and the *Festuca* species' groups the initial protein content of the constituent genotypes also differed significantly (P < 0.001). There was significant variation in the initial protein content between plants within the LpFg (P < 0.05) and LmFg (P < 0.001) F_1 hybrid groups. Despite their initial difference in protein content, following 24 h exposure to the temperature and oxygen conditions of the rumen, there remained no significant difference in protein content between *Lolium* and *Festuca* parent and their respective hybrid genotypes.

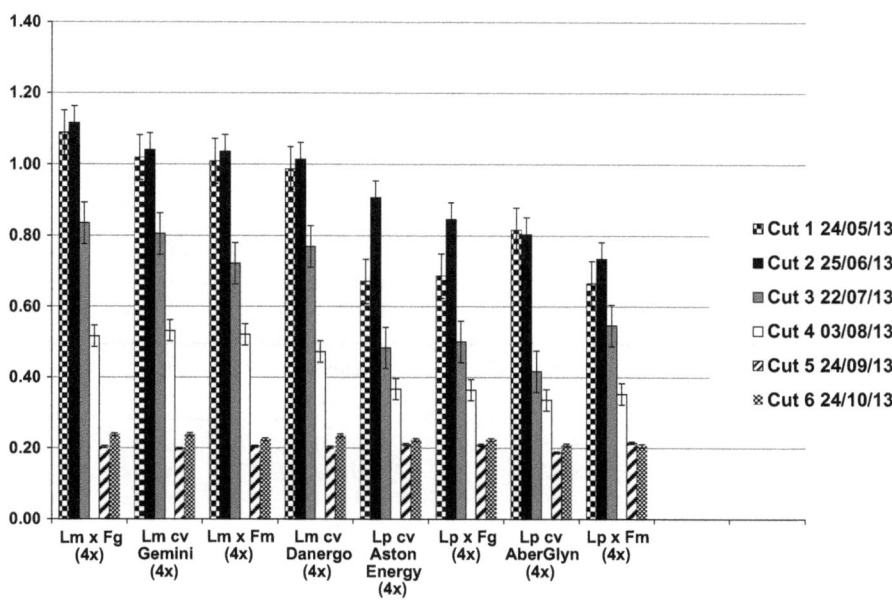

Figure 1. Dry matter yield (DMY kg) of field plot trials of *Festulolium* populations LmFg, LmFm, LpFg, and LpFm (2n = 4x = 28) compared with tetraploid *Lolium multiflorum* (Lm) and *L. perenne* (Lp) cultivar controls. Ranked by total yields (DMY kg).

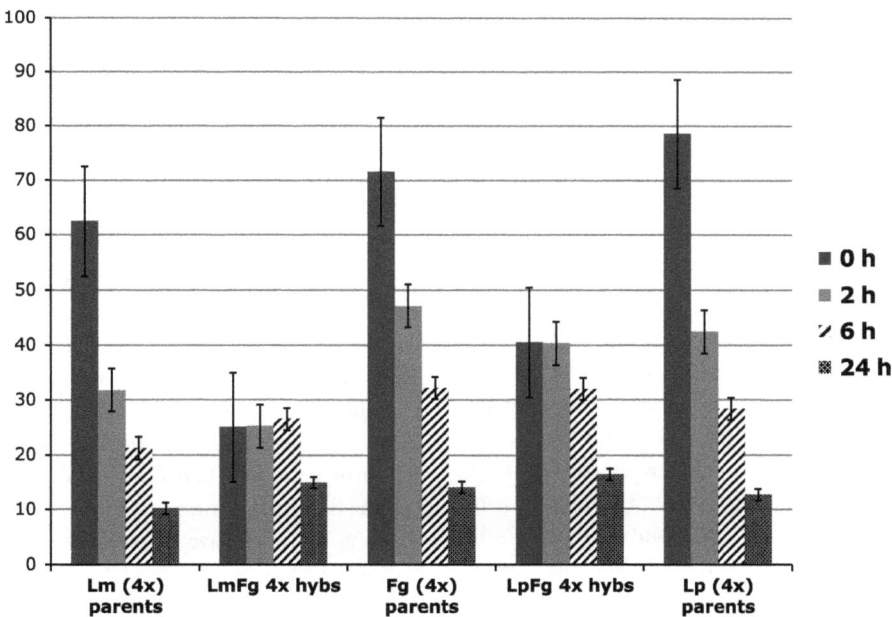

Figure 2. Protein content (mg/g DW) of *Lolium multiflorum*, *L. perenne*, *Festulolium arundinacea* var. *glaucescens* parent genotypes, and their *L. multiflorum* × *F. arundinacea* var. *glaucescens* and *L. perenne* × *F. arundinacea* var. *glaucescens* F_1 hybrid progeny when exposed in vitro to 0, 2, 6, and 24 h at 39°C under anoxia.

A comparison of $t^{1/2}$ of each Lm, Fg parent genotype, and their respective F1 progeny, and of Lp, Fg, and their respective F1 progeny is presented in Table 5. In addition, and using the combined data for all time points, replicates, and genotypes for each parent and hybrid, the total $t^{1/2}$ values were fitted (Fig. 3). Overall, the $t^{1/2}$ of the Fg parents

was significantly higher ($P < 0.05$) than the Lp (4x) genotypes (4.5 h) used to produce the LpFg F_1 hybrids. Similarly, $t^{1/2}$ of the Fg parents were higher than Lm but in this case, the difference was not significant ($P < 0.3$). However, $t^{1/2}$ for both F_1 hybrid groups (21 h and 18 h LmFg and LpFg, respectively) were significantly greater than their

Table 5. Protein half-lives ($t^{1/2}$) (SE) of (i) *Lolium multiflorum* (Lm), *Festulolium arundinacea* var. *glaucescens* (Fg), and their F1 progeny (all 2n = 4x = 28), and (ii) *L. perenne* (4x), *F. arundinacea* var. *glaucescens* (Fg), and their F1 progeny (all 2n = 4x = 28) when exposed in vitro to 24 h of anoxia at 39°C.

L. multiflorum (Lm) (4x)	Lm half-life ($t^{1/2}$) (SE)	X *F. glaucescens* (Fg) (4x)	Fg half-life ($t^{1/2}$) (SE)	= F1 LmFg (4x)	LmFg half-life ($t^{1/2}$) (SE)
Lm cv Roberta/1[1]	3.23 h (1.53)	X *Fg Bn354/4*[1]	8.12 h (2.41)	= LmFg1[1]	9.20 h (2.15)
Lm cv Gemini/2	8.70 h (3.26)	X *Fg Bn354/8*	13.70 (3.57)	= LmFg2	44.1 h ([2])
Lm cv Gemini/2	8.70 h (3.26)	X *Fg Bn354/8*	13.70 (3.57)	= LmFg3	21.15 h (10.19)
Lm cv Gemini/5	18.55 h (18.93)	X *Fg Bn354/8*	13.70 (3.57)	= LmFg4	25.29 h (10.86)
Lm cv Danergo/6	3.06 h (0.82)	X *Fg Bn354/35*	4.68 (0.92)	= LmFg5	37.99 h ([2])
Lm cv Danergo/9	1.53 h (0.40)	X *Fg Bn354/35*	4.68 (0.92)	= LmFg6	61.65 h ([2])

L. perenne (Lp) (4x)	Lp half-life ($t^{1/2}$) (SE)	X *F. glaucescens* (Fg) (4x)	Fg half-life($t^{1/2}$) (SE)	= F1 LpFg (4x)	LpFg half-life ($t^{1/2}$) (SE)
Lp cv AberDell/5	6.04 h (1.65)	X *Fg Bn354/15*	7.40 h (3.77)	= LpFg1	16.96 h (5.91)
Lp cv AberDell/5	6.04 h (1.65)	X *Fg Bn354/8*	13.70 (3.57)	= LpFg2	21.94 h (13.26)
Lp cv Dunluce/5	9.15 h (3.35)	X *Fg Bn354/17*	3.71 (1.02)	= LpFg3	26.63 h (12.19)
Lp cv Dunluce/5	9.15 h (3.35)	X *Fg Bn354/17*	3.71 (1.02)	= LpFg4	20.63 h (7.52)
Lp cv Dunluce/6	1.96 h (0.41)	X *Fg Bn354/8*	13.70 (3.57)	= LpFg5	9.91 h (2.46)

[1]Lm, Fg, and LmFg genotypes used previously in Shaw (2005).
[2]high SE due to protein $t^{1/2}$h values that far exceed the 0–24 h sample time points for protein measures.

respective *Lolium* parent groups (Lm, 4.1 h $P < 0.001$ and Lp 4.5 h, $P < 0.01$). LpFg hybrids had significantly greater $t^{1/2}$ than their Fg parent genotypes ($P < 0.05$) but the differences in $t^{1/2}$ between LmFg and Fg were not significant ($P < 0.15$). There were significant differences in $t^{1/2}$ within the Lm ($P < 0.001$), and the Lp ($P < 0.001$), but not between the Fg ($P < 0.25$) genotypes (Table 5). There were also significant differences in $t^{1/2}$ within both the LmFg ($P < 0.001$) and LpFg ($P < 0.05$) hybrid groups. While collectively Fg genotypes had greater $t^{1/2}$ than Lp ($P < 0.05$), they did not have greater $t^{1/2}$ than the Lm group ($P < 0.3$) Within the species' groups, individual genotypes of Lm and Lp were identified (Lm cv Gemini/5 and Lp cv Dunluce/5) as having superior $t^{1/2}$ compared to certain Fg genotypes (Bn354/17; Bn354/35).While the overall $t^{1/2}$ for LmFg F$_1$ was 21 h, genotypes with considerably higher $t^{1/2}$ values were identified (Table 5). All data on protein content were confined to time points within 0–24 h exposures to the in vitro stress conditions and the observation in three LmFg F$_1$ genotypes of $t^{1/2}$ calculations that extended well beyond the 24 h data set led to high SE for the genotypes concerned (Table 5) which carried forward when all LmFg F$_1$ were combined (Fig 3).

Discussion

Breeding for *Festulolium* varieties has over recent years gained increased importance as an aide to combat climate change and to achieve more sustainable grassland agriculture. This is because it is possible to capture in one variety the agronomic value of ryegrass and the resilience, water, and nutrient-use-efficiency found in different fescue species (Ghesquière et al. 2010). Initially, most advances and cultivars marketed involved Italian ryegrass (*L. multiflorum*) and meadow fescue (*F. pratensis*), a species combination known taxonomically as *Festulolium braunii*. However, in recent years especially at IBERS attention has moved more to employing alternative fescue species such as *F. arundinacea* var. *glaucescens* and more recently *F. mairei* (Humphreys et al. 2013). These have provided novel sources of genes for improved drought resistance, water-use-efficiency, and deep rooting for ryegrass (Humphreys et al. 2005, 2013). The plant–soil interactions generated by certain deep rooting *Festulolium* species combinations including those described herein and others such as *L. perenne* × *F. pratensis* (*Festulolium loliaceum*) can provide other benefits including ecosystem services such as flood mitigation (Humphreys et al. 2013; MacLeod et al. 2013) and have potential to increase soil organic carbon capture by grassland (Kell 2011).

While the potential of *Festulolium* to improve grassland persistency has long been recognized, for reasons such as genome instability (Canter et al. 1999), high costs of seed production compared to ryegrass, and inferior forage quality (Ghesquière et al. 2010) *Festulolium* has not yet been widely used. For these reasons, doubts surrounding its commercial development remain and these limitations must be overcome before *Festulolium* varieties become marketed widely and their full benefits realized. However, recent advances in *Festulolium* breeding technologies combined with an increased awareness of the need to find alternatives to ryegrass to better combat climate change and to

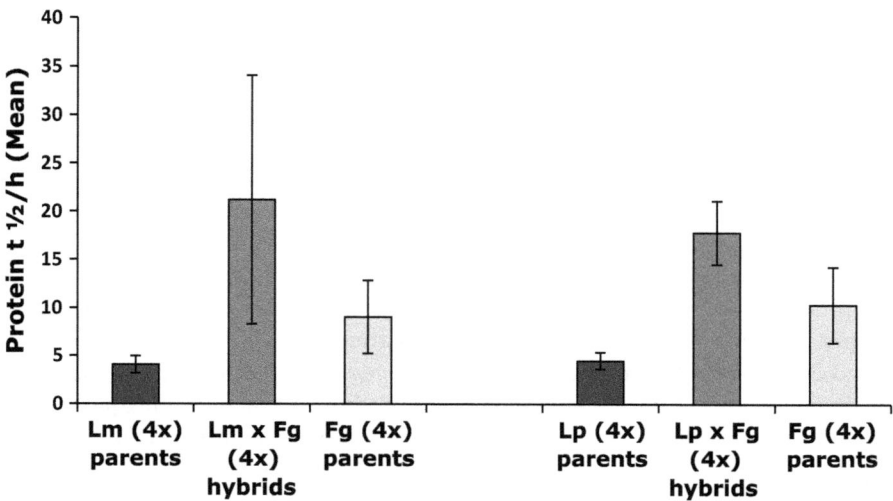

Figure 3. Comparisons of protein stability between tetraploid *Lolium perenne* (Lp), *L. multiflorum* (Lm), and *Festulolium arundinacea* var. *glaucescens* (Fg), and their respective F1 progeny LpFg and LmFg (2n = 4x = 28) when exposed in vitro to 24 h at 39°C under anoxia.

achieve more sustainable grassland systems have encouraged *Festulolium* development and led to the marketing of the cultivar AberNiche in the UK, an expected forerunner for others under development and trial. AberNiche is a synthetic form of *Festulolium braunii* comprising genome combinations of *L. multiflorum* and *F. pratensis*. It was exluded from the current study as in earlier work (Shaw, 2005), *F. pratensis* demonstrated a similar plant-mediated proteolysis to that recorded for *Lolium* spp., and was significantly inferior to the protein retention found in *F. arundinacea* var. *glaucescens* when exposed to the stress conditions applied in the current study. In an alternative field trial at IBERS AberNiche, *L. multiflorum* × *F. arundinacea* var. *glaucescens* (LmFg), and *L. multiflorum* × *F. mairei* (LmFm), all Lm-based tetraploid *Festulolium* hybrids, produced an equivalent forage yield throughout their first year harvests (Humphreys, unpubl.). The current work represents a comprehensive field and laboratory study of four new amphiploid *Festulolium* hybrid combinations, *L. multiflorum* × *F. arundinacea* var. *glaucescens* (LmFg), *L. perenne* × *F. arundinacea* var. *glaucescens* (LpFg), *L. multiflorum* × *F. mairei* (LmFm), and *L. perenne* × *F. mairei* (LpFm), and provides evidence to support their commercial development both in terms of their agronomic performance and their potential to improve ruminant nutrient-use-efficiency and thereby to reduce adverse environmental emissions by livestock. One of the species' combinations has already been developed for commercial use in France. A *L. multiflorum* × *F. arundinacea* var. *glaucescens* variety Lueur equivalent to the LmFg population used here has been developed commercially by INRA (Ghesquière et al. 1996, 2010) but in trials at IBERS has inferior yield to LmFg (Humphreys, unpubl.).

Two possible strategies for increasing efficiency of conversion of forage N to microbial N have been used to improve ruminant nutrition and decrease emissions of greenhouse gases by livestock and both have received considerable recent attention at IBERS over recent years. The first strategy aims at increasing the amount of readily accessible energy during the early part of the fermentation in the rumen. The second strategy aims to increase the protection of forage proteins, and thereby reducing the rate at which their breakdown products are made available to the colonizing microbial population. The high sugar grasses (HSG) are examples of the former where increased WSC has been shown to have a positive impact on meat yields (Lee et al. 2001) and milk production (Miller et al. 2001). The incorporation in legumes of protein protection methods such as increased PPO expression is an example of the second approach as applied to conserved forage (Lee et al. 2004) while increasing the $t^{1/2}$ for protein degradation in the rumen would increase N use efficiency by grazing ruminants, as discussed below.

Extensive nitrogen loss to the environment due to poor incorporation of dietary N by ruminants causes both pollution of ground water and contributes to nitrous oxide production. This is due to excessive proteolysis in the rumen, for which plant-mediated proteolysis is a contributory factor (Zhu et al. 1999; Wallace et al. 2001; Kingston-Smith et al. 2005) and hence current efforts to mitigate impact of ruminant farming through selection of improved forage genotypes. Shaw (2006) demonstrated that Fg was significantly more stable under rumen-like conditions than *Lolium* species. She also demonstrated that Fg was more stable than Fm when exposed to equivalent rumen-simulated trials. Although Shaw (2006) reported benefits in terms of improved protein stability in F_1 hybrids compared with Lm, these were lost during the course of a backcross breeding program identical to that described in Morgan et al. (2001). The current work sought to extend and verify the study described in Shaw (2006) by exploring the variation within Lp, Lm, and Fg for protein stability under rumen-like conditions and to determine the extent at which this trait is expressed in *Festulolium* F1 hybrids. Evidence is presented here to show (i) significant variation for protein stability within Lm, Lp, and Fg tetraploid genotypes, (ii) that the range of protein stability in Lm, Lp, and Fg genotypes has significant overlap, (iii) significant *Lolium-Festuca* genome interactions and transgressive segregation in F_1 hybrids between Fg and both Lm and Lp that gave rise to significantly higher protein stability than that expressed by either *Lolium* parent genomes.

Shaw (2006) proposed that protein-protection mechanisms that had evolved in Fg to combat the high temperatures experienced in Mediterranean locations were providing equivalent benefits when grass was exposed to the stresses encountered in the rumen. The current work indicates that the genetic control for protein stability is complex and demonstrates significant variation within and overlapping protein stability between Lm, Lp, and Fg populations. Heterosis between the *Lolium* and *Festuca* genomes in their F_1 hybrid forms provided improved protein stability compared with the *Lolium* and Fg parents. The initial protein content of the Lp, Lm, and Fg parent genotypes was consistently higher than that observed in their progeny. However, following exposure to 24 h of rumen-simulated conditions, despite significant difference in their protein half-lives, there was little or no actual difference in residual protein content between parent and hybrid genotypes. A similar relationship between initial low protein content and slow rates of protein degradation was seen previously with white clover (Kingston-Smith et al. 2006). While the outcome in terms of protein content in both *Lolium* and the *Festulolium* hybrid genotypes at 24 h was similar, the rate of protein decline was significantly greater in the *Lolium* genotypes. Protein degradation in the early time period following ingestion of forage is considered to be important. Despite the availability of peptide and amino acid substrates for microbial growth at this stage, the availability of energy will be determined (and possibly limited) by the extent of microbial colonization and cell wall degradation (Johnson 1976; Edwards et al. 2008). Hence, decreasing plant-mediated proteolysis has the potential to improve delivery of protein and energy by the feed. As protein building blocks are nonlimiting the main consequences of decreasing plant-mediated proteolysis would be predicted to be decreased activity of HAP bacteria and, on a whole animal level, increased N partitioning to product and away from urine. It will be important to now extend the current work by employing animal studies to fully assess the benefits of LmFg and LpFg hybrids as feed both in terms of their potential for improved ruminant nitrogen-use-efficiency and livestock gain and also for environmental gain by limiting N losses into the environment.

In field trials, genotypes of four *Festulolium* populations; LmFg, LpFg, LmFm, and LpFm were selected for seed multiplication; seed set from these and subsequent selections was far in excess of that required for the initial spaced plant trial and subsequent plot trial experiments. The two Lm-based populations were large and erect and contrasted with the two Lp-based populations which were large and high tillering and consistently prostrate. For the majority of foliar traits, it was the *Lolium* (Lm or Lp) rather than the *Festuca* (Fg or Fm) parent that was the determining factor in trait expression as reported previously in a root phenotype study (Humphreys et al. 2013). In the earlier work, root ontogeny was found to be more dependent on the presence of an Lm or an Lp genome rather than whether the accompanying genome complement was Fg or Fm. In the field plot trial, all four *Festulolium* populations compared favorably and were not significantly different in yield or forage quality from their respective Lm and Lp (4x) control varieties demonstrating the absence of any suggestion of transfers of deleterious forage characters from their fescue parent.

The combined field-based and in vitro proteolysis study provides evidence that both LpFg and LmFg amphiploid hybrids offer considerable potential for sustainable grassland agriculture. The LmFm and LpFm populations provided similar benefits and would be expected to be particularly drought and heat tolerant (Wang and Bughrara 2005). The root systems of all four populations combined the high growth rate and branching of *Lolium* with the root strength and depth of Fg and Fm (data not shown, Humphreys et al. 2013). The impact of these *Festulolium* hybrids in terms of plant–soil interactions is being assessed currently and will be compared with earlier research (MacLeod et al. 2013) for potential ecosystem service benefits. Taken together, the outcomes of the current research provide compelling evidence for benefits both for agriculture and the environment for future use of *Festulolium* hybrids.

Acknowledgments

We thank HCC and EBLEX for funding Sally O'Donovan's PhD program at IBERS some of whose research outcomes on use of *Festulolium* hybrids for improved ruminant nutrition are presented here. We also acknowledge BBSRC for funding the research inputs from Alison Kingston-Smith, Mike Humphreys, and Alan Gay. The IBERS field trials and seed multiplication were funded through an industrial DefraLINK Consortium (DefraLink Programme LK0688) led by Mike Humphreys and undertaken by Markku Farrell.

Conflict of Interest

None declared.

References

Canter, P. H., I. Pasakinskiene, R. N. Jones, and M. W. Humphreys. 1999. Chromosome substitutions and recombination in the amphiploid *Lolium perenne* × *Festuca pratensis* cv Prior (2n = 4x = 28). Theor. Appl. Genet. 98:809–814.

Durand, J. L., T. Bariac, M. Ghesquiere, P. Brion, P. Richard, M. W. Humphreys, et al. 2007. Ranking of the depth of water extraction by individual grass plants using natural ^{18}O isotope abundance. Environ. Experi. Bot. 60:137–144.

Edwards, J. E., S. A. Huws, E. J. Kim, A. H. Kingston-Smith, and N. D. Scollan. 2008. Advances in microbial ecosystem concepts and their consequences for ruminant agriculture. Animal 2:653–660.

Ghesquière, M., J.-C. Emile, J. Jadas-Hécart, C. Mousset, R. Traineau, and C. Poisson. 1996. First in vivo assessment of feeding value of *Festulolium* hybrids derived from *Festuca arundinacea* var. *glaucescens* and selection for palatability. Plant Breed. 115:238–244.

Ghesquière, M., M. W. Humphreys, and Z. Zwierzykowski. 2010. *Festulolium*(Chapter 12). Pp. 293–316 *in* F. Veronesi, U. Posselt, B. Beart, eds. Handbook on plant breeding, Eucarpia fodder crops and amenity grasses section. Springer-Verlag New York Inc., ISBN: 9781441907592, 5.

Humphreys, M. W., P. J. Canter, and H. M. Thomas. 2003. Advances in introgression technologies for precision breeding within the *Lolium - Festuca* complex. Ann. Appl. Biol. 143:1–10.

Humphreys, J., J. A. Harper, I. P. Armstead, and M. W. Humphreys. 2005. Introgression-mapping of genes for drought resistance transferred from *Festuca arundinacea* var. *glaucescens* into *Lolium multiflorum*. Theor. Appl. Genet. 110:579–587.

Humphreys, M. W., A. H. Marshall, R. P. Collins, and M. A. Abberton. 2012. Exploiting genetic and phenotypic plant diversity in grasslands. (Chapter 16) Pp. 148–157 *in* G. Lemaire, J. Hodgson, A. Chabbi, eds. Grassland Productivity and Ecosystem Services. CAB International (Pub), Oxfordshire, UK, ISBN-13:978 1 84593 809 3.

Humphreys, M. W., M. S. Farrell, A. Detheridge, J. S. Scullion, A. Kingston-Smith, and S. A. O'Donovan. 2013.

Resilient and multifunctional grasslands for agriculture and environmental service during a time of climate change. Pp. 335–337 *in* A. Helgadottir, A. Hopkins, eds. The Role of Grasslands in a Green Future. Threats and Perspectives in Less Favoured Areas. Grassland Science in Europe Vol. 18. Agricultural University of Iceland, Hvanneyri Borgarnes, Iceland. ISBN 978-9979-881-20-9.

Johnson, R. R. 1976. Influence of carbohydrate solubility on non-protein nitrogen utilization in the ruminant. J. Anim. Sci. 43:184–191.

Kell, D. B. 2011. Breeding crop plants with deep roots: their role in sustainable carbon, nutrient and water sequestration. Ann. Bot. 108:407–418.

Kingston-Smith, A. H., R. J. Merry, D. K. Leemans, H. Thomas, and M. K. Theodorou. 2005. Evidence in support of a role for plant-mediated proteolysis in the rumens of grazing animals. Br. J. Nutr. 93:73–79. doi: 10.1079/BJN20041303

Kingston-Smith, A. H., A. L. Bollard, and F. R. Minchin. 2006. The effect of nitrogen status on the regulation of plant-mediated proteolysis in ingested forage; an assessment using non-nodulating white clover. Ann. Appl. Biol. 149:35–42.

Kingston-Smith, A., J. E. Edwards, S. A. Huws, E. J. Kim, and M. Abberton. 2010. Plant-based strategies towards minimising "livestocks' long shadow". Proc. Nutr. Soc. 2000:1–8.

Lee, M. R., E. L. Jones, J. M. Moorby, M. O. Humphreys, M. K. Theodorou, J. C. MacRae, et al. 2001. Production responses from lambs grazed on *Lolium perenne* selected for an elevated water-soluble carbohydrate concentration. Anim. Res. 50:441–450.

Lee, M. R. F., A. L. Winters, N. D. Scollan, R. J. Dewhurst, M. K. Theodorou, and F. R. Minchin. 2004. Plant-mediated proteolysis in red clover with different polyphenol oxidase activities. J. Sci. Food Agric. 84:1639–1645.

MacLeod, C. J. A., M. W. Humphreys, R. Whalley, L. B. Turner, A. Binley, C. W. Watts, et al. 2013. A novel grass hybrid to reduce flood generation in temperate regions. Scientific Reports 3; 1-7.

Miller, L., J. M. Moorby, D. R. Davies, M. O. Humphreys, N. D. Scollan, J. C. MacRae, et al. 2001. Increased concentration of water soluble carbohydrate in perennial ryegrass (*Lolium perenne* L.): milk production from late lactation dairy cows. Grass Forage Sci. 56:383–394.

Morgan, W. G., I. P. King, S. Koch, J. A. Harper, and H. M. Thomas. 2001. Introgression of chromosomes of *Festuca arundinacea* var. *glaucescens* into *Lolium multiflorum* revealed by genomic in situ hybridisation (GISH). Theor. Appl. Genet. 103:696–701.

Ripple, W. J., P. Smith, H. Haberl, S. A. Montzka, C. MvAlpine, and D. A. Boucher. 2014. Ruminants, climate change and climate policy. Nat. Climate Change 2014: 2–5.

Ross, G.J.S. 1987. Maximum likelihood program, version 3.08. Oxford: Numerical Algorithms Group.

Ross, G. J. S. 1990. Nonlinear estimation. Springer, New York, NY.

Shaw, R.K. 2006. Effect of gene transfer from *Festuca* to *Lolium* on plant-mediated proteolysis. In IBERS: animal and Microbial Sciences, Vol PhD. Aberystwyth University, Wales, UK.

Van Soest, P. J. 1967. Development of a comprehensive system of feed analyses and its application to forages. J. Anim. Sci. 26:119–128.

Wallace, R. J., C. J. Newbold, B. J. Bequette, J. C. MacRae, and G. E. Lobley. 2001. Increasing the flow of protein from ruminal fermentation – Review – Asian-Australasian. J. Anim. Sci. 14:885–893.

Wang, J. P., and S. Bughrara. 2005. Evaluation of drought tolerance for Atlas fescue, perennial ryegrass, and their progeny. Euphytica 164:113–122.

Wilkins, P. W., and M. O. Humphreys. 2003. Progress in breeding perennial forage grasses for temperate agriculture. J. Agric. Sci. 140:129–150.

Zhu, W.-Y., A. H. Kingston-Smith, D. Troncoso, R. J. Merry, D. R. Davies, G. Pichard, et al. 1999. Evidence of a role for plant proteases in the degradation of herbage proteins in the rumen of grazing cattle. J. Dairy Sci. 82:2651–2658. doi: 10.3168/jds.S0022-0302(99)75522-0

9

Food security and food self-sufficiency in China: from past to 2050

Bishwajit Ghose

Institute of Nutrition and Food Science, University of Dhaka, Dhaka, Bangladesh

Keywords
China, food price, food security, grain, self-sufficiency.

Correspondence
Bishwajit Ghose, Institute of Nutrition and Food Science, University of Dhaka, Dhaka, Bangladesh.
E-mail: brammaputram@gmail.com

Funding Information
No funding information provided.

Abstract

Reducing hunger and malnutrition and improving food security have come to the forefront of global political agenda. In the wake of recent spells of food price hike, national and supranational development organizations and governments have begun to express serious concerns about the world's capacity to feed its burgeoning population. In response to the target of increasing food production by 70% in 2050, many countries are formulating their agricultural policies to promote domestic food self-sufficiency and many are building international networks for outsourcing food supply beyond national borders. In the face of massive demand for food for its growing population, China is both strengthening its food self-sufficiency strategies and relying on large-scale imports from international market which has been a major driver of food price inflation in recent years. In China, increase in income and socioeconomic status on one hand have dramatically improved dietary intake and overall nutritional status of the population, and are creating an enormous pressure on land and water resources and natural environments on the other. Maintaining food and water security and for its huge population with its limited resources while at the same time sustaining the economic growth momentum are offering significant challenges to China's macroeconomic prospects. China's domestic food production and self-sufficiency status have certain repercussions on the volatility of global agri-food market and food security in food-import and food-aid-dependent countries. The objectives of this study is to provide an in-depth overview of China's food production and demand scenario with a particular focus on its challenging perspectives toward food secure 2050. The first half of the paper is designed to show recent trends in food production and consumption and the impacts on global food market. The second half describes the major challenges with a brief discussion on policy implication.

Introduction

Since the 2007–2008 global food price hike, which pushed around 400 million more people into poverty, increasing agriculture's capacity to feed burgeoning global population has become at the forefront of global development agenda. High food price and food scarcity have already caused severe social and political unrest in many countries in Asia, Africa, and South America and is likely remain a worldwide concern for the next 50 years and beyond (Rosegrant and Cline 2003). Many researchers have explained the event from many perspectives such as drought in major wheat-producing nations, ban on grain export by many countries, rising per capita income, depreciation of dollar against foreign currencies, demand for corn for ethanol in Europe and the United States. Rising demand for food in India and China was indicated by some for the increase in world food prices. Increasing dependence on import and ban on export of food grains by Asian countries, especially India and China is very likely to aggravate global food crisis globally and can have significant implications for the poverty and malnutrition in the world (Keijiro 2013). The foundation of the republic in 1949 and the economic reform in late 1980s have been two most phenomenal events in China's agricultural history. Economic reform commenced with agrarian reform, and the introduction of the "Household Responsibility System" (家庭聯產承包責任制: Jiā tíng lián chǎn chéng bāo zé rèn zhì) has brought dramatic improvements in agricultural production and productivity. Since then China has been able to grow most of the food it needs and has become a number one producer and exporter of many agricultural products. In recent years, China's socioeconomic progress coupled with rapid demographic and nutrition transition however have accelerated the demand for food and which contributed to a decline in self-sufficiency rate and a sudden rise in food price. Food prices in China present a similar pattern to the global food price inflation and interest rates were raised several times in 2011 to combat this inflation (Zhihao and Shida 2012). As of the end of September 2011, the overall consumer price index was 13% higher than in 2010, with sharp price increases of all type of foods- food grains (12%), meat (28%), eggs (14%), seafood (14%), fresh vegetables (2%), fresh fruits (6%), and oils and fats (18%) (Zhihao and Shida 2012). Increase in food price has considerable implications for consumers who spend a high proportion of income for food. Aside from maintaining a robust economic growth rate, China has made great strides in improving its food situation during last few decades. Feeding one fifth of the world's population from less than a tenth of its arable land and freshwater is a daunting task, but this too has come at an expense of severe land and water pollution and many other environmental externalities. Biodiversity and environmental concerns have already reached a point to influence agricultural, food and trade policy making, and resource allocation strategies. To feed its 1.3 billion population with a per capita cultivated land far below the world average, China is already facing a great challenge of land scarcity. Farmland declined by about 11% between 1978 and 2006 (Fleming 2009). And accelerated urbanization along with explosive economic growth has further worsened the shortage of agricultural land over the last two decades (Jie 2007). While some agronomists and environmentalists believe that China should import more land-intensive food to reduce pressure on its already strained land and water resources, others express fear over the fact that China's long-term dependency on foreign exports will fuel food price increase and worsen the food insecurity status in many resource poor countries who largely rely on foreign imports (Minghua and Yan 2009; Feng et al. 2010). However, till now, maintaining a grain self-sufficiency of 95% is the central theme of China's food security blueprint. Since its accession to WTO in 2001, as world's largest agri-food market, China has gradually become a key player in global agri-food policy making. Today, with 9% of world's arable land and 8% freshwater resources, China produces 18% of the world's cereal grains, 29% of the world's meat, and 50% of the world's vegetables (Jiang 2007). China's agriculture is supporting a population of over 1.3 billion people today, compared to about 500 million in 1950, is projected to feeding around 1.4 billion by 2050 with a shrinking land base and water resource. As the population's economic status is improving, the quantity and quality of food consumed has also increased and average household income almost tripled since 1989. Caloric intake now stands at 2830 kcal/person/day, putting China near the top of the developing countries and approaching the levels of high-income countries. On the one hand, remarkable income increase had been encouraging demands for more animal protein, high-quality vegetables and fruits in addition to sufficient subsistence grains, on the other hand, accelerated urbanization is enlarging the markets of high-quality food products because the urban population in China normally has a much richer, more diverse diet than the rural population (Jie 2007). Leveraging intensification, introducing better irrigation techniques, more transparent land administration, bioremediation, and better environmental protection strategies to curb pollution seem to be vital for ensuring sustainable food security for China in 2050.

Self-Sufficiency in Grain

The term "Food security" translates literally as grain security in Chinese (粮食安全: Liáng shí ān quán). Grain sufficiency has always been at the heart of national food security agenda in China as in most other Asian countries (Ghose and Sajeeb 2013). Figure 1 shows the trend in grain production and consumption in China. Its global importance is realized by the fact that Chinese agriculture supports staple food supply for around 22% of global population and produces 30%, 15%, and 17% of global production of rice, wheat, and corn, respectively (Xiong and Conway 2009). Rice is the most important grain followed by wheat and corn and together account for 99% of total grain production. While Corn enjoys highest productivity of all types of grain, soybean production is lowest in comparison with other major grains (Fig. 2).

China's economic reform began with restructuring of the agricultural sector with a great emphasis on cereal grain production. Since decollectivization began in China in the late 1970s, both production and productivity of the farms have increased significantly and regional grain self-sufficiency was a major driving factor behind such gains (Justin et al. 1995). China's history as a grain importer is long and it is also hard to know last when China was self-sufficient in grain, and the Grain import program of China has long been a controversial topic among many

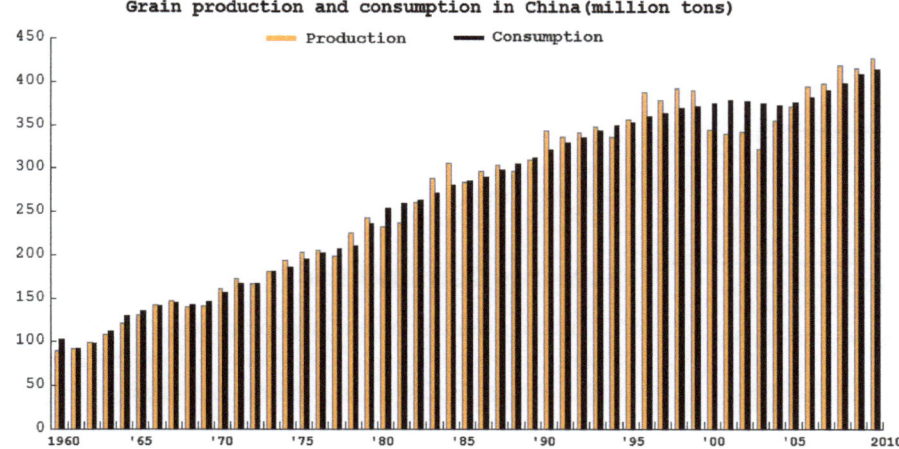

Figure 1. Trend in total domestic grain production in China since 1960. It reveals that total production remained higher than total consumption in most of the years except during the period between 2000 and 2005. Source: FAO STAT, 2010.

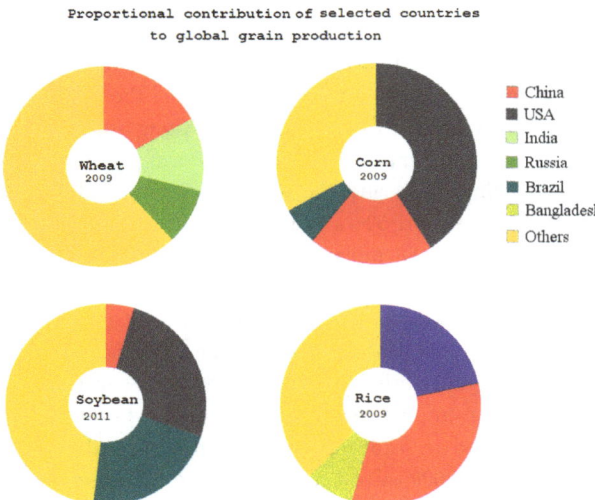

Figure 2. China's relative rice, wheat, corn, and soybean production status. It reveals that China is more self-sufficient in rice and wheat than corn and soybean. Comparatively lower production of soybean is a main reason why its which is shown in Figure 3. Source: Calculated from FAO and IFPRI data.

scholars (Xiao-yuan et al. 1995). In 1980s, grain import was attributed primarily to the inadequacy of domestic cereal production by some (Mah 1971) and to improved diet pattern by others (Ishikawa 1977). Today, when China boasts to be world's leading cereal grain producer, it still remains a net importer of grains and the causes behind this constant dependency despite huge domestic production have become more complex and inevitable in some cases.

Global urban population exceeded their rural counterpart in 2008 and China took only three more years to achieve this trend. Urbanization was a major driving factor behind the increased demand for food as it brought changes in demand for agricultural products both from increases in urban populations and from changes in their diets and demands (David et al. 2010). China is now world's second largest consumer of oil and the third largest producer and consumer of biofuel. Considering the need for energy to support the scale of industrialization China is undergoing, its decision based on trade-offs between outsourcing energy and achieving food self-sufficiency is a rather hard one. The situation is more complex for China compared to other major producers of biofuel because of its relatively the low per capita availability of agricultural land (Table 1) and a huge population burden. Though most of China's biofuel is produced from animal fat or waste vegetable oil, the cost of production is still higher compared to other sources of fuel (gasoline) and imported ones due to expensive land usage. Moreover, China's decision to produce biofuel has been criticized on the grounds that it will worse poverty (Qiu et al. 2012) and will have negative impact on food security and environment (Tatsuji 2013). It has been two decades now since some Chinese experts accepted that China will be forced to import 10% or more of its grain supply in the coming decades when it first became a net importer of oil for the first time in history in 1993 (Xiong and Conway 2009). China's grain production has increased from about 200 kg per capita in 1949 to about 400 kg in the early 1990s (Jianhua 2011) and per capita food supply rose from 2328 calories per day in 1980 to 3029 calories in 2000, an overwhelming 30% increase in a space of two decades (Carter 2011). Grain production in 2010 was 80% above the 1978 level and since the 1980s until 1999, self-sufficiency rate of grain in terms of weight was never below 95%. The rate dropped sharply in 2000, and went slightly below 90% during 2001 to 2003 period (Junichi and Jing 2013).

Table 1. Land statistics comparison of BRIC countries (km²/1000 population).

	Land area	Agricultural land	Arable land	Water surface	Forest area
China	7.0	4.1	1.1	0.2	1.5
Brazil	41.6	13.0	2.9	0.3	23.2
India	2.5	1.5	1.3	0.3	0.6
Russia	118.0	15.5	8.8	5	58.3

Table 1 shows different types land and water availability status of the BRIC countries. It is clear that China's land resources suitable for agricultural production and water availability is lower than average for most categories. Source: World Bank.

China is facing many far-reaching challenges to maintain its planned grain self-sufficiency of 95%, most prominent of them include loss of cultivated land, limited water resources, frequent natural disasters, impacts of climate change, vulnerable ecosystems, increased demand from population growth, improved standard of living, and outdated agricultural infrastructure (Minghua and Yan 2009). Some researchers suggest that to maintain a sustainable supply of cereal grains for its population, China may have to import increasing amounts of grain and this could trigger unprecedented rises in food prices globally (Lester and Linda 1995). In 2011, China harvested the largest grain crop of any country in history and 2 years later it became the largest importer of rice. Despite its grain output hit a record high of 601.94 million tons in 2013, China had to import another 15 million tons. China's population is projected to peak at 1.5 billion in 2033. If it is assumed that grain food per capita will increase from today's 400 to 470 kg in 2033, it will be necessary to increase grain production by at least 35% during the next 20 years (Jianhua 2011). A study reveals that the attainment of a 95% self-sufficiency rate would be quite challenging for China, unless the terms of trade in agriculture improve substantially in favor of producers (Junichi and Jing 2013).

Soaring Demand for Meat

To become world's largest meat-producing country from a state of limited ration of meat only for urban citizens, Chinese meat industry has gone through drastic changes in past three decades (Guanghong 2012). Although China has maintained a high degree of grain sufficiency, its seems to be influenced greatly by rising demand for animal feed. Though rice and wheat production is currently close to target, demand for corn and soybean has soared dramatically and account for bulk of the grain imports. In the 1960s about 80% of the corn was used for direct human consumption while in 1994, about 64% of the total corn production was used for animal feeds (Xiong and Conway 2009). Today, China's 70% of total corn production is used as animal feed and the demand is projected to rise by slightly less than 5% a year. China already reached its import limit of 2.88 million tons of corn this year and is not expected to get more corn until the end of the year. China is now world's largest single importer of soybean (Fig. 3) and the biggest customer of the United States who is currently the world's largest producer of soybeans. In 2013, soybean imports reached a record high of 60 million tons compared to 42.55 million tons in 2009 and 10 million tons in 2000. Corn used for feed increased from

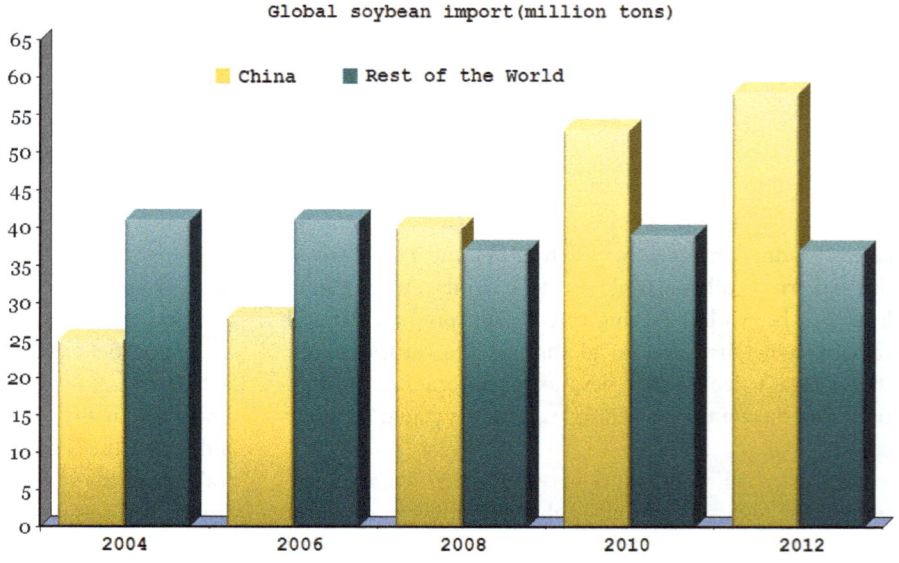

Figure 3. Trend in global soybean trade since 2004. China has been a major soybean importer since 2004 and since 2008 China's import amounted higher than those of all others nations combined. This explains the soaring demand for meat and meat products in China during last decade. Source: FAO, World Economic Outlook.

53 million metric tons (mmt) in 1990 to 93 mmt in 2002 at an annual growth rate of 4.7%. Use of soybean for feed grew even faster, from 1.03 mmt in 1990 to 16.65 mmt in 2002 at an annual growth rate of 30% (Steven et al. 2004). China has opened up soybeans imports to meet its soaring demand for animal feed in the late 1990s which brought about mass deforestation in Latin America and millions of local small-scale meat producers were moved out of the industry. In 2010, China replaced the United States as world's largest feed grain user and in 2011, its net import was five times as high as total domestic production (14 mmt). In 2012, China saw a 20-fold increase in soybean import since 1998 representing 64% of global soybean trade.

As people's socioeconomic status continues to rise, so does their demand for meat- and animal-based products. In 1978, China's meat consumption was one third of amount the consumed in the United States. Per capita meat consumption increased to 58.2 kg in 2009 from 14.6 kg in 1980. In 1990, total meat production was 30.42 million tons which was the largest in the world and per capita consumption reached world average in 1994 (Guanghong 2012). Between 1996 and 2007 total meat production went up by 50%, egg production by 30%, and milk production by 200%. In 1992, China overtook the United States as the world's leading meat consumer and today, around 27% of all the meat produced worldwide is consumed in China. Improving social and economic conditions, of course, leads to an ever higher consumption rate for both rural and urban people (Fleming 2009). Meeting increased demand for meat and dairy products will continue to be a challenge for China as these products require more land and water resources per unit production (Carter 2011). It takes about 16,000 L of water for 1 kg beef, 6000 and 3500 L, respectively, for same amount of

pork and chicken. Pork accounts for around 75% of total meat consumption in China (Fig. 4). In 2013, global pork consumption totaled 107 million tons and half of that occurred in China which is roughly six times as much pork as consumed in the United States. Many researchers have attributed the shocking rise in global grain prices in 2008 to China's increased demand for meat. Since the 1990s, rising incomes, especially in cities, have led to significant increases in demand for nonstaple animal-based foods such as meat, fish, and dairy products. Given that the number of higher income earners is expected to increase rapidly, demand for nonstaple food will also expand, increasing pressure on domestic production and import from other countries. Though meat industry is developing quickly, there is a significant lack of development in private enterprises, government, research institutes and the standards of fresh meat and meat products is still less than optimal (Guanghong 2012).

Fish

Globally, fish provides an important source of food and nutrition security. In 2009, fish accounted for 6.5% of all protein consumed. In china, fisheries sector has always played a major role in national food security. Even in the 1950s when fisheries output was just above 3 mmt, it used to contributed significantly to national food supply. Since 1980, China experienced the fastest growth in annual per capita fish consumption with its per capita consumption increased by more than 300% and at an annual average growth rate of 9%. Today, China is world's largest aquaculture and marine producer (Fig. 5) and has the highest number of fishers and fish farmers. China's share in world fish production grew from 7% in 1960 to around about 70% today. In 1988, China's annual fisheries output reached

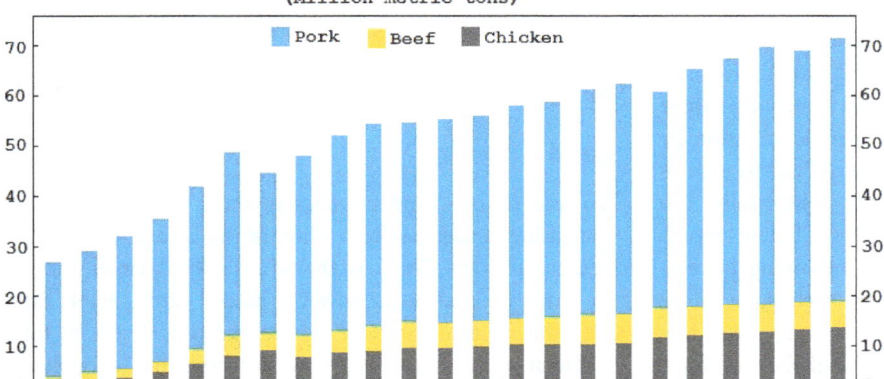

Figure 4. Demand for meat by type. Demand for all type of meat has been rising since 1990. Pork is the most widely consumed meat in China followed by beef and chicken.

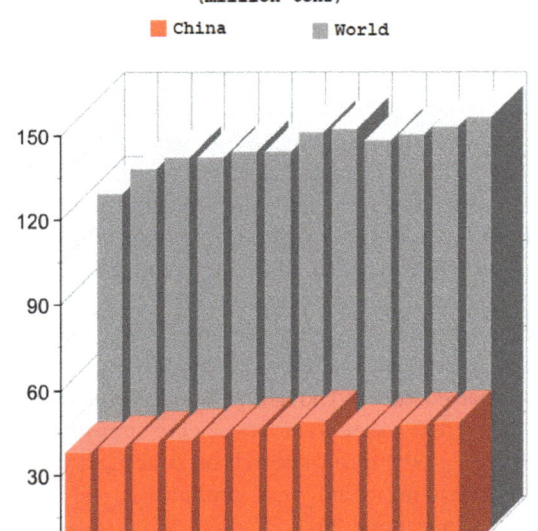

Figure 5. China's contribution in global fisheries (inland and marine) production. Total fish production has increased steadily from 1998 to 2005 but began to decline since then with an average 32% contribution to global production. Source: FAO (The State of World Fisheries and Aquaculture).

10 mmt mark for the first time in the country's history and was the third country in the world by then to achieve that scale of production. Globalization of the fish trade and subsequent relaxation of other regulatory barriers facilitated fish trade between China and the West. Following its accession to the WTO in 2001, China has been the world largest exporter of seafood since 2002. Since 1980, China experienced the fastest growth in annual per capita fish consumption with its per capita consumption increased by more than 300% and at an annual average growth rate of 9%. Total seafood consumption rose from 10 kg per capita in 1990 to 30 kg in 2009. China has a long history in aquaculture dates back to 3000 years ago and its mariculture dates back to the Song Dynasty about 1000 years ago. In Chinese, the word "fish" (鱼: yú) is homonymous with the word "abundance/surplus" (裕:yù). Though these words have slightly different accent, people occasionally prefer to pronounce fish as "yù" instead of "yú" to express a sense of abundance and/or prosperity. Historically, fish constitutes a major part of the diet for Chinese people and the role of fish in Chinese diet is becoming increasingly important. Besides its role in the diet, aquaculture also contribute to household food security by providing employment to a considerable proportion of the workforce (~15 million people) and also been an important component of the national economy. China is blessed with a

huge marine resource base with a coastline that extends more than 18,000 km. Fisheries and aquaculture have four subsectors- Inland (freshwater) catch, inland aquaculture, marine catch, and mariculture. Four seas (Bo Hai, Yellow Sea, East China Sea and South China Sea) surround the northeastern and southern territories and provide a great opportunity for marine catch and mariculture. Besides a huge marine water base, China is also rich in inland waters covering an area of 17.6 million ha, nearly one fiftieth of its land area. The country is crisscrossed by around 50,000 rivers and more than 2800 lakes and contains more than 4500 fish species. In the 1960s, marine capture fisheries accounted for the bulk of fisheries production. However, in recent years owing to rapid development in inland aquaculture, the proportion of marine capture has been decreasing. Total marine catch increased from 0.6 million tons in 1950 to 13.6 million tons in 2011 and marine culture increased almost 100-fold in the same period (Gongming and Mikko 2014). China's inland aquaculture is dominated by Tilapia- a tropical freshwater fish (*Oreochromis niloticus*) from Africa. It is the most widely grown of all farmed freshwater fish across the world and is cultivated all year round. China is the biggest producer and exporter of tilapia and accounts 56% for its global production.

As the world's biggest food producer, China also has the fastest growing share of fish and shellfish exports to EU and the United States (Qiu et al. 2012). Each year it processes a staggering 110,000 tons of cod products for the Europe and the United States markets which equates to approximately 15–20% of the global catch. It exported over USD 16 billion in seafood products in 2011. Besides being a major exporter, China also accounts for over a quarter of the world's seafood consumption and per capita consumption is expected to reach 35.9 kg by 2020. Despite such impressive production and export figures, China remains a major importer of seafood. In 2005, China exported 7.2 billion worth of seafood and imported $3.2 billion resulting in a huge fish trade surplus. And in 2011, China became the third largest importer of seafood.

Challenges for Food Security in China

Land

To feed China's huge population, small farmers are facing increasing challenges from diminishing land and water resource. Though China is the world's third largest nation in terms of land area. However its per capita arable land is around half the world's average and only 12.8% of total national terrestrial surface is available for agricultural activities (Jie 2007). Table 1 shows that China's land resource suitable for cultivation is comparatively lower than other major agri-food-producing countries. Pressure on agricultural land and land degradation has accelerated since 1978 due chiefly to rapid industrialization and urbanization (Yang and Li 2007). Between 1996 and 2008, cultivated land and grassland decreased by 6.4% and 0.59%, respectively, much of which was attributed to urbanization and industrialization (Ghose and Sajeeb 2013). Total cropland is expected to decline from 135 million ha in 2003 to 129 million ha in 2030 (Guanghong 2012). Since 2008, the central government has made urbanization, and agricultural and rural development a tactic to counter the global financial crisis and economic slowdown, as well as a long-term strategy for sustained economic growth (Ling 2011). The area of cereal production decreased from 97 million ha in 1978 to 93 million ha in 1995 due to the occupation of farmland for construction, and roads, and other developments (Xiong and Conway 2009). According to National Bureau of Statistics, China's urban population outnumbered rural dwellers for the first time in history in 2011 down from 81% in 1979. Nobel laureate Joseph E. Stiglitz maintained that, as the biggest developing country in the world, the urbanization of China and the high-tech development of the United States would be two important topics which would profoundly influence human development throughout this century (Hongxiang and Zhijun 2009). This urbanization spree has its footprints across various social and environmental spheres some of which are already being felt. Since 1990, over 8 million ha of its arable land have disappeared and per capita cropland has decreased from 0.18 ha in the 1950s to less than 0.1 ha today (Jianhua 2011). China completed the Three Gorges Dam where over 60,000 ha of farmland was lost, and according to a local news media (*China Daily Mail*, May 3, 2012), China lost 70% of its rice and 50% of its grain production with this one bad decision.

Besides declining in quantity, the quality of the survived land is also under huge threat. Urbanization is enhancing the degree of soil pollution through improper disposal of domestic and industrial waste, and acid deposition derived from urban air pollution (Jie 2007). About 2.5% of total arable land (3.3 million ha) have already become too polluted for cultivation. Contamination of croplands and rivers with heavy metal such as cadmium (Cd), lead (Pb), and arsenic (As) has received widespread attention due to their potential impact on public health and ecological systems (Zhiyuan and Zongwei 2014). Heavy metals are well-known for their carcinogenic properties in multiple organs (Matés et al. 2010). A test carried out by Guangzhou Food and Drug Administration in 2013 found that a majority of rice in local restaurants contained excessive levels of Cd. Another study on ten heavy metals in Hunan province showed that the tested soil samples were severely contaminated by the investigated metals (Xiangqin et al. 2010). As contamination

of food products especially of rice and vegetables has also raised serious food safety concerns nationwide as rice is the most widely consumed cereal in the country. Soil pollution by arsenic is the major source of As uptake by crops which is the main route for As to human food chain. Arsenic is widely recognized for its carcinogenic properties and is ranked first among all hazardous elements by the Agency for Toxic Substances and Disease Registry (Gao and Lai 2013). Though water is the main source of As contamination in most other Asian countries (Japan, Bangladesh, India, Thailand), a recent study showed that As intake for the Chinese population through rice (37.6%) is much higher than from drinking water (1.5%) (Feng et al. 2010), and suggested that issue must be dealt with more stringent policy making (Feng et al. 2010).

To avert such serious public health issues, in many areas where land and irrigation water is found to be highly polluted with toxic elements and pesticides are being discarded for cultivation. Thus, millions of hectares of agricultural land may become unsuitable for agriculture in near future and this will have grave impacts on agricultural output and domestic food self-sufficiency.

Aside from urbanization and pollution, land reform has been another heavily discussed topic in the context of agriculture and food security in China. Land administration is considered vital for food security (Georgina et al. 2013). Misappropriation of farmland in the rural areas is not a news in China. A nationwide campaign conducted in 2006 to regulate farmland allocation for commercial purposes detected 22395 illegal land appropriation cases (Jiang 2007). Since 1980s, prioritizing the need for land for nonagricultural purposes over the need for ones under agriculture has been serving as an implicit way to cater labor force for industrialization and urban development. And with the surge of more remunerative nonfarm market sectors, farmers are also becoming more unlikely to stick to the land to which their right is not firm.

Water

The concern over food and water insecurity is rising globally. Many countries have experienced decreasing agricultural yields recently due to diminishing stream flows and falling water tables. Drought in major food exporting regions like Australia and the United States was largely responsible for the global food price hike. Arab world is facing a huge food gap which is largely due to its incapacity to supply sufficient water for agriculture. And China appears to be no exception, and even has greater challenges owing to its fast ongoing urbanization and industrialization. According to WWF (World Wildlife Fund), 13% of China's lakes have disappeared in last 40 years along with half its coastal wetlands. The United Nations also identified China as one of

13 countries faced with extreme water shortages. The main underlying causes behind this scarcity include huge demand from agriculture and urbanization, uneven distribution of water resources, and extremely high level of pollution. Industry and agriculture combined account for 85% of all water use in China. Adequate availability of water is critical for optimum agricultural productivity and water scarcity is already seriously impacting grain production especially in northern parts of the country (Tatsuji 2013). China's annual per capita water resources is 2079 m³ compared to a global average of 6225 m³ which is expected to peak in 2030. While the Southern region enjoys 80% of the freshwater water resources, many regions in northern China are expected to run dry within 30 years which will have devastating impacts on the region's agricultural production and food supply. To prevent this catastrophic happening, Chinese government has initiated an astounding South-North Water Diversion Project which is set to become world's largest water diversion construction and the largest concrete structure on earth as well. The decision of going for this staggering $62 billion project in a situation where world economy is striving to recover from a deadly crisis speaks of the gravity of water insecurity in China (at least regionally). The channel is designed to divert about 45 billion m³ of water per year from the Yangtze River to the Yellow River basin in arid north once finished by 2050.

China's fisheries sector is confronted with a host of challenge arising from widespread water pollution from numerous sources and concern about the food safety standards. Most rivers and lakes especially near urban and industrial areas is severely affected by eutrophication and contamination by urban and industrial discharge, various agrochemicals, antibiotic contamination, and carcinogens. The Yangtze River, once the lifeblood of the country, is already heavily polluted and has become unusable for crop production in many parts due to eutrophication by agricultural run-off and heavy industrial effluent content. Out of 18 largest lakes in China, 14 are already entirely eutrophic, which is not only affecting aquaculture, but also causing shortage of drinking and industrial water for the surrounding communities (Peiqiao et al. 2013). Aside from eutrophication, aquaculture also suffers from degradation of genetic resources due to the lack of selective breeding, land reclamation dam building, etc. Overfishing and increasing pollution are destroying marine resources in the East China Sea, where 81% of the fishing regions are rated category 4 for pollution. The coastal regions like Shanghai and Liaoning have long been exposed to intense industrial activities and urbanization, and the people living in those areas are becoming more vulnerable to toxic contamination by consumption of local seafood. Mariculture is also facing a range of challenges including industrial pollution, loss of biodiversity, and increasing risks in food safety. Water quality of the Sanggou Bay, which is one of the most important mariculture regions in China, is facing severe deterioration due to overfishing which has resulted in reduced fish sizes and diminishing production (Tatsuji 2013).

Policy implications

Food security is a multidimensional concept which requires a range of factors to be considered for making long-term sustainable goals. China has been at the crossroads in its endeavor to design a long-term food security blueprint. For China, there exists a complex trade-off between outsourcing land and water intensive food for its population which makes it vulnerable to a volatile food market situation, and trying to stick to its self-sufficiency policy which is increasingly becoming more challenging and deleterious for the diminishing resources. Recent experience reveals that overdependence on imported food creates the preconditions for transmitting food price hikes in international market to the domestic market which has devastating impacts on low and middle-income families and also commonly leads to social disorder. Development specialists and think tanks around the world have been publishing extensively concerning the mammoth task of feeding 9.5–10 billion people in 2050 and suggest that food production needs to be increased by around 70%. Different countries have been adopting different measures to realize this nearly unachievable goal. China as a fast-growing country, and as a key player in global agri-food market is also faced with extraordinary challenges to meet its long-term food security goals. As China continues to rise as an economic superpower amid global crisis, this remarkable performance has in part overshadowed the critical challenges that came alongside this "rush development". Oversight of the negative externalities of unbalanced industrialization and urbanization has already begun to take huge toll on public health and environmental resources, and biodiversity. Though China's population is projected to be decreasing by 2050, its urban population will continue to grow and very likely will outstrip the country's catering capacity and increase strain on agricultural resources elsewhere. China's goal of grain self-sufficiency is already at a stake due to a great extent to huge demand for meat. The benefit of one child policy (独生子女: Dú shēng zǐ nǚ) on reducing food demand is being offset by rapid urbanization and changing demographics. This indicates that besides improving its capacity to maximize agricultural yield and reducing population growth rates, China's policy makers need to work on several other key areas which are not extant in the current food security framework. Food security specialists must innovate ways to feed the rapidly growing cities more safely and sustainably and curbing pollution at the same time to tackle the looming challenges of food security in the coming decades. Rapid urbanization has resulted in loss of arable land, environmental degradation, and new dietary demands on food production (Fleming 2009). Though population is projected to be decreasing by 2050, the demand for food can actually be multipled due to improved socioeconomic status and continued urbanization. Besides a declining ratio of food producers-to-consumers, urbanization increases the demand for more energy, land, water, and greenhouse gas emission-intensive food (David et al. 2010). Bringing a change in diet pattern can be effective in reducing pressure on meat industry which will also greatly help to achieve the grain self-sufficiency target. Many local experts have suggested that today's western type diet is largely responsible for China's chronic disease epidemic and Chinese people should revert to traditional high cereal and vegetable based diet with lower proportion of meat. On the supply side of the food security equation, sustainable intensification agricultural must be brought about by developing stress-resistant cultivars and climate-smart technologies. Better management of water resources and more transparent land administration are required, and community-based natural resource management (CBNRM) may also prove vital. Since China's food security relies on sustainable use of land and water resources, it faces great imperatives to innovate better strategies for meeting its demand with the given resources. Overall success in addressing the food security challenges will be determined by the ability to implement policies which will require greater coordination among policy makers, researchers, mass-media, and civil society.

Conclusion

Despite the fact that food production has doubled during the past three decades globally, demand has also accelerated from highly populated and fast-growing economies like China and India. From this study, it is clear that China continues to be a major contributor to global food production and consumption. China's past success in grain self-sufficiency doesn't reflect its future vulnerability to food insecurity since the demand for crops for nonfood use is rising enormously. China is likely to remain self-sufficient in rice and wheat but will continue to be dependent on other countries for soybeans and corn to meet its growing appetite for meat and dairy products. Food and nutrition security scenario can worse due to demographic pressures coupled with climate change, extreme pollution, diminishing arable land, and depleting aquifers. Proportion of the population not involved in food production will continue to increase, and so will urban consumers whose dietary choice is becoming more energy, water, and land intensive. Unbalanced industrialization and poor management of land

and water resources may cost dearly in future as soil and water pollution are already beginning to pose serious threats on food safety and on food security. Adoption of sustainable intensification techniques and developing better market strategies must be accompanied by high-level policy measures to strengthen environmental law enforcement and compliance to confront the challenges of food security in near future. Given the increasing phenomenon of global hunger and poverty, China, as a global economic powerhouse and major agrifood producer, is faced with an imperative of not only of feeding its own citizens, but also to contribute to structuring a more efficient and sustainable food and agricultural system and ensure better market balances, and a fairer allocation of resources and responsibilities among the global community so as to ensure food security for all and to improve the standard of living.

Acknowledgments

We express sincere thanks to our colleagues for their critical advices on the subject and to Luo Zhong for his contribution in data collection.

Conflict of Interest

None declared.

References

Carter, C. A. 2011. China's agriculture: achievements and challenges. AREUpdate 14:5–7.

David, S., M. Gordon, and T. Cecilia. 2010. Urbanization and its implications for food and farming. Phil. Trans. R. Soc. 365:2809–2820.

Feng, L., L. Yulan, Z. Guilin, et al. 2010. Total and speciated arsenic levels in rice from China. Food Addit. Contam. Part A 27:810–816.

Fleming, C. 2009. Food security, urbanization and social stability in China. J. Agrar. Change 9:548–575.

Gao, L. S., Q. L. Lai, et al. 2013. Arsenic, copper, and zinc contamination in soil and wheat during coal mining, with assessment of health risks for the inhabitants of Huaibei, China. Environ. Sci. Pollut. Res. 20:8435–8445.

Georgina, R., B. Rohan, and G. Liza. 2013. Land administration for food security: a research synthesis. Land Use Policy 32:337–342.

Ghose, B., S. Sajeeb, et al. 2013. Self-sufficiency in rice and food security: a South Asian perspective. Agric. Food Secur. 2:10.

Gongming, S., and H. Mikko. 2014. An overview of marine fisheries management in China. Mar. Policy 44:265–272.

Guanghong, Z., et al. 2012. China's meat industry revolution: challenges and opportunities for the future. Meat Sci. 92:188–196.

Hongxiang, W., and H. Zhijun. 2009. Study on the County-level City in China. J. Polit. Law 2: 50–54.

Ishikawa, S. 1977. China's food and agriculture: a turning point. Food Policy 3:90–102.

Jiang, Y. 2007. Chinese gov't uncovers officials who misappropriate farmland. Available at http://news.xinhuanet.com/english/2007-12/10/content_7226605.htm (accessed December 1, 2014).

Jianhua, Z. 2011. China's success in increasing per capita food production. J. Exp. Botany 32:1–5.

Jie, C. 2007. Rapid urbanization in China: a real challenge to soil protection and food security. CATENA 69:1–15.

Junichi, I. T. O., and N. I. Jing. 2013. Capital deepening, land use policy, and self-sufficiency in China's grain sector. China Econ. Rev. 24:95–107.

Justin, Y., Z. Guan, and W. James. 1995. China's regional grain self-sufficiency policy and its effect on land productivity. J. Comp. Econ. 21:187–206.

Keijiro, O. 2013. Food insecurity, income inequality, and the changing comparative advantage in world agriculture. Agricultural Economics 44:7–18.

Lester, R., and S. Linda. 1995. Who will feed China?: wake-up call for a small planet. W. W. Norton & Company, New York, ISBN-10 039331409X.

Ling, Z. 2011. Food security and agricultural changes in the course of China's urbanization. China World Econ. 19:40–59.

Mah, F. 1971. Why China imports wheat? China Quart. 45:128–129.

Matés, J. M., J. A. Segura, F. J. Alonso, and J. Marquez. 2010. Roles of dioxins and heavy metals in cancer and neurological diseases using ROS-mediated mechanisms. Free Radic. Biol. Med. 49:1328–1341.

Minghua, Z., and C. Yan. 2009. Problems, challenges, and strategic options of grain security in China. Adv. Agron. 103:101–147.

Peiqiao, J., Z. Wenbo, and L. Qigen. 2013. Lake fisheries in China: challenges and opportunities. Fish. Res. 140:66–72.

Qiu, H., L. Sun, and J. Huang. 2012. Liquid biofuels in China: current status, government policies, and future opportunities and challenges. Renew. Sust. Energy Rev. 16:3095–3104.

Rosegrant, M. W., and S. A. Cline. 2003. Global food security: challenges and policies. Science 302:1917–1919.

Steven, T. Y., F. Cheng, and S. Shew-Jiuan. 2004. Household food demand in urban China: a censored system approach. J. Comp. Econ. 32:564–585.

Tatsuji, K. 2013. Biofuel and food security in China and Japan. Renew. Sust. Energy Rev. 21:102–109.

Xiangqin, W., H. Mengchang, X. Jun, X. Jianhong, and L. Xiaofei. 2010. Heavy metal pollution of the world largest antimony mine-affected agricultural soils in Hunan province (China). J. Soils Sediments 10:827–837.

Xiao-yuan, D., S. Terrence, and M. Michele. 1995. China's grain imports: an empirical study. Food Policy 20:323–338.

Xiong, W., D. Conway, et al. 2009. Future cereal production in China: the interaction of climate change, water availability and socio-economic scenarios. Global Environ. Change 19:34–44.

Yang, H., and W. Li. 2007. Cultivated land and food supply in China. Land Use Policy 17:73–88.

Zhihao, Z., and R. H. Shida. 2012. Estimating the impacts of rising food prices on nutrient intake in urban China. China Econ. Rev. 23:1090–1103.

Zhiyuan, L., M. Zongwei, et al. 2014. A review of soil heavy metal pollution from mines in China: pollution and health risk assessment. Sci. Total Environ. 468–469:843–853.

10

Pesticides, environment, and food safety

Fernando P. Carvalho

Laboratório de Protecção e Segurança Radiológica, Instituto Superior Técnico/Universidade de Lisboa, Estrada Nacional 10, km 139, 2695-066 Bobadela LRS, Portugal

Keywords

agrochemicals, environmental health, food production, pesticides, residues

Correspondence

Fernando P. Carvalho, Laboratório de Protecção e Segurança Radiológica, Instituto Superior Técnico/Universidade de Lisboa, Estrada Nacional 10, km 139, 2695-066 Bobadela LRS, Portugal.

E-mail: carvalho@itn.pt

Abstract

Agrochemicals have enabled to more than duplicate food production during the last century, and the current need to increase food production to feed a rapid growing human population maintains pressure on the intensive use of pesticides and fertilizers. However, worldwide surveys have documented the contamination and impact of agrochemical residues in soils, and terrestrial and aquatic ecosystems including coastal marine systems, and their toxic effects on humans and nonhuman biota. Although persistent organic chemicals have been phased out and replaced by more biodegradable chemicals, contamination by legacy residues and recent residues still impacts on the quality of human food, water, and environment. Current and future increase in food production must go along with production of food with better quality and with less toxic contaminants. Alternative paths to the intensive use of crop protection chemicals are open, such as genetically engineered organisms, organic farming, change of dietary habits, and development of food technologies. Agro industries need to further develop advanced practices to protect public health, which requires more cautious use of agrochemicals through prior testing, careful risk assessment, and licensing, but also through education of farmers and users in general, measures for better protection of ecosystems, and good practices for sustainable development of agriculture, fisheries, and aquaculture. Enhanced scientific research for new developments in food production and food safety, as well as for environmental protection, is a necessary part of this endeavor. Furthermore, worldwide agreement on good agriculture practices, including development of genetically modified organisms (GMOs) and their release for international agriculture, may be urgent to ensure the success of safe food production.

Introduction

Pesticides and agrochemicals, in general, became an important component of worldwide agriculture systems during the last century, allowing for a noticeable increase in crop yields and food production (Alexandratos and Bruinsma 2012). Notwithstanding, the exponentially growing human population further stresses the need for enhancing food production. This need is aggravated by conflicts that paralyze food production and dislocate millions of refugees and, together with the effects of climate changes on agriculture, worsen scarcity of food in many

regions and call for renewed efforts in food production (UN 2015).

At the same time, during the last decades we realized that agrochemical residues did spread in the environment, causing significant contamination of terrestrial ecosystems and poisoning human foods (Carson 1962; EEA 2013). In addition, contamination of aquatic systems by pesticide residues around the world – illustrated herein with case studies in tropical coastal ecosystems – repeatedly compromised also aquatic food resources, fisheries, and aquaculture.

Paths, alternative to the intensive use of crop protection chemicals, are open to trial and assessment. However, the selection of future paths for enhanced food production shall be made through wise and science-based decision-making processes. Scientific research for developing food production and enhancing food safety, as well as environmental protection, is thus a necessary part of this process.

This article reviews the main issues related to pesticide residues, their environmental fate, and effects and discusses pathways for enhanced food safety.

The Role of Fertilizers and Pesticides in Agriculture

Agricultural production markedly increased since the beginning of the 20th century to cope with demographic growth. In about one century, population numbers exploded from 1.5 billion in 1900 to about 6.1 billion in 2000, which corresponds to an increase in world population three times greater than during the entire history of humanity. The world has added one more billion people since 2003, and at the current growth rates, it is estimated that world population will be of about 9.4–10 billion by 2050 (UN 2015).

The increase of world population in the 20th century would not have been possible without a parallel growth in food production, and this was achieved due to fertilizers. Organic fertilizers ("guano") were incipiently used by the end of the 19th century, but the introduction of mineral phosphate fertilizers took over in the beginning of the 20th century and continuously increased up to our days (Gilland 2015). The use of phosphates, together with development of improved crop varieties with higher yields, allowed for an unprecedented increase in agriculture productivity, the "green revolution," and the production of cereals more than duplicated per unit surface area of agriculture land (Brown 1995; Carvalho 2006). For example, in the USA from 1950 to 1990, the cereal production grew at 2.2% per year, although it has slowed down afterward (Brown 1995 2011). The growth of human population and the world production of phosphates for use as fertilizers were significantly and positively correlated over the last century (Roser and Ortiz-Ospina 2017), with a $R^2 = 0.97$ for the period 1900–1988 (Hendrix 2011) (Fig. 1).

From the 1940s onwards, further increase in food production was allowed by the introduction of synthetic crop protection chemicals. Worldwide pesticide production increased at a rate of about 11% per year, from 0.2 million tons in 1950s to more than 5 million tons by 2000 (FAO 2017; Fig. 2). Pesticides, or crop protection chemicals, include several groups of compounds, namely organochlorine, organophosphate, carbamate, pyrethroids, growth regulators, neonicotinoids, and now biopesticides, which have been developed one after the other. Pesticide sales

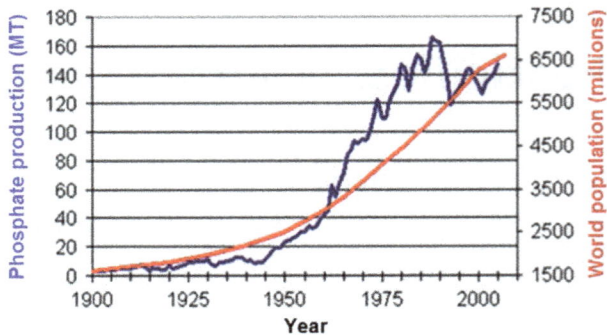

Figure 1. Increase of world population and phosphate rock production during last century (Modified from Roser and Ortiz-Ospina 2017).

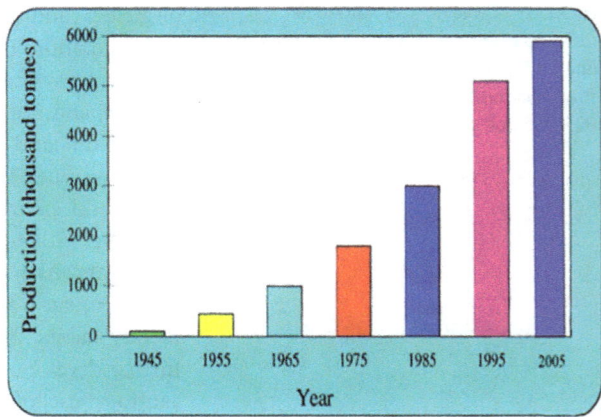

Figure 2. World production of formulated pesticides (based on FAO Statistics).

have increased for all types of pesticides, but herbicides were the group that expanded the most followed by insecticides and fungicides (Fig. 3).

The use of pesticides has not been the same across the world due to the cost of the chemicals (most of them patented), but also due to the cost of man power and the specific pests of each climatic/geographic region. Average application rates of pesticides per hectare of arable land have been computed by FAO and the highest average values, attaining 6.5–60 kg/ha, occurred in Asia and in some countries of South America (Fig. 4). While in North America and West Europe, the use of herbicides intensively applied in agriculture and in urban areas boomed in the last decades; in Asia, the use of herbicides remained low and contrasting with the use of insecticides that was very high (Fig. 4).

Early synthetic pesticides developed to control agriculture pests, such as DDT, were intensively used also for control of cattle ticks and human parasites in North America, Europe, and elsewhere (Fig. 5) and, although banned today, still are popular food preservatives of sun dry fish

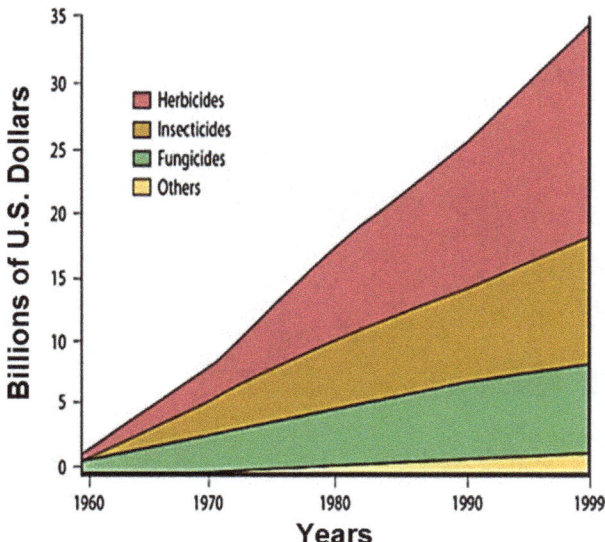

Figure 3. Estimated worldwide annual sales of pesticides (herbicides, insecticides, fungicides, and others in billion dollars; modified from Roser and Ortiz-Ospina 2017).

in South Asia and remain in use, sometimes illegally, to control malaria vectors and household pests in urban areas in the tropics (Taylor et al. 2003).

Environmental Fate and Effects of Pesticide Residues

Application of pesticides in agriculture has been made with the help of several techniques, from the manual spraying by workers on foot to truck- and airplane-based spraying techniques. At different times in different regions, some or all these techniques have been used.

Many cases of intoxication of farmers, rural workers, and their families did occur during pesticide applications and were documented in reports on poisoning and effects of synthetic chemicals on human health. It was reported that unintentional poisonings kill an estimated 355,000 people globally each year, and such poisonings are strongly associated with excessive exposure and inappropriate use of toxic chemicals (WHO 1990, 2012, Alavanja 2009; Alavanja and Bonner 2012).

Dispersion of pesticide residues in the environment and mass killings of nonhuman biota, such as bees, birds, amphibians, fish, and small mammals, were also reported (Köhler and Triebskorn 2013; Paoli et al. 2015; WHO 2017). Early reports and structured incident reporting systems certainly helped to develop regulations for pesticide applications, including dosage of chemicals and best periods of application (Hester and Harrison 2017). Over the years, a considerable research effort was developed also to understand the behavior of these chemicals in the environment, including their cycling and fate as well as their toxicity to biota.

Soon after the start of synthetic chemicals use, it was realized that the application of crop protection pesticides was causing contamination not only at local scale but also at global scale (Carson 1962; Fig. 6).

At local scale, chemicals applied on crops, as for example toxaphene applied in cotton crops in Nicaragua, remained in soils year after year and were carried by surface runoff to watersheds and coastal lagoons where residues contaminated aquatic biota (Carvalho et al. 1992, 2003). DDT applied to crops was often reported also to be transported to the aquatic environment where it is rapidly metabolized to DDE and bio-accumulated in aquatic food chains being returned eventually to humans (Kale et al. 1999). Endosulfan

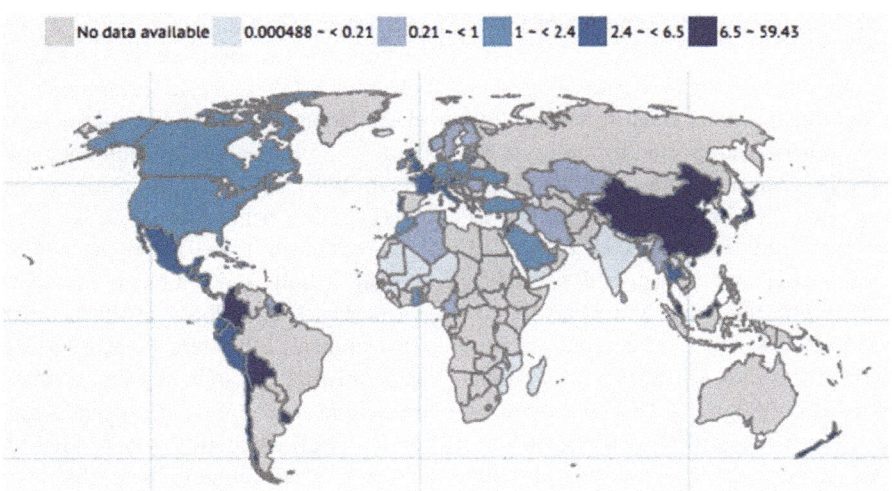

Figure 4. Use of pesticides per hectare of arable land, kg/ha, in the years 2005–2009 (FAO 2013).

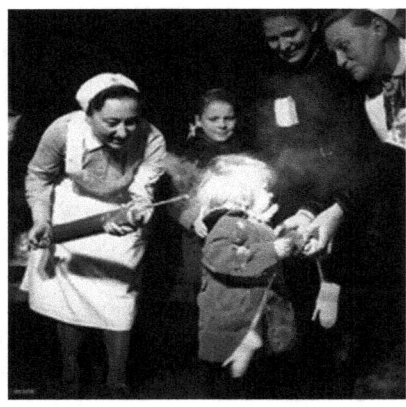

Figure 5. DDT application on humans and cattle, around the 1940s (photos from Internet).

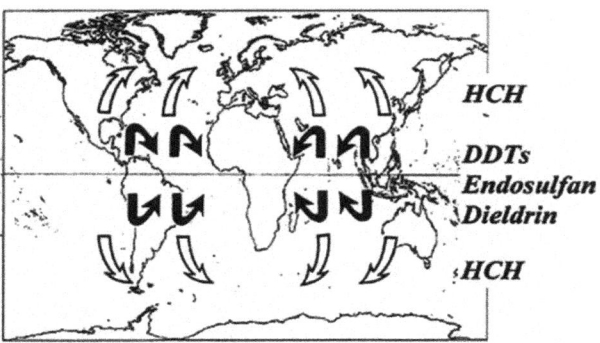

Figure 6. Volatilization and atmospheric transport of pesticides from tropical regions toward the poles.

was found to be metabolized by bacteria into endosulfan sulfate and could persist in soils and in aquatic sediments as a toxic chemical (Carvalho et al. 2002a,b). In general, these chemical compounds could undergo several chemical transformations and be transferred among environmental compartments, reaching other ecosystems outside the area of application and exerting toxic effects on nontarget species (Taylor et al. 2003).

At global scale, compounds such as hexacyclohexanes (HCH), chlordane, and toxaphene applied in fields in the south of USA were volatilized, transported by atmospheric processes, condensed in cooler climates, and deposited from the atmosphere onto the Great Lakes at Canada (Li and Jin 2013). The same did occur with HCH applied on rice fields in South Asia and transported to higher latitudes (Iwata et al. 1993; Simonich and Hites 1995). The most volatile compounds were more rapidly transported by atmospheric processes, reaching regions far away from the application areas (Fig. 6). This evaporation-condensation process was first observed with organochlorine compounds (OCs), but later was reported also for organophosphates (OPs), such as chlorpyrifos, that volatilized from application on banana plantations in the inter

tropical region of Central America and reached the ice pack in the Artic (Garbarino et al. 2002). This global scale dispersion process could have been predicted based on Henri's Law, which relates the volatility (fugacity) of compounds from liquid media to the air as inversely related to water solubility, and on van 't Hoff equation that parameterizes the effect of temperature on volatility of compounds (Rand 1995).

The organochlorine (OC) pesticides of first generation were soon reported as environmentally persistent, remaining long time in soils and sediments and accumulating in nonhuman organisms with devastating toxic effects at population level (Köhler and Triebskorn 2013). Organochlorine residues are generally transferred also in the food chains with impact on human health (discussed further in section Human Exposures to Residues and Public Health Concerns, below). Development of resistance by pests to these OC chemicals urged to replace them by new and less persistent chemicals, such as organophosphate (OP), carbamate, and pyrethroid compounds, supposedly more specific in the fight to pests too (The Agrochemicals Handbook 1991).

Research on all these chemicals, in particular using carbon-14 (^{14}C)-labeled compounds, shed light on the degradation rates in soils and in aquatic environments, and in accumulation by nontarget biota (e.g. Carvalho et al. 1992, 1997). Organochlorine compounds, such as DDT, HCH, heptachlor, toxaphene, and lindane, are in general, much more persistent and their residues may remain in soils and sediments over days, weeks, and even years (Fig. 7; Carvalho et al. 2002a,b, 2003). In the aquatic environment, OPs were expected to degrade rapidly, but experimental research has shown that they persist days/ weeks and are accumulated by crustacean and fish (Carvalho et al. 1992). Moreover, once released into the aquatic systems, these compounds are bio-accumulated in a few minutes and undergo also partitioning between water and particulate matter/sediment, with partitioning coefficients

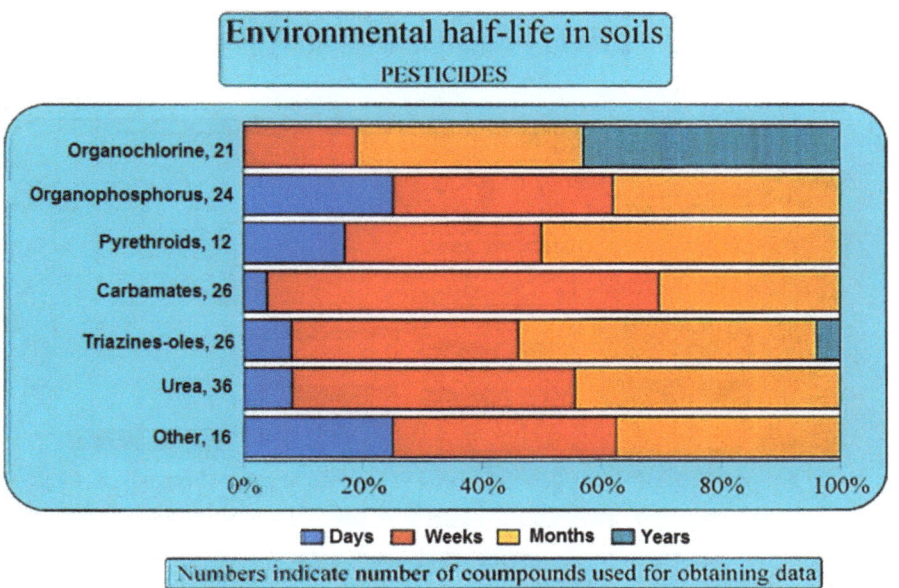

Figure 7. Environmental persistence (half-lives) of pesticides in soils by chemicals group. Numbers indicate number of compounds for which data are available (modified from Carvalho et al. 1997).

(K_p) that are positively correlated with the octanol-water partitioning coefficients (k_{ow}) of the compounds. Experimental studies in mesocosmos have shown that compounds, such as endosulfan, could persist long time as well and were accumulated up to the point to represent a toxicological risk to aquatic biota (Carvalho et al. 1999, 2002a,b; Nhan et al. 2002). Taking into account chemical properties and persistence in the environment, chemicals applied in agriculture fields may be transported and reach other ecosystems (Fig. 8). As predicted from results of experimental studies using ^{14}C-labeled compounds, the endosulfan applied in coffee and leguminous plantations and at the time seen as a nonpersistent compound, through field investigations was consistently found in aquatic systems near agriculture regions in central and North America countries (Carvalho et al. 2002b, 2009a,b). Later, it was verified that endosulfan residues are widespread in the environment and it is considered nowadays a "global pollutant" (Weber et al. 2010).

Compounds of different chemical groups have different toxic mechanisms and act on pest organisms in different ways. Organochlorine compounds (insecticides, e.g., aldrin, DDT, HCH, heptachlor, chlordane, endosulfan) are in general very effective contact insecticides, and they are structurally related to steroid hormones and act on the respective hormone receptor (Tebourbi et al. 2011). Organophosphates (mostly insecticides, e.g., parathion, malathion, chlorpyrifos, diazinon, dichlorvos) and carbamates (mostly herbicides and fungicides, e.g., aldicarb, carbofuran, ethienocarb, fenobucarb, methomyl) act as acetylcholinesterase (AchE) inhibitors causing

disruption of nervous impulse transmission at synaptic level. Pyrethroids (insecticides, e.g., cypermethrin, deltamethrin, esfenvalerate, fenvalerate) act on the voltage gated-sodium channels in cell membranes disrupting the Na^+ ion flux. The neonicotinoids (insecticides, e.g., acetamiprid, clothianidin, dinotefuran, imidacloprid) act as agonists at the nicotinic acetylcholine receptors (nAChRs), are neurotoxic, and act on the insect's nervous system, resulting in paralysis and death (Tomizawa and Casida 2005).

The mechanism for toxic action is not restricted to target pests, and toxicity is exerted also on nontarget similar organisms causing damage to biodiversity and ecosystems health. OCs impacted heavily the top predators in terrestrial food chains, as birds of prey, and accumulate in adipose tissues of animals and humans, being transferred to newborns with the milk fat, and act as endocrine disruptor (EEA 2013). Organophosphates were reported as highly toxic to arthropods in general, which includes insects but also shrimp, crabs and other crustacean, and also to vertebrates. Pyrethroids have also impact on insects and vertebrates. Many other compounds used, as herbicides have shown effects also on central nervous system and excretory system of mammals (Casida 2009; Singh et al. 2016).

Due to reports on contamination of the environment and toxic effects on biota, considerable efforts have been made to design new chemicals, improve pesticide formulations, application devices, and chemical delivery mechanisms such as the use of degradable nanoparticles as a vehicle to pesticides in an attempt to reduce exposure of biota and environmental contamination (De et al. 2014).

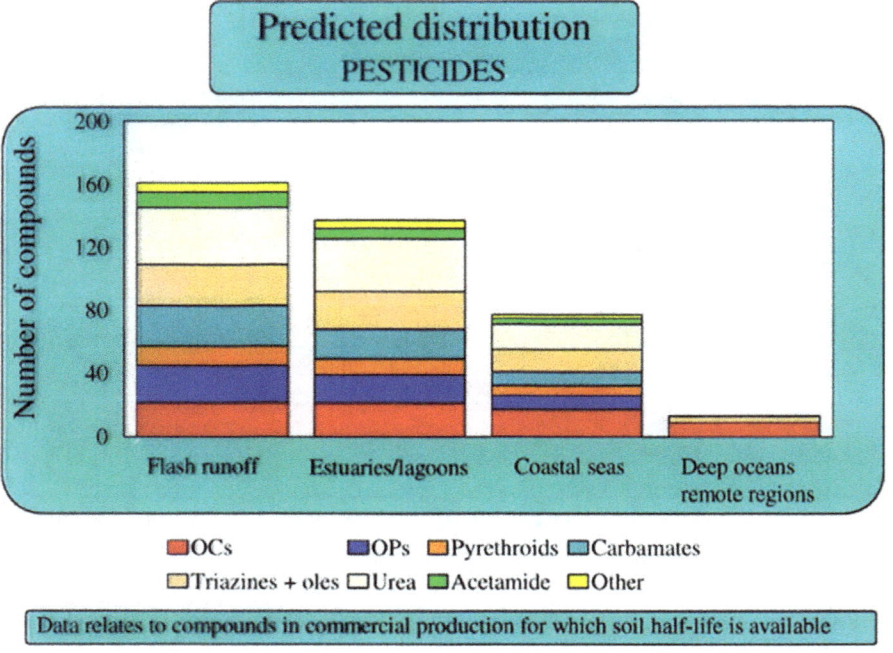

Figure 8. Potential for transport and dispersion of pesticides in the environment with ecosystems that they may reach. Data relate to compounds in commercial production for which soil half-life is available (modified from Carvalho et al. 1997).

Nevertheless, this has not resolved the collateral effects of pesticides and recurrent episodes with new chemicals, including neonicotinoids, have been reported (Bouwman et al. 2013; Hallmann et al. 2014; Park et al. 2015).

Residues in Soils and in Aquatic Environments

Persistent and bio-accumulative chemical compounds, such as DDT, HCH, toxaphene, aldrin, and dieldrin, were banned by the Stockholm Convention, approved in 2002, and have been replaced by environmentally friendly and less bio-accumulative chemicals. This has been the trend over the last decades, and it was driven by the toxicity of chemical residues present in food to humans as well as to chemicals' persistence in the environment and toxicity to nonhuman biota. However, from the massive application of OCs in the past, they are still present in soils, in sediments, and in the biosphere and are toxic. For example, toxaphene in cotton fields is not used anymore in Nicaragua but many years after cessation of applications, the deposit in agriculture soils was still a source of contaminants transported by surface runoff to aquatic environment and a threat to shrimp farming in coastal lagoons (Carvalho et al. 2002a,b, 2003).

Indeed, the ban of persistent OC compounds in agriculture abated application of OC pesticides in many regions but was not the end of concerns about toxic effects of these compounds. Today, we still find these OC compounds in environmental compartments as a legacy of past applications. Soils are the main reservoir of persistent OCs, and soil erosion, surface runoff, and river discharges carry and cycle significant amounts of persistent OCs in the environment. For example, results from the annual surveys of USA pollution trends reported pesticide residues in coastal sediments and biota (mussels and oysters) originated in river catchments. Many years after the ban of these compounds (e.g. DDT, chlordane), they were still present in the coastal environment where they degraded very slowly, as reflected by decreasing concentrations in biota over the years (Fig. 9). Similarly slow decrease of residue concentrations was also recorded in coastal environments of Mediterranean Sea in Europe (Villeneuve et al. 1999).

From a vast study carried out in tropical coastal ecosystems worldwide, it was concluded that pesticide residues were everywhere and were concentrated by marine fauna (Taylor et al. 2003). Other case studies showed similar conclusions, such as in the Manila Bay, Philippines, Mekong River Delta in Vietnam, coastal lagoons of NW Mexico, Laguna de Terminus, Caribbean Sea, Mexico, Todos-os-Santos Bay in Salvador, Brazil, coastal areas of Florida in USA, and North Sea and Baltic Sea in Europe (Carvalho et al. 1997, 1999, 2008, 2009a,b,c). In all these coastal areas, the residues of a large collection of crop protection chemicals, such as DDTs, HCHs, lindane, aldrin, toxaphene, and endosulfan, were determined (Carvalho et al. 2002a,b, 2003; Kimbrough et al. 2008;

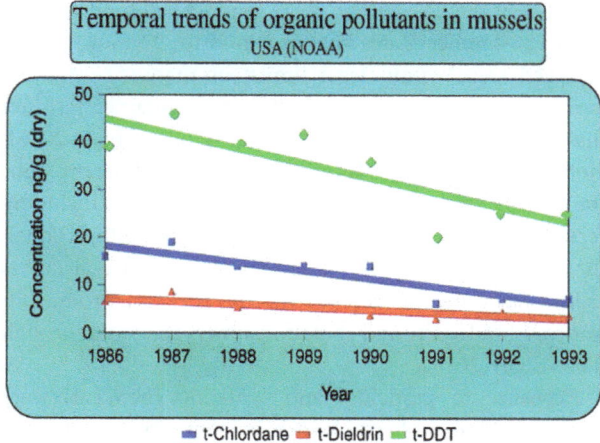

Figure 9. Temporal trends of organochlorine pesticides in coastal mussels. Average concentrations based on data from USA-NOAA Status and Trends Reports. Environmental half-lives in mussels are from 6 to 12 years depending on compound.

Moreno-Gonzalez and Leon 2017). More modern pesticides and other chemicals used in industry, such as PCBs, tributyltin, and pharmaceutical drugs, have been detected also in river waters and coastal areas and often originate in urban wastewater discharges (Barceló and Petrovic 2008).

Pesticide residues carried to the sea are also a threat to large marine ecosystems such as coral reefs. Agrochemical residues are currently monitored in sea water by the Great Barrier coral reef, Australia, where recent studies demonstrated the widespread contamination by pesticides, particularly herbicides which may impinge on symbiotic algae and destroy the coral reef (Lewis et al. 2009; Smith et al. 2012). Similarly, residues of persistent organic contaminants were found in biota in the deep sea, by many still seen as a remote and pristine environment (Jamieson et al. 2017).

Contamination of aquatic bodies by residues from periodic application of pesticides in crops was, thus, found in many environments, and this called for preventive measures. Not much was done at global scale, but near golf greens, where the use of herbicides and fertilizers is intensive, contamination of watercourses and groundwater by residues was of concern, and improved area management was advised and locally introduced through using constructed wetlands (Klaine et al. 1988).

Persistent OCs are not anymore used in Europe also, but HCH, DDT, and lindane still are present in rivers in Europe and are bioaccumulative (McKnight et al. 2015; Rasmussen et al. 2015). Organochlorine compounds residues are almost always present in environmental samples although in decreasing concentrations with the years, as reported, for example, in Denmark. There, the presence

of OC residues in river waters is from leaching of legacy applications persisting in soils, but their potential for toxic effects on aquatic fauna remains current (Rasmussen et al. 2015).

More recently, introduced and more degradable pesticides, such as chlorpyrifos, parathion, isoproturon, and mecroprop, are often detected also in river waters (Barceló and Petrovic 2008; Moreno-Gonzalez and Leon 2017). Residues of these new chemicals show the opposite concentration trend in surface waters, with concentrations often increasing over the years, such as for glyphosate (McKnight et al. 2015; Portier et al. 2016). This is very worrisome because this widespread presence of chemical residues compromises natural resources such as water for human consumption including groundwater and water for aquaculture activities.

Water Quality and Biodiversity in Freshwater and Coastal Ecosystems

The monitoring of water quality has been subject to stricter control with the EU Water Directive Framework (Directive 2000/60/EC of the European Parliament and of the Council) that required each contaminant concentration to be below 1 pg/L. Water quality for aquaculture is, however, more difficult to ensure because aquaculture itself makes use of chemicals. For example, salmon production, which in EU provided about 1/3 of fish for human consumption, uses antifouling agents, antibiotics, and chemicals for protection against lice, the main parasite of farmed salmon (SEP 2015). Residues from these chemicals, plus natural toxins from harmful algal blooms, dioxins, and PCBs, put the quality of coastal waters and the aquaculture production under pressure of contamination (SEP 2015). Although there has been a decrease of farmed salmon contamination by organic chemicals over the years, still the consumption of fish is a matter of concern and advice on intake limitation has been given to consumers (Nøstbakken et al. 2015; Ruzzina et al. 2015).

Discharges of industrial waste water and urban sewage into water lines and coastal zones have been a common procedure in most countries. The adoption of urban sewage treatment and their success has always been introduced, or attained late. One may recall the impact of pesticide residues on corals, fisheries, and shrimp aquaculture to understand that contamination has been always followed by ecological disasters and public health impacts, before regulations and mitigation measures were adopted. Increasing awareness of contamination multiplied the monitoring efforts that are continuously developing now. For example, the EU project Ocean of Tomorrow included a several initiatives to control chemical residues in sea food and development of a real-time monitoring system to respond to these challenges (e.g. Research Project Sea-on-a-Chip; http://www.sea-on-a-chip.eu). Removal of emerging contaminants from industrial waters and treatment of urban waters are also progressing (Barceló and Petrovic 2008).

The importance of controlling contamination of aquatic systems goes beyond the immediate need for water with quality for human consumption. Toxic residues in aquatic systems may eliminate aquatic species, reduce biodiversity, and compromise the functioning of ecosystems. A large research effort has been made in aquatic toxicology to understand bioaccumulation mechanisms and define toxicity levels to species selected as representative (plants, crustacean, fish) and elaborate guidelines for pollution control within tolerated limits (Rand 1995). However, toxic substances even in very low concentrations always bioconcentrate and may act on sensitive species or larval stages of biota impairing the ecosystem healthy functioning and compromising their services (Chagnon et al. 2015; Gilbert 2016). Dramatic examples are the reduction of pollinating insects, elevated concentrations of PCBs, and pesticides in farmed salmon, and the dying of the Great Barrier Coral Reef which may compromise entire ecosystems (Smith et al. 2012; Nøstbakken et al. 2015; Park et al. 2015). Eventually, instead refining toxicity testing and determination of LD_{50}, we should move the efforts to develop processes to remove contaminants from soils and effluents and prevent them to attain aquatic systems and bioaccumulate in food chains.

Human Exposures to Residues and Public Health Concerns

Worldwide, about 25 million agricultural workers experience unintentional pesticide poisonings each year, and it is estimated that approximately 1.8 billion people engage in agriculture and most use pesticides to protect food and commercial products that they produce. A few more are occupationally exposed during the use pesticides in sanitary campaigns and for lawn and garden applications (Alavanja 2009).

To reduce further exposure of population from widespread environmental contamination by these chemicals, it is not surprising that residues from both legacy applications and current agricultural, industrial, and household applications need to be controlled tightly in the environment and in the foods (EFSA 2016).

Currently, pesticide residues in North America and in EU are thoroughly monitored. In general, market foods are compliant with maximum permissible concentrations (MPC) and percentages of samples with detected residues exceeding MPCs fortunately are in small number. For

example in the EU, among more than 83,000 food samples from 28 Member States analyzed in 2014, 97% of samples analyzed were within legal limits; of these, 53.6% were free of quantifiable residues, and 43.4% contained residues that were within permitted concentrations (EFSA, 2016). Notwithstanding, in plant products, 154 different substances were found in measurable concentrations including recent and old crop protection chemicals and, although the food authority EFSA assessed the risk to consumers as low, recommendations were deemed necessary to further improve food safety and abate consumers exposure through diet (EFSA 2016).

Exposure to pesticides and synthetic chemicals were related to cancer, obesity, endocrine disruption, and other diseases in humans (Gorell et al. 1998; Bassil et al. 2007; George and Shukla 2011; Mrema et al. 2013; Araújo et al. 2016; WHO 2017). Phasing out persistent chemicals, as agreed in the Stockholm convention, contributed to reduce human exposure to toxic chemicals. Indeed, over the last decades, studies carried out in several countries have shown a consistent decrease of DDTs in human adipose tissues and milk (EEA 2013). Notwithstanding, exposure to chemical residues via water and food ingestion remains for the members of the public (i.e. without occupational exposure), a subject of concern and a burden to public health. Recent reviews of exposure and health impact of pesticides on human health have underlined the burden on human health and re-evaluated the current toxicity of legacy pesticide residues (Mrema et al. 2013). The WHO and IARC, among other organizations, keep under close scrutiny and revision the advisories on toxicity of new and old chemicals. Many agrochemicals were related to prostate cancer and other types of cancer and are increasingly regulated (Singh et al. 2016; ECA 2017).

At present, there is a widespread concern about effects of herbicides on human health, such as glyphosate that is of common use in agriculture and in cities to control weeds, and is a main carcinogenic agent (Araújo et al. 2016; Benbrook 2016). Glyphosate is the most widely applied pesticide worldwide, and in the USA, in 2014 farmers applied glyphosate at a rate of about 1 kg/ha in croplands (Benbrook 2016). The EU set the daily chronic reference dose for glyphosate to 0.5 mg/kg body weight per day, while the US EPA has set glyphosate daily chronic reference dose at 1.75 mg/kg body weight per day. However, recent compilation of toxicological data on glyphosate supports the need for reducing further the daily chronic reference dose to 0.1 mg/kg body weight per day (Antoniou et al. 2012).

In general, the maximum tolerated limits of residues in foods have been decreasing over the years, although exposure has not decreased sufficiently still due to legacy compounds in the environment and new chemicals introduced. Furthermore, it was recognized that most of the work done in the toxicity field has been reactive to problems and with marginal efficiency in anticipating and preventing the collateral toxic effects (EEA 2013).

Current Trends in Chemicals Control

The data base CAS Registry (www.cas.org) provided by the American Chemical Society includes more than 129 million unique organic and inorganic chemical substances and more than 67 million nucleotide sequences (by April 2017). More than 4000 new substances are added each day. The number of chemicals increased exponentially over the years with an average annual growth rate of about 15% in the last decades (Binetti et al. 2008). In this universe of chemicals, a small fraction is pesticides. In the data base of the US Pesticide Action Network (PAN), 6,400 pesticide active ingredients and their transformation products, as well as adjuvants and solvents used in pesticide products, were listed (www.pesticideinfo.org/). In the EU pesticide database, there are 1359 entries, not all approved for use, and about 700 registered chemicals are in use as pesticides (Eurostat, 2017). However, toxicological information about these chemicals is very poor for most of them.

In the USA, an EPA report of 1998 indicated that no information on toxicity was available for 43% of high production volume chemicals and a full set of toxicity data was available for 7% of them only (USEPA, 1998). A similar situation occurred also in the EU, and a study carried out in Denmark for 100,000 substances listed in the European Inventory of Existing Commercial Chemical Substances (EINECS) concluded that for 90% of them few toxicological data were available (Niemelä 1992).

The EU adopted in 2007 the new policy to control industrial chemicals called REACH (Registration, Evaluation, and Assessment of CHemicals), intended to create a central database on chemicals and entrusting the industry with the responsibility to evaluate and manage the risks of chemicals. In spite of large progress made in improving the knowledge about toxicity and environmental impact of chemicals, control of risks is far from being grasped and controlled (EUROSTAT 2012; EEA 2013). In a recent report, it was appreciated that, in the decade 2004–2013 in the EU, the production of environmentally harmful chemicals averaged about 150 million tons per year, representing about 40% of the total production of industrial chemicals (EUROSTAT, 2014). It was registered also a shift in production from more harmful to less harmful chemicals (based on aquatic toxicity and persistence), but still far from the objectives of sustainable development (EUROSTAT 2014).

EU objectives for 2020 foresee further action to implement REACH and achieve improvements in human life quality and environmental management regarding chemicals (7th EU Environment Action Plan). However, as pointed out before, experiments on hazards and risks cannot follow the same increasing trends for chemicals produced, because this would require very large amounts of expertise and very large amounts of human and laboratory resources to carry out complex tests (Binetti et al. 2008). Thus, timely risk assessment may be delayed.

Can we do better?

The need for producing more food to feed the growing human population is likely to increase (UN, 2015). To meet this goal, several options are open. One option might be to continue the path of intensive use of agrochemicals, including pesticides, with subsidiary research to produce more selective pesticides and improved application techniques. Other alternative options have been proposed and include the use of genetically modified organisms for better yield crops and crops resistant to pests, organic farming, development of new cultivars and recuperation of old cultivars, increased use of bio-pesticides and pheromone traps to control pests, and change of dietary habits of human populations.

The current pathway of applying synthetic crop protection chemicals has been walked through on a circular approach consisting of identification of a pest, development of a chemical, observation of collateral effects and rise of new problems, development of new chemicals, etc. We could consider this as an approach based on the trial and error method. There has been results temporarily achieved, certainly, but they always have come with an associated cost. Today, food and environment contamination with toxic chemicals impinging on public health over several human generations is considered unaffordable. We need to learn the lessons from the past and, desirably, this circle of trial and error should come to an end.

Probably, agriculture and intensive food production may not dispense the use of current agrochemicals in the next few years. Several measures could be introduced to better mitigate their collateral effects in the meantime. For example, introduction of precision application of agrochemicals (as well as precision irrigation) could reduce the amount of chemicals (and water) applied over the fields. Some other simple measures could be also immediately applied everywhere, such as: a) recovery and treatment of contaminated agriculture runoff with installation of wetland stripes suitable to clean up runoff and water drainage; b) reinforce education of farmers and the public in general about chemical hazards; and c) thorough toxicity testing and proper registration of chemicals and formulations. These measures may help to gain some extra time.

Meanwhile, we should look beyond the present time for sustainable solutions. There is a consensus that intensified research on better food production and production of food with better quality is needed. Furthermore, it is recognized that productive soil is a finite resource (as water) and, in order to ensure continued production of food, the agriculture must go side by side with soil and ecosystems preservation, restoration, and agronomic research on better yield cultivars. Therefore, it is urgent to achieve a generalized agreement on pesticide application and adoption of good agriculture practices, with consideration to Integrated Pest Management (IPM) techniques.

Consumers and the public in general have rejected already the environmental and health costs of hazardous chemicals, and awareness of chemical residues in foods created the demand for clean foods. More food and safer food is, therefore, required, but the human population and natural ecosystems may not survive longer to poor planning and poor agriculture practices. A systematic application of the precautionary principle in the introduction and application of all chemicals, including pesticides, is needed (EEA 2013). This requires thorough risk assessment of chemicals toxicity to environment and humans.

Emerging alternative paths in food production, such as development of GMO varieties and their release for international agriculture without application of the precautionary principle and satisfactory risk assessment, must be avoided. This issue deserves urgent international discussion. An agreement should be reached based on science and on ethical principles for ensuring food security and food safety. Moreover, alternative paths for food production should not repeat the mistakes of pesticide applications and must succeed in ensuring food safety and food security.

Conflict of Interest

None declared.

References

Alavanja, M. C. R. 2009. Pesticides use and exposure extensive worldwide. Rev. Environ. Health 24:303–309.

Alavanja, M. C. R., and M. R. Bonner. 2012. Occupational pesticide exposures and cancer risk: a review. J. Toxicol. Environ. Health B 15:238–263.

Alexandratos, N., and J. Bruinsma. 2012. World agriculture towards 2030/2050: the 2012 revision. ESA Working paper No. 12-03. FAO, Rome.

Antoniou, M., M. E. M. Habib, C. V. Howard, R. C. Jennings, C. Leifert, R. O. Nodari, et al. 2012.

Teratogenic effects of glyphosate-based herbicides: divergence of regulatory decisions from scientific evidence. J. Environ. Anal. Toxicol. S4:006. doi: 10.4172/2161-0525.S4-006.

Araújo, J., F. I. Delgado, and F. J. R. Paumgartten. 2016. Glyphosate and adverse pregnancy outcomes, a systematic review of observational studies. BMC Public Health 16:472.

Barceló, D., and M. Petrovic, eds. 2008. Emerging contaminants from industrial and municipal waste. Water Pollution Series, Springer.

Bassil, K. L., C. Vakil, M. Sanborn, D. C. Cole, J. S. Kaur, and K. J. Kerr. 2007. Cancer health effects of pesticides. Systematic review. Can. Fam. Physician 53:1704–1711.

Benbrook, C. M. 2016. Trends in glyphosate herbicide use in the United States and globally. Environ. Sci. Eur. 28:3.

Binetti, R., F. M. Costamagna, and I. Marcello. 2008. Exponential growth of new chemicals and evolution of information relevant to risk control. Ann. Ist. Super. Sanità 44:13–15.

Bouwman, H., R. Bornman, H. van den Berg, and H. Kylin. 2013. DDT: fifty years since Silent Spring. Pp. 240–259 in Late lessons from early warnings: science, precaution, innovation, Chapter 11. Environment and Health Environmental Scenarios. EEA Report No 1/2013. Environmental European Agency, Luxembourg.

Brown, L. R. 1995. Nature's limit (Chapter 1) from the state of the world 1995. World watch Institute, Washington, DC.

Brown, L. 2011. World on the edge. How to prevent environmental and economic collapse. Earth Policy Institute, Pub. W W Norton & Company, New York.

Carson, R. 1962. The silent Spring. Houghton Mifflin, New York.

Carvalho, F. P. 2006. Agriculture, pesticides, food security and food safety. Environ. Sci. Policy 9:685–692.

Carvalho, F. P., S. W. Fowler, J. W. Readman, and L. D. Mee. 1992. Pesticide residues in tropical coastal lagoons: the use of 14C-labelled compounds to study cycling and fates of agrochemicals. Pp. 637–653. in Applications of isotopes and radiation in conservation of the environment. Proceed. of an Int. Symposium, IAEA, Vienna.

Carvalho, F. P., S. W. Fowler, J. P. Villeneuve, and M. Horvat. 1997. Pesticide residues in the marine environment and analytical quality assurance of results. Pp. 35–57 in Environmental behaviour of crop protection chemicals. Proceed. of an FAO-IAEA Int. Symposium. International Atomic Energy Agency, Vienna.

Carvalho, F. P., S. Montenegro-Guillen, J. P. Villeneuve, C. Cattini, J. Bartocci, M. Lacayo, et al. 1999. Chlorinated hydrocarbons in coastal lagoons of the Pacific coast of Nicaragua. Arch. Environ. Contam. Toxicol. 36:132–139.

Carvalho, F. P., J.-P. Villeneuve, C. Cattini, I. Tolosa, S. Montenegro-Guillén, M. Lacayo, et al. 2002a. Ecological risk assessment of pesticide residues in coastal lagoons of Nicaragua. J. Environ. Monit. 4:778–787.

Carvalho, F. P., F. Gonzalez-Farias, J.-P. Villeneuve, C. Cattini, M. Hernandez-Garza, L. D. Mee, et al. 2002b. Distribution, fate and effects of pesticide residues in tropical coastal lagoons of the northwest of Mexico. Environ. Technol. 23:1257–1270.

Carvalho, F. P., S. Montenegro-Guillén, J.-P. Villeneuve, C. Cattini, I. Tolosa, J. Bartocci, et al. 2003. Toxaphene residues from cotton fields in soils and in the coastal environment of Nicaragua. Chemosphere 53:627–636.

Carvalho, F. P., J. P. Villeneuve, C. Cattini, I. Tolosa, D. D. Nhan, and N. V. Ahm. 2008. Agrochemical and polychlorobyphenyl (PCB) residues in the Mekong River delta, Vietnam. Mar. Pollut. Bull. 56:1476–1485.

Carvalho, F. P., J. P. Villeneuve, C. Cattini, J. Rendón, and J. M. Oliveira. 2009a. Pesticide and PCB residues in the aquatic ecosystems of Laguna de Terminos, a protected area of the coast of Campeche, Mexico. Chemosphere 74:988–995.

Carvalho, F. P., J. P. Villeneuve, C. Cattini, J. Rendón, and J. M. Oliveira. 2009b. Ecological risk assessment of PCBs and other organic contaminant residues in Laguna de Terminos, Mexico. Ecotoxicology 18:403–416.

Carvalho, F. P., J.-P. Villeneuve, C. Cattini, I. Tolosa, C. M. Bajet, and M. Navarro-Calingacion. 2009c. Organic contaminants in the marine environment of Manila Bay, Philippines. Arch. Environ. Contam. Toxicol. 57: 348–358.

Casida, J. E. 2009. Pest toxicology: the primary mechanisms of pesticide action. Chem. Res. Toxicol. 22:609–619.

Chagnon, M., D. Kreutzweiser, E. A. D. Mitchell, C. A. Morrissey, D. A. Noome, and J. P. Van der Sluijs. 2015. Risks of large-scale use of systemic insecticides to ecosystem functioning and services. Environ. Sci. Pollut. Res. Int. 22:119–134.

De, A., R. Bose, A. Kumar, and S. Mozumdar. 2014. Targeted delivery of pesticides using biodegradable polymeric nanoparticles. Springer, Berlin.

ECA. 2017. European Chemicals Agency. Available at https://echa.europa.eu/regulations/reach/legislation (accessed 10 February 2107).

EEA. 2013. Late lessons from early warnings: science, precaution, innovation. European Environment Agency, Report No 1/2013. EEA, Copenhagen.

EFSA. 2016. The 2014 European Union report on pesticide residues in food. European food safety authority. EFSA J. 14:4611 [139 pp.].

EUROSTAT. 2012. The REACH baseline study, 5 years update Summary report. Eurostat. Available at http://ec.europa.eu/eurostat/en/web/products-statistical-working-papers/-/KS-RA-12-024 (accessed 20 April 2017).

EUROSTAT. 2017. Chemicals production statistics. Data from September 2016. Eurostat. Available at http://

ec.europa.eu/eurostat/statistics-explained/index.php/Chemicals_production_statistics (accessed 20 April 2017).

FAO. 2013. FAO statistical yearbook 2013: World Food and Agriculture. Food and Agriculture Organization of the United Nations, Rome.

FAO. 2017. Available at http://www.fao.org/faostat/en/#home (accessed 23 January 2017).

Garbarino, J. R., E. Snyder-Conn, T. J. Leiker, and G. L. Hoffman. 2002. Contaminants in arctic snow collected over northwest Alaskan sea ice. Water Air Soil Pollut. 139:183–214.

George, J., and Y. Shukla. 2011. Pesticides and cancer: insights into toxicoproteomic-based findings. J. Proteomics. 74:2713–2722.

Gilbert, N. 2016. Global biodiversity report warns pollinators are under threat. Nat. News May 2017, https://doi.org/10.1038/nature.2016.19456.

Gilland, B. 2015. Nitrogen, phosphorus, carbon and population. Sci. Prog. 2015; 98(Pt 4): 379–390.

Gorell, J. M., C. C. Johnson, B. A. Rybicki, E. L. Peterson, and R. J. Richardson. 1998. The risk of Parkinson's disease with exposure to pesticides, farming, well water, and rural living. Neurology 50:1346–1350.

Hallmann, C. A., R. P. B. Foppen, C. A. M. van Turnhout, H. Kroon, and E. Jongejans. 2014. Declines in insectivorous birds are associated with high neonicotinoid concentrations. Nature 511:341–343.

Hendrix, C. S. 2011. Applying hubbert curves and linearization to rock phosphate. Working Paper Series WP 11-18. Peterson Institute for International Economics, Washington, DC.

Hester, R. E., and R. M. Harrison (Ed.), 2017. Agricultural chemicals and the environment: issues and potential solutions, 2nd edn. Issues in Environmental Science and Technology No.43. The Royal Society of Chemistry, London.

Iwata, H., S. Tanabe, N. Sakal, and R. Tatsukawa. 1993. Distribution of persistent organochlorines in the oceanic air and surface seawater and the role of ocean on their global transport and fate. Environ. Sci. Technol. 27:1080–1098.

Jamieson, A. J., T. Malkocs, S. B. Piertney, T. Fujii, and Z. Zhang. 2017. Bioaccumulation of persistent organic pollutants in the deepest ocean fauna. Nat. Ecol. Evol. 1:0051.

Kale, S., N. B. K. Murthy, K. Raghu, P. D. Sherkane, and F. P. Carvalho. 1999. Studies on degradation of 14C-DDT in the marine environment. Chemosphere 39:959–968.

Kimbrough, K. L., W. E. Johnson, G. G. Lauenstein, J. D. Christensen, and D. A. Apeti. 2008. An Assessment of Two Decades of Contaminant Monitoring in the Nation's Coastal Zone. Silver Spring, MD. NOAA Technical Memorandum NOS NCCOS 74. 105 pp.

Klaine, S. J., M. L. Hinman, D. A. Winkelmann, K. R. Sauser, J. R. Martin, and L. W. Moore. 1988. Characterization of agricultural nonpoint pollution: pesticide migration in a West Tennessee watershed. Environ. Toxicol. 7:609–614.

Köhler, H. R., and R. Triebskorn. 2013. Wildlife ecotoxicology of pesticides: can we track effects to the population level and beyond? Science 341:759–765.

Lewis, S. E., J. E. Brodie, Z. T. Bainbridge, K. W. Rohde, A. M. Davis, B. L. Masters, et al. 2009. Herbicides: a new threat to the Great Barrier Reef. Environ. Pollut. 157:2470–2484.

Li, R., and J. Jin. 2013. Modeling of temporal patterns and sources of atmospherically transported and deposited pesticides in ecosystems of concern: a case study of toxaphene in the Great Lakes. J. Geophys. Res. Atmos. 118:11863–11874.

McKnight, U. S., J. J. Rasmussen, B. Kronvang, P. J. Binning, and P. L. Bjerg. 2015. Sources, occurrence and predicted aquatic impact of legacy and contemporary pesticides in streams. Environ. Pollut. 200:64–76. https://doi.org/10.1016/j.envpol.2015.02.015.

Moreno-Gonzalez, R., and V. M. Leon. 2017. Presence and distribution of current-use pesticides in surface marine sediments from a Mediterranean coastal lagoon (SE Spain). Environ Sci Pollut Res Int. 24:8033–8048. https://doi.org/10.1007/s11356-017-8456-0.

Mrema, E. J., F. M. Rubino, S. Mandic-Rajcevic, E. Sturchio, R. Turci, A. Osculati, et al. 2013. Exposure to priority organochlorine contaminants in the Italian general population. Part 1. Eight priority organochlorinated pesticides in blood serum. Hum. Exp. Toxicol. 32:1323–1339.

Nhan, D. D., F. P. Carvalho, and B. Q. Nam. 2002. Fate of 14C-Chlorpyrifos in the tropical estuarine environment. Environ. Technol. 23:1229–1234.

Niemelä, J. 1992. EU project; priority setting for the purpose of future classification and labelling of dangerous substances (Contract No. B91/B4.3044/12200), Danish Environmental Protection Agency, Copenhagen, November 1992 and Jay Niemelä, 1994, Danish EPA.

Nøstbakken, O., H. Hove, A. Duinker, A. Lundebye, M. Berntssen, R. Hannisdal, et al. 2015. Contaminant levels in Norwegian farmed Atlantic salmon (Salmo salar) in the 13-year period from 1999 to 2011. Environ. Int. 74:274–280.

Paoli, D., F. Giannandrea, M. Gallo, R. Turci, M. S. Cattaruzza, F. Lombardo, et al. 2015. Exposure to polychlorinated biphenyls and hexachlorobenzene, semen quality and testicular cancer risk. J. Endocrinol. Invest. 38:745–752. https://doi.org/10.1007/s40618-015-0251-5.

Park, M. G., E. J. Blitzer, J. Gibbs, J. E. Losey, and B. N. Danforth. 2015. Negative effects of pesticides on wild bee communities can be buffered by landscape context.

Proc. Biol. Sci. 282:20150299. https://doi.org/10.1098/rspb.2015.0299.

Portier, C. J., B. K. Armstrong, B. C. Baguley, X. Baur, I. Belyaev, R. Bellé, et al. 2016. Differences in the carcinogenic evaluation of glyphosate between the International Agency for Research on Cancer (IARC) and the European Food Safety Authority (EFSA). J. Epidemiol. Community Health. https://doi.org/10.1136/jech-2015-207005.

Rand, G. 1995. Fundamentals of aquatic toxicology: effects, environmental fate and risk assessment. CRC Press, Boca Raton, Florida, USA.

Rasmussen, J. J., P. Wiberg-Larsen, A. Baattrup-Pedersen, N. Cedergreen, U. S. McKnight, J. Kreuger, et al. 2015. The legacy of pesticide pollution: an overlooked factor in current risk assessments of freshwater systems. Water Res. 84:25–32.

Roser, M., and E. Ortiz-Ospina. 2017. 'World Population Growth'. Published online at OurWorldInData.org. Available at https://ourworldindata.org/world-population-growth/ (accessed 15 February 2017).

Ruzzina, J., C. Bethuneb, A. Goksøyra, K. Hyllandc, D. H. Leed, D. R. Jacobs Jr, et al. 2015. Comment on "Contaminant levels in Norwegian farmed Atlantic salmon (Salmo salar) in the 13-year period from 1999 to 2011" by Nøstbakken et al. Environ. Int. 80:98–99.

SEP. 2015. Sustainable aquaculture. Sci. Environ. Policy. Future Brief 11. http://ec.europa.eu/science-environment-policy

Simonich, S. L., and R. A. Hites. 1995. Global distribution of persistent organochlorine compounds. Science 269:1851–1854.

Singh, Z., J. Kaur, R. Kaur, and S. S. Hundal. 2016. Toxic effects of organochlorine pesticides: a review. Am. J. Biosci. 4:11–18.

Smith, R., R. Middlebrook, R. Turner, R. Huggins, S. Vardy, and M. Warne. 2012. Large-scale pesticide monitoring across Great Barrier Reef catchments – Paddock to reef integrated monitoring, modelling and reporting program. Mar. Pollut. Bull. 65:117–127.

Taylor, M. D., S. J. Klaine, F. P. Carvalho, D. Barcelo, and J. Everaarts (Eds). (2003). Pesticide residues in coastal tropical ecosystems. Distribution, fate and effects. Taylor & Francis Publ., CRC Press, London. 576 pp., (ISBN: 0-415-23917-6).

Tebourbi, O., M. Sakly, and K. B. Rhouma. 2011. Molecular Mechanisms of Pesticide Toxicity. in M. Stoytcheva (Ed.). Pesticides in the modern world – Pests control and pesticides. exposure and toxicity assessment. InTech Publ. http://www.intechopen.com/books/pesticides-in-the-modernworld-pests-control-and-pesticides-exposure-and-toxicity-assessment.

The Agrochemicals Handbook. 1991. H. Kidd, D. R. James (Ed.) Royal Society of Chemistry (Great Britain), 3rd edn. London, UK.

Tomizawa, M., and J. E. Casida. 2005. Neonicotinoid insecticide toxicology: mechanisms of selective action. Annu. Rev. Pharmacol. Toxicol. 45:247–268.

UN. 2015. United Nations, Department of Economic and Social Affairs, Population Division (2015). World Population Prospects: The 2015 Revision, Key Findings and Advance Tables. Working Paper No. ESA/P/WP.241. United Nations New York, 2015.

USEPA. 1998. US Environmental Protection Agency – Office of Pollution Prevention and Toxics. Chemical hazard data. Availability study. What do we really know about the safety of high production volume chemicals? Washington DC. Available at www.epa.gov/HPV/pubs/general/hazchem.htm (accessed 20 April 2017).

Villeneuve, J. P., F. P. Carvalho, S. W. Fowler, and C. Cattini. 1999. Levels and trends of PCBs, chlorinated pesticides and petroleum hydrocarbons in mussels from the N.W. Mediterranean coast. Comparison of concentrations in 1973/74 and 1988/89. Sci. Total Environ. 237/238:57–65.

Weber, J., C. J. Halsall, D. Muir, C. Teixeira, J. Small, K. Solomon, et al. 2010. Endosulfan, a global pesticide: a review of its fate in the environment and occurrence in the Arctic. Sci. Total Environ. 408:2966–2984.

WHO. 1990. Public health impact of pesticides used in agriculture. World Health Organization, Geneva.

WHO. 2012. The WHO recommended classification of pesticides by hazard and guidelines to classification. World Health Organization, Geneva.

WHO. 2017. Agrochemicals, health and environment: directory of resources. Available at http://www.who.int/heli/risks/toxics/chemicalsdirectory/en/index1.html (accessed 10 February 2017).

11

An efficient antioxidant system and heavy metal exclusion from leaves make *Solanum cheesmaniae* more tolerant to Cu than its cultivated counterpart

Simão Branco-Neves[1,*], Cristiano Soares[1,*] (iD), Alexandra de Sousa[1], Viviana Martins[2], Manuel Azenha[3], Hernâni Gerós[2,4,5] & Fernanda Fidalgo[1]

[1]BioISI – Biosystems and Integrative Sciences Institute, Departamento de Biologia, Faculdade de Ciências, Universidade do Porto, Rua Campo Alegre s/n, 4169-007 Porto, Portugal
[2]CITAB-UM – Centre for the Research and Technology of Agro-Environmenal and Biological Sciences, Universidade do Minho, Campus de Gualtar, 4710-057 Braga, Portugal
[3]CIQ-UP, Departamento de Química e Bioquímica, Faculdade de Ciências, Universidade do Porto, Rua Campo Alegre 687, 4169-007 Porto, Portugal
[4]CBMA – Centre of Molecular and Environmental Biology, Universidade do Minho, Campus de Gualtar, 4710-057 Braga, Portugal
[5]CEB – Centre of Biological Engineering, Department of Biological Engineering, Universidade do Minho, Campus de Gualtar, 4710-057 Braga, Portugal

Keywords
Antioxidant system, biometric parameters, Cu accumulation, oxidative stress, tomato plants, tomato wild species

Correspondence
Cristiano Soares, Departamento de Biologia, Faculdade de Ciências, Universidade do Porto, Rua Campo Alegre s/n, 4169-007 Porto, Portugal.

E-mail: up201003798@fc.up.pt

Funding Information
This research was partially supported by national funds provided by Foundation for Science and Technology (FCT) through PEst-OE/BIA/UI4046/2014 (FCT through BioISI) and through the research project PTDC/AGR-PRO/7028/2014.

*These authors contributed equally to this work.

Abstract

Copper (Cu) is an abundant metal in the environment coming from anthropogenic activities and natural sources that, in excess, easily becomes phytotoxic to most species, being its accumulation in plants considered an environmental threat. This study aimed to compare the physiological and molecular responses of *Solanum lycopersicum* and its wild counterpart *Solanum cheesmaniae* to Cu stress. In particular, we wanted to address the hypothesis that *S. cheesmaniae* is more adapted to Cu stress than *S. lycopersicum*, since the former is equipped with a more efficient antioxidant defense system than the latter. Biomarkers of oxidative status (lipid peroxidation, hydrogen peroxide (H_2O_2) and superoxide anion (O_2^-) levels) revealed a more pronounced imbalance in the redox homeostasis in shoots of *S. lycopersicum* than in *S. cheesmaniae* in response to Cu. Furthermore, the activity of key antioxidant enzymes clearly differed in both species in response to Cu. Catalase (CAT) activity increased in *S. cheesmaniae* shoots but decreased in the domestic species, as well as ascorbate peroxidase (APX). Both species preferentially accumulated Cu in the radicular system, although a great increase in the aerial parts of *S. lycopersicum* was measured, while in leaves of Cu-treated *S. cheesmaniae*, the levels of Cu were not changed. Overall, results validated the hypothesis that *S. cheesmaniae* is more tolerant to excess Cu than *S. lycopersicum* and the data provided will help the development of breeding strategies toward the improvement of the resistance/tolerance of cultivated tomato species to heavy metal stress.

Introduction

As result of fluctuations in the abiotic environment, agricultural crops are frequently exposed to stress conditions, including drought, salinity, and pollution. Indeed, the steadily global industrialization is greatly increasing the incidence of metals in biosphere, which are already considered as serious environmental pollutants, disturbing the normal physiology of different animal and plant species (Nagajyoti et al. 2010).

Copper (Cu), a transition metal, is one of the oldest known metals and a component of the structure of earth's crust (Alloway 1995). The origin of Cu in soil may arise from both natural and anthropogenic activities. Cu naturally derives from rocks disintegration, parent material, minerals dissolution, and volcanic eruptions. The anthropogenic input of Cu results from livestock production, industrial activities, and intensive agriculture. In fact, in order to protect cultures against fungal diseases, several crops are treated with Cu-containing fungicides (Adrees et al. 2015), leading to an increase of Cu content in soils (Yruela 2005; Micó et al. 2006).

Some heavy metals (HM), like Cu, are essential micronutrients for higher plants and fundamental to different physiological processes, like photosynthesis, respiration, cell wall remodeling, and reactive oxygen metabolism (Burkhead et al. 2009; Marschner & Marschner 2012). However, when above a threshold level, they can easily become phytotoxic, impairing the normal growth and decreasing the nutritional quality and productivity of important crops. Phytotoxicity symptoms driven by Cu include the reduction of root growth prior to shoot growth, since roots are the preferred Cu accumulation site (Burkhead et al. 2009), inhibition of seed germination, anatomic alterations in diverse organs (Adrees et al. 2015), and induction of damages in the photosynthetic apparatus (Yruela 2005). In addition, Cu excess in plant tissues causes overproduction of reactive oxygen species (ROS). In this way, Cu excess often induces oxidative stress (Moller 2001; Yruela 2005), causing inhibition of enzymatic activities at both protein and gene expression level. Therefore, a common response of plants to Cu toxicity is the activation of the enzymatic and non-enzymatic antioxidant system (Fidalgo et al. 2013).

Tomato (*Solanum lycopersicum*) is one of the greatest produced agricultural products in the world being the second most important vegetable, constituting an excellent source of health-promoting compounds (Dorais et al. 2008). The importance of tomato is not restricted to its fresh consumption because ca. 80% of cultivated tomatoes are consumed in the form of processed products like sauce, puree, juice, or ketchup (Kaur et al. 2008). Consistent previous studies have already explored the problem of Cu stress in *S. lycopersicum*, as well as the response of the antioxidant system (Mazhoudi et al. 1997; Liao et al. 2000; Martins and Mourato 2006; Mediouni et al. 2006; Chamseddine et al. 2009; İşeri et al. 2011; Al Khateeb and Al-Qwasemeh 2014; Wang et al. 2015). However, in this study, although we have completed and performed some analyses in this species, the main purpose was to compare the responses of *S. lycopersicum* to its wild counterpart *Solanum cheesmaniae*, potentially more tolerant to Cu because it is considered a salt-tolerant species (Rajasekaran et al. 2000; Peralta and Spooner 2006), and

the induction of oxidative stress by metal toxicity is often associated with secondary water stress (Poschenrieder and Barceló 1999).

Solanum cheesmaniae is a wild tomato species, endemic from Galápagos Island (Rick, 1956), where it evolved in segregation from the continental wild tomato species, getting unique morphological characteristics, such as yellow-orange fruits and small-sized seeds. Since it can also be easily crossed with domestic tomato to produce fertile offspring (Rick 1979) and produce edible fruit (Darwin 2009), *S. cheesmaniae* seems to be an appropriate candidate species for tomato breeding.

In this study, *S. lycopersicum* and *S. cheesmaniae* plants were grown in a nutrient solution with up to 250 μmol L^{-1} Cu, and several parameters of both enzymatic and non-enzymatic components of the antioxidant system were analyzed and compared between the two species.

Materials and Methods

Plant material, growth conditions, and treatments

Solanum lycopersicum cv. Ciliegia and *Solanum cheesmaniae* seeds were surface-sterilized with 70% (v/v) ethanol for 5 min and 20% (v/v) commercial bleach (3.5% (v/v) of active chlorine) for 5 min and thoroughly washed with sterilized deionized water. Seeds of both tomato species were hydroponically cultured with a mixture of vermiculite:perlite (2:1) in a growth chamber at 24°C and 16 h-light/8 h-dark photoperiod, with a photosynthetic active radiation (PAR) of 65 μmol m^{-2} sec^{-1}. At the beginning of the experiment, seeds of both species were divided into two sets and allowed to germinate. In control conditions, watering was performed with 25% modified Hoagland solution (HS; Taiz et al. 2015), while under Cu stress, watering solution contained 25% HS supplemented with 250 μmol L^{-1} CuSO$_4$. This Cu concentration was set according to previous reports (Kopittke and Menzies 2006; Zhang et al. 2008; Choudhary et al. 2010; Fidalgo et al. 2013) and preliminary studies (see Supporting Information). For each experimental condition (control and Cu-treated plants from each species), a total of four biological replicates were considered, with five plants in each replicate. After 28 days of growth, plants from each biological replicate were collected and separated into shoots and roots, and the material was carefully processed for different biometric, biochemical, and molecular assays.

Cu concentration

Samples of dried material (0.1 g) from both experimental conditions were digested with a mixture of HCl:HNO3, (1:3) and then dissolved in a rigorous deionized water volume. Five aliquots of each digested sample were used to prepare solutions for the Cu quantification via multiple standard addition procedure. Cu levels of each sample were measured by flame-atomic absorption spectroscopy (AAnalyst 200 model; Perkin Elmer, Waltham, Massachusetts, USA). A few samples were fortified at the digestion step in order to check for possible losses or contamination during this critical operation. The recovery levels oscillated from 95% to 106%.

Photosynthetic pigments

Photosynthetic pigments were extracted from frozen plant samples (0.2 g) in 80% (v/v) acetone and quantified according to Lichtenthaler (1987), after reading the absorbance at 470, 647, and 663 nm. The results were expressed in mg g^{-1} fresh weight.

Lipid peroxidation

Malondialdehyde (MDA) content was used to quantify lipid peroxidation, as described by Heath and Packer (1968), using frozen aliquots of around 0.250 g. The concentration of MDA was calculated by using the extinction coefficient of 155 mM^{-1} cm^{-1} and expressed as nmol MDA g^{-1} fresh weight.

H$_2$O$_2$ levels

The quantification of hydrogen peroxide (H$_2$O$_2$) content was performed in shoots and roots (c.a. 0.250 g) as previously described by Jana and Choudhuri (1982). After measuring the absorbance at 410 nm, the H$_2$O$_2$ content was calculated using the extinction coefficient of 0.28 μM^{-1} cm^{-1} and expressed as nmol g^{-1} of fresh weight.

O$_2^-$ levels

Superoxide anion (O$_2^-$) levels were quantified according to Gajewska and Sklodowska (2007). Samples of plant fresh material (c.a. 0.3 g) were cut in small equal pieces of 1 cm^2 width and incubated in a mixture containing 0.01 mol L^{-1} sodium phosphate (pH 7.8), 0.05% (w/v) nitroblue tetrazolium (NBT), and 10 mmol L^{-1} azide (NaN$_3$). The NBT reducing activity (indicating O$_2^-$ generation) was expressed as the increase in A$_{580}$ h^{-1} g^{-1} fresh weight.

Total phenolics

The total content of phenolic compounds was determined following the method described by Singleton and Rossi (1965). Samples of fresh material (0.3 g) were cut in small equal pieces of 1 cm^2 and were immersed in a mixture of 1% (v/v) methanol in 1% (v/v) HCl. The total content of phenolic compounds was expressed in terms of μg gallic acid equivalents (GAE) g^{-1} fresh weight, calculated from a calibration curve prepared with gallic acid.

Total flavonoids

The total flavonoid content was determined with a colorimetric method, as previously described (Chang et al. 2002). Samples of fresh material (0.3 g) were cut in small equal pieces of 1 cm^2 and immersed in 4.5 mL of reaction solution containing 1.5 mL methanol, 0.1 mL of 10% (w/v) AlCl$_3$, 0.1 mol L^{-1} CH$_3$COOK, and 2.8 mL of water. After incubation during 30 min in the dark, the absorbance of the reaction mixture was read at 415 nm. The concentration of total flavonoid content was calculated from a calibration curve prepared with up to 100 μg of quercetin per mL of ethanol. The results were expressed as μg g^{-1} fresh weight.

Extraction and activity of the antioxidant enzymes

The extraction of superoxide dismutase (SOD – EC.1.15.1.1), catalase (CAT – EC.1.11.1.16), and ascorbate peroxidase (APX – EC.1.11.1.11) enzymes was performed according to de Sousa et al. (2013), and the soluble protein concentration in the extracts was determined by the method of Bradford (1976). The total activity of SOD was spectrophotometrically assayed by measuring the inhibition of the photochemical reduction of NBT at 560 nm (Donahue et al. (1997). The results were expressed as units of SOD mg^{-1} protein, with one SOD unit being defined as the amount of enzyme that inhibits by 50% the photochemical reduction of NBT at 560 nm. The total activity of CAT was spectrophotometrically assayed according to Rao et al. (1996), by monitoring H$_2$O$_2$ degradation at 240 nm over 1 min. The H$_2$O$_2$ extinction coefficient of 39.4 mM^{-1} cm^{-1} was used to express the activity of CAT as nmol H$_2$O$_2$ min^{-1} mg^{-1} protein. Total activity of APX was spectrophotometrically assayed according to the method described by Amako et al. (1994), based in the oxidation rate of ascorbate at 290 nm. The reaction was initiated by the addition of H$_2$O$_2$ and the variation in absorbance at 300 nm was immediately recorded for 30 sec. The total activity of APX was calculated using the ascorbate extinction coefficient of 0.49 mM^{-1} cm^{-1} and expressed as μmol oxidized ascorbate min^{-1} mg^{-1} protein.

Statistical analysis

In all biometric, physiological, and biochemical measurements, at least three biological replicates were used, each one with three technical replicates and the results expressed as mean ± standard deviation (STDEV) of the mean. The statistical analysis was accomplished by performing a two-way ANOVA, with Cu treatment and species defined as fixed factors and a significance level of 0.05, after checking the homogeneity of variances by using the Levene's test. In cases of significant P-values found for Cu treatment, a one-way ANOVA analysis was performed in order to detect the differences between control and Cu-treated plants for each species. When a significant interaction was recorded between both factors tested in the two-way ANOVA, the one-way ANOVA was performed with correction for simple main effects. When the homogeneity of variances was not accomplished, a nonparametric test (Mann–Whitney) was executed in order to discriminate differences between control and Cu-treated plants of each species. All ANOVA results, with respective F and P-values, can be found in Supporting Information. All statistical data were generated by GraphPad® Prism 6 (GraphPad Software Inc., La Jolla, California, USA).

Results

Accumulation of Cu by roots, stems and leaves

As reported in Materials and Methods, preliminary experiments were performed to find the appropriate Cu concentration in medium to evaluate plant growth and antioxidant response. The concentration of 250 μmol L^{-1} was adopted as it enabled the growth of both plant species, while inducing several toxic symptoms and antioxidant responses (see Supporting Information). Also, this concentration falls in the range of those tested in other reports for several plant species (Choudhary et al. 2010; Fidalgo et al. 2013; Kopittke and Menzies 2006; Zhang et al. 2008a,b).

Table 1 summarizes the values of Cu concentration in leaves, stems and roots measured by flame-atomic absorption spectroscopy in 28-day-old S. cheesmaniae and S. lycopersicum plants. As can be seen, both species preferentially accumulated Cu in roots. In S. cheesmaniae, a 6.3-fold increase was observed over the control, while in S. lycopersicum, Cu concentration increased threefold. In Cu-treated plants, Cu levels also increased in stems of both species (1.7- and 3.4-fold increase in S. cheesmaniae and S. lycopersicum, respectively) and in leaves of S. lycopersicum (threefold increase). Thus, leaves of S. cheesmaniae seem to be protected from the harmful effects of Cu because no increase in Cu levels was observed in Cu-treated plants in comparison with control plants.

Effects of Cu on growth and photosynthetic pigments of S. lycopersicum and S. cheesmaniae

Tables 2 and 3 summarize the effects of 250 μmol L^{-1} Cu on different biometric, physiological, and oxidative stress markers of S. cheesmaniae and S. lycopersicum plants cultivated in hydroponic conditions during 28 days.

Regarding growth-related parameters, root and shoot growth of S. lycopersicum was inhibited by 73% and 41%, respectively, while S. cheesmaniae revealed to be more tolerant to Cu, because the same growth parameters were inhibited by only 36% and 24%. Regarding biomass production, S. cheesmaniae revealed even less sensitive to the harmful effects of Cu than S. lycopersicum (Tables 2 and 3). While in S. cheesmaniae, the final shoot and root biomass were reduced by 64% and 48%, respectively, over the control, in S. lycopersicum, shoot and root biomass suffered a strong reduction higher than 90%. No changes in the final content of both chlorophylls and carotenoids were observed in both species in response to 250 μmol L^{-1} Cu (Table 2).

Based on these results, it can be assumed that Cu effects on the plant growth were dependent on the species, with a more pronounced effect on S. lycopersicum than on S. cheesmaniae.

Table 1. Cu concentration in leaves, stems and roots of 28-day-old S. cheesmaniae and S. lycopersicum plants. Cu partition in plants exposed to 250 μmol L^{-1} Cu, expressed in % relative to the total amount of Cu accumulated, is also represented. The concentration is expressed in μg of Cu per g of dry weight.

Parameter	S. cheesmaniae			S. lycopersicum		
	Control	250 μmol L^{-1} Cu	Cu partition (%)	Control	250 μmol L^{-1} Cu	Cu partition (%)
Leaves	14 ± 1.2	14 ± 1.2; $P > 0.05$	4.0	16 ± 1.2	42 ± 2.2; **$P < 0.001$**	9.5
Stems	7 ± 1.2	12 ± 2.2; **$P < 0.05$**	3.5	5.3 ± 2.0	18 ± 6.8; **$P < 0.05$**	4.0
Roots	51 ± 1.6	319 ± 3.8; $P < 0.001$	92.5	126 ± 4.0	384 ± 7.8; $P < 0.001$	86.5

Data presented as mean ± STDEV ($n = 4$).
Significant results are bold with the respective P value.

Table 2. Effect of 250 μmol L^{-1} Cu on root length, root fresh mass, lipid peroxidation, H_2O_2 and O_2^- contents and total phenols and flavonoids in roots of *S. cheesmaniae* and *S. lycopersicum* plants.

Parameter	*S. cheesmaniae*		*S. lycopersicum*	
	Control	250 μmol L^{-1} Cu	Control	250 μmol L^{-1} Cu
Root length (cm)	41.33 ± 8.898	24.68 ± 1.719; $P > 0.05$	45.71 ± 6.346	12.36 ± 0.6843; **$P < 0.05$**
Root fresh mass (g)	1.261 ± 0.1843	0.4562 ± 0.03679; **$P < 0.05$**	1.712 ± 0.3749	0.1141 ± 0.01423; **$P < 0.05$**
MDA (nmol g^{-1} FW)	11.16 ± 0.469	12.82 ± 1.228; $P > 0.05$	14.668 ± 1.383	12.663 ± 0.415; $P > 0.05$
H_2O_2 (nmol g^{-1} FW)	888.9 ± 72.80	1429 ± 4.606; $P > 0.05$	692.757 ± 29.86	967.9 ± 54.06; $P > 0.05$
O_2^- (Abs h^{-1} g^{-1} FW)	1.556 ± 0.07676	0.8707 ± 0.05950; $P > 0.05$	1.648 ± 0.04323	1.03 ± 0.080; **$P < 0.05$**
Total phenols (μg GAE g^{-1} FW)	61.78 ± 1.236	76.28 ± 1.788; $P < 0.05$	61.59 ± 2.446	58.48 ± 2.325; $P < 0.05$
Flavonoids (mg QE g^{-1} FW)	44.27 ± 2.494	91.64 ± 6.095; $P < 0.05$	183.3 ± 19.31	201.3 ± 11.30; $P > 0.05$

Data presented as mean ± STDEV ($n \geq 3$).
Significant results are bold with the respective P value.

Oxidative status of *S. lycopersicum* and *S. cheesmaniae* organs

Regarding lipid peroxidation, no differences were found among both all groups of plants. However, in *S. lycopersicum* shoots, MDA levels tended to increase from control to Cu-treated plants, although the differences were not statistically significant. In what concerns ROS levels, Cu treatment induced significant differences in O_2^- and H_2O_2 in both organs, with significant interaction between treatment and species for shoot O_2^- (Supporting Information). The content of O_2^- strongly decreased in roots of both plant species (up to 58%) treated with Cu, but a strong increase of O_2^- content was found in shoots of *S. lycopersicum* (121%), in contrast to *S. cheesmaniae* where O_2^- levels decreased by 37% over the control. Cu treatment did not significantly affect H_2O_2 levels in both organs of *S. cheesmaniae*, but this ROS increased up to 52% in response to Cu in shoots of *S. lycopersicum* (Tables 2 and 3).

Antioxidant response of *S. lycopersicum* and *S. cheesmaniae* in response to Cu

Non-enzymatic component

Statistical data revealed that, in general, Cu concentration differentially affects the non-enzymatic antioxidant system of *S. cheesmaniae* and *S. lycopersicum*. Furthermore, based on the single effects of Cu treatment, differences were found for both phenols and flavonoids (Supporting Information). Thus, results showed that in Cu-treated plants of *S. cheesmaniae*, total phenols content increased by 43% and 24% in shoots and roots, respectively. Also, flavonoids levels were increased by 107% in roots, but a tendency for decreased values was found in shoots. Conversely, in *S. lycopersicum*, the flavonoid content was only increased in shoots (Tables 2 and 3).

Enzymatic component

Regarding the enzymatic component of the antioxidant response, only for CAT activity in shoots was found a significant interaction between Cu treatment and species; however, significant differences for Cu treatment were registered for SOD in roots and CAT both in roots and leaves (Supporting Information). A strong increase of SOD activity (100%) was observed in roots of Cu-treated *S. cheesmaniae*, while a 60% increase was observed in *S. lycopersicum* in response to Cu. In shoots of both species, SOD activity did not change in response to Cu over the controls (Fig. 1).

In shoots of Cu-treated *S. cheesmaniae*, CAT activity was stimulated by 68%, but decreased by 29% in roots, while in *S. lycopersicum* exposed to Cu excess, CAT activity decreased both in roots and shoots (Fig. 2). Regarding APX, although not statistically different, a tendency for an increased activity was observed in shoots of Cu-treated *S. cheesmaniae*, while in *S. lycopersicum* APX activity was not affected in shoots, but decreased by 33% in roots (Fig. 3).

Discussion

The inhibitory effect of Cu on plant growth has been described in different reports for both monocotyledons, including rice (Lin et al. 2013), rye grass (*Lolium perenne*; Verdejo et al. 2015), maize (*Zea mays*; Ali et al. 2002; Aly and Mohamed 2012; Barbosa et al. 2013; Benimeli et al. 2010) and wheat (Gajewska and Skłodowska 2010; Gang et al. 2013), and dicotyledon species, like *Brassica juncea* (Ansari et al. 2013), and cucumber (İşeri et al. 2011). Different studies have also shown that excess of Cu negatively affects the physiological performance and growth of *S. lycopersicum* (Mazhoudi et al. 1997; Liao et al. 2000; Martins and Mourato 2006; Mediouni et al.

Table 3. Effect of 250 μmol L^{-1} Cu on shoot height, shoot fresh mass, total chlorophyll and carotenoid contents, lipid peroxidation, H_2O_2 and O_2^- contents and total phenols and flavonoids in shoots of *S. cheesmaniae and S. lycopersicum* plants.

Parameter	S. cheesmaniae		S. lycopersicum	
	Control	250 μmol L^{-1} Cu	Control	250 μmol L^{-1} Cu
Shoot height (cm)	9.755 ± 0.3895	7.381 ± 0.2797; **$P < 0.001$**	9.716 ± 0.5859	5.698 ± 0.2001; **$P < 0.001$**
Shoot fresh mass (g)	5.315 ± 0.9211	2.792 ± 0.2067; **$P < 0.05$**	7.625 ± 0.9657	0.7546 ± 0.1028; **$P < 0.001$**
Chlorophyll content (mg g^{-1} FW)	0.6473 ± 0.02315	0.8630 ± 0.09390; $P > 0.05$	0.7229 ± 0.02875	0.7645 ± 0.04515; $P > 0.05$
Carotenoids (mg g^{-1} FW)	0.1168 ± 0.004656	0.1460 ± 0.01635; $P > 0.05$	0.1323 ± 0.004102	0.1335 ± 0.008281; $P > 0.05$
MDA (nmol g^{-1} FW)	13.05 ± 1.977	14.373 ± 0.355; $P > 0.05$	14.94 ± 0.482	17.04 ± 0.442; $P > 0.05$
H_2O_2 (nmol g^{-1} FW)	2239 ± 150.2	2453 ± 45.88; $P > 0.05$	1432 ± 105.1	2174 ± 215.6; **$P < 0.05$**
O_2^- (Abs h^{-1} g^{-1} FW)	0.7216 ± 0.008102	0.4576 ± 0.02067; **$P < 0.001$**	1.003 ± 0.1526	2.216 ± 0.1616; **$P < 0.05$**
Total phenols (μg GAE g^{-1} FW)	221.4 ± 29.06	315.5 ± 20.89; $P > 0.05$	174.3 ± 12.63	165.0 ± 10.17; $P > 0.05$
Flavonoids (mg QE g^{-1} FW)	265.0 ± 13.44	193.1 ± 8.434; $P > 0.05$	576.78 ± 7.29	898.2 ± 43.15; $P > 0.05$

Data presented as mean ± STDEV ($n \geq 3$).
Significant results are bold with the respective P value.

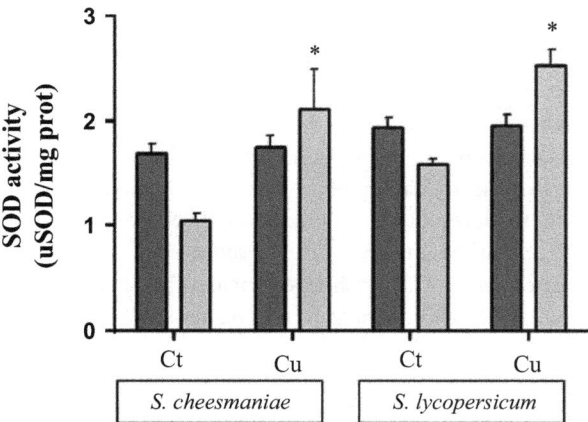

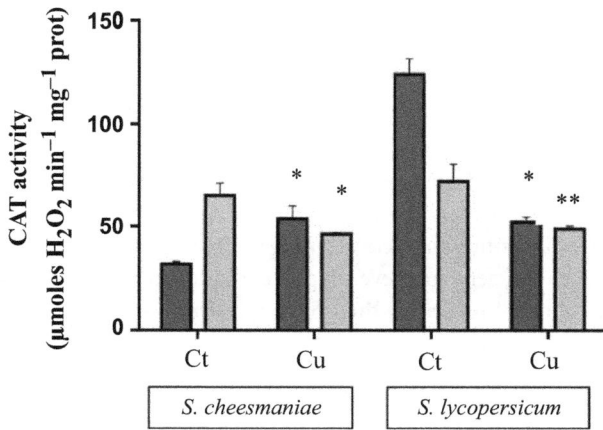

Figure 1. Superoxide dismutase activity in shoots (dark gray) and roots (light gray) of *S. cheesmaniae* and *S. lycopersicum* plants cultivated in nutritional medium supplemented with basal Cu levels (Ct, control) and Cu excess (Cu, 250 μmol L^{-1} Cu). * above bars represent significant differences at $P \leq 0.05$.

Figure 2. Catalase activity in shoots (dark gray) and roots (light gray) of *S. cheesmaniae* and *S. lycopersicum* plants cultivated in nutritional medium supplemented with basal Cu levels (Ct, control) and Cu excess (Cu, 250 μmol L^{-1} Cu). * and ** above bars represent significant differences at $P \leq 0.05$ and $P \leq 0.001$, respectively.

2006; Chamseddine et al. 2009; İşeri et al. 2011; Al Khateeb and Al-Qwasemeh 2014; Wang et al. 2015). However, an integrative study focused in the interplay of oxidative stress and antioxidant defense in response to Cu was still missing in *S. lycopersicum,* and the effect of Cu on the wild species *S. cheesmaniae* was so far unknown. In this study, we compared the physiological and biochemical mechanisms underlying the antioxidant responses of *S. lycopersicum* and its wild counterpart *S. cheesmaniae* to Cu stress. After 28 days in hydroponic culture, the presence of 250 μmol L^{-1} Cu promoted a decrease in root and shoot length in both *S. cheesmaniae* and *S. lycopersicum* plants and inhibited biomass production, but these effects were more pronounced in *S. lycopersicum* individuals, clearly showing that *S. cheesmaniae* is more tolerant to the toxic effects of the heavy metal.

The more pronounced decrease of biomass and organ length observed in *S. lycopersicum* in response to Cu is likely correlated with the higher levels of Cu found in *S. lycopersicum* organs.

Moreover, excess Cu may interfere with other nutrients' accumulation. Thus, a decrease in the content of calcium (Ca), iron (Fe), and zinc (Zn) in leaves and Mg in roots of tomato plants exposed to Cu was previously reported (Martins and Mourato 2006). The higher accumulation of Cu observed in roots of both plant species than in stems and leaves is in line with previous results in tomato and chicory plants under Cu stress (Liao et al. 2000). Also, this behavior was already found in stone pine (*Pinus pinea*), maritime pine (*Pinus pinaster*), and ash (*Fraxinus angustifolia*) exposed to Cd (Arduini et al. 1996) and Cu in plants of white lupin (*Lupinus albus* L.) growing in

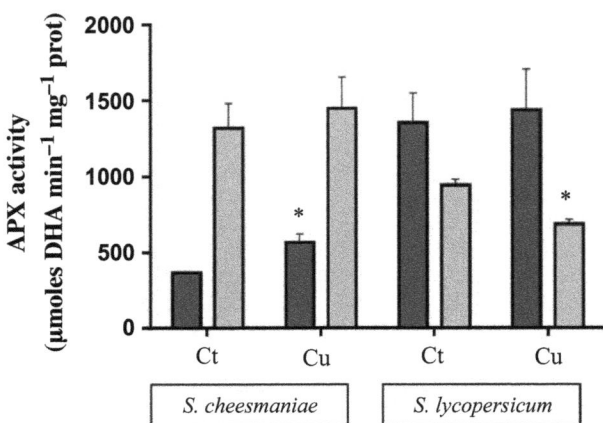

Figure 3. Ascorbate peroxidase activity in shoots (dark gray) and roots (light gray) of *S. cheesmaniae* and *S. lycopersicum* plants cultivated in nutritional medium supplemented with basal Cu levels (Ct, control) and Cu excess (Cu, 250 μmol L^{-1} Cu). * above bars represent significant differences at $P \leq 0.05$.

soils contaminated with Cu, Zn, and nickel (Ni; Fumagalli et al. 2014) and in guava seedlings (*Psidium guajava*) exposed to high concentrations of Ni (Bazihizina et al. 2015). However, the results of this study are particularly relevant, once they clearly suggest that *S. cheesmaniae* is much efficient in preventing Cu translocation from roots to shoots and leaves than *S. lycopersicum*, which support the observed higher capacity of *S. cheesmaniae* to grow under Cu stress than *S. lycopersicum*. These results also validated our hypothesis that being the wild species salt-tolerant (Rush and Epstein 1976; Knapp and Darwin 2006; Peralta and Spooner 2006), it could also be more tolerant to HM than its cultivated counterpart, further supporting that the response to both stresses may share conserved mechanisms.

Being Cu a redox-active transition element, it is able to catalyze the overproduction of ROS (Halliwell and Gutteridge 1984), which in turn can lead to harmful effects in proteins and nucleic acids and the peroxidation of lipids. It is widely accepted that lipid peroxidation occurs as a consequence of oxidative stress, being one of the most damaging process to all organisms (Gill and Tuteja 2010). However, in this study results showed that Cu only induced significant changes in MDA levels in shoots of *S. lycopersicum*, possibly because lipid peroxides were efficiently neutralized or its production avoided by a positive response of the antioxidant (AOX) system. MDA levels also did not change in response to HM stress in some previous reports (Gajewska et al. 2006; Gajewska and Sklodowska 2007; Soares et al. 2016a).

Several studies have reported increased ROS accumulation in several plants species under HM stress (Li et al. 2012; Thounaojam et al. 2012; Lukatkin et al. 2014; Soares

et al. 2016b). In this study, we have observed that O_2^- and H_2O_2 levels in response to excess Cu were tissue- and species-dependent: the exposure of *S. cheesmaniae* to Cu led to a decrease of O_2^- levels in roots and shoots, but H_2O_2 levels increased in roots. In contrast, in *S. lycopersicum*, both O_2^- and H_2O_2 content suffered significant increases in consequence of Cu excess, with the exception of O_2^- levels in roots which decreased. In general, *S. cheesmaniae* showed a higher capacity to control and/or to scavenge both O_2^- and H_2O_2 than *S. lycopersicum* in response to Cu.

Plants, as sessile organisms, have developed a complex antioxidant system, to withstand the toxic effects of ROS and oxidative stress (Sharma et al. 2012). In agreement, when the activities of SOD, CAT, and APX were measured in different tissues of both species, our results showed a less efficient AOX system in *S. lycopersicum* than in its wild counterpart. As reported, roots of both species exposed to Cu showed an increased SOD activity and, in accordance to this SOD behavior, roots of Cu-treated plants exhibited a reduction in O_2^- levels and an increase of H_2O_2 content. In shoots, SOD activity remained unaltered in response to Cu in both plant species and, in agreement, the levels of O_2^- suffered a decrease in *S. cheesmaniae* and a marked increase in *S. lycopersicum*, which accumulated more Cu in shoots. Apparently, this reduction of O_2^- in *S. cheesmaniae* is not related to SOD activity. Actually, taking into account that the total Cu accumulated in the shoots of this species did not change between control and Cu-treated plants, we can hypothesize that other AOX mechanism or metabolite is responsible for the decrease of O_2^- levels. Therefore, it seems that SOD showed a differential organ response, with a more pronounced and active protection role in roots. With respect to H_2O_2 detoxifying enzymes, our results showed that CAT activity was downregulated by Cu excess in both species, with decreased activity in all experimental conditions, excepting *S. cheesmaniae* shoots, where a positive response was recorded. In agreement, the inhibition of CAT activity as a consequence of HM stress has been well reported in literature and specifically in what regard Cu stress in several plants species like sunflower (*Helianthus annuus*; Gallego et al. 1996), durum wheat (*Triticum durum*; Sgherri et al. 2001), and in vitro grown plants of Indian ginseng (*Withania somnifera*; Khatun et al. 2008). This decrease in CAT activity can be a consequence of its auto reduction (De Vos et al. 1992; Gallego et al. 1996; Weckx and Clijsters 1996; Mazhoudi et al. 1997; Yamamoto et al. 1997) and/or the autoxidation and Fenton reactions which can also, in turn, cause oxidative injury in defense enzymes (Schutzendubel and Polle 2002). Regarding APX activity, a species- and organ-dependent response was observed. In fact, higher APX activity was observed in shoots of Cu-treated *S. cheesmaniae* than in control, while it did not change in shoots of *S. lycopersicum*, but decreased in roots. Several studies report a decrease or maintenance in APX activity in response to Cu stress (Mazhoudi et al. 1997; Teisseire and Guy 2000; Bankaji et al. 2015), and a previous work of our group with *S. nigrum* clearly showed that APX was downregulated in response to 100 and 200 μmol L^{-1} Cu in a dose-dependent manner (Fidalgo et al. 2013). Overall, the results of this study showed that the H$_2$O$_2$ levels in both plant species in response to excess Cu correlated with the observed APX and CAT activity, which is in accordance with the main role of both enzymes in the cellular detoxification of H$_2$O$_2$ (Sharma et al. 2012).

A more effective response to excess Cu of the nonenzymatic AOX system was also observed in *S. cheesmaniae* than in *S. lycopersicum*, when total phenolics and flavonoids were quantified. In what concerns total phenols, Cu exposure led to changes in *S. cheesmaniae*, with a significant increase in both plant organs in plants cultivated with Cu excess, suggesting that they may have a role in the observed higher tolerance of *S. cheesmaniae* to Cu toxicity. Particularly relevant was the observation that flavonoids increased substantially in roots of *S. cheesmaniae* in response to excess Cu and decreased in shoots, data which go in agreement with several references that state that these metabolites can be translocated from one organ to another, specifically from shoots to roots (Saslowsky and Winkel-Shirley 2001; Buer and Muday 2004; Buer et al. 2007). Also, it was also evident from our results that the basal level of flavonoids was higher in the domestic species, although its content did not changed in response to Cu. In this way, and bearing in mind the previously considered hypothesis, it seems that although flavonoids are less produced in *S. cheesmaniae*, their increase in roots may help to prevent oxidative damage and limit the translocation of Cu to the aerial parts of the plants.

Overall, our results validated the hypothesis that *S. cheesmaniae* is more tolerant to excess Cu than its domesticated counterpart, in part, due to its high capacity to limit the translocation of Cu to the aerial parts and its enhanced AOX performance. Studies in progress involving the identification and functional characterization/tissue localization of Cu transporters in both plant species may have important repercussion on the understanding of the molecular basis for the observed capacity of *S. cheesmaniae* in limiting the toxic effects of Cu on leaf tissues and will open new avenues for plant breeding.

Acknowledgments

This research was partially supported by national funds provided by Foundation for Science and Technology (FCT) through PEst-OE/BIA/UI4046/2014 (FCT through BioISI) and through the research project PTDC/AGR-PRO/7028/2014. Professor Ruth Pereira is greatly acknowledged for her support with the statistical analysis of the obtained data.

Conflict of Interest

The authors also declare that there is no conflict of interest.

References

Adrees, M., S. Ali, M. Rizwan, M. Ibrahim, F. Abbas, M. Farid, et al. 2015. The effect of excess copper on growth and physiology of important food crops: a review. Environ. Sci. Pollut. Res. Int. 22:8148–8162.

Al Khateeb, W., and H. Al-Qwasemeh. 2014. Cadmium, copper and zinc toxicity effects on growth, proline content and genetic stability of *Solanum nigrum* L., a crop wild relative for tomato; comparative study. Physiol. Mol. Biol. Plants 20:31–39.

Ali, N. A., M. P. Bernal, and M. Ater. 2002. Tolerance and bioaccumulation of copper in *Phragmites australis* and *Zea mays*. Plant Soil 239:103–111.

Alloway, B. J. 1995. Heavy metals in soils. Blackie Academic & Professional, Glasgow, U.K.

Aly, A. A., and A. A. Mohamed. 2012. The impact of copper ion on growth, thiol compounds and lipid peroxidation in two maize cultivars (*Zea mays* L.) grown 'in vitro'. Aust. J. Crop Sci. 6:541–549.

Amako, K., G.-X. Chen, and K. Asada. 1994. Separate assays specific for ascorbate peroxidase and guaiacol peroxidase and for the chloroplastic and cytosolic isozymes of ascorbate peroxidase in plants. Plant Cell Physiol. 35:497–504.

Ansari, M. K. A., E. Oztetik, A. Ahmad, S. Umar, M. Iqbal, and G. Owens. 2013. Identification of the phytoremediation potential of indian mustard genotypes for copper, evaluated from a hydroponic experiment. CLEAN 41:789–796.

Arduini, I., D. L. Godbold, and A. Onnis. 1996. Cadmium and copper uptake and distribution in Mediterranean tree seedlings. Physiol. Plant. 97:111–117.

Bankaji, I., I. Cacador, and N. Sleimi. 2015. Physiological and biochemical responses of Suaeda fruticosa to cadmium and copper stresses: growth, nutrient uptake, antioxidant enzymes, phytochelatin, and glutathione levels. Environ. Sci. Pollut. Res. Int. 22:13058–13069.

Barbosa, R. H., L. A. Tabaldi, F. R. Miyazaki, M. Pilecco, S. O. Kassab, and D. Bigaton. 2013. Foliar copper uptake by maize plants: effects on growth and yield. Cienc. Rural 43:1561–1568.

Bazihizina, N., M. Redwan, C. Taiti, C. Giordano, E. Monetti, E. Masi, et al. 2015. Root based responses account for *Psidium guajava* survival at high nickel concentration. J. Plant Physiol. 174:137–146.

Benimeli, C., A. Medina, C. Navarro, R. Medina, M. Amoroso, and M. Gómez. 2010. Bioaccumulation of copper by *Zea mays*: impact on root, shoot and leaf growth. Water Air Soil Pollut. 210:365–370.

Bradford, M. M. 1976. A rapid and sensitive method for the quantitation of microgram quantities of protein utilizing the principle of protein-dye binding. Anal. Biochem. 72:248–254.

Buer, C. S., and G. K. Muday. 2004. The transparent testa4 mutation prevents flavonoid synthesis and alters auxin transport and the response of Arabidopsis roots to gravity and light. Plant Cell 16:1191–1205.

Buer, C. S., G. K. Muday, and M. A. Djordjevic. 2007. Flavonoids are differentially taken up and transported long distances in Arabidopsis. Plant Physiol. 145: 478–490.

Burkhead, J. L., K. A. Reynolds, S. E. Abdel-Ghany, C. M. Cohu, and M. Pilon. 2009. Copper homeostasis. New Phytol. 182:799–816.

Chamseddine, M., B. Wided, H. Guy, C. Marie-Edith, and J. Fatma. 2009. Cadmium and copper induction of oxidative stress and antioxidative response in tomato (*Solanum lycopersicon*) leaves. Plant Growth Regul. 57:89–99.

Chang, C., M. Yang, H. Wen, and J. Chern. 2002. Estimation of total flavonoid content in propolis by two complementary colorimetric methods. J. Food Drug Anal. 10:178–182.

Choudhary, S. P., R. Bhardwaj, B. D. Gupta, P. Dutt, R. K. Gupta, S. Biondi, et al. 2010. Epibrassinolide induces changes in indole-3-acetic acid, abscisic acid and polyamine concentrations and enhances antioxidant potential of radish seedlings under copper stress. Physiol. Plant. 140:280–296.

Darwin, S. C.. 2009. The systematics and genetics of tomatoes on the Galápagos Islands (Solanum, Solanaceae). Genetics, Evolution and Environment, University College London, London.

De Vos, C. H. R., M. J. Vonk, R. Vooijs, and H. Schat. 1992. Glutathione depletion due to copper induced phytochelatin synthesis causes oxidative stress in *Silene cucubalus*. Plant Physiol. 98:853–858.

Donahue, J. L., C. M. Okpodu, C. L. Cramer, E. A. Grabau, and R. G. Alscher. 1997. Responses of antioxidants to paraquat in pea leaves: relationships to resistance. Plant Physiol. 113:249–257.

Dorais, M., D. L. Ehret, and A. P. Papadopoulos. 2008. Tomato (*Solanum lycopersicum*) health components: from the seed to the consumer. Phytochem. Rev. 7:231–250.

Fidalgo, F., M. Azenha, A. F. Silva, A. de Sousa, A. Santiago, P. Ferraz, et al. 2013. Copper-induced stress in *Solanum nigrum* L. and antioxidant defense system responses. Food Energy Secur. 2:70–80.

Fumagalli, P., R. Comolli, C. Ferrè, A. Ghiani, R. Gentili, and S. Citterio. 2014. The rotation of white lupin (*Lupinus albus* L.) with metal-accumulating plant crops: a strategy to increase the benefits of soil phytoremediation. J. Environ. Manage. 145:35–42.

Gajewska, E., and M. Sklodowska. 2007. Effect of nickel on ROS content and antioxidative enzyme activities in wheat leaves. Biometals 20:27–36.

Gajewska, E., and M. Skłodowska. 2010. Differential effect of equal copper, cadmium and nickel concentration on biochemical reactions in wheat seedlings. Ecotoxicol. Environ. Saf. 73:996–1003.

Gajewska, E., M. Skłodowska, M. Słaba, and J. Mazur. 2006. Effect of nickel on antioxidative enzyme activities, proline and chlorophyll contents in wheat shoots. Plant Biol. 50:653–659.

Gallego, S. M., M. P. Benavídes, and M. L. Tomaro. 1996. Effect of heavy metal ion excess on sunflower leaves: evidence for involvement of oxidative stress. Plant Sci. 121:151–159.

Gang, A., A. Vyas, and H. Vyas. 2013. Toxic effect of heavy metals on germination and seedling growth of wheat. J. Environ. Res. Dev. 8:206–213.

Gill, S. S., and N. Tuteja. 2010. Reactive oxygen species and antioxidant machinery in abiotic stress tolerance in crop plants. Plant Physiol. Biochem. 48:909–930.

Halliwell, B., and J. M. Gutteridge. 1984. Oxygen toxicity, oxygen radicals, transition metals and disease. Biochem. J. 219:1–14.

Heath, R. L., and L. Packer. 1968. Photoperoxidation in isolated chloroplasts. Arch. Biochem. Biophys. 125:189–198.

İşeri, Ö., D. Körpe, E. Yurtcu, F. Sahin, and M. Haberal. 2011. Copper-induced oxidative damage, antioxidant response and genotoxicity in *Lycopersicum esculentum* Mill. and *Cucumis sativus* L. Plant Cell Rep. 30:1713–1721.

Jana, S., and M. A. Choudhuri. 1982. Glycolate metabolism of three submersed aquatic angiosperms during ageing. Aquat. Bot. 12:345–354.

Kaur, D., A. A. Wani, D. P. S. Oberoi, and D. S. Sogi. 2008. Effect of extraction conditions on lycopene extractions from tomato processing waste skin using response surface methodology. Food Chem. 108:711–718.

Khatun, S., M. B. Ali, E. J. Hahn, and K. Y. Paek. 2008. Copper toxicity in *Withania somnifera*: growth and antioxidant enzymes responses of in vitro grown plants. Environ. Exp. Bot. 64:279–285.

Knapp, S., and S. C. Darwin. 2006. (1736) Proposal to conserve the name *Solanum cheesmaniae* (L. Riley)

Fosberg against *S. cheesemanii* Geras. (Solanaceae). Taxon 55:806–807.

Kopittke, P., and N. Menzies. 2006. Effect of cu toxicity on growth of cowpea (*Vigna unguiculata*). Plant Soil 279:287–296.

Li, X., H. Ma, P. Jia, J. Wang, L. Jia, T. Zhang, et al. 2012. Responses of seedling growth and antioxidant activity to excess iron and copper in *Triticum aestivum* L. Ecotoxicol. Environ. Saf. 86:47–53.

Liao, M. T., M. J. Hedley, D. J. Woolley, R. R. Brooks, and M. A. Nichols. 2000. Copper uptake and translocation in chicory (*Cichorium intybus* L. cv. Grasslands Puna) and tomato (*Lycopersicon esculentum* Mill. cv. Rondy) plants grown in NFT system. I. Copper uptake and distribution in plants. Plant Soil 221:135–142.

Lichtenthaler, H. K. 1987. Chlorophylls and carotenoids: pigments of photosynthetic biomembranes. Methods in Enzymology. 149:350–382.

Lin, C. Y., N. Trinh, S. F. Fu, Y. C. Hsiung, L. C. Chia, C. W. Lin, et al. 2013. Comparison of early transcriptome responses to copper and cadmium in rice roots. Plant Mol. Biol. 81:507–522.

Lukatkin, A., I. Egorova, I. Michailova, P. Malec, and K. Strzalka. 2014. Effect of copper on pro- and antioxidative reactions in radish (*Raphanus sativus* L.) in vitro and in vivo. J. Trace Elem. Med Biol. 28:80–86.

Marschner, P. 2012. Marschner's mineral nutrition of higher plants, 3rd ed. Academic Press, London.

Martins, L. L., and M. P. Mourato. 2006. Effect of excess copper on tomato plants: growth parameters, enzyme activities, chlorophyll, and mineral content. J. Plant Nutr. 29:2179–2198.

Mazhoudi, S., A. Chaoui, M. Habib Ghorbal, and E. El Ferjani. 1997. Response of antioxidant enzymes to excess copper in tomato (*Lycopersicon esculentum*, Mill.). Plant Sci. 127:129–137.

Mediouni, C., O. Benzarti, B. Tray, M. H. Ghorbel, and F. Jemal. 2006. Cadmium and copper toxicity for tomato seedlings. Agron. Sustain. Dev. 26:227–232.

Micó, C., M. Recatalá, M. Peris, and J. Sánchez. 2006. Assessing heavy metal sources in agricultural soils of an European Mediterranean area by multivariate analysis. Chemosphere 65:863–872.

Moller, I. M. 2001. Plant mitochondria and oxidative stress: electron transport, NADPH turnover, and metabolism of reactive oxygen species. Annu. Rev. Plant Physiol. Plant Mol. Biol. 52:561–591.

Nagajyoti, P. C., K. D. Lee, and T. V. M. Sreekanth. 2010. Heavy metals, occurrence and toxicity for plants: a review. Environ. Chem. Lett. 8:199–216.

Peralta, I. E., and D. M. Spooner. 2006. History, origin and early cultivation of tomato (Solanaceae). Pp. 1–24 *in* M. K. Razdan and A. K. Mattoo, eds. Genetic improvement of solanaceous crops Volume 2: tomato. Science Publishers, Enfield, NH.

Poschenrieder, C., and J. Barceló. 1999. Water relations in heavy metal stressed plants, heavy metal stress in plants. Pp 207–229. Springer, Berlin Heidelberg.

Rajasekaran, L. R., D. Aspinall, and L. G. Paleg. 2000. Physiological mechanism of tolerance of *Lycopersicon* spp. exposed to salt stress. Can. J. Plant Sci. 80:151–159.

Rao, M. V., G. Paliyath, and D. P. Ormrod. 1996. Ultraviolet-B- and ozone-induced biochemical changes in antioxidant enzymes of *Arabidopsis thaliana*. Plant Physiol. 110:125–136.

Rick, C. M., 1956. Genetic and systematic studies on accessions of Lycospersicon from the Galapagos Islands. American Journal of Botany. 43:687–696.

Rick, C. M. 1979. Biosystematic studies in Lycopersicon and closely related species of *Solanum*. Pp. 667–678 *in* J. G. Hawkes, R. N. Lester and A. D. Skelding, eds. The biology and taxonomy of the solanaceae. Academic Press, London.

Rush, D. W., and E. Epstein. 1976. Diferences betwen salt-sensitive and sal-tolerant genotypes of the tomato. Plant Physiol. 57:162–166.

Saslowsky, D., and B. Winkel-Shirley. 2001. Localization of flavonoid enzymes in Arabidopsis roots. Plant J. 27:37–48.

Schutzendubel, A., and A. Polle. 2002. Plant responses to abiotic stresses: heavy metal-induced oxidative stress and protection by mycorrhization. J. Exp. Bot. 53:1351–1365.

Sgherri, C., M. T. A. Milone, H. Clijsters, and F. Navari-Izzo. 2001. Antioxidative enzymes in two wheat cultivars, differently sensitive to drought and subjected to subsymptomatic copper doses. J. Plant Physiol. 158:1439–1447.

Sharma, P., A. B. Jha, R. S. Dubey, and M. Pessarakli. 2012. Reactive oxygen species, oxidative damage, and antioxidative defense mechanism in plants under stressful conditions. J. Bot 26:26.

Singleton, V. L., and J. A. Rossi. 1965. Colorimetry of total phenolics with phosphomolybdic-phosphotungstic acid reagents. Am. J. Enol. Vitic. 16:144–158.

Soares, C., A. de Sousa, A. Pinto, M. Azenha, J. Teixeira, R. A. Azevedo, et al. 2016a. Effect of 24-epibrassinolide on ROS content, antioxidant system, lipid peroxidation and Ni uptake in *Solanum nigrum* L. under Ni stress. Environ. Exp. Bot. 122:115–125.

Soares, C., S. Branco-Neves, A. de Sousa, R. Pereira, and F. Fidalgo. 2016b. Ecotoxicological relevance of nano-NiO and acetaminophen to *Hordeum vulgare* L.: combining standardized procedures and physiological endpoints. Chemosphere 165:442–452.

de Sousa, A., J. Teixeira, M. T. Regueiras, M. Azenha, F. Silva, and F. Fidalgo. 2013. Metalaxyl-induced changes in the antioxidant metabolism of *Solanum nigrum* L. suspension cells. Pestic. Biochem. Physiol. 107:235–243.

Taiz, L., E. Zeiger, I. M. Moller, and A. Murfy. 2015. Plant physiology and development, 6th ed. Sinauer Associates Inc, Sunderland, MA.

Teisseire, H., and V. Guy. 2000. Copper-induced changes in antioxidant enzymes activities in fronds of duckweed (*Lemna minor*). Plant Sci. 153:65–72.

Thounaojam, T. C., P. Panda, P. Mazumdar, D. Kumar, G. D. Sharma, L. Sahoo, et al. 2012. Excess copper induced oxidative stress and response of antioxidants in rice. Plant Physiol. Biochem. 53:33–39.

Verdejo, J., R. Ginocchio, S. Sauve, E. Salgado, and A. Neaman. 2015. Thresholds of copper phytotoxicity in field-collected agricultural soils exposed to copper mining activities in Chile. Ecotoxicol. Environ. Saf. 122:171–177.

Wang, L., X. Yang, Z. Ren, X. Hu, and X. Wang. 2015. Alleviation of photosynthetic inhibition in copper-stressed tomatoes through rebalance of ion content by exogenous nitric oxide. Turk. J. Botany 39:10–22.

Weckx, J. E. J., and H. M. M. Clijsters. 1996. Oxidative damage and defense mechanisms in primary leaves of *Phaseolus vulgaris* as a result of root assimilation of toxic amounts of copper. Physiol. Plant. 96:506–512.

Yamamoto, Y., A. Hachiya, and H. Matsumoto. 1997. Oxidative damage to membranes by a combination of aluminum and iron in suspension-cultured tobacco Cells. Plant Cell Physiol. 38:1333–1339.

Yruela, I. 2005. Copper in plants. Braz. J. Plant. Physiol. 17:145–156.

Zhang, H., Y. Xia, G. Wang, and Z. Shen. 2008. Excess copper induces accumulation of hydrogen peroxide and increases lipid peroxidation and total activity of copper–zinc superoxide dismutase in roots of *Elsholtzia haichowensis*. Planta 227:465–475.

12

Stress resilience in crop plants: strategic thinking to address local food production problems

William J. Davies[1] & Jean-Marcel Ribaut[2]

[1]The Lancaster Environment Centre, Lancaster University, Bailrigg, Lancaster LA1 4YQ, UK
[2]Generation Challenge Programme (GCP) c/o CIMMYT, Carretera Mexico-Veracruz, El Batan, Texcoco, Estado de Mexico, Mexico

Keywords
crop plants, local food production problems, strategic thinking, Stress resilience.

Correspondence
William J. Davies, The Lancaster Environment Centre, Lancaster University, Bailrigg, Lancaster LA1 4YQ, UK.

E-mail: w.davies@lancaster.ac.uk

Funding Information
The authors are grateful to the Society for Experimental Biology for financial support.

Abstract

There are many ways to assess or define the stress resilience of crop production, but ultimately the resilience of systems (and communities), i.e., an ability to survive and prosper, is driven by profitability. Here we review challenges for those who seek to bring about beneficial change in practice or policy as we translate novel crop science research findings into impacts on the food supply chain. While advances in plant and crop science are relevant to this challenge, the context of application is crucial here and this will mean that many other considerations, discussed below, will potentially moderate the impact on crop growth and yield of what could be the introduction of very significant breakthroughs in genetic gain. This paper considers opportunities for plant scientists seeking to address the world's growing food security challenge by exploiting new understanding of the basis of crop stress resilience. Ultimately the local challenge is to increase the resilience of cropping systems and rural communities.

Introduction: The Challenge and a Local Response

There are many ways to assess or define the stress resilience of crop production, but ultimately the resilience of systems (and communities), i.e., an ability to survive and prosper, is driven by profitability. Here, we review challenges for those who seek to bring about beneficial change in practice or policy as we translate novel crop science research findings into impacts on the food supply chain. While advances in plant and crop science are relevant to this challenge, the context of application is crucial here and this will mean that many other considerations, discussed below, will potentially moderate the impact on crop growth and yield of what could be the introduction of very significant breakthroughs in genetic gain This paper considers opportunities for plant scientists seeking to address the world's growing food security challenge by exploiting new understanding of the basis of crop stress resilience. Ultimately the local challenge is to increase the resilience of cropping systems and rural communities.

Even though advances in plant and crop science understanding have helped us make considerable progress toward meeting the food-related Millennium Development Goals and the more recent Sustainable Development Goals, there is still a very significant "Global Food Security Challenge." This is a multidisciplinary challenge which depressingly now also involves a necessity to address the fact that for the first time in history, there are more obese people in the world than there are hungry people. We recognize that both hunger and obesity are promoting significant health problems associated with unhealthy and/or inadequate diets. While stress resilience is of less relevance to those addressing this set of issues, stress effects on crop and food quality can be appreciable and there are opportunities here for crop science to deliver change for the better.

We need to increase the availability of food in many regions of the world and also increase peoples' access to this food but the food should also be healthy. There are many social cultural and economic considerations that contribute to local differences in food availability. These considerations can be captured effectively in the following identity which describes major influences which can determine the impact of a change in a food production system:

$$G \times E \times M \times S \text{ (Genetics} \times \text{Environment} \times \text{Management} \times \text{People/Society)}$$

This interaction between a multitude of factors effectively tells us that a "local" approach to addressing many food challenges must be important. Crop science is well aware of the importance of G × E interactions in determining how effective new traits may be in particular locations/environments. Probably not surprisingly, some traits can have very positive effects on crop yield in some stress environments but the same traits can have neutral or even negative effects when environmental conditions are varied (Tardieu 2012; Bonneau et al. 2013). Often crop production is most profitable in good years (optimal conditions) and it is these profitable years that help to sustain farmers through suboptimal years when different stresses are present. Breeding for resilience, requires assessment of performance under optimal and suboptimal conditions to ensure that genetic gain under abiotic stress is not associated with a yield penalty in the absence of stress (Ribaut 2006). One of the major consequence of climate changes is the increasing unpredictability of climatic conditions and an increase in the stress intensity. As a result improved rice cultivars in some regions of southeast Asia need to be resistant to flooding during the first part of the crop cycle, but at the same time being drought tolerant as water limited conditions might occur during flowering or grain filling stages; the good news is that surprisingly those "opposite" stresses might have some common genetic basis (Fukao et al. 2011;. Rubaiyath et al. 2016).

Recent work by agronomists at CSIRO (Kirkegaard and Hunt 2010) in collaboration with breeders in the same organization shows the importance of even the most basic of crop management options (M in above equation) and many other studies show that social considerations (S) are also very important in determining whether an innovation is taken up and whether it impacts on peoples' lives. Even in the most general consideration of the Food Security challenge it is apparent that peoples' access to diets dominated by poorly nutritious, often unsafe food can cause massive health problems for many. Price et al. (2013) show how novel plant stress biology implemented through genetics and crop management can have very beneficial effects on the safety of food but this crop-specific challenge requires a local "solution."

Some Targets for Plant Scientists in the Delivery of "Sustainable Intensification"

Crop scientists who focus on the interaction between the genetic basis of their crop of choice and the environment are mostly concerned with the impact of the environment (stress) on the genetic potential yield. Increasingly however we are concerned with the impact of agriculture (the crop/food production process) on the environment. There is particular concern for the overuse of the input resources required for crop production and excessive water use is a major problem on several continents, with falling water tables due to over extraction of water for irrigation having a particularly significant effect on natural vegetation and ultimately promoting desertification (Kang et al. 2008). Overuse of fertilizer impacts adversely on soil quality (e.g., Guo 2010) and on quality of ground water and surface water which can create important health risks (e.g., Campbell et al. 2016)). The stress biology at issue here is variation in water and nutrient availability and there is now much information to show how these stress variables can be exploited to the benefit of both resource use and crop production. Stress is effectively being used as a crop growth regulator. Among the best example is alternate wetting and drying irrigation (AWD) which saves water while sustaining yield and can have beneficial effects on greenhouse gas emissions and crop quality (Yang and Zhang 2010)

It goes without saying that we should seek wherever possible to minimize the damaging effects of agriculture such as those detailed above, while still seeking new ways of increasing productivity. Exploitation of understanding of the genetic basis of crop stress resilience, or how to mitigate it such as through crop diversification (Lin 2011), can be key here. International Initiatives such as the Generation Challenge Programme (GCP) have demonstrated that translational research in crop improvement is not only achievable but can be highly successful with the right combination of technical and "soft" science skills and expertise. The GCP was able to demonstrate that harnessing plant genetic diversity and applying modern biology to the development of new crop varieties that meet the needs of smallholder farmers is both an efficient and effective means of conducting translational research. This Programme promoted a way of working based on "true" partnerships by assembling the right combination of expertise into teams, by providing these teams with adequate resources- including budget- and managing their evolution toward synergy and delivery of outputs while, at the same time, encouraging and enforcing information sharing (Ribaut 2014).

Recently, the term "sustainable intensification" has been coined to describe a target for future food production methodology. This may be a useful development but most are well aware that this term is highly location-specific and even in meta-environments, techniques for sustainable use of water and nutrients in agriculture will be context-specific, depending on for example the nature of the soils and the hydrology of the region. Local "solutions" need to consider agricultural, environmental and social factors which will differ in importance, again with location and land use objective. Pollock (2016) has highlighted the fact that the preservation of viable rural communities intimately linked to local agricultural needs to be given more attention if we are to also preserve/achieve rural social stability. We will see below how crop genetics and management techniques based on understanding of the basis of crop resilience can be influential in climate-stressed communities.

Crop Science to Ameliorate the Impact of a 4 Degree World on Food Production

Projections of climate change impacts produced by a number of different modeling approaches indicate near certainty that global crop production will be negatively affected by climate change (Challinor et al. 2014). Most predictions also suggest reduced crop quality and nutritional value (i.e., decreases in leaf and grain N, protein and nutrient (Fe, Zn, Mn, Cu) concentrations) associated with warmer climates and increased CO_2 levels. (Stress effects that need to be overcome).

To date, only a relatively few studies have delivered estimates of climate change effects for different regions of the world. Lobell et al. (2011) have identified South Asia and southern Africa as two regions that, in the absence of significant crop adaptation, would suffer the most negative impacts on important food crops (some of which have received little attention from stress biologists). The expectation is that future climate will be on average both warmer and wetter. Crop seasonality is affected by both the intensity and the distribution of the rains over time and both are affected by climate change (Feng et al. 2013). Increases in the inter-annual variability of yields are also likely to become more pronounced and will potentially affect stability of food availability and access (Porter et al. 2014).

Hochman et al. at CSIRO (2017) analyzed data from 50 weather stations located throughout Australia's wheat-growing areas and found that, on average, the amount of rain falling on growing crops declined by 2.8 mm per season, or 28% over 26 years, while maximum daily temperatures increased by an average of 1.05°C. By modeling these data using APSIM they calculate that the national wheat yield will fall from the recent average of 1.74 tonnes per hectare to 1.55 tonnes per hectare in 2041.

Plant science now has the capacity to develop crop varieties that are better suited to contrasting and new climatic conditions more rapidly than has previously been the case. Increases in the incidence of water deficits,

chronically high temperatures and an increase generally in mean temperature can sensitively affect different stages of reproductive crop development while also accelerating crop development, resulting in shorter crop durations and reduced time to accumulate biomass and grain yield. The time from trait identification, through breeding, local availability and adoption of a new variety can be up to 30 years and although revised breeding strategies and new methodologies, such as double haploids or genomics (Varshney et al. 2012), can reduce the cycle significantly, there are many other factors that determine the adoption of new varieties by farmers. In addition to market demands that might determine profitability, new varieties require efficient regulatory processes and distribution networks and will likely be accompanied by improved management practices that enhance yield and quality potential.

Challinor et al. (2016) have identified this chain of developments through to impact as the BDA process (Breeding, Development, Adoption). These authors show that for maize in Africa both adaptation and mitigation can reduce loss of yield due to shortening cropping duration and they argue that climate projections have the potential to provide target elevated temperatures for regional breeding operations. They also stress that while options for reducing BDA time are highly context-dependent, there are common threads.

Many recent reports on the global food security challenge have stressed the need for enhanced knowledge exchange strategies in many parts of the world, including the developed world (e.g., UK Foresight). This may particularly be the case in the developing world as highlighted by Challinor et al. (2016). As many of those living in poverty in the developing world depend on agriculture for their income, vibrant agricultural systems are the key to development. The five countries in the world with the greatest problems with agricultural production and hence the greatest food and nutrition needs are all found in sub-Saharan Africa. Agricultural development can feed more people in the region and can also link to more general economic growth and reduction of poverty by generating employment. GPC (http://globalplantcouncil.org/) can help focus the attention of plant science and scientists in the developed world on this region of the developing world.

In recent years, crop yields in many African countries have begun to rise and this is early evidence, that African agriculture may now be generating its own "Green Revolution." Progress has been driven by a number of factors, including increased investment in infrastructure, introduction of policies to enhance both local and international markets, and some development of extension programs to help farmers take profit from new knowledge which can enhance crop productivity (Foresight Africa). As is the case with many aspects of food systems around the world, there is no single silver bullet which will "solve" the problem of food and nutrition insecurity. There is, however, a general view that with appropriate focus upon regional constraints, capacity development, investment and partnerships, many African countries have the potential to address the problem of substantial crop yield gaps that historically have held back development on the continent (Van Ittersum et al. 2016).

Evidence for the considerable potential of African agriculture may be found by looking at recent or intended investments by the African Development Bank. Africa currently imports one-third of all calories consumed (USD 77 Billion pa) and with widespread poverty (49% of the population in Africa lives on <USD 1.25/day) and high youth unemployment (40–60%), the imperative for an agricultural transformation that will result in broader impacts is very obvious (Chianu, 2016).

The challenges are many. Up to 60% of all famers are non-commercial or semi-commercial. Markets are under-developed and in many instances value chains are very weak. However, the Feed Africa Initiative has set ambitious goals for the period to 2025. It will aim to substantially eliminate extreme poverty, end hunger and malnutrition, enhance the performance of value chains in agriculture and turn Africa into a net food exporter.

To achieve these ambitious aims will require a commitment by governments and many others, especially to invest in human capital; the researchers and practitioners who will drive the development and sustainability of agricultural commodities and processes. A key challenge will be to retain the best and brightest young minds and to create a cadre of innovative scientists, including plant breeders, who see a future in African agriculture. This will not be easy. Budding young scientists often see a future in developed countries or international agencies where their talents will be well-rewarded. However, we are optimistic, the potential is there (Diop et al. 2013). We see a future where agriculture and agricultural research play an important part in national economies; where science and education will be key to economic development and resourced accordingly; and, where regional initiatives and international organizations all have a role to play in creating an enabling and rewarding environment for young African researchers.

The development of African agriculture will be both global and local; globally, the biophysical potential is huge- about 60% of the world's un-utilized but potentially available cropland is in Africa. Locally, the vast migration of populations from rural to urban areas is creating new market opportunities.

New developments in KE with small holder farmers that might be applied globally with regional tuning have recently been described by Zhang et al. (2016). Here agronomy students from a range of regional Universities and from China Agricultural University (the project coordinator) are assigned to "Science and Technology Backyards" (STBs) in rural China. Often these are single villages or groups of small communities where the students work to develop farmers' co-operatives and to introduce new technology and changed farming practice. Increases in water and nutrient use productivity and yielding that have been achieved in these villages are impressive (Zhang et al. 2016).

Campbell et al. (2016) have recently argued that given the serious threats to food security posed by climate change, attention should shift to an action-oriented research agenda. He and co-authors see four key challenges:

(a) changing the culture of research;
(b) deriving stakeholder-driven portfolios of options for farmers, communities and countries;
(c) ensuring that adaptation actions are relevant to those most vulnerable to climate change;
(d) combining adaptation and mitigation strategies.

The emphasis here is to increase stakeholder engagement in research and by definition general principles and strategies to mitigate climate change impact must be implemented at the local level. In reality the BDA catena defined by Challinor et al. (2016), also termed the research to implementation gap, or the science-policy gap, is often substantial. Action is needed to address this shortcoming and GPC may have a role to play here.

Adoption rate of technologies with the potential to reduce risks in agriculture has traditionally been slow. For example, despite a global shortage of water for most purposes, the adoption of improved water management practices has been slow, even in agriculture, where around 70% of the world's available fresh water is used. There seems to be a clear case here for enhanced knowledge exchange between farmers, scientists and regional policy makers. How can stress resilience biology help us produce 'more crop per drop'?

Three Examples of Possible Local Interventions to Increase Food Security, Health and Well-being at Decreasing Scale of Operation

(a) The Community Scale: Eco- and Climate-Smart Villages

Some years ago, the EU funded the development of so-called eco-villages in different regions of sub-Saharan Africa. Introduction of technological innovation on a village scale resulted in enhancement of social sustainability of the communities as a component of enhanced environmental sustainability, the importance of which was highlighted by Pollock (2016). In particular, introduction of solar arrays generated significant increases in health and well-being of children as a result of phasing out of kerosene-based lighting system and their adverse effect on air quality in the home. Energy was also used to great effect for water pumping for irrigation and deficit irrigation techniques were applied. In the Chinese STB communities described above, crop scientists have shown villagers how to grow crops with reduced nutrient and water input. Crop geneticists have also played a part.

In what appears to be a very successful collaboration between CGIAR-CCAFS and several national programmes in Africa, rural communities are encouraged to develop Climate Sensitive Villages (CSVs) as platforms where researchers, local partners, farmers' groups and policymakers collaborate to select and trial a portfolio of technologies and institutional interventions. The focus is on the objectives of climate-smart agriculture (Campbell et al. 2016): namely, enhancing productivity, incomes, climate resilience and mitigation. Importantly, context-specific objectives are established by the stakeholders.

The Campbell paper notes that a broad range of adaptation technologies are introduced into the CSVs. These include water-smart practices, weather-smart activities, nutrient-smart practices, carbon-and energy-smart practices and knowledge-smart activities, all of which have been discussed above.

(b) The Farming System Scale: Conservation Agriculture

Conservation Agriculture (CA) has been widely adopted with some success throughout the Americas, where the effects of tillage had previously resulted in loss of soil structure, soil erosion with the loss of large quantities of good quality soil. CA is said to increase yields, to improve soil fertility, reduce soil water loss, control weed growth and reduce erosion. There may also be savings on use of tractor fuel and reduced C emissions, all changes resulting in a much more stress resilient agricultural system.

However, Giller et al. (2009) have suggested that CA can leave farmers with a heavy dependence on herbicides and fertilizers. The same group has highlighted particular concerns for use of conservation agriculture in Africa. These include: decreased yields often observed with CA, increased labor requirements when herbicides are not used, an important gender shift of the labor burden to women and a lack of mulch due to poor productivity and due to the priority given to feeding of livestock with crop residues. This appears to be an excellent example of different regional manifestations of the interaction between G × E × M × S (above).

(c) The Crop Scale: Putting Nitrogen Fixation to Work for Smallholder Farmers in Africa (N2Africa) http://www.n2africa.org/

Here, the crop stress which is a major problem in much of sub Saharan Africa, is a shortage of nitrogen for crops. N2Africa, a Gates-funded long term project directed by Ken Giller at Wageningen University, is focused on enabling African smallholder farmers to benefit more fully from symbiotic N2-fixation by grain legumes. The thrust of the project is a locally-focused knowledge exchange and capacity-building effort and the development of effective production technologies including inoculants and fertilizers. The capacity that is built will sustain the pipeline and deliver continuous improvement in legume production technologies tailored to local settings.

Discovery research is aimed at the identification of new elite strains of rhizobium for the several major grain legumes other than soybean – common bean, cowpea and groundnut. New elite strains will be made available to inoculant producers for scaling up the technology. The project website stresses that delivery and dissemination approaches will be tailored to local needs. New, innovative tools for monitoring and evaluation will allow "best fit technologies" to be developed at the field and farm-scale to be translated into "best-fit approaches" at the country or regional scale. In the first phase, N2Africa reached more than 230,000 farmers who evaluated and employed improved grain legume varieties, rhizobium inoculants and basal (P) fertilizers. The impact on the family of the increased utilization of legumes is particularly large as the crop is largely grown by women and used within the home.

Introduction of N fixation biology into non-legume crops may also be a game-changer if these new seeds can be made available to the very large numbers of smallholders in developing countries who can benefit from this stress resilience technology (Charpentier and Oldroyd, 2010).

It is clear from the above examples that there is much action-orientated research underway in farming communities around the world. It is equally clear that there is much still to do within the framework of the BDA pipeline (above) or the research to outcome catena. One size interventions will not "fit all" across the globe and we ask now what the Global Plant Council can do to facilitate progress in implementation as plant science and scientists seek to address a mounting number of global food challenges.

Food security is a global issue; by 2050 food production must increase by at least 60% to meet the demands of a growing population and changing diets. Meeting this challenge will require global and strategic thinking and planning. We have outlined some of the challenges to be addressed and presented examples of models that work. The key is stakeholder engagement at all levels and, in this regard, we submit that challenges for crop production will be best addressed at the local level either through the adoption and adaptation of generic solutions or through the development of local solutions through knowledge exchange and with the benefit of indigenous knowledge. There are encouraging signs that governments and regional bodies understand the importance of increasing agricultural productivity to meet the growing demands and we highlight the importance of international collaboration as a major element in increasing crop productivity and food production.

Actions for the Global Plant Council

1. Help facilitate partnerships in research to implementation projects across disciplinary boundaries and geographic borders (the right technology in the right place)
2. Help develop partnerships with international agencies
3. Promote sharing of data and working practices
4. Promote development of Knowledge Exchange resources and international training courses (novel science must be freely available to policy makers and importantly to the large numbers of practitioners producing food in the developing world)
5. Lead in the provision of advocacy for policy, practice, funding change
6. Lead in reducing the science-policy gap
7. Encourage a "bottom-up" approach to intervention
8. Lead in promoting regionally relevant interventions at a range of scales (understand the local landscape)
9. Encourage introduction of initiatives along the delivery chain.

Acknowledgments

This paper is based on outcomes from a Stress Resilience Symposium held in Brazil in October 2015 organized by the Global Plant Council and Society for Experimental Biology. The authors would like to thank the Society for Experimental Biology for funding support for this symposium.

Conflict of Interest

None declared.

References

Bonneau, J., J. Taylor, B. Parent, et al. 2013. Multi-environment analysis and improved mapping of a yield-related QTL on chromosome 3B of wheat. Theor. Appl. Genet. 126:747–761.

Campbell, B. M., S. J. Vermeulen, P. K. Agarwal, C. Corner-Dolloff, E. Girvetz, A. M. Loboguerrero, et al., 2016. Reducing risks to food security from climate change. Global Food Sec. 11:34–43.

Challinor, A., P. Martre, S. Asseng, P. Thornton, and F. Ewert. 2014. Making the most of climate impact ensembles. Nat. Clim. Change 4:77–80.

Challinor, A. J., A. K. Koehler, J. Ramirez-Villegas, S. Whitfield, and B. Das. 2016. Current warming will reduce yields unless maize breeding and seed systems adapt immediately. Nat. Clim. Chang. 6:954–958.

Charpentier, M., and G. Oldroyd. 2010. How close are we to nitrogen fixing cereals? Curr Opin Plant Biol. 13:556–564.

Chianu, J. 2016. Technologies for African Agricultural Transformation (TAAT) International Institute for Tropical Agriculture. Available via https://issuu.com/iita/docs/bulletin_taat_2341_special

Diop, N. N., F. Okono, and J.-M. Ribaut. 2013. Evaluating human resource capacity for crop breeding in national programs in Africa and South and Southeast Asia. Creat. Educ. 4:72–81.

Feng, X., A. Porporato, and I. Rodriguez-Iturbe. 2013. Changes in rainfall seasonality in the tropics. Nature Climate Change 3:811–815.

Foresight Africa. Available at https://www.brookings.edu/research/foresight-africa-top-priorities-for-the-continent-in-2016-2/. (accessed 17 March 2017).

Fukao, T., E. Yeung, and J. Bailey-Serres. 2011. The submergence tolerance regulator SUB1A mediates crosstalk between submergence and drought tolerance in rice. Plant Cell 23:412–427.

Giller, K. E., E. Witter, M. Corbeels, and P. Tittonell. 2009. Conservation agriculture and smallholder farming in Africa: the heretics' view. Field. Crop. Res. 114:23–34.

Guo, J. H. 2010. Significant acidification in major Chinese croplands. Science 327:1008–1010.

Hochman, Z., D. L. Gobbett, and H. Horan. 2017. Changing climate has stalled Australian wheat yields: study. The conversation, Environment + Energy. Available at https://theconversation.com/changing-climate-has-stalled-australian-wheat-yields-study-71411. (accessed 17 March 2017).

Lin, B. B. 2011. Resilience in agriculture through crop diversification: adaptive management for environmental change. BioScience 61:183–193.

Kang, S. Z., X. L. Su, L. Tong, J. H. Zhang, L. Zhang, and W. J. Davies. 2008. A warning from an ancient oasis: intensive human activities are leading to potential ecological and social catastrophe. Int. J. Sustain. Dev. World Ecol. 15:440–447, 9.

Kirkegaard, J. A., and J. R. Hunt. 2010. Increasing productivity by matching farming system management and genotype in water-limited environments. J. Exp. Bot. 61:4129–4143.

Lobell, D. B., W. Schlenker, and J. Costa-Roberts. 2011. Climate trends and global crop production since 1980. Science 333:616–620.

Pollock, C. J. 2016. Sustainable Farming: chasing a mirage? Food Energy Secur. 5:205–209.

Porter, J. R., L. Xie, J. Andrew, K. Cochrane, S. M. Howden, M. M. Iqbal, D. B. Lobell, and M. Travasso (2014) Food security and food production systems. In: Climate Change 2014: Impacts, Adaptation, and Vulnerability. Part A: Global and Sectoral Aspects. Contribution of Working Group II to the Fifth Assessment Report of the. Intergovernmental Panel on Climate Change. Available via http://www.ipcc.ch/pdf/assessment-report/ar5/wg2/WGIIAR5-Chap7_FINAL.pdf (accessed 17 March 2017).

Price, A. H., G. J. Norton, D. E. Salt, O. Ebenhöh, A. A. Meharg, C. Meharg, M. R. Islam, R. N. Sarma, T. Dasgupta, A. M. Ismail, K. L. McNally, H. Zhang, I. C. Dodd, and W. J. Davies 2013. Alternate wetting and drying irrigation for rice in Bangladesh: is it sustainable and has plant breeding something to offer? Food Energy Secur. 2:120–129.

Ribaut, J.-M. 2006. Drought adaptation in cereals, 642 pp. Haworth's Food Products Press, New York.

Ribaut, J.-M. 2014. How to build research partnership that benefit farmers. SciDev.Net. Available at http://www.scidev.net/global/r-d/opinion/build-research-partnerships-benefit-farmers.html. (accessed 17 March 2017).

Rubaiyath, A., A. N. M. Bin Rahman, and J. Zhang. 2016. Flood and drought tolerance in rice: opposite but may coexist. Food Energy Secur. 5:76–88.

Tardieu, F. 2012. Any trait or trait-related allele can confer drought tolerance: just design the right drought scenario. J. Exp. Bot. 63:25–31.

van Ittersum, M. K., L. G. J. van Bussel, J. Wolf, P. Grassini, J. van Wart, N. Guilpart, et al. 2016. Can sub-Saharan Africa feed itself? PNAS 113:14964–14969.

Varshney, R., J.-M. Ribaut, E. S. Buckler, R. Tuberosa, J. A. Rafalski, and P. Langridge. 2012. Can genomics boost productivity of orphan crops. Nat. Biotech. 30:1172–1176.

Xue, F., A. Porporato, and I. Rodriguez-Iturbe. 2013. Changes in rainfall seasonality in the tropics. Nat. Clim. Chang. 3:811–815.

Yang, J., and J. Zhang. 2010. Crop management techniques to enhance harvest index in rice. J. Exp. Bot. 61:3177–3189.

Zhang, W., G. Cao, X. Li, H. Zhang, C. C. Wang, Q. Liu, et al. 2016. Closing yield gaps in China by empowering smallholder farmers. Nature 537:671–674.

13

Photobiology in protected horticulture

Phillip A. Davis & Claire Burns

Stockbridge Technology Centre, Cawood, Selby, North Yorkshire YO8 3TZ, UK

Keywords
Horticulture, Light emitting diodes, crop production, photobiology, photomorphogenesis, photosynthesis.

Correspondence
Dr Phillip A. Davis, Stockbridge Technology Centre, Cawood, Selby, North Yorkshire YO8 3TZ, UK.

E-mail: phillip.davis@STC-nyorks.co.uk

Funding Information
Dr Davis' contribution to this review was funded by AHDB Horticulture Fellowship (CP085).

Abstract

The introduction of high power LED lighting systems for horticulture has stimulated substantial interest from both the research community and the protected horticulture industry. LED lighting systems have the potential to reduce electrical energy consumption compared to conventional high pressure sodium lights and their energy efficiency continues to improve. In addition to the potential of LEDs to reduce carbon footprints and reduce running costs, LED lighting also provides considerable opportunities to exploit the wealth of photobiological knowledge to produce horticultural benefits. The narrow emission spectra of LEDs allows lighting systems to be tightly designed to stimulate specific plant photoreceptors, allowing plants to be manipulated to produce desirable characteristics. Lighting systems can be designed to maximize growth, control morphology, and optimize flavor and pigmentation. This review outlines how the light spectrum influences photosynthesis and how plant photoreceptors sense light and control growth. The review then discusses the ways in which this knowledge is being implemented in commercial horticulture to improve factors such as yield, flavor, color, plant growth, and flowering as well as pest and pathogen management and control. Research in this area is moving rapidly as the LED systems improve and increase in efficiency and as the range of novel horticultural applications expands.

Glossary

Cryptochrome	A photoreceptor that is sensitive to blue and UVA light.	Photomorphogenesis	The processes that causes plant morphology and pigmentation to change following exposure to light. These processes are activated and controlled by several photoreceptors
DE	Day-length-extension lighting. Light treatments provided to extend the length of the photoperiod.		
EOD	End-of-day lighting. Light treatments provided for a short period at the end of the day to manipulation plant light responses.	Photon irradiance	A measurement of the number of photons incident on a surface, which has units of $\mu mol[photons]/m^2/s$.
HPS	High pressure sodium lighting.	Photoreceptor	Light-sensitive proteins that initiate light responses.
LED	Light-emitting diodes.		
NB	Night-break lighting. Light treatments provided during the middle of the night.	Phototropin	A photoreceptor that detects blue and UVA light.
PAR	Photosynthetically active radiation (PAR) is light with wavelengths in the range 400–700 nm that can be used by plants for the process of photosynthesis.	Phytochrome	A photoreceptor that can sense the red:far-red ratio of light.
		UVR8	A photoreceptor that is able to detect UVB light
PGR	Plant growth regulators.		

Introduction

Advances in technology are rapidly exploited by the protected horticulture industry, and have expedited the progression from structures that provide simple frost protection to sophisticated automated plant factories in which all environmental parameters are carefully regulated. Such advances have allowed crops to be produced year-round and can facilitate substantial yield increases. For example, lettuce production in plant factories increased crop yields by up to 100 times per unit area of land compared to conventional outdoor practices (Kozai 2013). As with glasshouse production in light-limited conditions such as winter months in northern latitudes, crop production in plant factories requires artificial lighting. In large-scale plant production systems, the efficiency of the lighting system is key to maintaining profitability. The introduction of LED lighting systems for horticultural use has attracted considerable attention for their potential to reduce electrical energy inputs, making winter crop production more financially viable. Horticultural LED lighting systems exhibit considerable diversity in their design, control systems, and the light spectra produced. LED lighting systems can be highly energy efficient; however, not all LEDs are more energy efficient than standard high-pressure sodium (HPS) lights (Pearson et al. 2015), and care is needed during design and implementation to ensure that the installed systems meet the needs of the crop production system.

The potential reduction in energy consumption that LED systems can deliver is highly desirable to the horticultural industry. However, the greatest potential of LED lighting systems to alter and improve crop production comes from the tight spectral control that LEDs provide. Light quality influences all aspects of plant biology, and much is known regarding the photobiological processes by which plants sense and respond to the light environment. Manipulation of the light spectrum using LED lighting allows the exploitation of this substantial body of knowledge to achieve improved crop production systems. Optimal lighting regimes have the potential to increase yields and improve plant quality, nutritional value, and flavor. In addition to enhancement of plant health and quality, the impacts of pests and pathogens can be reduced both as a result of elevated plant resistance and also by direct disruption of pest/pathogen biology. In this review, we briefly outline how plants use and respond to light before examining how LEDs are being used to manipulate crop photobiology and improve protected horticulture.

Optimizing the Use of Light to Maximize Photosynthesis

Plants are able to use light with wavelengths in the range 400–700 nm for photosynthesis. This waveband, which is often referred to as photosynthetically active radiation (PAR), contains 26% of the photons and 42% of the energy reaching the Earth's surface (calculated from ASTM G173-03 reference spectra). Plants must receive sufficient light to drive active growth and maintain plant quality and productivity. To maximize productivity and minimize energy inputs, artificially-supplied light must provide wavelengths that are used efficiently and meet plant needs. The light action spectrum for photosynthesis was first described by McCree (1971), who showed that red light was optimal for light-limited photosynthesis. More recently, research showed that changes in the concentration of plant accessory pigments such as carotenoids, which absorb predominantly in the blue region of the PAR spectrum, are responsible for the differences in light-use efficiency between a) the red and blue regions of the spectrum, and b) leaves grown under different conditions (Hogewoning et al. 2012). Although such pigments reduce the light-use efficiency of blue light, they are vital for protecting the photosynthetic machinery against UV damage (Middleton and Teramura 1993). While red light is utilized most efficiently for photosynthesis, red light alone is not sufficient to maximize photosynthesis. Blue light is required to prevent 'red light syndrome' (Trouwborst et al. 2016), which is characterized by suboptimal morphology and aberrant gene expression and biochemistry. Blue light is also needed to promote stomatal opening, improving access to CO_2, and driving transpiration and nutrient uptake (Hogewoning et al. 2010; van Ieperen et al. 2012; Nanya et al. 2012; Savvides et al. 2012). Other light wavelengths can further increase photosynthesis under certain circumstances. For example, green light can penetrate further into both canopies and individual leaves than red or blue light, and can drive photosynthesis in cells/leaves that are not reached by red and blue light (Sun et al. 1998; Terashima et al. 2009; Paradiso et al. 2011). The results of Terashima et al. (2009) also demonstrated the interactions of light quality and intensity with the greatest benefit of green light occurring at intermediate light intensities. The benefits of including green light in a customized spectrum for plant production would need to be evaluated with regard to the energy required to generate green wavelengths. Currently, red and blue LEDs are more energy efficient than green LEDs.

Unlike traditional lighting systems, LEDs can be turned on and off rapidly (hundreds of times per second). This creates the opportunity to potentially maximize the photosynthetic performance of crops while minimizing the energy inputs. In theory, it would be possible to pulse the light in such a way as to deliver the correct amount of light energy to excite every photosystem in a leaf without inducing the array of energy dissipation mechanisms that help protect plants from damage under natural conditions. This would help to maximize the light-use efficiency of plants. Tennessen et al. (1995) demonstrated that, provided intervals between light pulses were less than 200 μs, the amount of photosynthesis was proportional to the total amount of light provided to the plants. Jao and Fang (2004) observed that Potato plantlets grew fastest when light was pulsed at 720 Hz. and noted that 180 Hz provided the most energy-efficient system and would be appropriate where reduction in energy consumption was paramount. Shimada and Taniguchi (2011) found that photosynthetic rate and plant morphology were adversely affected in plants exposed to out-of-phase red and blue light pulses compared to plants exposed to in-phase light pulses. While this experiment provided interesting results from the perspective of how plants sense and use light, the most pronounced physiological effect observed was an increased shade avoidance response, which is of no benefit for most horticulture applications. In addition, pulsed light may not provide the desired effects in a glasshouse setting where natural light is present. Several research groups are developing sensor-controlled lighting systems that modulate the light regime to match ongoing plant needs. These technologies have the potential to maintain plant growth rates and quality during variable weather conditions while minimizing energy consumption.

Mobile lighting systems, which use fewer lamps and have correspondingly lower costs, have been trialed for crop production. Mobile systems have two major limitations: a) the amount of light supplied to plants is usually lower than with a static system, and b) systems to move lamps are needed, with associated installation and maintenance requirements. However, costs for mobile systems could be minimal if existing mobile irrigation booms were used to mount lamps. Li et al. (2014) showed that lettuce plants could be grown under mobile lights. However, this mobile system used half the number of lamps as were used for the fixed LED control treatment, and a more substantial reduction in lights would be needed to achieve viable savings in a commercial system. Due to the design of mobile lighting systems plants receive a variable intensity as the lights pass over the crop. However, plants can take up to 45 minutes of constant light to achieve maximum photosynthetic rates (Kirschbaum and Pearcy 1988), and much of the light provided during passage of a mobile light would therefore not be used for photosynthesis. Mobile lighting is therefore unlikely to be suitable for the majority of applications. However, in instances where only low doses of light are required (for example, end-of-day light treatments, or UVC/UVB treatments), mobile lights mounted on irrigation booms may provide an economically viable way of installing lamps.

Improving Yields with Interlighting

The light intensity within plant canopies decreases with depth as leaves absorb the light. Due to the light gradients within canopies leaves at the top may be light saturated when the canopy as a whole is light limited. Under these conditions adding artificial light at the top of the canopy can increase yield but any light absorbed by leaves that are already light saturated will provide no additional growth potential and those leaves may even become light stressed. By directly providing supplemental light to leaves that are shaded lower down in the canopy (interlighting) a greater proportion of the light can be used for photosynthesis without exceeding the point of photosynthetic light saturation. LEDs have made interlighting systems practical in commercial settings as they are cool to touch and can be placed close to crops without burning the plants. LEDs located within a cowpea (*Vigna unguicultata* L. Walp.) canopy were able to improve biomass production as well as reducing the senescence of older leaves deep in the canopy (Massa et al. 2008). Trouwborst et al. 2010 found that interlighting in cucumber crops increased leaf photosynthetic rate and photosynthetic potential of leaves lower in the canopy. However, the interlighting treatments caused extensive leaf curling. This led to a reduction in light interception by the canopy and prevented the interlighting treatment from increasing crop yields. Hao et al. 2012 also had mixed results when using interlighting with cucumber. Over the first two weeks of the experiment, visual quality and yield increased by more than the increase in total photon irradiance; however, these gains declined as the experiment progressed, especially in the blue interlighting treatment where some leaf curling was observed. Guo et al. (2016a) found that LED interlighting plus HPS top lighting increased the yield and concentration of health providing compounds of sweat peppers in comparison to HPS only light treatments. However, it should be noted that the total amount of light supplied in the interlighting treatments was greater than that provided by the HPS only treatments. Interlighting in tomato crops has proved highly successful and there are now a growing number of commercial installations, all of which are reporting significant increases in yields. Interlighting trials have also investigated the use of different light spectra at different locations with canopies. The addition of increasing amounts of blue light within the canopy was found to increase yields of cucumber but not tomato plants, though the blue light reduced the internode lengths of both species (Ménard et al. 2006). Guo et al. (2016b) found that applying far-red at the top of a cucumber canopy was able to increase the yield while providing blue at the top reduced yield. The best yields were achieved with far-red at the top of the canopy and blue light at the bottom. Many novel approaches to lighting crops with LEDs are now possible but trialing all these approaches would be costly and time consuming. To aid the development of lighting systems, modeling approaches will be increasingly important for identifying designs that should be trialed in the real world. de Visser et al. (2014) used 3D ray tracing models to assess the light interception of tomato canopies when illuminated with interlighting set at different angles of incidence. Their models predicted that light interception and photosynthesis could be maximized by positioning the interlights so they shone slightly upwards compared to horizontally as is the case in most commercial systems.

The Impact of LEDs of Overall Glasshouse Energy Budgets

One potential negative impact of LED lighting in glasshouses is the lack of radiative heat that is produced by LEDs. In experiments where the energy consumption of glasshouses has been monitored, LED-lit compartments required higher air temperatures to counteract the loss of radiative heat. This reduced the overall energy saving as there was a greater heating demand. Dueck et al. (2012) reported that the use of LEDs for tomato production increased energy consumption; however, this was attributed to the energy demands of the water cooling systems of the LEDs used in that particular system. The majority of current commercially available LEDs do not require water cooling systems. Gómez and Mitchell (2013) examined the use of LED towers in comparison to standard HPS lighting for tomato production. Their results indicated that the LEDs provided a significant energy saving but provided similar yields as the HPS lighting systems. In a more detailed analysis of the system, Gómez et al. 2013 measured the efficiency of electrical conversion into fruit biomass to be 75% greater for the LED lights compared to sodium lamps; however, this did not take heating requirements into consideration. In all experiments that compare HPS and LED light there is a need to assess the differences in plant temperature to ensure that any effect of temperature can be separated from the effects of light on plants responses. HPS light can increase leaf temperature by several degrees and this can increase plant growth rates. While the drop in crop temperature may have negative effects on crops in the colder months of the year, the lower temperature will benefit crops on warm days with low light levels. As with any significant change in crop environment, the switch from HPS to LED lighting will require a period of learning to develop protocols for correct management of plant irrigation and growth.

Overview of Plant Photoreceptors and Photomorphogenesis

Photomorphogenesis is the process by which photoreceptors drive changes in seedling morphology in response to light exposure after germination. Several processes responsible for efficient, healthy growth are stimulated during this process. For example, one of the prime ways in which photomorphogenesis optimizes plant performance is by stimulating seedlings to orient toward light, thereby maximizing light capture and photosynthesis. Photomorphogenesis is mediated by several types of photoreceptor, each of which is sensitive to distinct parts of the light spectrum. The different photoreceptors control several photomorphogenic processes that are important for plant survival, growth, and development.

The UVR8 photoreceptor responds to UVB light, with peak sensitivity at ~290 nm (Brown et al. 2009). Plants produce a range of pigments and other secondary metabolites in the presence of UVB light, and these act as sunscreens to provide protection against UV light damage (Chalker-Scott 1999). In particular, production of flavonoids and anthocyanins increases after exposure to UVB (Tevini and Iwanzik Wm Thoma 1981; Beggs and Wellman 1985), and the visual appeal of leaves and flowers is enhanced (Paul et al. 2006). Plants also retain a compact shape when exposed to UVB light (Gardner et al. 2009), and form tougher, more robust, leaves (Wargent et al. 2009). A further benefit of UVB light exposure is an increase in the concentration of essential oils in herbs (Kumari et al. 2009; Hikosaka et al. 2010).

Plants grown in the absence of blue light become etiolated and leaves tend to hang downwards and remain curled. Several photoreceptor families are responsible for sensing blue and UVA light, including the phototropins and cryptochromes. Phototropins control a wide range of plant responses such as stomatal opening, phototropism (bending toward light), chloroplast movement (Briggs and Christie 2002), leaf flattening (de Carbonnel et al. 2010), and de-etiolation of the hypocotyl (Folta and Spalding 2001). Phototropins regulate cellular processes while leaving gene expression unchanged. By contrast, cryptochromes regulate gene expression, resulting in downstream effects on secondary metabolism and pigment synthesis (Vlohr and Drumm-Herrel 1983), flowering (Giliberto et al. 2005), and inhibition of hypocotyl elongation (Folta and Spalding 2001). Cryptochromes also function to entrain circadian rhythms (Cashmore 2003). Phototropins and cryptochromes are similarly sensitive to UVA and blue light (Briggs and Christie 2002) but there is some evidence that cryptochromes can also be partially inactivated by green light (Sellaro et al. 2010). For a more detailed overview of the blue/UV light signaling networks see Huché-Thélier et al. (2016).

Red and far-red light are sensed by another family of photoreceptors, the phytochromes, which function similarly to the cryptochromes in that they mediate their effects by altering gene expression. Phytochromes are involved in the control of a number of photobiological responses such as germination (photoblasty), inhibition of hypocotyl elongation, apical hook straightening, leaf expansion, flowering time, circadian rhythm entrainment, and chlorophyll biosynthesis. Plants possess several phytochromes (e.g., arabidopsis has five phytochromes, rice has three, and maize has six), each of which has a different functional range. Two types of phytochrome have received extensive investigation: phytochrome A (phyA) and phytochrome B (phyB). PhyB is activated by red light (peak absorbance, 666 nm) and deactivated by far-red light (peak absorbance, 730 nm). A short pulse of red light is sufficient to activate phyB; however, if a red pulse is followed by a far-red pulse the red light responses do not occur. This is termed red, far-red reversibility. In contrast with phyB, phyA can be activated by both far-red and red light (Shinomura et al. 1996). However, in the light, phyA is downregulated both transcriptionally and post-transcriptionally (Chen and Chory 2011) and PHYA primarily accumulates in plant tissues during periods of darkness. Although phytochromes are mainly considered with regard to their ability to detect red:far-red ratios, phytochromes absorb light wavelengths from the full spectrum, including blue light. Green light, for example, can stimulate phyA- and phyB-mediated germination (Shinomura et al. 1996). For a more detailed overview of red:far-red responses in plants, see Demotes-Mainard et al. (2016).

No photoreceptor with specificity for green light has been identified to date. Inclusion of green light in illumination mixes has been reported to increase plant growth rates; however, it remains to be seen whether this occurs as a result of the direct effect of green light on photosynthesis or via some other photomorphogenic effects. For a detailed review of the influence of green light on plant production, see Wang and Folta (2013).

The cellular and plant physiological responses induced by different photoreceptors overlap to some extent, with many of these responses caused by alterations to the synthesis and transport of plant hormones (Lau and Deng 2010). The tight spectral control provided by LED lighting allows different photoreceptors to be selectively activated, permitting the light environment to be tuned to manipulate plant responses and enhance desirable plant qualities. Such manipulation is impossible with traditional light technology, and it is through the precise activation of different photoreceptors in diverse crops that LED lighting has the potential to revolutionize horticultural practices.

The Effect of Light Quality on Propagation

Plant propagation is the first stage of horticultural crop production, and maximizing the efficiency of seed germination and rooting of cuttings can have huge impacts on overall yields and profitability. Light is an important cue both for seed germination and for root development of cuttings and providing optimal lighting conditions can greatly improve both the speed and success of propagation. Jankowska-Blaszczuk and Daws (2007) found that seeds were less dependent on light for germination as they increased in size. Seed size was also found to correlate with the tolerance of germination to shade conditions, as determined by the response to red:far-red light ratios (high red:far-red corresponds to full sun conditions with little shade). Smaller seed needed higher red:far-red ratios for germination than larger seeds. Germination efficiency can also be tailored using other regions of the light spectrum. For example, germination of Chinese ladder brake fern (*Pteris vittata*) spores can be inhibited by blue light. Interestingly, Sommer and Franke (2006) observed that exposing seeds of cress, radish, and carrots to bright green laser light caused the plants to grow considerably larger. No biological explanation for this observation has been elucidated, but further investigation may identify some useful practical applications.

Many important horticultural crops are propagated from cuttings or using micropropagation techniques. Using spectral manipulation to improve propagation efficiency is of particular interest for high-value crops that are challenging to root. One of the challenges of taking cuttings is preventing dehydration. Plastic sheeting and fogging helps to reduce transpiration, but can also be reduced by manipulation of the light spectrum. Blue light drives stomatal opening, so removing blue light from the spectrum helps to reduce transpiration and improve cutting survival. Red light has been shown to be beneficial in promoting root development in several species. Rooting was improved in two of three varieties of grape (*Vitis ficifolia*) when illuminated with red light compared to fluorescent or blue light. In the third grape variety, rooting levels were high and similar in all light treatments (Poudel et al. 2008). When Wu and Lin (2012) propagated *Protea cynarodies* plantlets under red LED light, 67% rooted compared to 7% under conventional fluorescent tubes, and 13% rooted under blue light or a red:blue (50:50%) LED light combination. Root development was also found to be more extensive under the red light treatments. In *Protea cynarodies* cuttings, Wu (2006) observed that the concentration of phenolic compounds increased over time and that root development only occurred after their concentration reached a certain level. Further investigation demonstrated that the phenolic compound 3,4-dihydroxybenzoic acid could promote root formation up to a concentration of 100 mg/L but inhibited root formation at higher concentrations (Wu et al. 2007). The use of red light in the propagation phase caused the plants to generate phenolic compound concentrations favorable to rooting, while the inclusion of blue light elevated concentrations further, thus inhibiting rooting. Similar effects may be occurring in other species, though the active compounds are likely to vary between species.

Limiting the spectrum to 100% red light is not always optimal for propagation, and a range of red:blue light mixtures have been found to produce optimal rooting in several species. *In vitro* propagation of banana plantlets and subsequent transfer to a soil based growing substrate was found to be best when performed under 80% red: 20% blue light (Nhut et al. 2002). A 50% red: 50% blue light treatment was most effective for the propagation of Cotton plants (Li et al. 2010). Strawberry plantlets performed best when propagated under 70% red: 30% blue light (Nhut et al. 2003). In climbing Gentian, red light was found to promote rooting while blue light inhibited rooting, with optimal rooting observed using a 70% red: 30% blue mixture (Moon et al. 2006). Even in species where 100% red light treatments are thought to provide optimal rooting, plants must still be moved to a light treatment containing some blue light after the critical stage of root initiation in order to prevent etiolation of the young plants and help ongoing root development: blue light enhances both root and shoot development (Nhut et al. 2003).

Light treatments provided to parental stock plants prior to removal of cuttings may also influence rooting success rates. *Eucalyptus grandis* cuttings had greater rooting success when the stock plants were grown under low red:far-red ratios (Hoad and Leakey 1996). Cutting success is closely linked to cutting quality, and improving the quality of stock plants through changes to lighting or with spectral filters is expected to provide significant benefits, especially if combined with optimal postcutting light treatments.

The Influence of Light Quality on Plant Growth

The first attempts at producing food crops under LED lighting systems were limited by the intensity and color of the LED lamps (only red LEDs were available), and fluorescent tubes were used to provide the blue light required to maintain plant health and growth rates (Bula et al. 1991; Barta et al. 1992; Tennessen et al. 1994). LED technologies have advanced greatly since these early attempts, and LED lighting systems that provide multiple colors of light are now available for horticultural

production. The benefits of including blue light in the spectrum have been demonstrated on numerous occasions. Wheat and Arabidopsis plants produced more seeds (Goins et al. 1997, 1998), lettuce, radish, and spinach produced more biomass (Yorio et al. 2001), and frigo strawberries produced more fruit with higher sugar contents (Samuoliene et al. 2010) when grown in the presence of blue light.

While the need for blue light in artificial growth systems is clear, there is less consensus regarding optimal red:blue mixtures. Part of this uncertainty is due to differences in stomatal light responses between species. Ouzounis et al. (2014) showed that stomatal conductance in campanula showed no response to supplemental LED lighting, whereas stomatal conductance increased in both roses and chrysanthemums. Light quality affected stomatal development in *Withania somnifera* plantlets, with monochromatic light resulting in impaired stomatal development (Lee et al. 2007). Stomatal development can also be influenced by UVB light. For example, soybean produced fewer stomata after UVB exposure and, although this could improve drought tolerance, photosynthetic performance could be adversely affected (Gitz et al. 2005). Although stomata routinely open and close in response to light to regulate water use and CO_2 uptake, any influence of light quality on the development and density of stomata during leaf growth will have long-term impacts on stomatal conductance, photosynthetic performance, and water-use efficiency.

In addition to its effects on stomata, light quality also influences many other responses required for healthy plant growth through regulation of plant metabolism and morphology. Cucumber plants grown under blue and UVA light were found to have both higher photosynthetic potential and increased transcription of the genes required for carbon fixation compared to plants grown under red, green, or yellow light (Wang et al. 2009). In rice, addition of blue light to a red background led to higher photosynthetic and stomatal conductance rates and was associated with higher chlorophyll and Rubisco contents (Matsuda et al. 2004). In lettuce, growth rates (measured as biomass accumulation) decreased as UVA and blue light increased (Li and Kubota 2009; Son and Oh 2013). In contrast, an increase in rapeseed growth rate was observed as blue light percentage increased from 0% to 75% (in a red: blue mix; Li et al. 2013). Folta and Childers (2008) observed the greatest growth of strawberry plants under 34% blue light. However, Yoshida et al. (2012) found that Strawberry fruit yield was greatest in plants grown under continuous blue light, and that red light inhibited flowering.

The inclusion of green light from LEDs increased fresh and dry weight biomass accumulation in lettuce when green light replaced some of the blue or red light in a light treatment mixture (Kim et al. 2004a; Stutte et al. 2009). There are two possible reasons for these observed increases in growth rate. First, green light can penetrate deeper into the plant canopy and, therefore, drive more photosynthesis. Second, reducing the amount of blue reduces the restriction on leaf expansion imposed by plant photoreceptors (Dougher and Bugbee 2004), thus increasing leaf area, light capture and growth. Research to date suggests that addition of green light to the growth spectrum does not enhance crop performance in all cases. Contrasting with the study noted above, Li and Kubota (2009) found that addition of green light caused no increase in lettuce biomass, but that plant morphology was affected and that an increase in stem and leaf elongation was observed. Early in lettuce crop development, larger leaves may benefit crop performance by allowing greater light capture; however, larger leaves later in the crop cycle may reduce plant quality, especially if combined with stem extension. Some of the discrepancies between the different sets of published results may be due to differences in the total amount of light provided as well as the proportions of green light provided. Kim et al. (2004b) found that 24% green light boosted lettuce yields; however, yields were reduced when greater than 50% green light was used, probably as a result of lower overall photosynthetic rates. Kim et al. (2004a) also found that in lettuce green light could cause stomatal closure and that stomatal opening was greatest under broad spectrum lighting (suggesting that white light may be better in this case). In Tomato transplants, the addition of small amounts of green (520 nm), orange (622 nm), or yellow (595 nm) LED light was found to reduce plant growth rates (Brazaitytė et al. 2010), and some of the negative impacts on plant growth could still be observed one month after exposure to the different light treatments (Brazaitytė et al. 2009). Yellow light was also found to suppress the growth of lettuce plants (Dougher and Bugbee 2001); however, it should be noted that these experiments were not performed with LED lighting and spectral assessments were complex. Lu et al. (2012) examined the effect of supplemental LED light on Tomato production on the single truss system and concluded that white light would be more effective at driving canopy photosynthesis in dense canopies than red or blue light, because the green light component of the white light spectrum penetrates further into the canopy than red or blue.

Far-red light is important for plant development and performance throughout the life of the crop. For example, far-red light can inhibit germination of lettuce seeds (Borthwick et al. 1952; Shinomura et al. 1996) and generally counteracts the influence of red and blue light on plant morphology. This results in plant stretching and reduced chlorophyll content (Li and Kubota 2009), which can also reduce photosynthetic rates. Far-red can, however, be beneficial to crop growth by increasing leaf area in lettuce (Li and Kubota 2009; Stutte et al. 2009), potentially allowing greater light capture. Far-red light can also increase the yields of green beans (Davis 2013). Due to the sensing mechanisms of the phytochromes, the effects of far-red light can be achieved by providing light at the end of day (EOD), or during the night period with day-length-extension (DE) or night break (NB) lighting techniques, rather than during the day. For example, EOD-far-red treatments are effective in encouraging tomato plants to grow taller, which could be used optimize the production of seedlings for grafting, and EOD-red treatments can improve plant compactness (Kubota et al. 2012). These lighting techniques generally require lower light intensities (1–5 μmol/m^2/s can provide strong influences on crop responses) and durations than daytime far-red light treatments and can therefore be used to influence morphology with lower energy investments.

Improving Crop Morphology and Reducing the Use of Plant Growth Regulators

Spectral manipulation can maximize biomass production but, depending on the particular conditions and crop, larger plants may not be desirable. Morphology and quality may be negatively affected if plants are grown 'too soft'. Several methods are used in the industry to control plant morphology during crop production, including reduced irrigation, increased electrical conductivity (EC) of irrigation solutions, application of plant growth regulators (PGRs), and altering temperature profiles (e.g., negative DIF). The use of PGRs in the ornamentals sectors is particularly widely used to help maintain crop compactness during periods of low light. The ability to control the light spectrum with LEDs, or with spectral filters, provides the potential to manipulate plant morphology, reducing the need for chemical intervention. Poinsettias grown under 80% red: 20% blue supplemental LED lighting were 20-34% shorter than those grown under HPS (5% blue) lamps (Islam et al. 2012). Although leaves were smaller and plants accumulated less dry matter, there was no delay in bract color formation or postproduction performance, indicating that LEDs could be useful for reducing the use of PGRs for Poinsettia production. An increase in the blue light proportion of supplemental light was also found to cause roses and chrysanthemum to remain more compact during production: the most compact roses were observed under 40% supplemental blue light (the highest proportion examined in the study; Ouzounis et al. 2014). The quality of the supplemental light was also found to strongly influence leaf morphology, with 100% red light treatments causing rose leaves to become curled. In many species, stem elongation decreases as the proportion of blue light increases (Moon et al. 2006; Nanya et al. 2012). While higher percentages of blue light reduce plant height, the concomitant reduction in leaf size may have negative influences on growth and development that could influence production periods. The red:blue ratio, while important, is not solely sufficient to control plant morphology: light intensity is also critical. In tomato plants, the absolute blue light intensity rather than the blue percentage in the light recipe controlled hypocotyl length and stem extension (Nanya et al. 2012). While stem elongation was controlled by blue light, the position of the first flower truss developed in proportion to the total photosynthetic rate of the plant (more photosynthesis = earlier truss development). In principle, this would mean that a plant grown in 75% blue light of 100 μmol/m^2/s would have the same internode size as a plant grown in 38% blue light at 200 μmol/m^2/s, though it should be noted that plants grown at the higher light intensity would grow more quickly and flower earlier. Higher light intensities can reduce crop production time which, after an initial capital investment in lamps, has the potential to reduce production costs.

Other colors of light also influence plant morphologies. In chrysanthemums, blue light reduced leaf mass, green light reduced stem mass, and red and far-red light caused a reduction in root mass (Jeong et al. 2012). In contrast to roses and chrysanthemums, campanula height was unaffected by supplemental blue light and, in this case, the addition of red light provided the greatest effect on reducing plant height (Ouzounis et al. 2014). The addition of red light likely reduced plant height by changing the red:far-ratio, and reducing far-red light with spectral filters could have a similar influence on plant morphology. Differences between plant morphological responses to red/far-red and blue light are associated with differences in the relative contributions of phytochromes and blue-sensitive photoreceptors (cryptochromes and phototropins) to inhibition of stem extension. A better understanding of the light regulatory pathways between species will help further improve lighting strategies but may also highlight new areas for detailed scientific investigation.

The Use of Far-red Light in Control of Flowering

The effects of far-red light on plant morphology can be substantial; however, the greatest potential for far-red light in horticultural applications is in the control of flowering time. Runkle and Heins (2001) demonstrated that far-red light promotes flowering in several long-day ornamental species and that an absence of far-red light can even prevent flowering. This work examined plant light responses using spectral filters. With the use of LED light treatments, plants can be grown using even more extreme ranges of red:far-red ratios, providing the potential for either delaying or advancing flowering still further than has been seen with spectral filters. In petunias and pansies grown under red:blue:far-red light mixtures, flowering was induced up to two weeks earlier in plants treated with far-red light compared to plants grown without far-red light (Davis et al. 2015).

Adams et al. (2012) investigated the use of several types of LED with different spectra for use in NB and DE lighting to promote and delay flowering in several long- and short-day species. Spectral quality of the lights was found to have a significant impact on their effectiveness in controlling flowering. Far-red-only and red + white + far-red lamps promoted flowering at levels similar to incandescent lamps, while red + white lamps were less effective than incandescent light. For example, in short days, chrysanthemum flowering was delayed by NB and DE illumination with red + white and red + white + far-red lamps. Far-red-only lamps had no effect on any of the short-day plants. None of the LED light combinations were found to be as effective as incandescent lamps at delaying flowering in Christmas cactus. Begonia and poinsettia flowering times were advanced in response to red+white + far-red light treatments. LED treatments affected morphology as well as flowering time and plants grew taller as the amount of far-red in the treatments increased. Craig and Runkle (2016) showed that NB lighting had its greatest effect on flowering in several long-day plants when an intermediate red:far-red ratio was applied. If red:far-red ratios were too high or too low, flowering promotion was reduced. The influence on plant stretching was also greatest under the light treatments that had the greatest influence on flowering. Chrysanthemums normally flower when days are shorter than 13.5 h, but Jeong et al. (2012) observed that a 4 h day DE blue light treatment prompted flowering during a 16 h day. These experiments demonstrate that LEDs can be used effectively to control plant flowering and also highlight the importance of the spectral composition of lights used for this application.

Secondary Metabolites: Pigmentation, Flavor, and Aroma

Primary metabolites are the chemicals that are directly involved in normal growth, development, and reproduction, and loss of these compounds results in death. Plants also produce many other compounds, known as secondary metabolites, that act to improve the fitness of an organism and help it acclimate to a changeable environment (Lambers et al. 1998). Many of these compounds convey qualities that are desirable by humans such as color, flavor, and aroma. The production of many secondary metabolites is regulated by light (Samuoliene et al. 2013).

Red, far-red, and blue light have all been implicated in driving synthesis of the pigments required for photosynthesis (Tripathy and Brown 1995; Miyashita et al. 1997; Tanaka et al. 1998; Huq et al. 2004; Kim et al. 2004a; Moon et al. 2006; Li et al. 2010). Blue and red light cause an increase in chlorophyll levels, whereas far-red results in reduced chlorophyll contents. As well as influencing the appearance of plants, these changes can also alter the rate of photosynthesis and therefore impact plant growth rates. The link between secondary metabolites and photosynthesis comprises an additional layer of complexity that should be considered when designing light recipes.

Many crops have red-colored leaves or flowers that are distinctive and desirable, and maximizing pigmentation is important to retain quality for customers. Red pigmentation is mainly provided by two types of compound: anthocyanins and betacyanins. Anthocyanin synthesis is regulated by many different biochemical pathways, but blue-light via the cryptochromes (Ninu et al. 1999) is an important signal for driving synthesis. In lettuce, supplying supplemental LED lighting of different colors against a background of fluorescent white light resulted in increases in leaf anthocyanin, xanthophyll, and β-carotene concentrations (Li and Kubota 2009). UV-A and blue light both increased the anthocyanin concentration, with blue light prompting the largest increase. In contrast, far-red light and green light reduced anthocyanin concentration, and far-red light also reduced chlorophyll, xanthophyll, and β-carotene content. UVB was also shown to be a potent stimulator of anthocyanin production in lettuce (Park et al. 2007), and UV transparent spectral filters increased plant and flower pigmentation (Paul et al. 2006). Carotenoid concentration was found to be greater in buckwheat seedlings grown under white light compared to those grown with 100% blue or red light (Tuan et al. 2013). It should be noted that few plants perform well under 100% red or blue light, and a combination of red and blue light may produce carotenoids in similar quantities as white light. Polyphenol concentrations in chrysanthemum were at their highest levels when grown with red or green supplemental lighting and at their lowest levels when grown with blue supplemental lighting. However, the plants grown under blue light flowered, which may have influenced the production of secondary metabolites.

Betacyanins have replaced anthocyanins as red pigments in the Caryophyllales Order (excluding the families Caryphyllaceae, which contains *Dianthus,* and

Molluginaceae; Sakuta 2014). The Caryophyllales contains 6% of all eudicotes (~11,155 species) and includes the Amaranthaceae, (the family that contains spinach, swiss chard, and beetroot). Unlike anthocyanins, betacyanins do not appear to increase in concentration in response to blue/UVB light. These pigments instead appear to accumulate in response to red light and their synthesis is thought to be controlled by the phytochromes (Elliott 1979). This means that, while the red pigmentation of many species could be improved by increasing the amount of blue light, it is probable that plants in the Caryophyllales Order will improve their red pigmentation on provision of more red light.

Large differences in pigment contents can alter the flavor of crops, but light is also important in regulating the biosynthesis of many of the volatile compounds that create the flavor and aroma of leaves, fruits, and flowers. UVB light exposure has been linked to increased oil and volatile contents in a range of herb species including sweet flag (*Acorus calamus* L.; Kumari et al. 2009), japanese mint (*Metha arvensis* L var. piperascens; Hikosaka et al. 2010), lemon balm, sage, lemon catmint (Manukyan 2013), *Cynbopogon citratus* (Kumari and Agrawal 2010), and basil (Bertoli et al. 2013). In basil plants, blue light was also found to increase the oil content of leaves in comparison to white light treatments (Amaki et al. 2011). In the same study, green and red light were found to have little effect on oil contents, although green light was shown to increase crop biomass production compared to other light treatments. While more blue light can increase oil and other secondary metabolite contents, it is not always sufficient to simply provide more blue light. In basil plants grown under 100% blue light, rosmaric acid (RA) levels were 3 mg/L but under 100% red or white light the RA concentration reached 6 mg/L (Shiga et al. 2009). A possible reason for the lower level of secondary metabolite production observed in this study was that the photosynthetic rate under blue light was lower than under red or white light. Data from Manukyan (2013) indicated that increasing PAR led to an increase in production of secondary metabolites. It is important to provide plants with sufficient light to drive enough photosynthesis as this provides the metabolic building blocks for the various biosynthetic pathways as well as stimulating the biosynthetic pathways to maximize production of desirable compounds.

More recently, the effect of postharvest light treatments on secondary metabolites has been considered. Postharvest light treatments provide the potential to enhance crop qualities during transport, prior to sale or to delay the onset of senescence, thus extending shelf life. Costa et al. (2013) found that exposure to 2 hours of low intensity red light (30-37 μmol/m^2/s) delayed senescence of basil leaves for 2 days during storage at 20°C in the dark. The authors concluded that the effects were due to changes in gene expression mediated by phytochromes, which mediate cell senescence in low light conditions, rather than via photosynthetic carbon gain. Colquhoun et al. (2013) showed that postharvest light treatments in petunia, tomatoes, blueberries, and strawberries could alter the volatile compounds produced by the different crops. Eight-hour red and far-red light treatments increased levels of several volatile compounds in petunia that are known to be important components of flower scent. Fewer compounds were examined in strawberries and tomatoes, but both large increases and decreases were observed in amounts of volatile compounds in these crops following exposure to different light treatments.

Changing the light spectrum can cause amounts of some compounds to increase while others may decrease. While it is apparent that light treatments increase the concentration of certain compounds, it is not always understood how these changes may impact crop flavor. Many of the studies focus on just one or two compounds, but flavor is influenced by a large range of compounds. Due to the limits of our understanding regarding the ways in which secondary metabolites are influenced by light and how these influence flavor, it is currently more efficient to develop light treatments for improved flavor by trial-and-error. The compounds of importance, and their synthesis in response to light, can subsequently be elucidated.

Insect Management Under LEDs

Light is a highly important environmental cue for all insect species. Several aspects of the light environment influence insects, such as daylength, intensity, direction, polarization, spectrum, and contrast. These environmental cues influence many insect biological and behavioral responses including the circadian rhythm, host identification, take-off and landing frequency, reproductive success, phototaxis (movement toward or away from light), and feeding frequency. Improving our understanding of insect light responses will be important to ensure both pollination and pest control can be maintained under LED light sources. There are two main aspects to consider: 1) the direct effect of light quality on insect responses, and 2) the effect of host species responses to light quality in the insect of interest.

Our knowledge of the spectral sensitivity of insect vision is limited to a few species, but the diversity between species is considerable. For example, bees are able to see UV (peak absorbance ~350 nm), blue (peak absorbance ~450 nm), and green (peak absorbance ~550 nm) light but have low sensitivity for red light (Backhaus 1993). Many insects are also able to detect the polarization of light and use this information to navigate (Rossel 1993; Reppert et al. 2004). The pest *Caliothips phaseoli*, a thrip

species that attacks soy, has one photoreceptor that can only detect UV (UVA and UVB) light, and these insects are therefore blind to PAR. However, spectral sensitivity in this species is enhanced: some of the ommatidia of their compound eyes contain pigments that fluoresce under UVA light, and this acts as a UVA filter so those eyes only detect UVB light. This allows the insects to distinguish UVB from UVA even though they only have one photoreceptor (Mazza et al. 2010). Western flower thrips (*Frankliniella occidentalis*) see both visible and UV light. Males and females have similar visual responses but have different swarming behaviors, with males more likely to gather on flowers than females (Matteson et al. 1992). Behavioral responses to light, both innate and learned, provide added complexity to insect light responses that will provide added challenges to understanding and manipulating insect light responses.

A better understanding of pest light responses can be used to improve traps for monitoring insect populations. Making traps more attractive to insects can render them more effective, which enables earlier identification of pest issues. Green LEDs have been used to increase trap effectiveness for West Indian sweet potato weevils (*Eusceoes postfasciatus*; Nakamoto and Kuba 2004), Whitefly (*Bemisia tabaci*), Greenhouse whitefly (*Trialeurodes vaporarioum*), Fungus gnats (*Bradysia coprophila*), and Aphids (*Aphis gossypii*; Chu et al. 2004). In environments with no natural light, trap effectiveness will be more strongly influenced by the color of the traps and the color of the LEDs than in glasshouses where natural light dominates.

Indirect effects of light quality on pests are caused by plant responses to light. In a species of wild tomato (*Lycopersicon hirsutum*), seasonal changes in day length and quantity cause large changes in synthesis of 2-tridecanone, resulting in much greater concentrations in June than in January (Kennedy et al. 1981). When caterpillars of *Manduca sexta* were fed on tissue from plants grown in January, 8% died compared to 87% that perished when fed on plants grown in June. More subtle effects are likely to influence pest performance on crops grown in different light conditions. Plants produce a range of volatile organic compounds (VOCs) that act as attractants to both pests and beneficial insects. Changes in light quantity (Paré and Tumlinson 1999) and quality (Kegge et al. 2013) alter the production of VOCs and spectral manipulation may help enhance VOC production to maximize crop protection.

Plant Pathogens and Their Interactions with Light

The interactions between plants and their pathogens are also influenced by the light environment. Light affects many aspects of plant biology and many of these responses influence plant resistance to disease. The red:far-red ratio in particular has been shown to influence the expression of many genes, via the phytochromes, that are involved in disease resistance (Griebel and Zeier 2008). Low red:far-red ratios decrease the production of many secondary metabolites involved in disease resistance and thus reduce resistance (Ballaré et al. 2012). Salicylic acid (SA) and jasmonic acid (JA) both play important roles in mediating defenses against pathogens and low red:far-red ratios have been shown to reduce the response of both pathways to disease attack (de Wit et al. 2013).

Light will also have direct effects on fungal pathogens as they also possess an array of photoreceptors that modulate their gene expression (Corrochano 2007). Fungi have circadian rhythms (Liu and Bell-Pedersen 2006) and certain species sporulate at specific times of day to coincide with events that enable them to infect plants, such as during times when leaves are likely to be wet. Rose powdery mildew (*Podosphaera pannosa*) was found to release spores during the day and more spores were released with brighter light (Suthaparan et al. 2010). Light color was also important: compared to white light, more spores were released under blue and far-red light, and fewer spores were released under red light. Both day extension and night break light treatments with red light greatly reduced the release of mildew conidia, and such treatments may be useful in reducing the intensity and spread of mildew in crops. As powdery mildews are obligate pathogens, it is not possible to determine if the effect of the light treatments occurs as a direct effect on the pathogen or as a result of the plant responses. However, red light treatments have also been found to increase occurrence of two diseases in broad bean: *Alternaria tenuissima* and *Botrytis cinerea* (Islam et al. 1998). Spore germination rates were also affected by light color, with blue light reducing germination by 16.5% compared to other treatments. The spores of many plant pathogens are killed by exposure to solar radiation (Kanetis et al. 2010), with the UVB component of solar radiation being the most likely to cause spore death. Models of spore germination could be used to define the best time of day to provide a pulse of UV light that would maximize effectiveness. It may also be possible to use novel light strategies to increase disease control. Blue light inhibits spore germination, so if red light only is provided early in the day, spores will germinate and they can then be more easily killed with a UV pulse before the blue light is again turned on. UVC light has also been trialed for the control of plant diseases. For these treatments to be effective, it is important to make sure that treatments are applied when the pathogens are vulnerable. If the UV light is provided before sporulation or after infection then it will be ineffective at providing protection. If applied during the germination of the spores, then UVC can be effective at preventing infection. Designing the light scheme to co-ordinate UVC application with spore release or germination may be an effective method for controlling disease in controlled environment chambers. There are, however, health and safety issues regarding the use of UVC light in commercial settings where staff can be exposed.

Just as different plant species have different light responses, the light responses of different pathogens vary, as do the interactions in different plant/pathogen systems in response to light. Schuerger and Brown (1997) observed that in tomatoes infected with bacterial wilt (*Pseudomonas solanacearum*) and cucumber plants infected with powdery mildew (*Sphaerotheca fuliginea*), disease symptoms were at their lowest in plants grown under 100% red light. By contrast, for tomato mosaic virus (ToMV) on pepper plants, disease symptoms were slower to develop and less severe in plants grown in the presence of blue/UVA light. These data indicate that spectral modification could be used as part of an integrated disease management system, with the caveat that care must be paid to the development and achievement of appropriate light treatments.

Conclusions

The introduction of LEDs for use in horticulture is facilitating the application of photobiology at all stages of crop production from propagation to postharvest quality control. The diversity of LED applications is expected to increase in the near future as our understanding of plants, pest, and pathogens increases but also as LED technologies improve in efficiency and capital costs decrease. This technology has the potential to improve food quality, reduce energy consumption, and increase food security. This use of LEDs in horticulture is a rapidly evolving field and is expected to gradually revolutionize commercial crop production.

References

Adams, S., S. Jackson, V. Valdes, J. Akehurst, A. Hambidge, D. Fuller, et al. 2012. Protected ornamental: Assessing the suitability of energy saving bulbs for day extension and night break lighting. *HDC Project PC 296 Final report*.

Amaki, W., N. Yamazaki, M. Ichimura, and H. Watanabe. 2011. Effects of light quality on the growth and essential oil content in sweet basil. Acta Hortic. 907:91–94.

Backhaus, W. 1993. Color vision and color choice behavior of the honey bee. Apidologie 24:309.

Ballaré, C. L., C. A. Mazza, A. T. Austin, and R. Pierik. 2012. Canopy light and plant health. Plant Physiol. 160:145–155.

Barta, D. J., T. W. Tibbitts, R. J. Bula, and R. C. Morrow. 1992. Evaluation of light emitting diode characteristics for a space-based plant irradiation source. Space 12:141–149.

Beggs, C., and E. Wellman. 1985. Analysis of light-controlled anthocyanin formation in coleoptiles of *Zea mays* L.: The role of UV-B, blue, red and far-red light. Photochem. Photobiol. 41:481–486.

Bertoli, A., M. Lucchesini, A. Menuali-Sodi, M. Leonardi, S. Doveri, A. Magnabosco, et al. 2013. Aroma characterization and UV elicitation of purple basil from different plant tissue cultures. Food Chem. 141:776–787.

Borthwick, H. A., S. B. Hendricks, M. W. Parker, E. H. Toole, and V. K. Toole. 1952. A reversible photoreaction controlling seed germination. PNAS 38:662–666.

Brazaitytė, A., P. Duchovskis, A. Urbonavičiūtė, G. Samuolienė, J. Jankauskienė, V. Kazėnas, et al. 2009. After-effect of light-emitting diodes lighting on tomato growth and yield in greenhouse. Sci. Works Lithuanian Institute Hortic. Lithuanian University Agric. 28:115–126.

Brazaitytė, A., P. Duchovskis, A. Urbonavičiūtė, G. Samuolienė, J. Jankauskienė, J. Sakalauskaite, et al. 2010. The effect of light-emitting diodes lighting on the growth of tomato transplants. Zemdirbyste Agric. 97:89–97.

Briggs, W. R., and J. M. Christie. 2002. Phototropins 1 & 2: versatile plant blue-light receptors. Trends Plant Sci. 7:204–210.

Brown, B. A., L. R. Headland, and G. I. Jenkins. 2009. UV-B Action Spectrum for UVR8-Mediated HY5 Transcript Accumulation in Arabidopsis. Photochem. Photobiol. 85:1147–1155.

Bula, R. J., R. C. Morrow, T. W. Tibbitts, D. J. Barta, R. W. Ignatius, and T. S. Martin. 1991. Light-emitting Diodes as a Radiation Source for Plants. HortScience 26:203–205.

de Carbonnel, M., P. A. Davis, M. Rob, G. Roelfsema, S. Inoue, I. Schepens, et al. 2010. The *Arabidopsis* PHYTOCHROME KINASE SUBSTRATE 2 protein is a phototropin signalling element that regulates leaf flattening and leaf positioning. Plant Physiol. 152:1391–1405.

Cashmore, A. R. 2003. Cryptochromes: enabling plants and animals to determine circadian time. Cell 114:537–543.

Chalker-Scott, L. 1999. Environmental significance of anthocyanins in plant stress responses. Photochem. Photobiol. 70:1–9.

Chen, M., and J. Chory. 2011. Phytochrome signaling mechanisms and the control of plant development. Trends Cell Biol. 21:664–671.

Chu, C.-C., A. M. Simmons, T.-Y. Chen, A. P. Alexander, and T. J. Henneberry. 2004. Lime green light-emitting diode equipped yellow sticky card traps for monitoring whiteflies, aphids and fungus gnats in greenhouses. Entomologia Sinica 11:125–133.

Colquhoun, T. A., M. L. Schwieterman, J. L. Gilbert, E. A. Jaworski, K. M. Langer, C. R. Jones, et al. 2013. Light

modulation of volatile organic compounds from petunia flowers and select fruits. Postharvest Biol. Technol. 86:37–44.

Corrochano, L. M. 2007. Fungal photoreceptors: sensory molecules for fungal development ad behavior. Photochem. Photobiol. Sci. 6:725–736.

Costa, L., Y. M. Montano, C. Carrióna, N. Rolnya, and J. J. Guiamet. 2013. Application of low intensity light pulses to delay postharvest senescence of *Ocimum basilicum* leaves. Postharvest Biol. Technol. 86:181–191.

Craig, D. S., and E. S. Runkle. 2016. An intermediate phytochrome photoequilibria from night-interruption lighting optimally promotes flowering of several long-day plants. Environ. Exp. Bot. 121:132–138.

Davis, P. A. (2013) Securing skills and expertise in crop light responses for UK protected horticulture, with specific reference to exploitation of LED technology. AHBD Project CP085 Annual Report 2013.

Davis, P. A., R. Beynon-Davies, G. M. McPherson, J. Banfield-Zanin, D. George, C. O. Ottosen, et al. 2015. Understanding crop and pest responses to LED lighting to maximise horticultural crop quality and reduce the use of PGRs. AHDB Project CP125 Year One report

Demotes-Mainard, S., T. Péron, A. Corot, J. Bertheloot, J. Le Gourrierec, S. Pelleschi-Travier, et al. 2016. Plant responses to red and far-red lights, applications in horticulture. Environ. Exp. Bot. 121:4–21.

De Visser, P. H. B., G. H. Buck-Sorlin, and G. W. A. M. van der Heijden. 2014. Optimizing illumination in the greenhouse using a 3D model of tomato and a ray tracer. Front. Plant Sci. 5:48. doi:10.3389/fpls.2014.00048.

Dougher, T. A. O., and B. Bugbee. 2001. Differences in the response of Wheat, Soybean, and Lettuce to reduced blue radiation. Photochem. Photobiol. 73:199–207.

Dougher, T. A. O., and B. Bugbee. 2004. Long-term blue light effects on the histology of lettuce and soybean leaves and stems. J. Am. Soc. Hortic. Sci. 129:467–472.

Dueck, T. A., J. Janse, B. A. Eveleens, F. L. K. Kempkes, and L. F. M. Marcelis. 2012. Growth of tomatoes under hybrid LED and HPS lighting. Acta Hortic. 952:335–342.

Elliott, D. C. 1979. Temperature-sensitive responses of red light-dependent Betacyanin Synthesis. Plant Physiol. 64:521–524.

Folta, K. M., and K. S. Childers. 2008. Light as a growth regulator: controlling plant biology with narrow-bandwidth solid-state lighting systems. HortScience 43:1957–1964.

Folta, K. M., and E. P. Spalding. 2001. Unexpected roles for cryptochrome 2 and phototropin revealed by high-resolution analysis of blue light-mediated hypocotyl growth inhibition. Plant J. 26:471–478.

Gardner, G., C. Lin, E. M. Tobin, H. Loehrer, and D. Brinkman. 2009. Photobiological properties of the inhibition of etiolated Arabidopsis seedling growth by ultraviolet-B irradiation. Plant, Cell Environ. 32:1573–1583.

Giliberto, L., G. Perrotta, P. Pallara, J. L. Weller, P. D. Fraser, P. M. Bramley, et al. 2005. Manipulation of the blue light photoreceptor cryptochrome 2 in tomato affects vegetative development, flowering time, and fruit antioxidant content. Plant Physiol. 137:199–208.

Gitz, D. C., L. Liu-Gitz, S. J. Britz, and J. H. Sullivan. 2005. Ultraviolet-B effects on stomatal density, water-use efficiency, and stable carbon isotope discrimination in four glasshouse-grown soybean (*Glycine max*) cultivars. Environ. Exp. Bot. 53:343–355.

Goins, G. D., N. C. Yorio, M. M. Sanwo, and C. S. Brown. 1997. Photomorphogenesis, photosynthesis, and seed yield of wheat plants grown under red light-emitting diodes (LEDs) with and without supplemental blue lighting. J. Exp. Bot. 48:1470–1413.

Goins, G. D., N. C. Yorio, M. M. Sanwo-Lewandowski, and C. S. Brown. 1998. Life Cycle experiments with Arabidopsis grown under red light-emitting diodes (LEDs). Life Support Biosph. Sci. 52:143–149.

Gómez, C., and C. A. Mitchell. 2013. Supplemental lighting for greenhouse-grown tomatoes: intracanopy LED towers vs. overhead HPS lamps. Acta Hortic. 1037:855–862.

Gómez, C., M. C. Morrow, C. M. Bourget, G. D. Massa, and C. A. Mitchell. 2013. Comparison of intracanopy light-emitting diode towers and overhead high-pressure sodium lamps for supplemental lighting of greenhouse-grown tomatoes. Horttechnology 23:93–98.

Griebel, T., and J. Zeier. 2008. Light regulation and daytime dependency of inducible plant defenses in *Arabidopsis*: phytochrome signaling controls systemic acquired resistance rather than local defense. Plant Physiol. 147:790–801.

Guo, X., X. Hao, S. Khosa, K. G. S. Kumar, R. Cao, and N. Bennett. 2016a. Effect of LED interlighting combined with overhead HPS light on fruit yield and quality of year-round sweet pepper in commercial greenhouse. Acta Hortic. 1134:71–78.

Guo, X., X. Hao, J. M. Zheng, C. Little, and S. Khosa. 2016b. Response of greenhouse mini-cucumber to different vertical spectra of LED lighting under overhead high pressure sodium and plasma lighting. Acta Hortic. 1134:87–94.

Hao, X., J. M. Zheng, C. Little, and S. Khosl. 2012. LED inter-lighting in year-round greenhouse mini-cucumber production. Acta Hortic. 956:335–340.

Hikosaka, S., K. Ito, and E. Goto. 2010. Effects of Ultraviolet Light on Growth, Essential Oil Concentration, and Total Antioxidant Capacity of Japanese Mint. Environ. Control. Biol. 48:185–190.

Hoad, S. P., and R. R. B. Leakey. 1996. Effects of pre-severance light quality on the vegetative propagation of *Eucalyptus grandis* W. Hill ex maiden. Trees 10:317–324.

Hogewoning, S. W., G. Trouwborst, H. Poorter, W. van Ieperen, and J. Harbinson. 2010. Blue light dose–responses of leaf photosynthesis, morphology, and chemical composition of *Cucumis sativus* grown under different combinations of red and blue light. J. Exp. Bot. 61:3107–3117.

Hogewoning, S. W., E. Wientjes, P. Douwstra, G. Trouwborst, W. van Ieperen, R. Croce, et al. 2012. Photosynthetic quantum yield dynamics: from photosystems to leaves. Plant Cell 24:1921–1935.

Huché-Thélier, L., L. Crespel, J. Le Gourrierec, P. Morel, S. Sakr, and N. Leduc. 2016. Light signaling and plant responses to blue and UV radiations—Perspectives for applications in horticulture. Environ. Exp. Bot. 121:22–38.

Huq, E., B. Al-Sady, M. E. Hudson, C. Kim, K. Apel, and P. H. Quail. 2004. PHYTOCHROME-INTERACTING FACTOR 1 is a critical bHLH regulator of chlorophyll biosynthesis. Science 305:1937–1941.

van Ieperen, W., A. Savvides, and D. Fanourakis. 2012. Red and blue light effects during growth on hydraulic and stomatal conductance in leaves of young cucumber plants. Acta Hortic. 956:223–230.

Islam, S. Z., Y. Honda, and S. Arase. 1998. Light-induced resistance of broad bean against *Botrytis cinerea*. J. Phytopathol. 146:479–485.

Islam, M. A., G. Kuwar, J. L. Jihong, R. D. Blystad, H. R. Gislerød, J. E. Olsen, et al. 2012. Artificial light from light emitting diodes (LEDs) with a high portion of blue light results in shorter poinsettias compared to high pressure sodium (HPS) lamp. Sci. Hortic. 147:136–143.

Jankowska-Blaszczuk, M., and M. I. Daws. 2007. Impact of red: far red ratios on germination of temperate forest herbs in relation to shade tolerance, seed mass and persistence in the soil. Funct. Ecol. 21:1055–1062.

Jao, R. C., and W. Fang. 2004. Effects of frequency and duty ratio on the growth of potato plantlets in vitro using light emitting diodes. HortScience 39:375–379.

Jeong, SW, S Park, JS Jin, ON Seo, G-S Kim, H Bae, et al. 2012. Influences of four different light-emitting diode lights on flowering and polyphenol variations in the leaves of Chrysanthemum (*Chrysanthemum morifolium*). J. Agric. Food Chem. 60:9793–9800.

Kanetis, L., G. J. Holmes, and P. S. Ojiambo. 2010. Survival of *Pseudoperonospora cubensis* sporangia exposed to solar radiation. Plant. Pathol. 59:313–323.

Kegge, W., B. T. Weldegergis, R. Solerm, M. Vergeer-Van Eijk, M. Dicke, L. A. Voesenek, et al. 2013. Canopy light cues affect emission of constitutive and methyl jasmonate induce volatile organic compounds in *Arabidopsis thaliana*. New Phytol. 200:861–874.

Kennedy, G. G., R. T. Yamamoto, M. B. Dimock, W. G. Williams, and J. Bordner. 1981. Effect of day length and light intensity on 2-tridecanone levels and resistance in *Lycopersicon hirsutm* f. *glabratum* to *Maduca sexta*. J. Chem. Ecol. 7:707–716.

Kim, H. H., G. D. Goins, R. M. Wheeler, and J. C. Sager. 2004a. Green light supplementation for enhanced lettuce growth under red and blue light emitting diodes. HortScience 39:1617–1622.

Kim, H. H., G. D. Goins, R. M. Wheeler, and J. C. Sager. 2004b. Stomatal conductance of lettuce grown under or exposed to different light qualities. Ann. Bot. 94:691–697.

Kirschbaum, M. U. F., and R. W. Pearcy. 1988. Gas Exchange Analysis of the Relative Importance of Stomatal and Biochemical Factors in Photosynthetic Induction in *Alocasia macrorrhiza*. Plant Physiol. 86:782–785.

Kozai, T. 2013. Sustainable plant factory: closed plant production systems with artificial light for high resource use efficiencies and quality produce. Acta Hortic. 1004:27–40.

Kubota, C., P. Chia, Z. Yang, and Q. Li. 2012. Applications of far-red light emitting diodes in plant production under controlled environments. Acta Hortic. 952:59–66.

Kumari, R., and S. B. Agrawal. 2010. Supplemental UV-B induced changes in leaf morphology, physiology and secondary metabolites of an Indian aromatic plant Cymbopogon citratus (D.C.) Staph under natural field conditions. Int. J. Environ. Stud. 67:655–675.

Kumari, R., S. B. Agrawal, S. Singh, and N. K. Dubey. 2009. Supplemental ultraviolet-B induced changes in essential oil composition and total phenolics of *Acorus calamus* L. (sweet flag). Ecotoxicol. Environ. Saf. 72:2013–2019.

Lambers, H., F. S. III Chaplin, and T. L. Pons. 1998. Plant physiological ecology. Pp. 413–436. Springer, New York.

Lau, O. S., and X. W. Deng. 2010. Plant hormone signalling lightens up: integrators of light and hormones. Curr. Opin. Plant Biol. 13:571–577.

Lee, S.-H., R. K. Tewari, E.-J. Hahn, and K.-Y. Paek. 2007. Photon flux density and light quality induces changes in growth, stomatal development, photosynthesis and transpiration of *Withania somnifera* (L.) Dunal. plantlets. Plant Cell, Tissue Organ Cult. 90:141–15.

Li, Q., and C. Kubota. 2009. Effects of supplemental light quality on growth and phytochemicals of baby leaf lettuce. Environ. Exp. Bot. 67:59–64.

Li, H., Z. Xu, and C. Tang. 2010. Effect of light-emitting diodes on growth and morphogenesis of upland cotton (Gossypium hirsutum L.) plantlets in vitro. Plant Cell, Tissue Organ Cult. 103:155–163.

Li, H., C. Tang, and Z. Xu. 2013. The effects of different light qualities on rapeseed (*Brassica napus* L.) plantlet growth and morphogenesis in vitro. Sci. Hortic. 150:117–124.

Li, K., Q.-C. Yang, Y.-X. Tong, and R. Cheng. 2014. Using movable light-emitting diodes for electrical saving in a plant factory growing lettuce. Horttechnology 24:546–553.

Liu, Y., and D. Bell-Pedersen. 2006. Circadian Rhythms in Neurospora crassa and Other Filamentous Fungi. Eukaryot. Cell 5:1184–1193.

Lu, N., T. Maruo, M. Johkan, M. Hohjo, S. Tsukagoshi, Y. Ito, et al. 2012. Effects of supplemental lighting with light-emitting diodes (LEDs) on tomato yield and quality of single-truss tomato plants grown at high planting density. Environ. Control. Biol. 50:63–74.

Manukyan, A. 2013. Effects of PAR and UV-B Radiation on Herbal Yield, Bioactive Compounds and Their Antioxidant Capacity of Some Medicinal Plants Under Controlled Environmental Conditions. Photochem. Photobiol. 89:406–414.

Massa, G. D., H.-H. Kim, R. M. Wheeler, and C. A. Mitchell. 2008. Plant Productivity in Response to LED Lighting. HortScience 43:1951–1955.

Matsuda, R., K. Ohashi-Kaneko, K. Fujiwara, E. Goto, and K. Kurata. 2004. Photosynthetic characteristics of rice leaves grown under red light with or without supplemental blue light. Plant Cell Physiol. 45:1870–1874.

Matteson, N., I. Terry, A. Ascoli-Christensen, and C. Gilbert. 1992. Spectral efficiency of the western flower thrips, *Frankinella occidentalis*. J. Insect Physiol. 38:453–459.

Mazza, C. A., M. M. Izaguirre, J. Curiale, and C. L. Ballaré. 2010. A look into the invisible: ultraviolet-B sensitivity in an insect (*Caliothrips phaseoli*) revealed through a behavioural action spectrum. Proc. R. Soc. B 277:367–373.

McCree, KJ. 1971. The action spectrum, absorptance and quantum yield of photosynthesis in crop plants. Agric. Meteorol. 9:191–216.

Ménard, C, M Dorais, T Hovi, and A Gosselin. 2005. Developmental and physiological responses of tomato and cucumber to additional blue light. In V International Symposium on Artificial Lighting in Horticulture 711: 291–296.

Middleton, E. M., and A. H. Teramura. 1993. The role of flavonol glycosides and carotenoids in protecting soybean from ultraviolet-B damage. Plant Physiol. 103:741–752.

Miyashita, Y. T., Y. Kinura, C. Kitaya, C. Kubota, and T. Kozai. 1997. Effects of red light on the growth and morphology of potato plantlets in vitro using light emitting diodes (LEDs) as a light source for micropropagation. Acta Hortic. 418:169–173.

Moon, H. K., S.-Y. Park, Y. W. Kim, and C. S. Kim. 2006. Growth of Tsuru-rindo (*Tripterospermum japonicum*) cultured in vitro under various source of light-emitting diode (LED) irradiation. J. Plant Biol. 49:174–179.

Nakamoto, Y., and H. Kuba. 2004. The effectiveness of a green light emitting diode (LED) trap at capturing the West Indian sweet potato weevil, *Euscepes postfasciatus* (Fairmaire) (Coleoptera: Curculionidae) in a sweet potato field. Appl. Entomol. Zool. 39:491–495.

Nanya, K., Y. Ishigami, S. Hikosaka, and E. Goto. 2012. Effects of blue and red light on stem elongation and flowering of tomato seedlings. Acta Hortic. 956:264–266.

Nhut, D. T., L. T. A. Hong, H. Watanabe, M. Goi, and M. Tanaka. 2002. Growth of banana plantlets cultured in vitro under red and blue light-emitting diode (LED) irradiation source. Acta Hortic. 575:117–124.

Nhut, D. T., T. Takamura, H. Watanabe, K. Okamoto, and M. Tanaka. 2003. Responses of strawberry plantlets cultured in vitro under superbright red and blue light-emitting diodes (LEDs). Plant Cell, Tissue Organ Cult. 73:43–52.

Ninu, L., M. Ahmad, C. Miarelli, A. R. Cashmore, and G. Giuliano. 1999. Cryptochrome 1 controls tomato development in response to blue light. Plant J. 18:551–556.

Ouzounis, T., X. Fretté, E. Rosenqvist, and C. O. Ottosen. 2014. Spectral effects of supplementary lighting on the secondary metabolites in roses, chrysanthemums, and campanulas. J. Plant Physiol. 171:1491–1499.

Paradiso, R., E. Meinen, J. F. H. Snel, P. De Visser, W. Van Leperen, S. W. Hogewoning, et al. 2011. Spectral dependence of photosynthesis and light absorptance in single leaves and canopy in rose. Sci. Hortic. 127:548–554.

Paré, P. W., and J. H. Tumlinson. 1999. Plant volatiles as a defence against insect and herbivores. Plant Physiol. 121:325–331.

Park, J.-S., M.-G. Choung, J.-B. Kim, et al. 2007. Gene up-regulated during red colouration in UV-B irradiated lettuce leaves. Plant Cell Rep. 26:507–516.

Paul, N. D., J. M. Moore, and M. Huey. 2006. The potential benefits of three modified plastic crop covers in Hardy Ornamental Nursery Stock production: initial investigations on a grower holding (Garden Centre Plants, Preston). HDC Project PC19a Final report.

Pearson, S., P. A. Davis, and H. Kitchener 2015. Commercial review of lighting systems for UK Horticulture. AHDB Horticulture CP 139 Final Report.

Poudel, P. R., I. Kataoka, and R. Mochioka. 2008. Effect of red- and blue-light-emitting diode on growth and morphogenesis of grapes. Plant Cell, Tissue Organ Cult. 92:147–153.

Reppert, S. M., H. Zhu, and R. H. White. 2004. Polarized light helps monarch butterflies navigate. Curr. Biol. 14:155–158.

Rossel, S. 1993. Navigation by bees using polarized skylight. Comp. Biochem. Physiol. A Mol. Integr. Physiol. 104:695–708.

Runkle, E. S., and R. D. Heins. 2001. Specific functions of red, far red, and blue light in flowering and stem extension of long-day plants. J. Am. Soc. Hortic. 126:275–282.

Sakuta, M. 2014. Diversity in plant red pigments: anthocyanins and betacyanins. Plant Biotechnol. Rep. 8:37–48.

Samuolienė G, Brazaitytė A, Urbonavičiūtė A, Šabajevienė G and Duchovskis P (2010) The effect of red and blue light component on the growth and development of frigo strawberries, Zemdirbyste-Agriculture 97: 99–104.

Samuoliene, G., A. Brazaitytė, R. Sirtautas, A. Visile, J. Sakalauskaite, S. Sakalauskaite, et al. 2013. LED illumination affects bioactive compounds in romaine baby leaf lettuce. J. Sci. Food Agric. 93:3286–3291.

Savvides, A., D. Fanourakis, and W. van Ieperen. 2012. Co-ordination of hydraulic and stomatal conductance across light qualities in cucumber leaves. J. Exp. Bot. 63:1135–1143.

Schuerger, A. C., and C. S. Brown. 1997. Spectral quality affects disease development of three pathogens on hydroponically grown plants. HortScience 32:96–100.

Sellaro, P., M. Crepy, S. A. Trupkin, E. Karayekov, A. S. Buchovsky, C. Constanza Rossi, et al. 2010. Cryptochrome as a sensor of the blue/green ratio of natural radiation in Arabidopsis. Plant Physiol. 154:401–409.

Shiga, T., K. Shoji, H. Shimada, S. Hashida, F. Goto, and T. Yoshihara. 2009. Effect of light quality on rosmarinic acid content and antioxidant activity of sweet basil, Ocimum basilicum L. Plant Biotechnol. 26:255–259.

Shimada, A., and Y. Taniguchi. 2011. Red and blue pulse timing for pulse width modulation light dimming of light emitting diodes for plant cultivation. J. Photochem. Photobiol., B 104:399–404.

Shinomura, T., A. Nagatani, H. Hanzawa, M. Kubota, M. Watanabe, and M. Furuya. 1996. Action spectra for phytochrome A- and B-specific photoinduction of seed germination in Arabidopsis thaliana. PNAS 93:8129–8133.

Sommer, A. P., and R.-P. Franke. 2006. Plants grow better if seeds see green. Naturwissenschaften 93:334–337.

Son, K.-H., and M.-M. Oh. 2013. Leaf shape, growth, and antioxidant phenolic compounds of two lettuce cultivars grown under various combinations of blue and red light-emitting diodes. HortScience 48:988–995.

Stutte, G. W., S. Edney, and T. Skerritt. 2009. Photoregulation of bioprotectant content of red leaf lettuce with light-emitting diodes. HortScience 44:79–82.

Sun, J., J. N. Nishio, and T. C. Vogelmann. 1998. Green Light Drives CO2 Fixation Deep within Leaves. Plant Cell Physiol. 39:1020–1026.

Suthaparan, A., S. Torre, A. Stensvand, M. L. Herrero, R. I. Pettersen, D. M. Gadoury, et al. 2010. Specific Light-emitting diodes can suppress sporulation of Podosphaera pannosa on greenhouse roses. Plant Dis. 94:1105–1110.

Tanaka, M., T. Takamura, H. Watanabe, M. Endo, T. Yanagi, and K. Okamoto. 1998. In vitro growth of Cymbidium plantlets cultured under super bright and blue light emitting diodes (LEDs). J. Hortic. Sci. Biotechnol. 73:39–44.

Tennessen, D. J., E. L. Singsaas, and T. D. Sharkey. 1994. Light-emitting diodes as a light source for photosynthesis research. Photosynth. Res. 39:85–92.

Tennessen, D. J., R. J. Bula, and T. D. Sharkey. 1995. Efficiency of photosynthesis in continuous and pulsed light emitting diode irradiation. Photosynth. Res. 44:261–269.

Terashima, I., T. Fujita, T. Inoue, W. S. Chow, and R. Oguchi. 2009. Green light drives leaf photosynthesis more efficiently than red light in strong white light: Revisiting the enigmatic question of why leaves are green *Plant Cell*. Physiology 50:684–697.

Tevini, M., and U. Iwanzik Wm Thoma. 1981. Some effects of enhanced UV-B irradiation on the growth and composition of plants. Planta 153:388–394.

Tripathy, B. C., and C. S. Brown. 1995. Root-shoot interaction in the greening of wheat seedlings grown under red light. Plant Physiol. 107:407–411.

Trouwborst, S. W., G. Trouwborst, H. Maljaars Hm Poorter, W. van Ieperen, and J. Harbinson. 2010. Blue light dose-responses of leaf photosynthesis, morphology and chemical composition of *Cucumis sativus* grown under difference combination of red and blue light. J. Exp. Bot. 121:75–82.

Trouwborst, G., S. W. Hogewoning, O. van Kooten, and J. Harbinson. 2016. Plasticity of photosynthesis after the 'red light syndrome' in cucumber. Environ. Exp. Bot. 61:3107–3117.

Tuan, P. A., A. A. Thwe, Y. B. Kim, J. K. Kim, S.-J. Kim, S. Lee, et al. 2013. Effects of white, blue, and red light-emitting diodes on carotenoid biosynthetic gene expression levels and carotenoid accumulation in sprouts of Tartary Buckwheat (*Fagopyrum tataricum* Gaertn.). Journal of Agricultural and Food Chemistry 61:12356–12361.

Vlohr, H., and H. Drumm-Herrel. 1983. Coaction between phytochrome and blue/UV light anthocyanin synthesis in seedlings. Physiol. Plant. 58:408–414.

Wang, Y., and K. M. Folta. 2013. Contributions of green light to plant growth and development. Am. J. Bot. 100:70–78.

Wang, G., M. Gu, J. Cui, K. Shi, Y. Zhou, and J. Yu. 2009. Effects of light quality on CO_2 assimilation, chlorophyll-fluorescence quenching, expression of Calvin cycle genes and carbohydrate accumulation in *Cucumis sativus*. J. Photochem. Photobiol., B 96:30–37.

Wargent, J. J., J. P. Moore, A. R. Ennos, and N. D. Paul. 2009. Ultraviolet radiation as a limiting factor in leaf expansion and development. Photochem. Photobiol. 85:279–286.

de Wit, M., S. H. Spoel, G. F. Sanchez-Perez, C. M.M. Gommers, C. M. J. Pieterse, L. A. C. J. Voesenek, et al. 2013. Perception of low red:far-red ratio compromises both salicylic acid- and jasmonic acid-dependent pathogen defences in Arabidopsis. Plant J. 75:90–103.

Wu, H. C. 2006. Improving in vitro propagation of *Protea cynaro*ides L. (King Protea) and the roles of starch and phenolic compounds in the rooting of cuttings. PhD thesis. University of Pretoria, Pretoria.

Wu, H.-C., and C.-C. Lin. 2012. Red light-emitting diode light irradiation improves root and leaf formation in difficult-to-propagate *Protea cynaroides* L. plantlets in vitro. HortScience 47:1490–1494.

Wu, H. C., E. S. du Toit, C. F. Reinhardt, A. M. Rimando, F. van der Kooy, and J. J. M. Meyer. 2007. The phenolic, 3,4-dihydroxybenzoic acid, is an endogenous regulator of rooting in *Protea cynaroides*. Plant Growth Regul. 52:207–215.

Yorio, N. C., G. D. Goins, H. R. Kagie, R. M. Wheeler, and J. C. Sager. 2001. Improving spinach, radish, and lettuce growth under red light emitting diodes (LEDs) with blue light supplementation. HortScience 36:380–383.

Yoshida, H., S. Hikosaka, E. Goto, H. Takasuna, and T. Kudou. 2012. Effects of light quality and light period on flowering of everbearing strawberry in a closed plant production system. Acta Hortic. 956:107–112.

14

Photosynthesis and growth in diverse willow genotypes

P. John Andralojc[1], Szilvia Bencze[1], Pippa J. Madgwick[1], Hélène Philippe[1], Stephen J. Powers[2], Ian Shield[3], Angela Karp[3] & Martin A. J. Parry[1]

[1]Plant Biology and Crop Science Department, Rothamsted Research, Harpenden, Hertfordshire, AL5 2JQ, United Kingdom
[2]Computational and Systems Biology Department, Rothamsted Research, Harpenden, Hertfordshire, AL5 2JQ, United Kingdom
[3]AgroEcology Department, Rothamsted Research, Harpenden, Hertfordshire, AL5 2JQ, United Kingdom

Keywords
A/Ci, carbon isotope ratio, CO_2 assimilation, photosynthesis, Rubisco, Salix

Correspondence
P. John Andralojc, Plant Biology and Crop Science Department, Rothamsted Research, Harpenden, Hertfordshire, AL5 2JQ, United Kingdom.

E-mail: john.andralojc@rothamsted.ac.uk

Present address
Szilvia Bencze, Agricultural Institute, Centre for Agricultural Research, Hungarian Academy of Sciences, P.O.Box 19, Martonvásár, H-2462, Hungary

Funding Information
This work was funded by Institute Strategic Programme Grants from the Biotechnology and Biological Sciences Research Council (BBSRC; BB/J/00426X/1, BB/I002545/1, BB/I017372/1, BB/J004278/1) of the United Kingdom.

Abstract

During a study of the contribution of photosynthetic traits to biomass yield among 11 diverse species of willow, the light and CO_2 dependence of photosynthesis were found to differ, with absolute rates at ambient and saturating CO_2, together with maximum rates of Rubsico-limited and electron-transport-limited photosynthesis (V_{cmax} and J, respectively) varying by factors in excess of 2 between the extremes of performance. In spite of this, the ratio, J/V_{cmax} – indicative of the relative investment of resource into RuBP regeneration and RuBP carboxylation – was found to fall within a narrow range (1.9–2.5) for all genotypes over two successive years. Photosynthetic rate (μmol CO_2 fixed m^{-2} sec^{-1}) showed a strong, inverse correlation with total leaf area per plant. Photosynthetic capacity, expressed on a leaf area basis, showed a strong, positive correlation with yield among some of the species, but when expressed on a whole plant basis all species indicated a positive correlation with yield. Thus, both leaf area per plant and photosynthetic rate per unit leaf area contribute to this relationship. The abundance and kinetic characteristics of Rubisco play a pivotal role in determining photosynthetic rate per unit leaf area and so were determined for the chosen willow species, in parallel with Rubisco large subunit (LSU) gene sequencing. Significant differences in the rate constants for carboxylation and oxygenation as well as the affinity for CO_2 were identified, and rationalized in terms of LSU sequence polymorphism. Those LSU sequences with isoleucine instead of methionine at residue 309 had up to 29% higher carboxylase rate constants. Furthermore, the A/Ci curves predicted from each distinct set of Rubisco kinetic parameters under otherwise identical conditions indicated substantial differences in photosynthetic performance. Thus, genetic traits relating specifically to Rubisco and by implication to photosynthetic performance were also identified.

Introduction

Food and nonfood crops are increasingly being used as renewable sources of energy and fixed carbon for industrial processes. Such crops are the subject of intense investigation, the aim of which is to provide compelling alternatives to the use of environmentally compromising fossil fuels.

Among these, short rotation coppiced (SRC) willow is finding increasing favor within temperate regions, including the United Kingdom, Northern and Eastern Europe, and North America. A recent life cycle analysis, whose objective was to identify the most resource-efficient energy crops, found that willow was second to sugar cane on the combined bases of land- and N-use efficiency (Miller 2010), making willow among the choices of preference in regions where more tropical-adapted species, such as sugar cane, do not grow. Yields in the order of 14 tons DM ha^{-1} year^{-1} have been recorded for SRC willow and are set to increase further as more productive varieties emerge (Christersson 1987; Karp et al. 2011).

All plant organic matter is derived from photosynthetic assimilation. Even so, when comparing the productivity of closely related species, leaf photosynthetic performance per se may not always be the principal cause for differences in yield. Indeed, evidence has been presented which suggests that yield and photosynthetic rate of trees may only be weakly correlated (Ericsson et al. 1996; Taylor et al. 2001). Changing requirements for development of SRC willow over the course of a single season may confine any clear correlation between photosynthetic rate and yield to periods of intense stem growth, for example, following bud burst and primary canopy establishment, typically in the late Spring/early Summer period (Cannell et al. 1987; Neergaard et al. 2002). In the field, differing degrees of competition, water stress, and pathogen interaction may also obscure any such correlation (Karp and Shield 2008).

The current research was undertaken to identify leaf characteristics which impact upon willow growth and yield. This study assessed photosynthetic capacity and contributory processes in a broad range of genotypes to determine whether sufficient natural variation relevant to yield exists to warrant the initiation of genetic improvement strategies targeting the underlying genes. The National Willow Collection at Rothamsted Research encompasses a diverse range of willow species (~100) and genotypes (~1300), exhibiting a broad range of growth habits, yield characteristics, and disease resistance traits. As such, it represents a unique resource for the identification of the genetic basis of agronomically desirable qualities, and accessions from this collection were chosen for the present study. Analyses of photosynthetic parameters of *Salix* and other woody species have been performed previously, in combination with leaf anatomical and light harvesting processes (e.g., Patton and Jones 1989; Liu et al. 2003; Robinson et al. 2004; Manter and Kerrigan 2004; Merilo et al. 2006). Similar to more recently reported work (Bouman and Sylliboy 2012), this study not only differs from these in its breadth of species diversity and range of measurements chosen but also differs from the latter work by an additional focus on the species-specific investment of resource in components of RuBP carboxylation and regeneration.

The contribution of plant leaf area to photosynthesis and yield has been extensively studied (Tharakan et al. 2005; Merilo et al. 2006; Bouman and Sylliboy 2012). The existence of two strategies among high-yielding *Salix* have been demonstrated (Tharakan et al. 2005), namely, relatively low leaf area index, but high foliar nitrogen on one hand, and high leaf area index, but low foliar nitrogen on the other, which have clear correlates in the genotypes studied here. *Salix* varieties with higher photosynthetic rates on a leaf area basis, relative to plant leaf area, have been shown to result in greater yields (Bouman and Sylliboy (2012)). Thus, in common with other commercially important plant species, improvements to photosynthetic efficiency in *Salix* has significant potential for yield improvement (Long et al. 2006).

The 11 willow accessions used in the current study included many widely used in breeding willows for bioenergy (e.g., *Salix viminalis*, *Salix eriocephala*, and *Salix schwerinii*), the parents (R13 and S3) of a key mapping population (K8) used extensively for quantitative trait locus (QTL) mapping (Hanley and Karp 2013), and a representation of diverse, pure species. Of the latter, *Salix triandra* ("Baldwin") is of particular interest as previous studies, of genetic diversity using molecular markers, have challenged the current taxonomy of this species showing it to be as distinct from most other willows species as poplar (Trybush et al. 2008). Also interesting is *Salix exigua* – a North American accession which is also distinct in terms of amplified fragment length polymorphism (AFLP) analysis and has diverse morphological traits (e.g., stomata on the upper leaf surfaces). *Salix daphnoides* was included as it is of potential interest in drought tolerance studies.

Our work aimed to determine whether a clear correlation between estimated photosynthesis and yield across a broad range of *Salix* genotypes could be demonstrated, and to identify relationships between photosynthetic capacity and the abundance of easily measured leaf components including starch, soluble carbohydrate, chlorophyll, protein, and Rubisco. The material for this study comprised pot-grown cuttings, reared under glass, during a single season of growth (2008) after which a destructive harvest took place. The rationale behind this approach was to ensure that all genotypes were subjected to identical, controlled, disease-free, growth conditions and were developmentally equivalent. This experimental approach was repeated in a second consecutive growing season (2009) in order to complement and extend the observations made in the first year. A strong correlation between Rubisco and photosynthetic performance, evident in both growing seasons, was followed up by determining the associated Rubisco large subunit (LSU) gene sequences as well as the kinetic characterization of the purified Rubisco holoenzymes, which led to the identification of three sequence polymorphisms with significant kinetic consequences.

Materials and Methods

The 11 accessions studied here are shown in Table 1 and are all present in the National Willow Collection maintained

Table 1. Identities and GenBank accession numbers of Rubisco LSU sequences.

Accession	Identity	Landrace or isolate	Abbreviation
HE610660	*Salix triandra*	Landrace Baldwin	*tri*
HE610661	*Salix dasyclados*	Isolate 77056	*das*
HE610662	*Salix udensis* (formerly *sacchalinensis*)	Landrace Sekka	*ude*
HE610663	*Salix viminalis*	Isolate bowes hybrid	*vim*
HE610664	*Salix eriocephala*	Isolate R632	*eri*
HE610665	*Salix viminalis* × *Salix schwerinii*	Isolate R13	*R13*
HE610666	*Salix viminalis* × *Salix schwerinii*	Isolate S3	*S3*
HE610667	*Salix purpurea* × *Salix viminalis*	Landrace Ulbrichtweide	*pxv*
HE610668	*Salix schwerinii*	Isolate K3 hilliers	*sch*
HE610670	*Salix daphnoides*	Isolate Fastigate	*dap*
HE610669	*Salix exigua*	–	*exi*

The abbreviations used for each genotype in subsequent figures are also shown.

at Rothamsted Research, Harpenden, UK (51°48'30"N, 0°21'22"W; 125 m AOD). The three hybrid accessions (R13, S3, and "Ulbrichtweide") are parents of mapping population families, of which R13 and S3 are full sibs of *S. viminalis* L. × *S. schwerinii* E. Wolf. All accessions are maintained as a coppiced collection and their identity has been previously verified, unless stated (Trybush et al. 2008). The pure species were chosen to represent diversity across the genus and include many that are commonly used in breeding programs worldwide.

Experimental design and growth conditions

A randomized block design with three blocks was used to conduct an experiment to assess the differences in photosynthetic performance and the yield between the 11 willow genotypes in each of two growing seasons (2008 and 2009). A set of 25–30 cm long cuttings of each genotype were collected from the National Willow Collection 1–2 months beforehand and stored in polythene bags at −4°C. In March, each was planted in a 25 cm diameter pot, containing Rothamsted prescription mix compost with added nutrients (75% medium grade [L&P] peat, 12% screened sterilized loam, 3% medium grade vermiculite, 10% grit [5-mm screened, lime free], 3.5 kg "Osmocote® Exact 3–4 month" per m³ [Scotts (UK) Ltd., Godalming, Surrey], 0.5 kg PG mix/m³ [Hydro Agri (UK) Ltd., Bury St. Edmunds, Suffolk], lime [approximately

3 kg/m³ to pH 5.5–6.0], Vitax Ultrawet [wetting agent: 200 mL/m³]) to a depth of 18–20 cm, and the design was set up in our glasshouse facilities. The pots were placed in large circular dishes which contained water at all times, to avoid drought stress. Supplementary lighting was provided to ensure an irradiance of at least 400 μmol m^{-2} sec^{-1} throughout the light period (16 h day/8 h night). Bud emergence did not take place until April. The experiment was terminated by shoot harvest in late December.

Leaf gas exchange measurements

Light and CO_2 dependence of photosynthesis was measured for all genotypes in 2008 and 2009. Leaf photosynthesis was measured at midday ±3 h, on young, fully expanded leaves attached to the dominant stem of each plant, using a gas exchange analyzer equipped with an LED light source (Li-6400, Li-Cor Inc., Lincoln, Nebraska, USA). Instrument start-up/calibration according to the manufacturer's instructions was performed at the start of each day. Leaf chamber conditions, unless stated otherwise, consisted of a photosynthetic photon flux density of 1250 μmol photons m^{-2} sec^{-1} (with 10% blue light) and a CO_2 concentration of 385 μmol mol^{-1}. During measurement, leaf temperature was maintained at 25 ± 1.5°C and relative humidity at 60%. For A/Ci curves, reference CO_2 values of 450, 300, 150, 75, 450, 600, 800, and 1100 μmol mol^{-1} were applied. For A/Q curves, the sample CO_2 concentration was set to 385 μmol mol^{-1} and photosynthesis measured at light levels of 1500, 1000, 750, 500, 250, 125, 50, and 0 μmol photons m^{-2} sec^{-1}. Each data point was logged when the change in each of CO2S, H2OS, Flow, Photo, and Cond were no more than 1% CV (coefficient of variation) over a 20-sec period. At low deltas (<15 ppm CO_2), the IRGAS were matched immediately before measurement. The resulting A/Ci data were processed using a published algorithm (Sharkey et al. 2007) to determine values – normalized to the reference temperature of 25°C – of the maximum carboxylation rate allowed by Rubisco, V_{cmax} (i.e., the product of Rubisco carboxylase activity and Rubisco abundance per unit leaf area), and the rate of photosynthetic electron transport, J (based on NADPH requirement), while also providing estimates for day respiration and mesophyll conductance. To increase accuracy, values for the relevant Rubisco kinetic parameters used by this algorithm (the K_M for carboxylase [K_c] and oxygenase [K_o] activities and the specificity factor [$S_{c/o}$]), were determined using Rubisco purified from the genotypes under study.

The gas exchange data describing the light dependence of photosynthesis were used to determine the light-saturated rate of photosynthesis (A_{maxQ}), the irradiance required to achieve half this value ($Q_{1/2}$), and the light compensation point (Γ_Q). This was made possible, since the (A/Q) paired data points were faithfully described by the function:

$$A = A_{maxQ}(Q - \Gamma_Q)/[Q_{1/2} + (Q - \Gamma_Q)], \qquad (1)$$

where A is the observed rate of CO_2 assimilation at light intensity Q. An Excel solver routine was constructed to find the best fit of the experimental data to this equation, by varying the values of A_{maxQ}, $Q_{1/2}$, and Γ_Q.

Measurement of growth and leaf area per plant

The number, length, and diameter (at 0.5-m intervals) of all shoots and branches were systematically tabulated, together with the number and area of all the associated leaves. Leaf area was measured using a Li-Cor LI-3000 leaf area meter, which enabled the rapid and nondestructive measurement of leaf area, length, and width. It took approximately 10 days to process all 33 samples (11 × 3 replicates) and this task was performed on two separate occasions: from mid-June and from early August (2009). As soon as this task had been completed, the associated gas exchange measurements were made.

Leaf sampling and extraction for quantification of starch, chlorophyll, soluble protein, and Rubisco

Leaf samples (10 cm²) taken in the early Summer, 3–4 h after midday, were snap-frozen in liquid N_2 immediately after excision from the plant by means of a 3.6 cm diameter cork borer and stored at −80°C until extraction. Extraction buffer consisted of: 50 mmol/L MES-NaOH, pH 7.0; 10 mmol/L MgCl₂; 1 mmol/L ethylene glycol tetraacetic acid (EGTA); 1 mmol/L ethylenediamine-tetraacetic acid (EDTA); 50 mmol/L 2-mercaptoethanol; 1% (v/v) Tween 80; 2 mmol/L benzamidine; 5 mmol/L ε-aminocaproic acid; 10 mmol/L D,L-dithiothreitol; 1% (v/v) Sigma plant protease inhibitor cocktail (P 9599); and 1 mmol/L PMSF – the last three components being added immediately before extraction. Three milliliters of this buffer was used per 10 cm² of leaf material, as follows: 0.5 mL of extraction buffer was first ground to a fine powder in liquid N_2 by means of a precooled pestle and mortar. To this was added the frozen leaf tissue together with 0.2 g acid washed sand, which were ground to a fine powder. A further 0.5 mL of extraction buffer was added and ground-in, followed by two more aliquots of 1 mL with accompanying grinding. The homogeneous paste was allowed to thaw, accompanied by frequent grinding. Once thawed, a 0.70 mL subsample was taken for immediate chlorophyll and starch determination, and the remaining homogenate was clarified by centrifugation (5 min, 4°C, 14,250 × g). The clarified supernatant was divided into 0.40 mL aliquots and snap-frozen until needed. This procedure releases large

quantities of soluble protein which (in the case of Rubisco) retains full catalytic activity and can be scaled-up, as necessary, for the purification of sufficient Rubisco for specificity factor assays. To determine the soluble protein in these Tween-containing samples, we found the 2-D Quant Kit (GE Healthcare, Hatfield, Hertfordshire, UK) to be ideal, on account of its sensitivity and the incorporation of an acid-precipitation step, enabling removal of interfering solutes (including Tween 80) prior to protein-dependent color development. The homogenized samples taken prior to centrifugation for chlorophyll and starch determination were added immediately to ethanol (giving 90% [v/v] ethanol) and the chlorophyll content of the clarified solution determined spectrophotometrically as described by Wintermans and De Mots (1965). The resulting insoluble, decolorized pellet could then be assayed sequentially for soluble carbohydrate then starch content, using anthrone – a reagent specific for soluble carbohydrate – by a modification of the approach of Hansen and Møller (1975) in which the soluble carbohydrate was first removed from the insoluble starch by repeated extraction/sedimentation first using 80% (v/v) and then 50% (v/v) ethanol. The sugar content of the combined ethanolic supernatants was then determined, while the resulting starch-containing pellet was dried under vacuum, and starch digestion initiated by addition of 0.5 mL of 100 mmol/L sodium acetate (pH 4.5) containing 10 units mL⁻¹ of purified amyloglucosidase (Product A 7420, Sigma Aldrich Co Ltd., Poole, Dorset, UK). Digestion proceeded at 37°C for 24 h, after which the digests were clarified by centrifugation and the sugar content determined by the anthrone method. A parallel digestion control was always performed, to enable correction for traces of soluble carbohydrate present in the amyloglucosidase itself.

Rubisco extraction for kinetic assays

This was achieved by a small-scale extraction procedure very similar to that described earlier, except that by the time these were performed it had been found that excellent Rubisco activities could also be obtained by replacing Tween 80 with PEG 4000. An important consequence of this (the absence of Tween 80) was that the Rubisco could then be quantified by the quicker ¹⁴CABP-binding method of Yokota and Canvin (1985) rather than by band quantification following SDS-PAGE. Leaf material for this purpose was from youngest, fully expanded leaves, sampled in late Spring/early Summer (2008) and mid/late Summer (2009). The complete extraction buffer for this purpose was: 50 mmol/L MES-NaOH, pH 7.0; 5 mmol/L MgCl₂; 1 mmol/L EGTA; 1 mmol/L EDTA; 50 mmol/L 2-mercaptoethanol; 10 mmol/L NaHCO₃; 2 mmol/L benzamidine; 5 mmol/L ε-aminocaproic acid; 5% (w/v) PEG 4000; 10 mmol/L D,L-dithiothreitol; 1% (v/v) Sigma plant protease inhibitor cocktail (P 9599); insoluble

PVPP (150 mg gFW^{-1}), and 1 mmol/L PMSF – the last four ingredients being added immediately before extraction. The resulting, clarified supernatant was then immediately applied to a PD-10 desalting column (GE Healthcare) at 4°C, pre-equilibrated, and subsequently developed with the following ice-cold buffer: 100 mmol/L Bicine-NaOH, pH 8.2; 10 mmol/L MgCl$_2$; 1 mmol/L EDTA; 1 mmol/L benzamidine; 1 mmol/L ε-aminocaproic acid; 1 mmol/L Na$_2$HP$_i$; 10 mmol/L NaHCO$_3$; 2% (w/v) PEG 4000; 10 mmol/L D,L-dithiothreitol. Fractions of 0.5 mL were collected and two or three fractions containing the protein peak were pooled, supplemented with 1% (v/v) Sigma plant protease inhibitor cocktail (P 9599) and snap-frozen in liquid N$_2$, and stored in liquid N$_2$ until assayed. Pilot experiments established that freezing in this way had no effect on subsequent catalytic activity.

Rubisco kinetic characterization

Rates of Rubisco ^{14}CO$_2$-fixation using rapidly extracted and desalted leaf protein extracts were measured in 7 mL septum-capped scintillation vials, containing reaction buffer (yielding final concentrations of 100 mmol/L Bicine-NaOH, pH 8.0, 20 mmol/L MgCl$_2$, 0.4 mmol/L RuBP, and about 100 W-A units of carbonic anhydrase) and sodium [^{14}C] bicarbonate to give one of six different concentrations of CO$_2$ (4–100 μmol/L, each with a specific radioactivity of 3.7 × 10^{10} Bq mol^{-1}), each at four concentrations of O$_2$ (0%, 21%, 60%, and 100% [v/v]), as described previously (Parry et al. 2007). Assays (1.0 mL total volume) were started by the addition of activated leaf extract, and the V_{max} for carboxylase activity, together with the Michaelis–Menten constant (K_m) for CO$_2$ (K_c) determined by application of a curve optimization program (Enzfitter, Elsevier Biosoft, PO Box 98, Cambridge, UK). The K_m for the oxygenase activity (K_o) was calculated from the relationship, $K_{c(21,61,100\%o_2)} = K_{c(0\%o_2)}(1 + [O_2]/K_o)$ using the corresponding apparent K_c values at 21%, 60%, and 100% O$_2$. Replicate measurements (n = 2–8) were made using protein preparations from leaves of different individuals. For each sample, the maximum rate of carboxylation (k^c_{cat}) was extrapolated from the corresponding V_{max} value after allowance was made for the Rubisco active site concentration, as determined by [^{14}C]CABP binding (Yokota and Canvin 1985). Rubisco CO$_2$/O$_2$ specificity ($S_{c/o}$) was measured as described (Galmés et al. 2005) using enzyme purified by PEG precipitation and ion-exchange chromatography and the values given for each species were the mean of 5–6 replicate determinations. The maximum oxygenation rate (k^o_{cat}) was calculated using the equation $S_{c/o} = (k^c_{cat}/K_c)/(k^o_{cat}/K_o)$. All kinetic measurements were performed at 25°C. All radiochemicals and associated instruments and consumables were purchased from PerkinElmer (Seer Green, Buckinghamshire, UK). All other chemicals were of analytical grade and supplied by Sigma.

As described in the Discussion section, the maximum carboxylation rate of Rubisco (V_{cmax}) for each willow species could be calculated after determination of the leaf Rubisco concentration and the corresponding rate constant for carboxylation, using the equation:

$$V_{cmax} = \text{Rubisco concentration } (g \cdot m^{-2}) \times k^c_{cat}(s^{-1})$$
$$\times 14.5 \mu\text{mol active sites} \cdot g\,\text{Rubisco}^{-1}, \quad (2)$$

(assuming a molecular weight (MW) of 550 kDa for Rubisco and eight active sites per Rubisco holoenzyme). [Correction added on 15 December 2014 after initial publication on 10 October 2014. In Equation 2, the final term Rubisco^{-1} should be g Rubisco^{-1}. This is now corrected.]

Biomass measurement and yield estimation

The number, length, and diameter (at 0.5 m intervals) of all shoots and branches were systematically tabulated immediately prior to harvest. Aboveground material from each sample, excluding any remaining leaves, was cut into 25 cm lengths, weighed, then oven dried at 80°C for 96 h, and reweighed immediately. The resulting dry mass (DM) data are regarded as the biomass yield. A nondestructive estimate of yield in early and again in mid/late Summer (2009) was provided by measurement of shoot and branch numbers, length, and diameters at 0.5 m intervals, from which the total shoot volume could be calculated, which was highly correlated to wood DM (Table 3, col 1, row A).

LSU gene sequencing

DNA was extracted from two young leaves of each willow variety, by use of the Wizard™ Genomic DNA Isolation Kit (Product A1120, Promega, Southampton, UK). These preparations included chloroplast DNA. rbcL was amplified from primers binding within the genes on either side of rbcL in the chloroplast genome, ATP synthase β (atpB3: GTGTCAATC ACTTCCATTCCTCTC), and acetyl-CoA carboxylase β (accDR2: CCATTGATTTAYTTCRCCYACACCTG). DNA sequencing of purified PCR products was carried out from these primers and additional primers SF1, SF2, SR1, and SR2 (CGAGTAGACCTTGTTGCTGTGAG; CTTCTACTGGTACA TGGACAACTG; CTTTTAGTAAAAGATTGTTCCTAT; CAT CTTTGGTAAAATCAAGTCCACC).

Statistical analysis

Association between mean values (three biological replications per genotype) of the indicated genotype-specific parameters (11 genotypes per parameter set) were calculated using the Pearson product–moment correlation (Tables 2, 3), as implemented by SigmaPlot 12.0 (Systat Software, Inc., Hounslow, London, UK) highlighting correlation coefficients

Table 2. Pearson product-moment correlations between mean values for a variety of yield- and photosynthesis-related parameters, determined for each genotype in 2008.

							Leaf size	SLA	Leaf area	A_{amb}	C_i	ΣA_{amb}	A_{max}	V_{cmax}	Rubisco	J	$Q_{1/2}$	A_{maxQ}	Γ_Q	CHO	Starch	Chl	a/b	Protein
							1	2	3	4	5	6	8	9	10	11	12	13	14	15	16	17	18	19
A	Shoot DM	g/plant					0.36	0.57	0.61	−0.48	0.45	0.71	−0.64	−0.67	−0.66	−0.62	−0.66	−0.77	−0.46	0.22	0.55	−0.31	0.47	−0.59
B	Leaf size	cm²/leaf						0.12	0.65	−0.86	−0.16	0.53	−0.68	−0.69	−0.65	−0.70	−0.84	−0.77	−0.34	0.61	0.52	−0.41	0.66	−0.71
C	SLA	cm²/g							0.12	−0.09	0.19	0.16	−0.19	−0.21	−0.20	−0.22	−0.41	−0.47	−0.72	−0.10	0.15	−0.19	0.53	−0.27
D	Leaf area	m²/plant								−0.75	−0.15	0.96	−0.72	−0.67	−0.67	−0.71	−0.85	−0.88	−0.52	0.63	0.75	−0.30	0.39	−0.90
E	A_{amb}	μmol m⁻² sec⁻¹									0.07	−0.63	0.77	0.77	0.71	0.78	0.81	0.80	0.30	−0.51	−0.71	0.61	−0.77	0.73
F	C_i	μmol mol⁻¹										−0.08	−0.32	−0.40	−0.28	−0.31	0.18	0.00	−0.20	−0.51	0.22	−0.38	0.12	0.06
G	ΣA_{amb}	μmol/m²/plant											−0.65	−0.61	−0.67	−0.63	−0.81	−0.86	−0.48	0.66	0.65	−0.20	0.27	−0.83
I	A_{max}	μmol/m²/sec												0.98	0.95	1.00	0.73	0.83	0.47	−0.33	−0.89	0.68	−0.53	0.85
J	V_{cmax}	μmol/m²/sec													0.94	0.98	0.68	0.79	0.39	−0.27	−0.81	0.72	−0.57	0.78
K	Rubisco	g/m²														0.95	0.76	0.86	0.40	−0.47	−0.79	0.67	−0.45	0.80
L	J	μmol/m²/sec															0.73	0.83	0.51	−0.34	−0.90	0.70	−0.58	0.86
M	$Q_{1/2}$	μmol/m²/sec																0.97	0.43	−0.80	−0.69	0.28	−0.65	0.86
N	A_{maxQ}	μmol/m²/sec																	0.52	−0.73	−0.78	0.38	−0.68	0.91
O	Γ_Q	μmol/m²/sec																		−0.30	−0.61	0.15	−0.56	0.70
P	CHO	g/m²																			0.29	0.08	0.22	−0.60
Q	Starch	g/m²																				−0.60	0.47	−0.84
R	Chl	g/m²																					−0.49	0.36
S	a/b																							−0.50

Each measurement was the mean of three biological replications. $P < 0.05$ ($r = 0.60 - 0.72$) is indicated by a yellow background and $P < 0.01$ ($r = 0.73+$) by green. A_{amb}, photosynthesis at ambient CO_2, with 1250 μmol PAR m⁻² sec⁻¹, 25°C, 60% relative humidity. A, net rate of CO_2 uptake per unit leaf area (μmol m⁻² sec⁻¹); A_{amb}, A_{max}, A as measured in air containing ambient (385 μmol mol⁻¹) or saturating (1100 μmol mol⁻¹) concentrations of CO_2, respectively; A_{maxQ}, theoretical value of A prevailing at light saturation; C_i, intercellular CO_2 concentration (μmol mol⁻¹) at ambient external CO_2); $Q_{1/2}$, irradiance at which ½ A_{maxQ} is attained (μmol photons m⁻² sec⁻¹); ΣA_{amb}, net rate of CO_2 uptake per plant (μmol m⁻² plant⁻¹); Γ_Q, light compensation point (μmol photons m⁻² sec⁻¹); V_{cmax}, maximum rate of RuBP carboxylation (μmol CO_2 fixed m⁻² sec⁻¹); J, rate of whole chain electron transport (μmol electrons m⁻² sec⁻¹); SLA, specific leaf area (cm² g⁻¹); CHO, soluble carbohydrate (g m⁻²); Chl, total chlorophyll concentration (g m⁻²); a/b, ratio of Chl a: Chl b (dimensionless); protein, total soluble leaf protein (g m⁻²).

Table 3. Pearson product-moment correlations between mean values for a variety of yield- and photosynthesis-related parameters, determined for each genotype in 2009.

Section groupings: columns 1–2 = **At harvest**; columns 3–15 = **Early Summer**; columns 16–20 = **Mid/late Summer**.

Code	Variable	Unit	1 Shoot V	2 Root DM	3 Leaf size	4 A_{amb}	5 $A_{amb}M$	6 Ci	7 A_{max}	8 V_{cmax}	9 J	10 $Q_{½}$	11 A_{maxQ}	12 $Γ_Q$	13 Shoot V	14 Leaf area	15 $ΣA_{amb}$	16 Shoot V	17 Leaf area	18 Rubisco	19 A_{amb}	20 $ΣA_{amb}$
A	Shoot DM	g/plant	0.97	0.88	0.34	-0.40	-0.46	-0.07	-0.28	-0.45	-0.26	-0.69	-0.61	-0.32	0.63	0.73	0.87	0.88	0.48	-0.26	-0.53	0.38
B	Shoot V	cm³ plant⁻¹		0.88	0.36	-0.37	-0.40	-0.13	-0.19	-0.34	-0.17	-0.61	-0.57	-0.33	0.68	0.71	0.88	0.94	0.35	-0.13	-0.49	0.28
C	Root DM	g/plant			0.40	-0.49	-0.53	-0.06	-0.40	-0.51	-0.40	-0.69	-0.69	-0.28	0.43	0.76	0.90	0.72	0.54	-0.38	-0.44	0.49
D	Leaf size	cm²/leaf				-0.68	-0.57	-0.05	-0.69	-0.54	-0.69	-0.64	-0.66	-0.29	0.33	0.63	0.42	0.38	0.50	-0.44	-0.48	0.19
E	A_{amb}	μmol m⁻² sec⁻¹					0.90	0.42	0.81	0.80	0.82	0.64	0.96	0.57	-0.50	-0.89	-0.68	-0.37	-0.78	0.61	0.68	-0.27
F	$A_{amb}M$	μmol m⁻² sec⁻¹						0.25	0.89	0.87	0.88	0.62	0.86	0.36	-0.27	-0.86	-0.69	-0.30	-0.82	0.77	0.60	-0.45
G	Ci	μmol mol⁻¹							-0.05	0.16	0.00	-0.25	0.35	0.65	-0.40	-0.40	-0.37	-0.33	-0.15	-0.28	0.49	0.41
H	A_{max}	μmol m⁻² sec⁻¹								0.81	0.95	0.69	0.77	0.10	-0.07	-0.66	-0.43	-0.05	-0.81	0.88	0.49	-0.49
I	V_{cmax}	μmol m⁻² sec⁻¹									0.91	0.58	0.77	0.48	-0.12	-0.73	-0.56	-0.23	-0.82	0.81	0.59	-0.54
J	J	μmol m⁻² sec⁻¹										0.61	0.75	0.27	-0.03	-0.67	-0.43	-0.06	-0.88	0.91	0.52	-0.48
K	$Q_{½}$	μmol m⁻² sec⁻¹											0.77	0.20	-0.48	-0.68	-0.62	-0.47	-0.58	0.60	0.33	-0.62
L	A_{maxQ}	μmol m⁻² sec⁻¹												0.51	-0.58	-0.93	-0.81	-0.52	-0.79	0.58	0.69	-0.37
M	$Γ_Q$	μmol m⁻² sec⁻¹													-0.55	-0.56	-0.48	-0.51	-0.39	0.07	0.47	0.14
N	Shoot V	cm³ plant⁻¹														0.60	0.63	0.83	0.20	0.09	-0.59	-0.20
O	Leaf area	m² plant⁻¹															0.93	0.67	0.69	-0.49	-0.64	0.35
P	$ΣA_{amb}$	cm³ plant⁻¹																0.81	0.57	-0.32	-0.59	0.34
Q	Shoot V	cm³ plant⁻¹																	0.26	0.05	-0.59	-0.02
R	Leaf area	m² plant⁻¹																		-0.84	-0.75	0.41
S	Rubisco	g m⁻²																			0.40	-0.53
T	A_{amb}	μmol m⁻² sec⁻¹																				0.17

Each measurement was the mean of three biological replications. $P < 0.05$ ($r = 0.61 – 0.70$) is indicated by a yellow background and $P < 0.01$ ($r = 0.71+$) by green. A, net rate of CO_2 uptake per unit leaf area (μmol m⁻² sec⁻¹); A_{amb}, A_{max}, A as measured in air containing ambient (385 μmol mol⁻¹) or saturating (1100 μmol mol⁻¹) concentrations of CO_2, respectively; $A_{amb}M$, A_{amb} derived by biochemical modelling; A_{maxQ}, net rate of theoretical value of A prevailing at light saturation; Ci, intercellular CO_2 concentration (μmol mol⁻¹) at ambient external CO_2; $Q_{½}$, irradiance at which ½ A_{maxQ} is attained (μmol photons m⁻² sec⁻¹); $ΣA_{amb}$, net rate of CO_2 uptake per plant (μmol m⁻² plant⁻¹); J, rate of whole chain electron transport (μmol electrons m⁻² sec⁻¹); $Γ_Q$, light compensation point (μmol photons m⁻² sec⁻¹); V_{cmax}, maximum rate of RuBP carboxylation (μmol CO_2 fixed m⁻² sec⁻¹); V, volume (cm³ plant⁻¹); SLA, specific leaf area (cm² g⁻¹); DM, dry mass (g plant⁻¹).

Table 4. Comparing Rubisco from the chosen *Salix* genotypes.

Species	Polymorphic residues						Rubisco kinetic constants									
Spinach	9	142	230	255	309	363	k_{cat}^c (sec^{-1})		k_{cat}^o (sec^{-1})		K_c(µM)		K_o(µM)		$S_{c/o}$	
	A	V	A	V	M	Y	Mean	LSD	Mean	LSD	Mean	LSD	Mean	LSD	Mean	LSD
S. triandra	A	P	T	I	I	Y	3.45	0.35	2.24	0.54	7.2	1.46	409	104	88.3	3.1
R13	T	T	A	I	I	Y	2.78		1.17		11.0		408		88.4	
S. exigua	T	V	A	V	M	F	2.67		1.78		6.4		379		88.5	
Other *Salix*	T	T	A	I	I	Y	3.02	0.27	1.62	0.42	7.4	1.15	352	82	88.7	2.4

Amino acid differences between the LSU sequences of *Salix* Rubisco and the corresponding Rubisco kinetic parameters, describing the Michaelis–Menten constants for CO_2 (K_c) and O_2 (K_o), the maximum rate of carboxylation k_{cat}^c and oxygenation (k_{cat}^o), and the specificity factor ($S_{c/o}$). Only those residues which differ between the 11 species of *Salix* species are shown. Kinetic constants for *S. triandra*, R13, and *S. exigua* are the means of two biological replications, while those for the remaining Rubisco isolates – which were considered to be identical – were combined, giving (eight biological replicates). Values of k_{cat}^o were calculated using the equation $S_{c/o} = (k_{cat}^c/K_c)/(k_{cat}^o/K_o)$. Two values for the LSD at the 5% confidence interval are shown; one for comparison between the top three genotypes ($n = 2$) and one for comparison between the value at the bottom ($n = 8$) and those above. In all cases, the degrees of freedom for calculation of LSD was 10. [Correction added on 15 December 2014, after first online publication: The first 4 headings of the 5 categories of parameter under Rubisco kinetic constants were previously incorrect and these have now been replaced.]

with $P < 0.05$ and $P < 0.01$. The one-way analysis of variance (ANOVA) treatments (SigmaPlot 12.0) of Figure 4 state mean values and the standard error of difference (SED) between the means, with 20–22 degrees of freedom depending on the response variable. The ANOVA of Table 4 is similar to that of Figure 4, except values for the least significant difference (LSD) at the 5% confidence interval are given, and the associated degrees of freedom stated in the legend. Other figures show specific comparisons between parameter sets, indicating mean values ± standard errors ($n = 3$) together with the derived correlation coefficient, r.

Results

Measuring leaf photosynthesis

In both years, *S. daphnoides* and *S. exigua* represented the upper and *S. dasyclados* the lower extremes of A/Ci performance, respectively (Fig. 1). For clarity, the dependence of A on intercellular CO_2 concentration (Ci) is shown for four (rather than all 11) of the genotypes, chosen to illustrate the range of observed responses. Values of V_{cmax} and J were determined (Fig. 2) which – when substituted into the equations of Farquhar and von Caemmerer describing RuBP carboxylation- and regeneration-limited CO_2 assimilation (von Caemmerer 2000) – described curves which faithfully followed the observed A/Ci data points for each genotype (Fig. 1, solid lines). Collectively, the values of V_{cmax} and J for all genotypes from both years could be described by a linear relationship, with a gradient (J/V_{cmax}) of approximately 2. The distribution of values differed between the 2 years – showing a broader range of values in 2008 (□) than 2009 (◇). Even so, the four genotypes with the highest values, and the three genotypes with the lowest values were the

same in each year, although data points from 2008 appeared to have marginally higher rates of electron transport for any given V_{cmax} than those of 2009, the mean values for J/V_{cmax} being 2.24 ± 0.30 and 2.05 ± 0.10, respectively.

Diverse responses to light intensity were also found. The light dependence of photosynthesis was faithfully described for all genotypes in both years by a series of hyperbolic functions (eq. 1, Material and Methods), from which values for the light-saturated rates of CO_2 assimilation (A_{maxQ}), the photosynthetically active radiation required to achieve half these rates (effectively the K_M for photon flux density, $Q_{1/2}$), and the light compensation point (Γ_Q) could be derived. A linear relationship was found between light-saturated

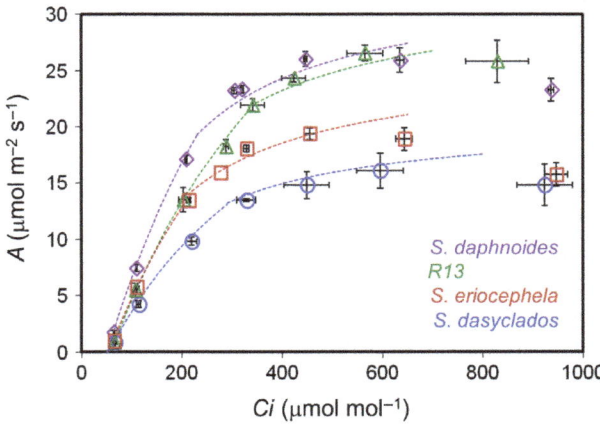

Figure 1. Dependence of leaf CO_2 assimilation (A) on intercellular CO_2 concentration (Ci). Data from a representative selection of genotypes in 2009. Light intensity 1250 µmol PAR m^{-2} sec^{-1}; leaf temperature 25 ± 1°C; relative humidity 60%. Each measurement was the mean of three biological replications, for which mean values and standard errors are indicated. The dotted lines show the modeled curves, based on the average values of V_{cmax} and J for each genotype.

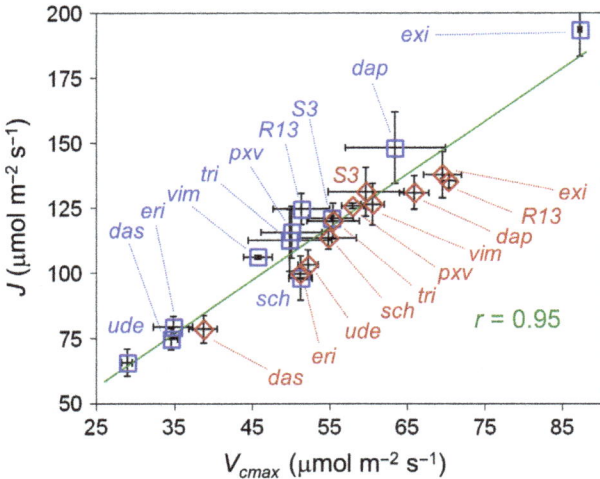

Figure 2. Correlation between maximum rates of Rubsico-limited and electron-transport-limited photosynthesis (V_{cmax} and J, respectively). Values derived by an iterative procedure (Sharkey et al. 2007) using A/Ci data for each genotype from both years. During this procedure, V_{cmax} and J were varied to minimize the deviation of the modeled curve from the observed data points. The resulting values were then normalized to 25°C. Each measurement was the mean of three biological replications, for which mean values and standard errors are indicated: 2008 (blue) and 2009 (red).

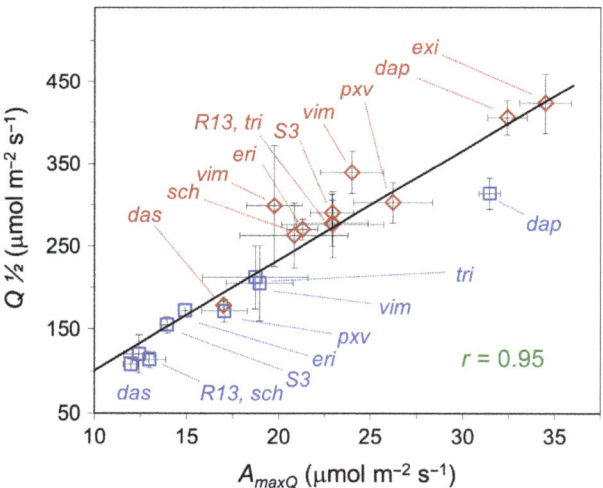

Figure 3. Correlation between rates of CO_2 assimilation at light saturation (A_{maxQ}) and the irradiance required to achieve 50% of these rates ($Q_{\frac{1}{2}}$). Data derived from a study of the light dependence of photosynthesis in both years, with chamber conditions set to 385 μmol CO_2 mol^{-1}, 25°C, 60% relative humidity. 2008 (blue) and 2009 (red). Each measurement was the mean of three biological replications, for which mean values and standard errors are indicated.

A_{maxQ} and $Q_{\frac{1}{2}}$ among the chosen genotypes from both years (Fig. 3). Values for the same genotype between years showed greater variation that the A/Ci data, although the relative positions of the extremes of performance were similar.

Identifying relationships across a wider range of parameters

In addition to those above, a selection of additional biochemical, physiological, and yield-related parameters were measured in the first year of growth (2008). Those relating to gross leaf morphology (size and specific leaf area) and photosynthetic performance were made in early/mid-Summer (late June) in parallel with leaf sampling for constituent analysis. Multiple pair-wise correlations between all the measured and derived parameters identified/confirmed likely positive and negative relationships (at $P > 99\%$ and $P > 95\%$) between many of these. The pattern and significance of the derived correlations were virtually identical when either the ungrouped (not shown) or the grouped (mean) data for each genotype were treated in this way (Table 2). In particular, a consistent, positive correlation was evident between total leaf protein and all parameters relating to leaf level photosynthesis (Table 2, col 19) and especially between leaf Rubisco and both light and CO_2 saturated photosynthesis (Table 2, col 10 and also row K). The same was also indicated for total leaf chlorophyll. In contrast, DM and leaf starch both showed strong negative correlations with maximum leaf photosynthetic rate, both at ambient (A_{amb}) and saturating CO_2 (A_{max}) or saturating light (A_{maxQ}). However, as elaborated below, when the total leaf area per plant was estimated from leaf number and size data, and whole plant photosynthetic capacity estimated by combining this with the ambient rates per unit leaf area (as an estimate of the whole plant photosynthetic rate [ΣA_{amb}]), then a significant positive correlation between ΣA_{amb} and both DM and starch was indicated (Table 2). Surprisingly, although total leaf area (m²/plant) was positively correlated to DM, leaf starch, and leaf size (cm²/leaf) and negatively with all parameters relating to leaf level photosynthesis, and specific leaf area (SLA, cm²/g DM) showed little correlation to any of the measured parameters, except light compensation point (Γ_Q).

A similar collection of paired correlations was constructed using data from the second season (2009). As described previously, the pattern and significance of the derived correlations were virtually identical when either the ungrouped (not shown) or grouped (mean) data for each genotype were analyzed (Table 3). In this season, emphasis was placed on detailed measurement of shoot biomass and total leaf area from mid-June ("early Summer") and again from early August ("mid/late Summer"), together with leaf photosynthesis.

In early Summer (Table 3, col 14 and 15) total plant leaf area and whole plant photosynthetic potential (ΣA_{amb}) showed strong positive correlation to final yield (shoot DM, shoot volume, and root DM – rows A, B, and C) and equally strong negative correlations to a variety of photosynthetic characteristics expressed per unit leaf area (A_{amb}, A_{max}, A_{maxQ}, V_{cmax} Table 3, col 14 and 15). In mid/late Summer, the negative correlations between total plant leaf area and the same leaf level photosynthetic parameters were even more pronounced (Table 3, col 17) although photosynthetic potential – expressed on a whole plant basis – showed a weaker correlation to final yield (Table 3, col 20).

Identifying relationships between specific parameters

Mean (and SED) values for some of the parameter sets obtained in 2008 and 2009 and ranked according to magnitude are shown (Fig. 4) to illustrate the range and significance of a collection of the measured values. Although there are clear differences between the extremes of performance, the intermediate genotypes showed some degree of overlap. However, when the correlation between pairs of data sets was examined, likely relationships emerged. Some noteworthy examples are elaborated here.

Biomass yield and photosynthetic rate

Biomass yield and photosynthetic rate under ambient CO_2 and near-saturating light in the early Summer (A_{amb}) showed a weak, negative correlation (Table 3, row A, columns 5 and 20). However, graphical illustration of this relationship (using the early Summer yield data) highlighted a subset of genotypes which demonstrated a highly significant, positive correlation (Fig. 5A, red vs. black line). When the correlation between leaf area per plant and yield was examined (Fig. 5B), a significant correlation to which all species broadly conformed was found. An even more significant correlation was observed when both leaf area per plant and assimilation rate per unit leaf area were integrated (Fig. 5C), which again was a correlation to which all genotypes conformed. Thus, both leaf area per plant and assimilatory capacity per unit leaf area appear to make significant contributions to biomass. V_{cmax} and leaf area per plant were inversely correlated (Fig. 6). Since V_{cmax} is a function of the leaf content of Rubisco, we also compared leaf area and leaf Rubisco concentration, and found that they were most strongly (negatively) correlated in the mid/late Summer – as would be expected since (in 2009) the samples for Rubisco analysis had been taken over this period (Table 3, column 18).

Leaf starch content and maximum photosynthetic rate

Leaf starch content – measured in leaf samples taken in the mid-afternoon in early Summer (3–4 h after midday) – showed a strong negative correlation with parallel measurements of maximum leaf photosynthetic rate at saturating CO_2 (A_{max}) (Fig. 7) as well as with numerous other photosynthetic processes (Table 2, col 16 and row Q).

The correlation between common data sets measured in 2008 and 2009, while not identical, nonetheless indicate similar relative rankings between the diverse genotypes between successive years (Fig. S2).

Probing Rubisco diversity among the willow genotypes

Rubisco was sampled from collections of young, fully expanded leaves of each genotype in the late Spring. These samples provided material for subsequent extraction and/or purification of Rubisco, for determination of the relative specificity for CO_2 and O_2 ($S_{c/o}$) as well as the rate constants for carboxylase and oxygenase activities (k_{cat}^c, k_{cat}^o, respectively), and the Michaelis constants for CO_2 and O_2 (K_c and K_o, respectively). In parallel, DNA was extracted and the LSU gene sequences determined. The GenBank accession numbers for these sequences are given in Table 1, and the integrated data in Table 4, with additional kinetic data given in Table S1. Significant differences were identified between the genotypes with respect to (k_{cat}^c), (k_{cat}^o), and K_c (Table 4).

For nine of the genotypes, the LSU gene sequences were identical. Compared to these, the *S. triandra* sequence contained three different codons, representing amino acid substitutions T9A, T142P, and A230T, while *S. exigua* contained three distinct residue substitutions, these being, I255V, I309M, and Y363F together with an alternative substitution, T142V (Table 4).

A comparison of the maximum carboxylation rate of Rubisco (V_{cmax}) for each willow species, as deduced from gas exchange measurements, with the corresponding parameters deduced from direct measurement of the leaf Rubisco concentration and the rate constant for carboxylation is shown in Figure S4. This reveals that the activity of Rubisco during the photosynthesis measurements was considerably lower than would be expected based on the measured Rubisco content and catalytic characteristics.

Discussion

Leaf photosynthesis

The mean rates of photosynthesis at ambient CO_2 reported here for *S. viminalis* "Bowles hybrid" (19.5 ± 1.0 [SD, $n = 3$]) and *S. dasyclados* (13.7 ± 0.1 [SD, $n = 3$]) are very similar to those reported by Patton and Jones (1989) (i.e., 17.2 ± 2.3 and 13.9 ± 1.3, respectively). The values of V_{cmax} and J (Fig. 2) derived from the A/C_i analyses (Fig. 1) define the maximum rates of Rubisco-dependent (A_c) and

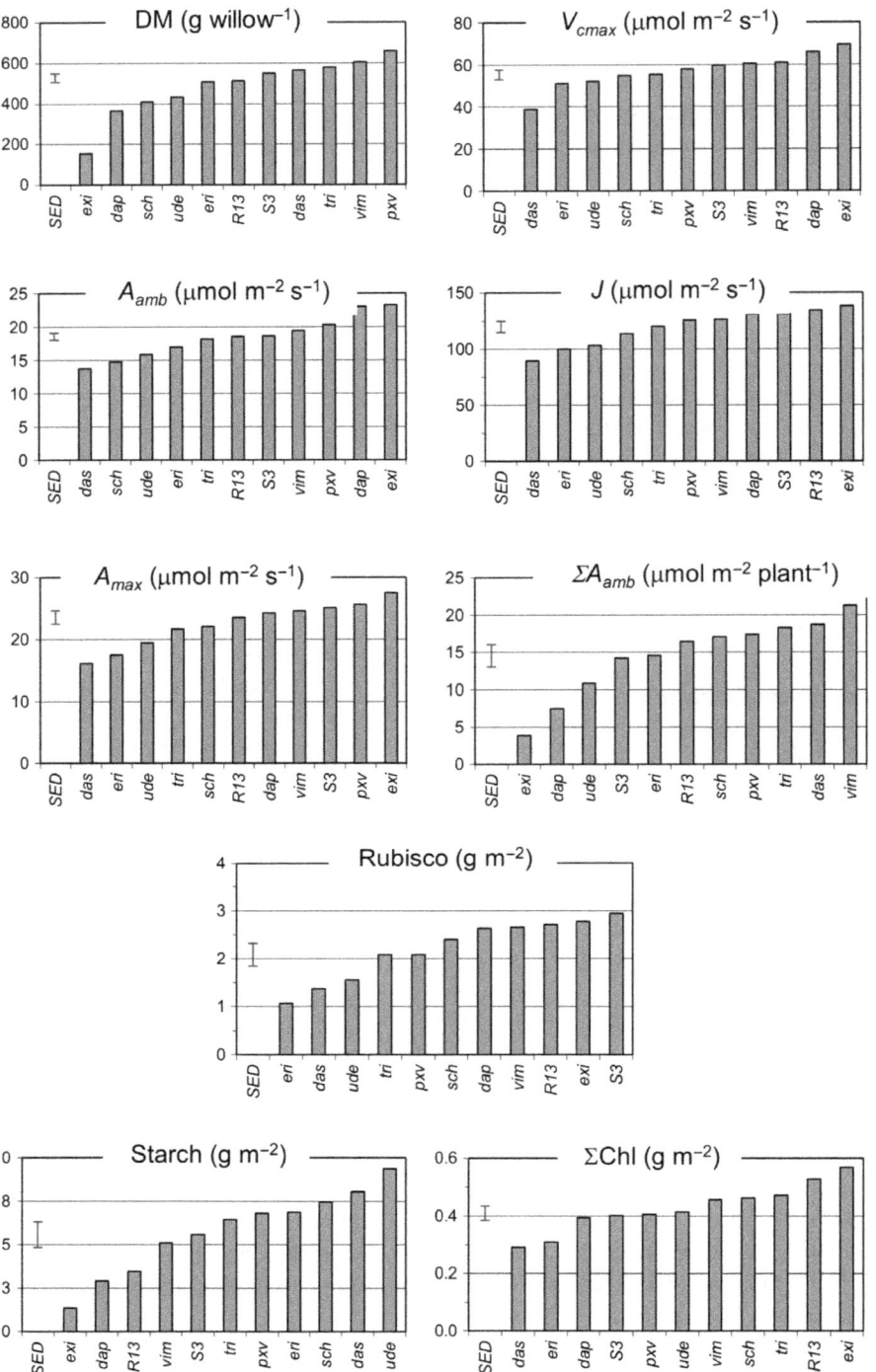

Figure 4. Means and SED values for an assortment of genotype properties. The data were from the 2009 experiment, except for starch and total chlorophyll (ΣChl) which were from 2008. Photosynthetic capacity on a whole plant basis (ΣA_{amb}) was measured in early Summer (end June/early July). Each measurement was the mean of three biological replications. Each data set was analyzed for statistical significance using a one-way analysis of variance (SigmaPlot 12.0) from which the mean values and standard error of difference (SED) between the means (with 20–22 degrees of freedom depending on the response variable) are shown.

RuBP-regeneration-dependent (A_j) CO_2 assimilation, respectively, as well as the initial gradient of the A/Ci relationship and the maximum attainable assimilation rates. The ratio,

J/V_{cmax} (at 25°C) typically ranges between 1.5 and 2.0 (von Caemmerer 2000) which are similar to the values determined here (2.0–2.5). In an analysis across a range of woody plant

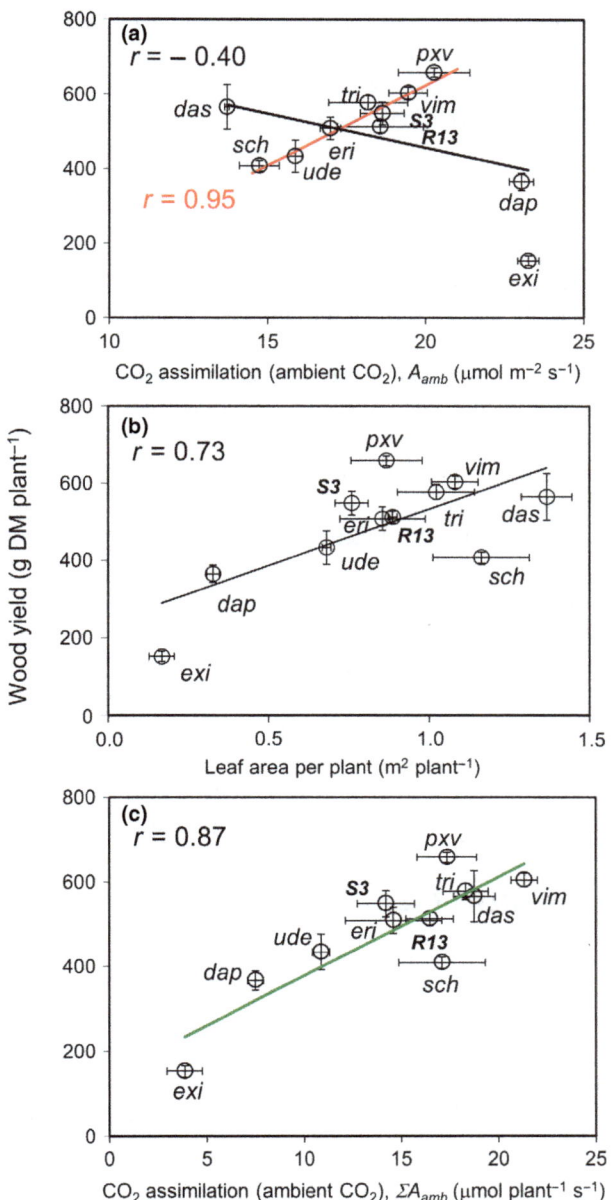

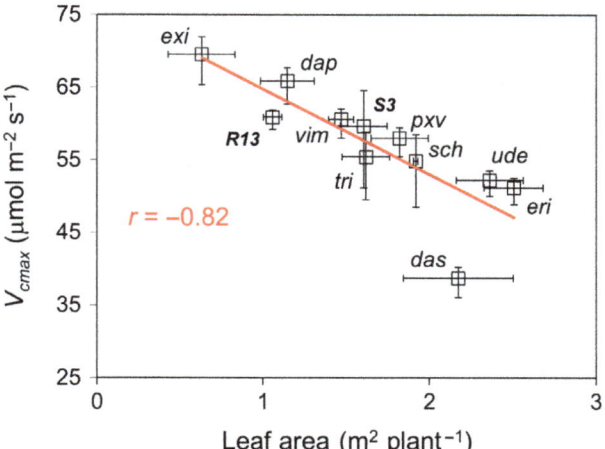

Figure 6. Correlation between V_{cmax} and leaf area per plant. Values shown are means ± standard error for 2009 data set ($n = 3$).

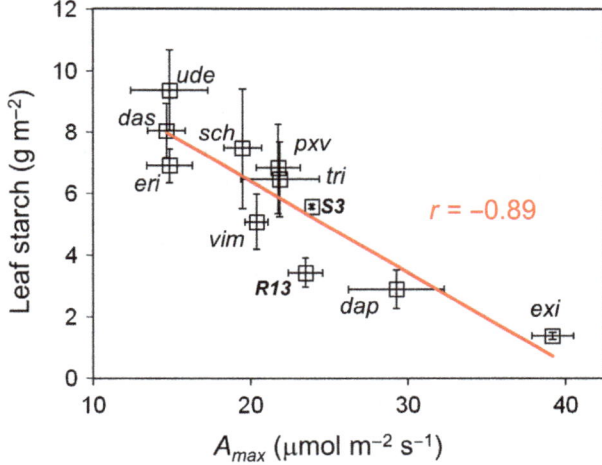

Figure 7. Correlation between photosynthesis at CO_2 saturation (A_{max}) and leaf starch, 3–4 h after midday. Values shown are means ± standard error from 2008 ($n = 3$).

Figure 5. (A) Correlation between yield and photosynthesis at ambient (385 μmol mol⁻¹) CO_2, 25°C, 60% relative humidity. The overall regression line is shown in black, while that shown in red applies to the cluster of points in the center. Values shown are means ± standard error ($n = 3$). (B) Correlation between yield and whole plant leaf area. (C) Correlation between yield and whole plant photosynthesis (calculated as: μmol CO_2 fixed m⁻² sec⁻¹ × leaf m² plant⁻¹). Coefficients derived from Pearson product–moment correlations, as implemented by SigmaPlot 12.0, are shown. All data were from the 2009 experiment.

species, Manter and Kerrigan (2004) also found a linear correlation between J and V_{cmax} analogous to that of Figure 2, with a mean ratio (J/V_{cmax}) of 2.5. By contrast, Robinson et al. (2004) measured ratios of 4.5 and 5.2 for high- and low-yielding willow species, respectively. Our data are

consistent with a narrow range of J/V_{cmax} ratios among the willows investigated, irrespective of yield capacity. The absolute amounts of the components which determine J and V_{cmax} will also impact on the observed rates of photosynthesis. Thus, *S. exigua* and *S. daphnoides*, which in both years showed the highest rates of photosynthesis at both ambient and saturating CO_2 (Figs. 1, 4) also had the highest V_{cmax} and J, while *S. dasyclados*, which had among the lowest rates in both years, also had the lowest V_{cmax} and J (Figs. 1, 2, and 4). These observations were complemented by the total leaf protein, chlorophyll (Table 2), and Rubisco (Tables 2, 3) measurements, all of which correlated positively with leaf photosynthetic performance. The process of parameter optimization to obtain the best values of V_{cmax} and J (Sharkey et al. 2007) also generated complementary values for light-independent respiration (R_d) and mesophyll conductance (*gm*). However, no significant correlations were evident using these values (not shown). In addition, the *Ci* at which the RuBP carboxylation limited rate of CO_2 assimilation is identical to the RuBP regeneration limited rate of CO_2 assimilation (the "tipping point") was generally in the vicinity of the observed *Ci* when the leaves were in normal air, and showed no consistent trend between the genotypes (not shown).

Considerable diversity was also apparent in the light dependence of photosynthesis among the willow genotypes; the estimated light-saturated rates of photosynthesis (at ambient CO_2) differing by up to 2.5-fold between the extremes of performance (Fig. 3). The genotypes with the highest photosynthetic rates at both ambient and saturating CO_2 also possessed the highest rates at light saturation and the highest irradiances to attain half this value.

Identifying relationships between parameters at specific periods

One important observation is that photosynthetic rate – either at ambient or saturating CO_2 – together with many of the underlying processes contributing to leaf photosynthetic performance (Rubisco, total protein, chlorophyll) were negatively correlated to yield. In contrast, total leaf area per plant was positively correlated to yield, particularly in the early Summer period (Tables 2, 3, Fig. 5A). However, yield is a function of both total leaf area and leaf photosynthetic rate (per unit leaf area). This is evident from Figure 5 where the rate of assimilation of eight of the species was very significantly correlated to yield (Fig. 5A). However, the total leaf area per plant for all genotypes was also significantly correlated to yield (Fig. 5B, $P < 0.01$). When these two components were integrated, the resulting whole plant photosynthetic capacity was also very significantly correlated to yield ($P < 0.001$). Yield and the corresponding parameters from mid/late Summer were less significantly correlated. Since the average factor by which leaf area increased over this period was 2.6 ± 0.5 (SD), the poorer correlation later in the season may have been due to leaf capacity being in increasing excess over available radiation. In other words, a slowly diminishing amount of solar radiation (since the solar angle and day length both diminish after June 21) was shared across an increasing total plant leaf area.

The relationship between starch, soluble carbohydrate, yield, and photosynthesis

The immediate products of photosynthesis – soluble carbohydrate and starch – measured in the late afternoon, showed a weak, positive correlation to yield, a stronger positive correlation to total leaf area per plant, and (particularly starch) a strong negative correlation to many components of the photosynthetic process (Table 2, columns 15 and 16). The latter observation prompts the question as to whether accumulation of leaf starch over this period, triggers processes which diminish photosynthetic rate. However, since starch is insoluble, the effects of starch accumulation may need to be relayed by a soluble factor whose concentration is responsive to starch. The obvious candidate for this would be a soluble sugar, which may explain the similar (albeit weaker) negative correlations between soluble carbohydrate and an assortment of photosynthetic components. More detail on the correlation between high starch content in leaves and low assimilation rate can be found elsewhere (Paul and Foyer 2001), although a strong negative correlation between net photosynthetic rate and starch concentration, independent of carbohydrate concentration, has been attributed to an increase in mesophyll (liquid phase) CO_2 diffusion resistance, suggesting that starch accumulation may reduce net photosynthetic rate by impeding intracellular CO_2 transport (Nafziger and Koller 1976). A key role for starch in the integration of plant growth has been reported (Selbig et al., 2009) although in that study (using Arabidopsis) it was found to correlate negatively with biomass, while the current study reports a positive correlation with biomass – of both shoot and leaf (Table 2).

A question of scale

The barely significant, negative, correlations between yield and many indicators of leaf photosynthetic performance, expressed per m² of leaf (A_{amb}, A_{max}, V_{cmax}, J, $Q_{1/2}$, A_{maxQ}, Γ_Q, and Rubisco; Table 3, rows A, B, and C) were all seen in fact to be significantly ($P > 0.05$) and positively correlated (with yield) when (like leaf photosynthesis itself, Fig. 5) these parameters were expressed on a whole plant basis (Fig. S3). The same was also true of the corresponding parameters of Table 2 (not shown). Evidently, the nature of an observed correlation depends on the way (or units) in which the parameters in question are expressed. That being so, in future it may be informative to relate photosynthetic parameters determined for much larger populations of plants obtained by measurements made over a similarly large area (e.g., by application of Eddy Covariance techniques) to biomass yields per hectare.

Assumptions, limitations, and shortcomings

The study of pot-grown material under glass was not intended to be a realistic substitute for field-grown material. However, in an attempt to ensure healthy, nutritionally replete, and otherwise unstressed plants in the same soil substrate at identical developmental stages, this approach was considered to be justified. It is unlikely that differences in the mass of the planted cuttings were responsible for the differences in willow growth recorded here (Hangs et al., 2011), neither are the differences in leaf level photosynthesis likely to be due to N availability – as identical growth media were used throughout and as fertilization has been shown to have little effect on leaf photosynthesis (Merilo et al. 2006). However, the absence of lateral shading – which would normally be present in dense plantations – may have led to unrealistic growth rates (and possibly other characteristics). Additional problems associated with restriction of root development may also have been present –and would undoubtedly have become more severe if the experiment had continued for more than a single season.

Furthermore, the photosynthetic performance of the youngest fully expanded leaves, measured under ideal, light-saturated conditions, together with a careful assessment of leaf area per plant, have been used to estimate the actual whole plant photosynthetic capacity. While this is no doubt on oversimplification, we did not have sufficient resources to conduct a more thorough investigation which, for example, might have investigated gradients of photosynthetic performance in leaves at different positions along the axis of the main stem and/or could have incorporated more stages in the season. Even so, we believe that the comparisons of estimated performance between genotypes attempted in this study are of value.

The extent of yearly variation between the photosynthetic parameters measured was surprising, although the correlation for yield remained strong (Fig. S1). Such differences are presumably due to the variation in summer irradiance and temperature between years.

Rubisco diversity among willow genotypes

Three of the six polymorphic LSU residues identified in Table 4 – namely 142, 255, and 309 – are positions which are among those most frequently found to be positively selected in evolutionary adaptation of the LSU (Kapralov and Filatov 2007). Residue identity at position 309 (typically methionine or isoleucine) correlates with kinetic differences between C_3 and C_4 forms of *Flaveria*, respectively (Kapralov et al. 2011). In the context of the *Flaveria* LSU sequence, isoleucine 309 has been shown to act as a catalytic switch that increases the Rubisco carboxylation rate (k_{cat}^c). An analogous situation is reported here between *S. triandra* (residue 309 = isoleucine) and *S. exigua* (residue 309 = methionine). In fact, all the LSU sequences determined, except that of *S. exigua*, had isoleucine at this position – and correspondingly had (4–29%) higher values for (k_{cat}^c) (Whitney et al., 2011) than *S. exigua*. But the extent of this difference and the relative magnitude of the other kinetic parameters differed. This is presumably due to the

effects of the other LSU residue changes which were observed. It is puzzling that R13 – whose LSU sequence was identical to that of the willow "consensus" sequence – was found to have distinct (k_{cat}^o) and K_c values (Table 4). This is presumably due to diverse properties of the accompanying nuclear-encoded small subunits, of which there are likely to be multiple, distinct, copies in each genotype, and which have been shown to differentially influence the kinetic properties of Rubisco (Ishikawa et al. 2011).

Making use of the modest differences in (k_{cat}^c), (k_{cat}^o), and K_c (Table 4) between the polymorphic willow LSUs, is beyond the scope of the current study, although these values generate distinct modeled A/Ci curves, assuming the leaves contained equal amounts of Rubisco and equal amounts of the components of RuBP regeneration, per unit leaf area (Fig. S2). All else equal, such differences could have a significant impact on performance over a range of Ci values, especially at lower Ci. Stomatal closure induced by water deficit will cause Ci to decline, as photosynthesis proceeds (Farquhar and Sharkey 1982). Under these conditions, species – like *S. triandra* – whose Rubisco can support higher rates of CO_2 assimilation (up to twofold more than R13 at $Ci = 200$ μmol mol^{-1}) would have a clear advantage. However, the modeled behavior was not reflected in the observed A/Ci curves or in the derived values for V_{cmax} and J (Figs. 1, 2), most likely owing to the actual leaf concentrations of Rubisco – which were greater in R13 and *S. exigua* than in many of the other genotypes (Fig. 4).

Mismatch between Rubisco activity and abundance

The species-specific, maximum carboxylation capacity of Rubisco (V_{cmax}) predicted from A/Ci analyses (e.g., Figs. 2, 4), may underestimate the actual V_{cmax}, as deduced from direct measurement of species-specific Rubisco abundance and catalytic capacity. Figure S4 presents a comparison between these alternative measures of V_{cmax} from 2009 (although the same was also apparent in 2008). One explanation for such a mismatch would be that the A/Ci measurements were not made with adequate (near saturating) light. This is very unlikely, since the mean photon flux density ($Q_{\frac{1}{2}}$) at which half the predicted light-saturated rate of photosynthesis was evident – was approximately 250 μmol photons m^{-2} sec^{-1} (Fig. 3), while the measuring intensity was 5× higher than this, and represented a similar level of exposure as that on a cloudless day under glass. A more likely explanation would be either that Rubisco was present in excess of requirement and that the excess catalytic capacity had been downregulated, or else that limitations existed in the Rubisco regulatory mechanism, preventing full expression of the available Rubisco activity. This possibility may be worth pursuing in future.

Concluding Remarks

This study assessed photosynthetic capacity and contributory processes in a broad range of genotypes to determine whether sufficient natural variation relevant to yield existed to warrant the initiation of genetic improvement strategies targeting the underlying genes. Significant differences in photosynthetic parameters and yield between the genotypes studied were identified and imply strong genotypic control of all these properties. A variety of parameters – expected to positively impact upon photosynthetic performance – were measured for each genotype and were collectively found to correlate positively with photosynthesis. Although a positive correlation between photosynthetic rate and biomass was only described by a subset of genotypes, at a specific period in the growing season (Fig. 5A), when leaf area was taken into account (Fig. 5B) and photosynthetic rate expressed on a whole plant basis (Fig. 5C), then it emerged as being positively correlated to yield: strongly during early Summer, but also later on, albeit less significantly (Table 3). It also emerged that when a variety of other performance-related photosynthetic parameters were expressed on a per plant basis (taking into account the leaf area per plant), they were also found to be positively correlated with yield (Fig. S3). This indicates that plant leaf area has been a more significant adaptive criterion for controlling growth among *Salix* genotypes, than the mechanistic capacity in a given leaf area. The kinetic properties of Rubisco make a significant contribution to V_{cmax} and J (the former through (k_{cat}^c), K_c, and K_o, than latter through specificity factor). These properties were shown to be distinct between certain willow genotypes, consistent with differences in their LSU gene sequences, also identified in this work. Hence, distinct, heritable traits relating specifically to Rubisco (and therefore photosynthetic) performance have been identified, consistent with our overall objectives. *Salix* varieties with higher photosynthetic rates per unit leaf area, relative to whole plant leaf area, have been shown to result in greater yields (Bouman and Sylliboy, 2012). Thus, in common with other commercially important plant species, improvements in photosynthetic efficiency in *Salix* has significant potential for yield improvement (Long et al. 2006).

Acknowledgments

We acknowledge the advice of Elina Vapaavuori (of the Finnish Forest Research Institute) during our initial attempts to extract protein (particularly active Rubisco) from our leaf material. We also thank William Macalpine for assisting in the identification of willow genotypes in the field; March Castle and Tim Barraclough for helping with the harvest shoot volume measurements of the 2009 experiment; Jérémy Guinard for assisting with the gas exchange measurements in 2008; and the glasshouse staff at Rothamsted Research for technical support throughout. P. J. A., P. J. M., and M. A. J. P. are supported by the UK Biotechnological and Biological Sciences Research Council (BBSRC) 20:20 Wheat® Institute Strategic Program (BBSRC BB/J/00426X/1) and BBSRC BB/I002545/1 and BB/I017372/1, and A. K. and I. F. S. are supported by the BBSRC Cropping Carbon Institute Strategic Program (BB/J004278/1). Rothamsted Research is an Institute supported by the BBSRC.

Conflict of Interest

None declared.

References

Bouman, O. T., and J. Sylliboy. 2012. Biomass allocation and photosynthetic capacity of willow (*Salix* spp.) bio-energy varieties. Forstarchiv 83:139–143.

van Caemmerer, S. 2000. P. 45 *in* Biochemical models of leaf photosynthesis. Chapter 3 – modelling C$_3$ photosynthesis, CSIRO Publishing, Canberra, Australia.

Cannell, M. G. R., R. Milne, L. J. Sheppard, and M. H. Unsworth. 1987. Radiation interception and productivity of willow. J. Appl. Ecol. 24:261–278.

Christersson, L. 1987. Biomass production by irrigated and fertilized *Salix* clones. Biomass 12:83–95.

Ericsson, T., L. Rytter, and E. Vapaavuori. 1996. Physiology of carbon allocation in trees. Biomass Bioenergy 11:115–127.

Farquhar, G. D., and T. D. Sharkey. 1982. Stomatal conductance and photosynthesis. Annu. Rev. Plant Phys. Plant Mol. Biol. 33:317–345.

Galmés, J., J. Flexas, A. J. Keys, J. Cifre, R. A. C. Mitchell, P. J. Madgwick, et al. 2005. Rubisco specificity factor tends to be larger in plant species from drier habitats and in species with persistent leaves. Plant Cell Environ. 28:571–579.

Hangs R. D., Schoenau J. J., Van Rees K. C. J., Steppuhn, H. 2011. Examining the salt tolerance of willow (Salix spp.) bioenergy species for use on salt-affected agricultural lands. Can. J. Plant Sci. 91:509–517.

Hanley, S. J., and A. Karp. 2013. Genetic strategies for dissecting complex traits in biomass willows (*Salix* spp.). Tree Physiol. doi: 10.1093/treephys/tpt089

Hansen, J., and I. Møller. 1975. Percolation of starch and soluble carbohydrates from plant tissue for quantitative determination with anthrone. Anal. Biochem. 68:87–94.

Ishikawa, C., T. Hatanaka, S. Misoo, C. Miyake, and H. Fukayama. 2011. Functional incorporation of sorghum small subunit increases the catalytic turnover rate of Rubisco in transgenic rice. Plant Physiol. 156:1603–1611.

Kapralov, M. V., and D. A. Filatov. 2007. Widespread positive selection in the photosynthetic Rubisco enzyme. BMC Evol. Biol. 7:73.

Kapralov, M. V., D. S. Kubien, I. Andersson, and D. A. Filatov. 2011. Changes in rubisco kinetics during the evolution of C4 photosynthesis in flaveria (Asteraceae) are associated with positive selection on genes encoding the enzyme. Mol. Biol. Evol. 28:1491–1503.

Karp, A., and I. Shield. 2008. Bioenergy from plants and the sustainable yield challenge. New Phytol. 179:15–32.

Karp, A., S. J. Hanley, S. O. Trybush, W. Macalpine, M. Pei, and I. Shield. 2011. Genetic improvement of Willow for bioenergy and biofuels. J. Integr. Plant Biol. 53:151–165.

Liu, M. Z., G. M. Jiang, Y. G. Li, L. M. Gao, S. L. Niu, H. X. Cui, et al. 2003. Gas exchange, photochemical efficiency, and leaf water potential in three *Salix* species. Photosynthetica 41:393–398.

Long, S. P., X. G. Zhou, S. L. Naidu, and D. R. Ort. 2006. Can improvement in photosynthesis increase crop yields? Plant Cell Environ. 29:315–330.

Manter, D. K., and J. Kerrigan. 2004. A/Ci curve analysis across a range of woody plant species: influence of regression analysis parameters and mesophyll conductance. J. Exp. Bot. 55:2581–2588.

Merilo, E., K. Heinsoo, and O. Kull. 2006. Leaf photosynthetic properties in a willow (*Salix viminalis* and *Salix dasyclados*) plantation in response to fertilization. Eur. J. Forest Res. 125:93–100.

Miller, S. A. 2010. Minimizing land use and nitrogen intensity of bioenergy. Environ. Sci. Technol. 44:3932–3939.

Nafziger, E. D., and H. R. Koller. 1976. Influence of leaf starch concentration on CO_2 assimilation in soybean. Plant Physiol. 57:560–563.

Neergaard, A. D., J. R. Porter, and A. Gorissen. 2002. Distribution of assimilated carbon in plants and rhizosphere soil of basket willow (*Salix viminalis* L.). Plant Soil 245:307–314.

Parry, M. A. J., P. J. Madgwick, J. F. C. Carvalho, and P. J. Andralojc. 2007. Prospects for increasing photosynthesis by overcoming the limitations of Rubisco. J. Agric. Sci. 145:31–43.

Patton, L., and M. B. Jones. 1989. Some relationships between leaf anatomy and photosynthetic characteristics of willows. New Phytol. 111:657–661.

Paul, M. J., and C. H. Foyer. 2001. Sink regulation of photosynthesis. J. Exp. Bot. 52:1383–1400.

Robinson, K. M., A. Karp, and G. Taylor. 2004. Defining leaf traits linked to yield in short-rotation coppice *Salix*. Biomass Bioenergy 26:417–431.

Selbig, J., A. R. Fernie, T. Altmann, and M. Stitt. 2009. Starch as a major integrator in the regulation of plant growth. Proc. Natl Acad. Sci. USA 106:10348–10353.

Sharkey, T. D., C. J. Bernacchi, G. D. Farquhar, and E. L. Singsaas. 2007. Fitting photosynthetic carbon dioxide response curves for C3 leaves. Plant Cell Environ. 30:1035–1040.

Taylor, G., K. P. Beckett, K. M. Robinson, K. Stiles, and A. M. Rae. 2001. Identifying QTL for yield in biomass poplar. Aspects Appl. Biol. 65 (Biomass and energy crops II):173–182.

Tharakan, P. J., T. A. Volk, C. A. Nowak, and L. P. Abrahamson. 2005. Morphological traits of 30 willow clones and their relationship to biomass production. Can. J. For. Res. 35:421–431.

Trybush, S., S. Jahodova, W. Macalpine, and A. Karp. 2008. Genetic studies of a germplasm resource reveal new insights into relationships among *Salix* subgenera, sections and species. BioEnergy Res. 1:67–79.

Whitney, S. M., R. E. Sharwood, D. Orr, S. J. White, H. Alonso, and J. Galmés. 2011. Isoleucine 309 acts as a C_4 catalytic switch that increases ribulose-1,5-bisphosphate carboxylase/oxygenase (rubisco) carboxylation rate in *Flaveria*. Proc. Natl Acad. Sci. USA 108:14688–14693.

Wintermans, J. F. G. H., and A. De Mots. 1965. Spectrophotometric characteristics of chlorophylls a and b and their phaeophytins in ethanol. Biochim. Biophys. Acta 109:448–453.

Yokota, A., and D. T. Canvin. 1985. Ribulose bisphosphate carboxylase/oxygenase content determined with [^{14}C] carboxypentitol bisphosphate in plants and algae. Plant Physiol. 77:735–739.

Supporting Information

Additional supporting information may be found in the online version of this article at the publisher's web-site.

Figure S1. The correlation between common data sets measured in 2008 and 2009. The nature of the data and the associated units are given in the title to each correlation.

Figure S2. Theoretical *A/Ci* curves, generated using the four distinct sets of kinetic parameters identified among the willow genotypes (shown in Table 4). To facilitate comparison, it was assumed that all leaves had the same Rubisco concentration (the mean value across both years, namely 29.7 ± 9.2 μmol active sites m^{-2}) and that the ratio of J to V_{cmax} was 2.16 (the mean ratio from data in Fig. 2).

Figure S3. Revealing significant positive correlations between components of CO_2 assimilation and willow biomass. (A) Mean leaf area in each willow species (error bars show SEM, $n = 3$); (B) ($\bullet$) Correlation between wood yield (abscissa axes) and stated parameters (ordinate axes) expressed on a leaf area basis (leaf m^{-2}); (C) ($\circ$) As in (B) except stated parameters expressed on a whole plant basis (plant^{-1}) using data from (A). In (B) and (C) mean values (only) are shown. All data from early Summer (2009).

Figure S4. Species-specific carboxylase rate constants (k_{cat}^c) and Rubisco content (upper graph); and species-specific

V_{cmax} derived either by *A/Ci* analyses or by integration of rate constant (k_{cat}^c) and Rubisco content data. Each measurement was the mean of three biological replications, for which mean values and standard errors are indicated (nd, not determined). All data from 2009.

Table S1. Individual data sets used to generate the data of Table 4. Each line shows data obtained from a different leaf protein extract and plant. Each value was derived from a series of technical replications, each data set resulting from Rubisco assays at (a) six CO_2 concentrations, at each of four O_2 concentrations ((k_{cat}^c), (k_{cat}^o), K_c and K_o); or (b) five or six total (RuBP) consumption assays in an oxygen electrode ($S_{c/o}$), as described in Materials and Methods. Values are means ± standard deviation.

15

China's food security is threatened by the unsustainable use of water resources in North and Northwest China

Taisheng Du[1], Shaozhong Kang[1], Xiying Zhang[2] & Jianhua Zhang[3]

[1]Center for Agricultural Water Research in China, China Agricultural University, Beijing 100083, China
[2]Center for Agricultural Resources Research, Institute of Genetics and Developmental Biology, Chinese Academy of Sciences, Shijiazhuang 050021, China
[3]School of Life Sciences and State Key Laboratory of Agrobiotechnology, The Chinese University of Hong Kong, Hong Kong, China

Keywords
China, crop production, food security, sustainability, water resources, water-saving agriculture.

Correspondence
Jianhua Zhang, School of Life Sciences and State Key Laboratory of Agrobiotechnology, The Chinese University of Hong Kong, Hong Kong, China.

E-mail: jhzhang@cuhk.edu.hk

Funding Information
We are grateful for research grants from the National Natural Science Foundation of China (51222905, 51079147, 51321001, 50939005), the National High-Tech 863 Project of China (2011AA100502), the Ministry of Water Resources of China (201001061), China–EU Int'l Collaboration Projects (S2010GR0692), National Basic Research Program of China (2012CB114300) and Shenzhen Overseas Talents Innovation & Entrepreneurship Funding Scheme (The Peacock Scheme).

Abstract

China has a growing population of 1.35 billion at the moment and has been largely self-sufficient in its food production. In consideration of the major agricultural resources (land, available fresh water, and ambient temperature), China is actually very vulnerable in maintaining its food security. Due to the unfavorable distributions of its water resources and temperature conditions, almost all the land that can be cropped has been utilized. Over cropping and over irrigation has led to some serious problems in two major areas of this country. In Northwest China where most of the land belongs to the inland river system, over expanding of irrigated area has resulted in some serious ecological problems, such as shrinking of oasis and desertification of grasslands. In North China Plain where about half of the country's wheat and maize are produced, over cropping has been supplemented with underground water for several decades, which has led to fast drop of underground water table, for example, 0.88 m per year in the recent 30 years at Luancheng County. China's food security is threatened by its diminishing and unsustainable use of water resources. Integrative agronomic, biological, engineering, and administrative practices that can sustainably use the water resources will be the key ways out to secure the country's future food production.

Introduction

Globally, 2.6 billion people lack improved sanitation, 800 million people lack safe drinking water, 1 billion people go to bed hungrily, 2 billion people are undernourished, and 60% of our ecosystem services are deteriorating according to the numbers released by the World Water Week in Stockholm, 26–31 August 2012. The theme of

World Water Day this year is *Water and Food Security*, showing us that the whole world food security will not be guaranteed if we do not take integrated water resources management to make our water resources more sustainable (Watson 2004; Biswas 2008; García 2008). It is expected that the world population will reach 8.3 billion by 2030 and food and fiber demand will be increased by 70%. However, we are already using 70% of our fresh water resources for crop production and have resulted in many serious environmental problems due to overexploitation of the water resources (Von Braun 2007; FAO 2012). Apparently any further increase in water consumption in agriculture is impossible. We are facing the inevitable changes in our focuses in agriculture: a greater emphasis on water productivity, defined as the ratio between crop yield and water consumption, rather than the yield per se is required.

China still has a growing population. The latest census shows that the country has a population at 1.35 billion at the moment and will reach 1.5 billion at 2030. If the grain food per capita increases from today's 400 to 470 kg in 2030, which is realistic with the rapid increase in living standard, we will need at least 35% increase in grain production in less than 20 years. China's total food production has surpassed 500 million tons over the last several years. Technical advancements, such as extensive breeding programs and applications of large amount of chemical fertilizers and agrochemicals, have helped maintaining an increasing trend over the last several decades. More importantly, the agricultural policy changes over the last two decades, things like the "Household Responsibility System", which distributes land into individual households, the government subsidies to grain crops and the phasing-out of agricultural land taxes, are the major driving forces for the food production increases since 1980s (Zhang 2011).

In terms of the water consumption for crop production in China, we should note that the food production increase has resulted from an extensive increase in irrigated areas in this country (Fig. 1). There is a linear correlation between the total grain production in China and the effective irrigated area (National Bureau of Statistics of China 2010a,b). However, this correlation does not indicate that further increase in food production needs to be achieved with the expansion of irrigated area, because we have reached the limits of water resources in many parts of China (discussed in the following sections). We really have to find other ways to maintain the increase in food production with the same or even less water resources.

China is a big country but its water resources per capita are only about 25% of the world average (Jiang 2009). In addition, the distribution of the water resources is very

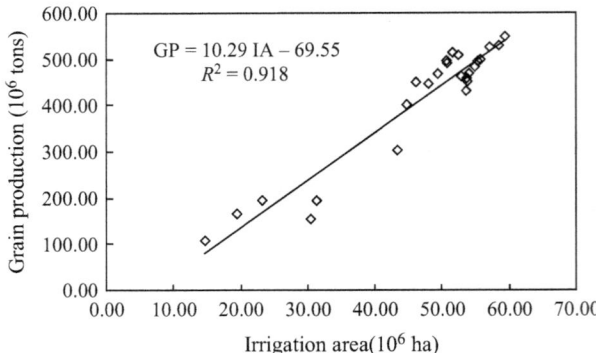

Figure 1. Relationship between irrigation area and grain production in China during the last 60 years. Data showing selected years are from National Bureau of Statistics of China (2010a,b).

uneven by locations and by timing. The southeast part of China near the coastline has an annual rainfall over 1500 mm but the inner northwest part with a third of the country's land area only has less than 200 mm per year. China's rainfall is also affected by typhoons in the summer. From June to August, about 70% of the annual precipitation is delivered by typhoons. Northern China, with 40% of the country's population and 51% of the country's cropped land, has only 20% of its water resources. North China's water availability is only about 271 m^3 of per capita value, which is only 1/25 of the world's average (Guan and Hubacek 2008). The water scarcity is as serious as in areas such as Israel which is 290 m^3 water resources per capita. The average agricultural water consumption accounts for 68% in China, but the value varies in different region. For example, in North China Plain (NCP) which is the important political, cultural, and industrial center, the value is between 71.6% and 76.9%, while in Northwest China the value is above 90% (Fig. 2).

In recent decades, developing irrigation agriculture has changed the transformation and relationship among water cycles, crop water use, and ecology, especially in arid areas where the water resources are very limited and agriculture heavily depends on irrigation. The desert areas have expanded, but river, lake, pasture, and grassland acreage have shrunk. Therefore, competition for the limited water resources between agriculture and ecosystem is increasingly becoming a serious problem. Due to the lack of surface water, groundwater has become a major source of irrigation water in most parts of North and Northwest China. Groundwater levels are persistently declining in many parts of the areas. Overutilization of groundwater resources has led to some serious consequences, for example, gradually falling groundwater table, shrinking of vegetation areas, soil salinization, and desertification in oasis region (Kang et al. 2008). Increased pumping costs and

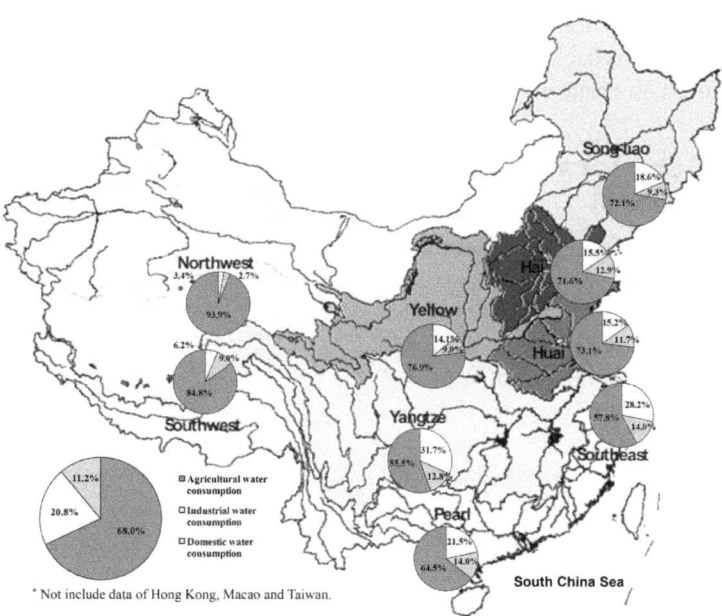

Figure 2. Agricultural water consumption ratio in different regions of China. Data are from Ministry of Water resource (2012).

saltwater intrusion have forced farmers to abandon thousands of wells in NCP in recent years (Kendy et al. 2004). It should be noted that this area accounts for about 50% of wheat and 33% of maize production in China. The wheat crop uses more than 70% of irrigation water resources in this area (Hu et al. 2010).

When compared to North and Northwest China, the other parts of China should have enough water resources for their crops but unique problems for their water resources still exist. For example, in Northeast China annual rainfall reaches 600 mm, which should be enough for the only one crop per year there. Periodic spring and summer droughts are still reported in recent years. They are largely attributed to abnormal atmospheric circulation, air–sea interaction (Kang et al. 2013) or local climatic fluctuations resulted from land use and land cover change mainly by rapid expansion of irrigated areas. It is believed that enhanced water engineering efforts and adoption of precise and water-saving irrigation technologies should help solve the issues. In fact, the Chinese Government has initiated a campaign to "Increase grain production with less water" there with an aim to boost the food production by 10% within 4 years. This area currently produces 22.0% of total grains with only 23.5% of the land in this country. It is believed that it has the potential to produce more if more and better irrigation can be provided.

In Southwest China where rainfall usually exceeds 1000 mm per year, agriculture should not lack of water if the water resources are properly managed. In recent years, however, severe spring droughts are almost always the headline news in China. Lack of proper water

conservation measures or projects plus the abnormal weathers, arguably the result of climate change are the causes (Cao et al. 2012; Patterson et al. 2013; Zhang et al. 2013a). Traditional or indigenous water storage ponds for the households have largely been abandoned by converting them to crop land and relying on large reservoirs for irrigation during the last several decades. This has actually accelerated the drought damage and made the agriculture more vulnerable in this largely mountainous area in abnormally dry years when the reservoirs quickly dries up due to the ever increasing demand of water in all the sectors of society. Agricultural irrigation becomes a low priority in the dry years.

This review focuses on the North and Northwest China (Table 1) where water scarcity is most serious and simple

Table 1. Comparison of key data between Northwest China and North China Plain.

	Northwest China[1]	North China Plain[2]
Area (km^2)	3.43 × 10^6	3.20 × 10^5
Average annual temperature (°C)	10–15	10–16
Annual rainfall (mm)	~50–400	~400–800
Frost-free days per year	15–256	175–220
Principal crops	Wheat, cotton, potato	Wheat, maize, cotton
Total land cultivated (ha)	1.21 × 10^7	2.65 × 10^7
Total land irrigated (ha)	6.20 × 10^6	1.62 × 10^7
Population	8.58 × 10^7	2.14 × 10^8

[1]Data from Zhang and Lu (2002).
[2]Data from Kendy et al. (2004) and Song et al. (2011).

water engineering approaches will not solve it. We believe the water-saving agriculture is the suitable option left to make these two areas sustainable in terms of water use.

Northwest China: Deteriorating Ecology and Environment

Northwest China belongs to the typical inland river system which means rivers end up within the land in lakes or wetland. The size of the area is about 29% of the country's land area, largely classified as arid or semiarid climate. Most of this area has an annual rainfall below 200 mm. In recent 30 years, 25 million hectares of farmland have suffered from drought disasters, which account for 20% of cropping acreage, with grain yield reduction at 57 billion kg (National Bureau of Statistics of China 2010a). The drought as a stress to crops exists almost all the time, regardless the seasons and accounts for more than 10% yield loss in most years, for example in drought-prone Gansu Province, an arid and semiarid area (Fig. 3). In recent 50 years, temperature has increased at 0.325, 0.339, and 0.360°C per decade in the mountain, oasis, and the desert areas, respectively. Precipitation has increased at 10.15, 6.29, and 0.87 mm per decade. It should be noted that only the increasing trend of temperature and precipitation in desert area was not significant (Li et al. 2013). In other areas of the watershed, such increases are significant. For example, in Wuwei, the average temperature only increased 1.2°C during last 50 years (Fig. 4).

The major water resources in this area come from glaciers in high mountains. Irrigation is therefore the lifeline for agriculture. The total area of this region is an estimated 2.78 million km², about 29% of the national total,

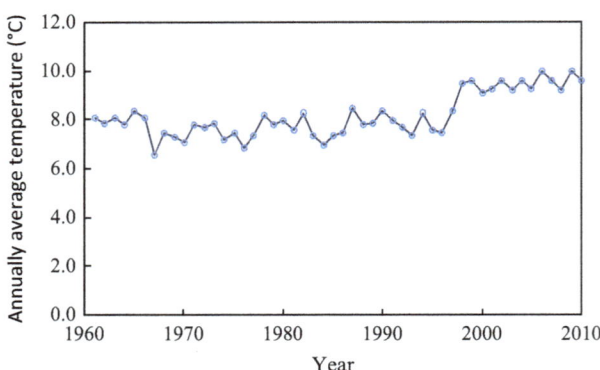

Figure 4. Annually average temperature change in Wuwei city during last 50 years. Data are from China Meteorological Administration National Meteorological Information Center (2013).

but almost 90% of people concentrate in the oasis which accounts for only 4% of the land surface there. Its water resources are estimated at only 3.8% of the country's total. Agriculture is in fact conducted in a very vulnerable ecological condition and almost totally relies on irrigation. As a consequence, agricultural water use is over 90% of the total water usage.

A typical story can be told by the example of Shiyang River basin (Kang et al. 2008). It is located in the eastern part of the Hexi Corridor of Gansu Province in Northwest China and on the northern slope of the Qilian Mountain. The river originates from the northern part of the mountain and ends at the Minqin oasis that is sandwiched between the Badanjilin and Tenggeli deserts (Kang et al. 2004). The whole river basin has an area of 41,600 km² with eight subcatchments from east to west. The annually average water resources in the basin are estimated to be 1.661×10^9 m³ between 1995 and 2000 (Fig. 5). The population there is 2.3 million with cropped land at 373,000 ha. The total river flow in the upper reach did not change much during the last six decades. The flow to the lower reach, the Minqin oasis, however, has been greatly reduced during the later 50 years of the last century, largely due to the intensive human activities, mainly the expanded irrigation, in the upper and middle reaches (Fig. 5). The oasis only received an annual flow at about 60 millions m³ of water in the early 2000s, a drastic reduction if it is compared to the annual flow over 500 million m³ of water in the 1950s. The Minqin oasis, once nurtured by lakes, is now largely an oasis mainly relying on the underground water. It is consistently under the invasions by the Badanjilin and Tenggeli deserts from the east, north, and west and there is the risk that it will disappear altogether. The Minqin oasis currently prevents fusion of the two large deserts and its disappearance will threaten the two neighboring cities of Jinchang and

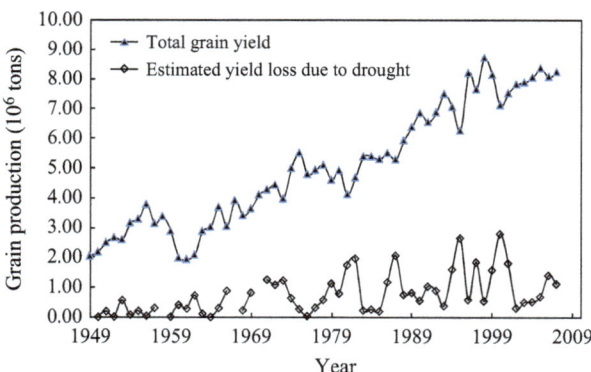

Figure 3. Total grain yield and estimated yield loss due to drought in Gansu province of Northwest China during the past 58 years. Data during 1950–1990 are from Editorial Committee of Flood and Drought Disasters in Northwest Inland River Basin (1999) and data during 1991–2008 are from Water Resource Bureau of Gansu Province (2010).

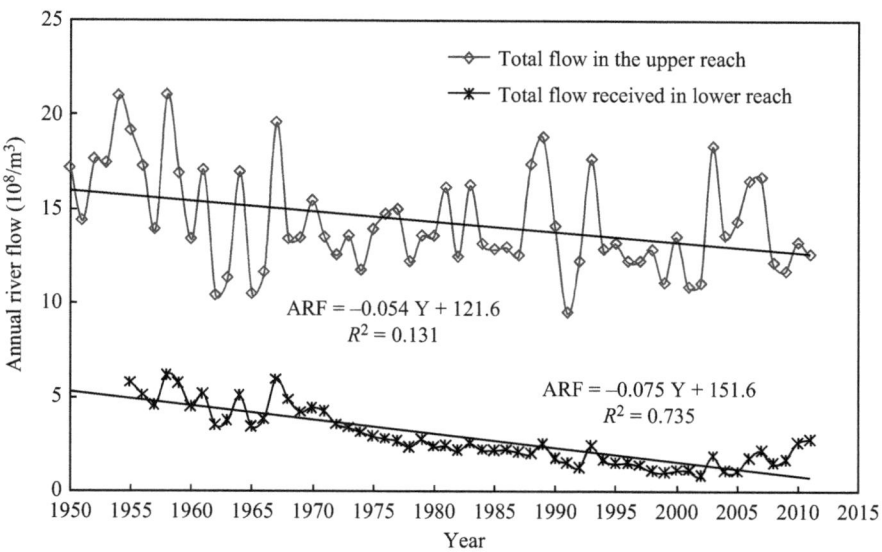

Figure 5. Annual river flows (ARF) of Shiyang River at the foot of Qilian Mountain (upper plot) and at the lower reach of Minqin Oasis (lower plot). Data are from Kang et al. (2009).

Liangzhou with desertification and block the Hexi Corridor, the ancient Silk Road, which connects Xinjiang Uyghur Autonomous Region to central China. The ecology and climate of the whole northern China may also be affected if such a mega desert is formed.

During the later 50 years of last century, irrigation in the Shiyang River region has increased steadily and rapidly. The area of irrigated land started from 185,900 hm^2 in the 1950s to 293,600 hm^2 in 2000, nearly 60% increase. Seventeen large irrigation complexes, each with an area of over 100 hm^2 have been constructed. An extensive system of lined canals has been built and the water transport efficiency of the canal system (ratio of canal inflow at the headwork to outflow of the end canal in an irrigation district) has increased from 0.30–0.35 in the 1950s to 0.54–0.72 in 2000. The total annual water usage in this area has reached 2.854×10^9 m^3. Agricultural water use was 2.455×10^9 m^3 which accounted for 80% of the total water use. The gross water utilization ratio (ratio of gross water utilization to the total water resources of the basin) reached 172% with water repeatedly used and combined with excessive groundwater extraction. The net water consumption by irrigation, industry, and domestic utilization increased from 0.867×10^9 m^3 in the early 1950s to 1.718×10^9 m^3 in 2000, virtually it has doubled in 50 years. The net water consumption ratio (ratio of net water consumption to the total water resources of the basin) reached 103% in 2000 (Kang et al. 2008).

During the last decade, much effort has been made in the Shiyang River basin to reduce the water use. The central government implemented the layout of a comprehensive reparation scheme of Shiyang River Basin and more

than 4.7 billion RMB has been spent since 2003 in repairing the damaged environment and improving water use efficiency and water productivity at different scales from crops, farmland, irrigation-covered areas, and watershed. Integrative approaches including agronomic water-saving measures, cropping system adjustment, water resource allocation, water engineering projects, ecosystem, and water resources conservation projects, inter-watersheds water diverting projects, and water resources information system, have been implemented. Rational and optimal allocation of water resources based on crop production and ecological demand has been put forward (Kang et al. 2009). A water-saving experimental station has been set up to demonstrate all sorts of water-saving irrigation techniques and facilities. The local government has been advocating the adoption of water-saving practices, setting the policies of water ration, and regulating the irrigation water use and decommissioning the irrigation areas. For example, they have rationed the underground water exploitation by trimming down the outputs of many "luxury" wells using IC card (carrying the user's identity and the amount of water available) on the pumps, subsidized the building of sun-warming plastic greenhouses (much less water consumptions for the vegetable crops inside due to the closed environment and high humidity) and adopted crops that require less irrigation. These measures are complemented by the extensive use of plastic mulching and advanced irrigation techniques such as improved furrow irrigation or partial root zone irrigation. There are also many reports that partial root zone drip irrigation increases crop water use efficiency more than conventional deficit irrigation (Kang et al. 2004; Du et al. 2006, 2008a,b, 2010; Fereres and Soriano 2007; Dodd Ian 2009; Sadras 2009). Furthermore, classification of water prices has been carried out, that is, greenhouse crops or drip irrigation crops enjoy 20% discount within the water quota. In contrast, for wheat and maize under traditional flood irrigation, the water price will be increased by 20%, and the cost per cubic meter of water increased from 1.93 Yuan RMB in 2006 to 5.7 Yuan in 2012. As a result, in the downstream area, the planting structure was adjusted, reducing the high water use crops (wheat and maize) and increasing cash crops (cotton, sunflowers, and vegetables) that can bring more profit (Fig. 6). The net income per farmer has increased from 4665.5 Yuan in 2006 to 7035 Yuan in 2012. The total water use in the middle reaches of the river basin has been reduced by 15.7%, and exploitation of groundwater in the whole area has been reduced by 55.3% at 2006 when compared to 1 year earlier at 2005. The water flow to Minqin has been gradually increased since 2006 (Fig. 5). The trend of ecological deterioration has been gradually reversed and groundwater table at Qingtu Lake in Minqin oasis has shown some increase. For example, the down trend of groundwater level became slow in recent years and even increased by 0.17 m at 2012 when compared to 1 year earlier.

Minqin oasis's story can be visually shown by the landscape changes of the Qingtu Lake, which used to receive the final flows of Shiyang River. In the history books, the lake of 2000 years ago was described as a water surface of several hundred square kilometers with numerous kinds of fishes and crowds of wild ducks. Before the 1958, the lake area was largely a wetland with an area of about 100 km^2. In 1958, a reservoir was built at Minqin to divert the water for irrigation. After that, the lake area has gradually become a desert with moving sand dunes. Surprisingly, due to the last 10 years of water conservation efforts, for example, in Wuwei city which is mainly in Shiyang River basin, the water distribution ratio of domestic, industry, ecology, and agriculture increased from 2:4:2:91 in 2006 to 5:14:8:73 in 2012, the annual river flow through Caiqi section reached 3.48 × 10^8 m^3 during 2012, the groundwater exploitation is controlled within 0.86 × 10^8 m^3, and the Qingtu Lake has resumed a water surface of 3 km^2! Especially in the Huangantan village, Jiahe Town, Minqin County near Tenggeli desert, seven wells which were closed in 2008 became usable in the spring of 2012. As a result of the above described integrative measures, the declining trend of water table in the whole river basin was suppressed, and a restoration of 2.05 × 10^4 ha man-made forest and 1.0 × 10^4 ha grassland expansion has been achieved. In addition, 1.1 × 10^4 ha sand dune has been controlled by wheat straw grid in the lower reach.

The NCP: Over Exploitation of Underground Water

NCP is located in the eastern part of China. It covers the area from Bohai Sea at its east to the Taihang Mountain at its west, and from Yellow River at its south to the Yanshan Mountain at its north. It includes the flat plain area of Beijing, Tianjin, and Hebei Province, and parts of the plain area of Henan and Shandong provinces. The total area is approximately estimated as 320,000 km^2, mainly including the Yellow, Huai, and Hai river basins. In the NCP, the area of agricultural land is about 8.85 × 10^7 ha with an irrigated area of 6.53 × 10^7 ha, and consumes about 70% total water resources (Fang et al. 2010; Zhang et al. 2011). The mean annual accumulated temperature has increased significantly by 348.5°C·day due to global warming during last half century (Song et al. 2011).

The flat landscape of NCP means it is best suitable for agriculture. In fact this area today produces half the wheat, nearly half the maize and one-third of the cotton for the country, a remarkable achievement if we consider its area is only 3.3% of the country's total (Zhang 2011). There are not many rivers to deliver water from other areas to NCP. The Yellow River has an average runoff

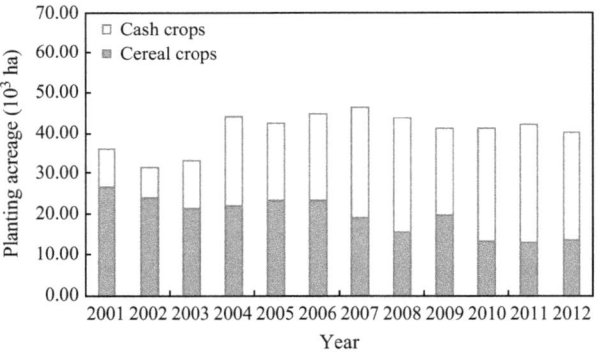

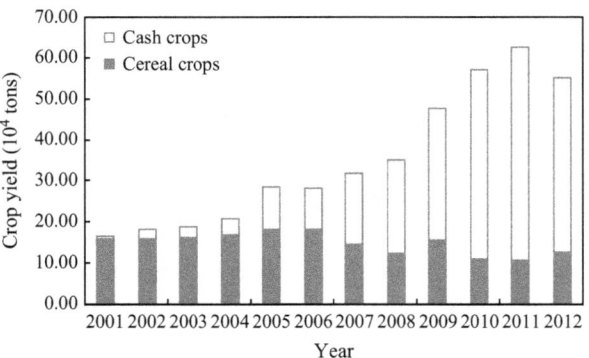

Figure 6. Planting acreage and crop yield change in Minqin county at the downstream of Shiyang River Basin during past 12 years. Data are from Kang et al. (2009) and local government statistic report.

only at 1175 m^3 s^{-1} and almost dries up when it reaches the sea. In fact the Yellow River dried up seasonally for consecutive 10 years from later 1990s to early 2000s. From 2003 to 2012, the annual water withdrawal and water consumption from Yellow River increased linearly (Fig. 7), which is correlated with the expansion of irrigated area. The annual rainfall for the area ranges from 376 to 750 mm. The average water resources per capita in this area are only about 20% of the national average. It is estimated that the crop land in this area only has an average of 668 mm water per year, enough for one crop but short for two crops per year (Guan and Hubacek 2008).

Traditionally, NCP has two crops per year, mostly winter wheat plus summer maize or cotton. In the past, water requirement in this area is supplemented by rivers that come from the mountains at west and north and from the Yellow River at south. The agricultural development in the upper reaches of these rivers during the last several decades, however, has used much of the water resources. Approximately since 1980s, NCP has begun to extensively exploit underground water to supplement the water shortage for their two crops per year. The availability of electricity supply to the countryside and the cheap electric pumps have made this trend possible and much accelerated the process. Ironically, the initial rapid exploitation of underground water and quick decline of the water table has brought an unexpected benefit to the agriculture in this area. It solved the long-lasting problem of soil salinity (Li et al. 2012)! The salt is deposited down the soil profile and cannot come up to the surface because of the lack of the capillary movement of salt toward the top soil. As a consequence, major crops such as wheat now yield many folds more than before the 1980s, namely from about 1 or 2 tons per hectare to over six tons per hectare (Zhang et al. 1998). According to a long-term field study conducted from 1979 to 2012 at the Luancheng Agro-Eco-Experimental Station of the Chinese

Academy of Sciences, which is located in the northern part of the NCP, the winter wheat annual yield increase was ~193 kg/ha per year from 1992 to 2001 (stage 2). The average yield increased rapidly during the first stage (from 1979 to 1992) and still maintained increase in the third stage (2001–2012). Other measures, such as the improved cultivars may also have helped (Zhang et al. 2013b), but salinity as a major problem existing in NCP for many centuries has largely disappeared (Yun and Wang 1997; Li et al. 2003; Chen et al. 2006; Ouyang et al. 2011).

The rapid and extensive exploitation of underground water indeed has helped the rapid increase in total wheat and maize production in NCP. For example, as the data released from Hebei Province, wheat and maize production increased by almost sixfolds when compared to 1970s. However, this fantastic achievement also accompanied a groundwater exploitation increase by the same sixfolds (Fig. 8)!

During the recent three decades, it has been increasingly realized that the agriculture at the NCP is not sustainable in terms of water consumption. The over exploitation of underground water has led to the rapid expansion of irrigation to large areas. The Hebei Province has seen this expansion by threefolds over the last 50 years (Fig. 9). Apparently the consequence of such rapid expansion is the rapid declining of underground water table. As one of the experimental stations at Luancheng has reported, the groundwater table steadily dropped from about 10 m in 1970s to over 40 m today (Fig. 10).

The data shown in Figure 9 are typical for NCP but far from the worst ones. There are many reports in the newspapers and other public media that the land sinking is accelerating in many parts of NCP (Wang et al. 2009). Also the sea water infiltration near the coastline area has also been frequently reported. More and more serious water problems such as continuous decline of water table and worsening of water quality have also been reported

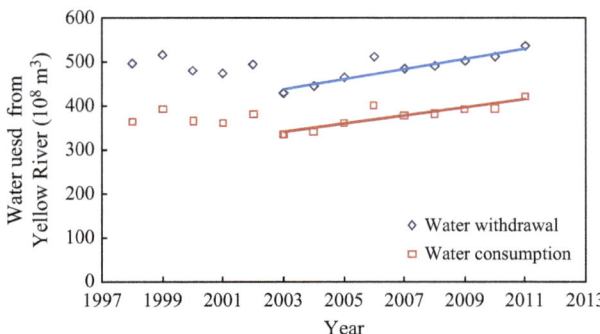

Figure 7. Annual water withdrawal and water consumption from Yellow River. Lines indicate recent significant increases in water consumption and water withdrawal. Data are from Yellow River Conservancy Commission, Ministry of Water Resources (2013).

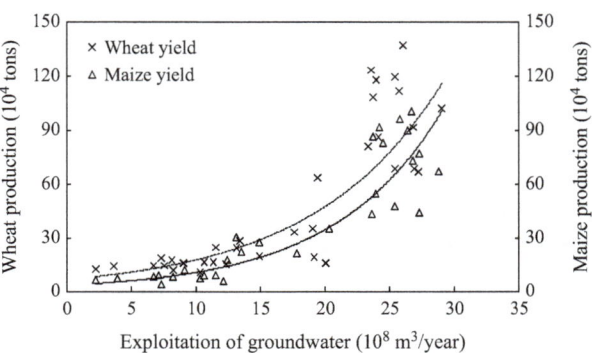

Figure 8. Total wheat and maize production at Hebei Province, North China Plain, in relation to the total exploitation of groundwater from 1970s to 2000s. Data are from Zhang et al. (2012).

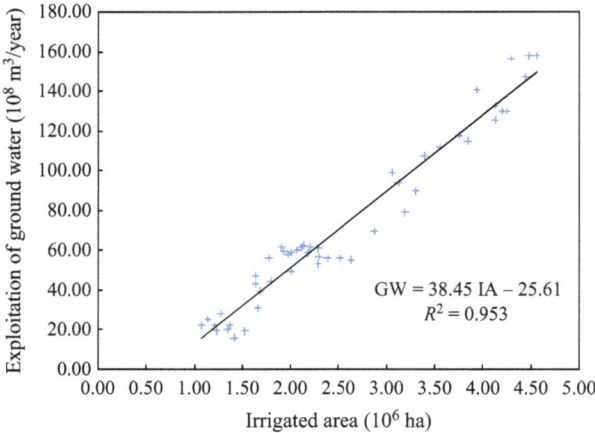

Figure 9. Relationship between groundwater exploitation (GW) and irrigation area (IA) in Hebei plain of North China during last 50 years. Data are from Zhang et al. (2009).

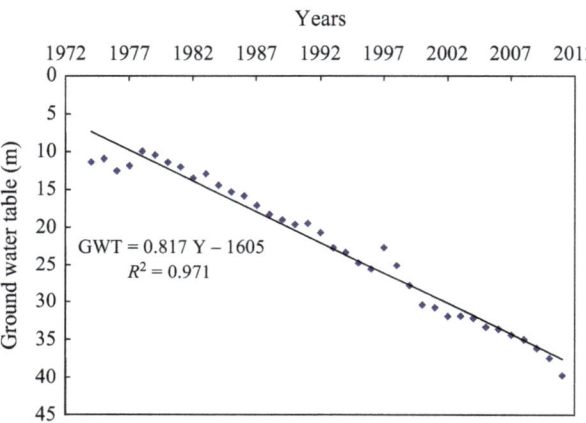

Figure 10. Ground water table (GWT) changes during the last 40 years at Luancheng, Hebei Province, North China Plain. Data are from Zhang et al. (2009) and Wu et al. (2012).

(Liu et al. 2001; Jia and Liu 2002). The impact of such deteriorating situation of water depletion on social and economic developments in this area is huge and immense (Zhang et al. 2004, 2009). We expect more serious consequences will appear before long. There is no sign that the trend is slowing at the moment.

The Way Out: Integrative Water-Saving Agriculture

Water-saving agriculture is a technology system that focuses on improving the efficient utilization of crop water, field water, channel water, and recycled water, and the benefits from agricultural production. It also includes key products and equipments characterized by high-efficiency, low-cost, eco- and environment-friendly advanta-

ges, and setting up a water-saving engineering and management pattern suited to the different needs of different areas. Water-saving irrigation should be the effective way to cope with the ever increasing food demand in China. Is it possible to make the water use sustainable for Northwest China and NCP where a crisis of water shortage is developing? The goal should be to reduce the agricultural water use to a level that can sustain itself. To achieve this goal, we need to set the target that we should reduce the water use by half if compared to the "luxury" water use during 1980s and still maintain the productivity of crop land in these areas. The way out for the sustainable use of agricultural water resources is to increase the water use efficiency and much more effort should be made to develop water-saving techniques for the field crops. In general, high efficient use of agricultural water resources can be achieved by water engineering, agronomic, and biological water-saving methods, and administrative policies.

The conventional engineering approaches for water saving in China are the rational exploitation of water resources such as temporal and spatial optimization and allocation of water resources and coordinative dispatch of multiple water resources including surface water and groundwater. Furthermore, in arid regions where crop production is heavily dependent on irrigation and water resources are scarce, unconventional water engineering approaches should be also encouraged and developed. For instances, rainwater harvesting and storage for agricultural use in arid mountainous areas of Gansu and Ningxia Provinces have been practiced. Some simple and new improvements have made this indigenous method more efficient and more widely accepted. Utilization of saline or brackish water with total dissolved solid up to 3–7 g/L or water electrical conductivity up to 3200–6500 μs/cm for irrigation on salt-resistant crops such as cotton and sunflower in the ecologically fragile oasis areas in Xinjiang or Gansu has been successfully developed. Use of recycled sewage water for irrigation has been promoted in the suburbs of Beijing city. Desalination of sea water or sea ice for irrigation in the seaside agricultural areas near the Bohai Bay has also been tested with some successes. These approaches have been complemented with traditional water-saving irrigation and engineering technologies such as canal lining, irrigation with low-pressure pipe conveyance, improved surface irrigation by shortening or narrowing border or furrow, surge flow irrigation based on laser-controlled land leveling technology and so on. In recent decades, sprinkling irrigation, micro-irrigation, drip irrigation under plastic film mulch, and subsurface irrigation have also been rapidly popularized in arid areas of Northwest China. One good example is the drip irrigation under plastic film mulch on cotton in Xinjiang

Uygur Autonomous Region, where more than one-third of the country's cotton, or one-eighth of world's cotton, is grown. Nearly half of the irrigated water can be saved in comparison to the traditional surface irrigation with this new irrigation method.

In recent years, more widely used approach of water saving in agriculture is the agronomic approach by ways such as crop rotation, water conserving through minimum tillage or no tillage, plastic film mulching or straw mulching, watering coupled with fertilization, breeding for drought tolerant varieties, chemical methods such as using superabsorbent polymers, soil amendments and antitranspirants. There are many reports for the successful use of these methods (Akhtar and Malik 2000; Puoci et al. 2008; He et al. 2009).

The most challenging and promising approach is arguably the biological water saving. By definition, the meaning of biological water saving is that water is saved directly from the reduced water consumption by the plants. This may be achieved through ways such as crop genetic improvement and physiological regulations through reduced stomatal opening or reduced light interception. One method that has been practiced and reported is the alternate partial root zone irrigation (Kang and Zhang 2004; Tang et al. 2005; Du et al. 2006, 2008a, b; Fereres and Soriano 2007; Davies et al. 2011). The practice in Northwest China showed that partial root zone irrigation can reduce irrigation, up to 30% reduction, and still maintain economic yield in crops such as cotton and maize (Kang et al. 1998; Tang et al. 2005, 2010). An additional benefit of partial root zone irrigation is that the plant growth pattern can be changed by continuously exposing some roots in drying soil. It has been reported that grapevine can have reduced vegetative growth and therefore reduced demand for pruning (Loveys et al. 2004). However, partial root zone irrigation can have some contrasting effects on vegetative vigor which may be agronomically advantageous according to total soil water availability (Romero et al. 2012).

Based on the widely used agronomic and promising biological water-saving approach, it should be emphasized that government policies on water allocation and government subsidies for water-saving practices should never be ignored in areas where the allocation of water resources is the key decision to allow balanced developments in different social sectors and maintain sustainable environment. As mentioned earlier in Northwest China, governments at different levels have set up special bureaus for specific watershed areas to allocate water resources. They are responsible to ration the water resources, set water-saving targets for different industries, monitor and charge the water consumptions at canals as well as at pumping wells. Water engineering projects and water-saving equipment are subsidized from central government or provincial governments. All of these administrative approaches should be the effective approach to reverse the deteriorating trend of ecological environment in the Northwest China. After more than a decade of effort, we have already seen some improving effects.

Concluding Remark

It can be foreseen that water will be the most limiting factor for food production in the world in the near future. As many have suggested, we shall need a "Blue Revolution", to increase the agricultural productivity per unit of water, or "more crop per drop" (UNIS 2000). Norman Borlaug said in 2000, "how can we continue to expand food production for a growing world population within the parameters of likely water availability? The inevitable conclusion is that humankind in the 21st century will need to bring about a "Blue Revolution -more crop for every drop" to complement the so(sic)-called "Green Revolution" of the 20th Century. Water use productivity must be wedded to land use productivity. Science and technology will be called upon to show the way".

In China, with a huge population and only limited agricultural resources, such a Blue Revolution is more urgent than ever. Specifically we should first of all identify our unique problems in water scarcity in different areas, analyze the possible consequences if the business-as-usual way of water use in food security and other social and economical developments, design-specific and strategic plans to cope with the challenges, and finally set the specific targets in water-saving and water usage. As discussed earlier, a 50% reduction in irrigated water use in North and Northwest China is the target that must be achieved in the near future to make our water resources sustainable. Water-saving agriculture is the only way to achieve the target. As we have discussed, the irrigated agriculture should be controlled in a suitable scope in the arid and ecologically fragile areas of NCP and Northwest China, and the allocation of water resources must include the ecological water demand. Optimized and integrated technologies for increasing water use efficiency, yield and quality are available. However, it needs cooperation from all aspects, such as scientific research, effective technology, optimum organization, and strong market and government applications.

Acknowledgments

We are grateful for research grants from the National Natural Science Foundation of China (51222905, 51079147, 51321001, 50939005), the National High–Tech 863 Project of China (2011AA100502), the Ministry of Water Resources of China (201001061), China–EU Int'l

Collaboration Projects (S2010GR0692), National Basic Research Program of China (2012CB114300) and Shenzhen Overseas Talents Innovation & Entrepreneurship Funding Scheme (The Peacock Scheme).

Conflict of Interest

None declared.

References

Akhtar, M., and A. Malik. 2000. Roles of organic soil amendments and soil organisms in the biological control of plant-parasitic nematodes: a review. Bioresour. Technol. 74:35–47.

Biswas, A. K. 2008. Integrated water resources management: is it working? Water Resour. Dev. 24:5–22.

Cao, C., J. Zhao, P. Gong, G. Ma, D. Bao, K. Tian, et al. 2012. Wetland changes and droughts in southwestern China. Geomatics Nat. Hazards Risk 3:79–95.

Chen, J., Z. Yu, J. Ouyang, and M. E. F. van Mensvoort. 2006. Factors affecting soil quality changes in the North China Plain: a case study of Quzhou County. Agric. Syst. 91: 171–188.

China Meteorological Administration National Meteorological Information Center. 2013. China meteorological data sharing service system. Available at http://cdc.cma.gov.cn (accessed 12 September 2013).

Davies, W. J., J. Zhang, J. Yang, and I. C. Dodd. 2011. Novel crop science to improve yield and resource use efficiency in water-limited agriculture. J. Agric. Sci. 149:123–131.

Dodd Ian, C. 2009. Rhizosphere manipulations to maximize 'crop per drop' during deficit irrigation. J. Exp. Bot. 60:2454–2459.

Du, T., S. Kang, J. Zhang, F. Li, and X. Hu. 2006. Yield and physiological responses of cotton to partial root-zone irrigation in the oasis field of northwest China. Agric. Water Manag. 84:41–52.

Du, T., S. Kang, J. Zhang, F. Li, and B. Yan. 2008a. Water use efficiency and fruit quality of table grape under alternate partial root-zone drip irrigation. Agric. Water Manag. 95:659–668.

Du, T., S. Kang, J. Zhang, and F. Li. 2008b. Water use and yield responses of cotton to alternate partial root-zone drip irrigation in the arid area of north-west China. Irrig. Sci. 26:147–159.

Du, T., S. Kang, J. Sun, X. Zhang, and J. Zhang. 2010. An improved water use efficiency of cereals under temporal and spatial deficit irrigation in north China. Agric. Water Manag. 97:66–74.

Editorial Committee of Flood and Drought Disasters in Northwest Inland River Basin. 1999. Pp. 119–131 in Flood and drought disasters in Northwest Inland river basin. Yellow River Water Resource Press, Zhengzhou, China.

Fang, Q. X., L. Mab, T. R. Green, Q. Yu, T. D. Wang, and L. R. Ahuj. 2010. Water resources and water use efficiency in the North China Plain: current status and agronomic management options. Agric. Water Manag. 97:1102–1116.

FAO. 2012. FAO Statistical Yearbook 2012. Available at http://www.fao.org/docrep/015/i2490e/i2490e00.htm (accessed 8 September 2013).

Fereres, E., and M. A. Soriano. 2007. Deficit irrigation for reducing agricultural water use. J. Exp. Bot. 58:147–158.

García, L. E. 2008. Integrated water resources management: a 'Small' step for conceptualists, a giant step for practitioners. Water Resour. Dev. 24:23–36.

Guan, D., and K. Hubacek. 2008. A new and integrated hydro-economic accounting and analytical framework for water resources: a case study for North China. J. Environ. Manage. 88:1300–1313.

He, S., M. Li, J. Song, and D. Wang. 2009. Research advances on anti-transpirant. Chin. J. Agrometeorol. 30(S1):77–81.

Hu, Y., J. P. Moiwo, Y. Yang, S. Han, and Y. Yang. 2010. Agricultural water-saving and sustainable groundwater management in Shijiazhuang Irrigation District, North China Plain. J. Hydrol. 393:219–232.

Jia, J., and C. Liu. 2002. Groundwater dynamic drift and response to different exploitation in the North China Plain: a case study of Luancheng County, Hebei Province. Acta Geogr. Sin. 57:201–209.

Jiang, Y. 2009. China's water scarcity. J. Environ. Manage. 90:3185–3196.

Kang, S., and J. Zhang. 2004. Controlled alternate partial root-zone irrigation: its physiological consequences and impact on water use efficiency. J. Exp. Bot. 55:2437–2446.

Kang, S., Z. Liang, W. Hu, and J. Zhang. 1998. Water use efficiency of controlled alternate irrigation on root-divided maize plants. Agric. Water Manag. 38:69–76.

Kang, S., X. Su, L. Tong, P. Shi, X. Yang, Y. ABE, et al. 2004. The impacts of human activities on the water–land environment of the Shiyang River basin, an arid region in northwest China. Hydrol. Sci. J. 49:413–427.

Kang, S., X. Su, L. Tong, J. Zhang, L. Zhang, and W. J. Davies. 2008. A warning from an ancient oasis: intensive human activities are leading to potential ecological and social catastrophe. Int. J. Sustain. Dev. World Ecol. 15:440–447.

Kang, S., X. Su, and T. Du, eds. 2009. P. 773 in Water resource transformation and efficient water-saving pattern in Arid Northwest China-a case study in Shiyang River Basin. Water Resources and Hydro Power Press, China.

Kang, S., B. Yang, C. Qin, J. Wang, F. Shi, and J. Liu. 2013. Extreme drought events in the years 1877–1878, and 1928, in the southeast Qilian Mountains and the air-sea coupling system. Quatern. Int. 283:85–92.

Kendy, E., Y. Zhang, C. Liu, J. Wang, and T. Steenhuis. 2004. Groundwater recharge from irrigated cropland in the North

China Plain: case study of Luancheng County, Hebei Province, 1949–2000. Hydrol. Process. 18:2289–2302.

Li, B., R. Li, and Y. Shi. 2003. Thirty years (1973–2003)-research on soil water and salt movement. J. China Agric. Univ. 8(S):5–19 (in Chinese with English abstract).

Li, J., J. Pu, M. Zhu, and R. Zhang. 2012. The present situation and hot issues in the salt-affected soil research. J. Geog. Sci. 67:1233–1245.

Li, B., Y. Chen, X. Shi, Z. Chen, and W. Li. 2013. Temperature and precipitation changes in different environments in the arid region of northwest China. Theor. Appl. Climatol. 112:589–596.

Liu, C., J. Yu, and E. Kendy. 2001. Groundwater exploitation and its impact on the environment in the North China Plain. Water Int. 26:265–272.

Loveys, B. R., M. Stoll, and W. J. Davies. 2004. Physiological approaches to enhance water use efficiency in agriculture: exploiting plant signaling in novel irrigation practice. Pp. 113–141 in M. A. Bacon, ed. Water use efficiency in plant biology. Blackwell Publishing, Oxford.

Ministry of Water Resources. 2012. Pp. 15–18 in Statistic bulletin on China water activities. China Waterpower Press, Beijing, China.

National Bureau of Statistics of China. 2010a. Pp. 35–60 in China compendium of statistics 1949–2008. China Statistics Press, Beijing, China.

National Bureau of Statistics of China. 2010b. P. 850 in China statistical year book. China Statistics Press, Beijing, China.

Ouyang, Z., L. Wu, C. Wang, and Y. Li. 2011. Practice and experience for modern agricultural development mode with resources-saving type at Yucheng City, Shandong Province, China. Bull. Chin. Acad. Sci. 26:383–389 (in Chinese with English abstract).

Patterson, R. T., A. S. Chang, A. Prokoph, H. M. Roe, and G. T. Swindles. 2013. Influence of the Pacific decadal oscillation, El Niño-southern oscillation and solar forcing on climate and primary productivity changes in the northeast Pacific. Quatern. Int. 310:124–139.

Puoci, F., F. Iemma, U. G. Spizzirri, G. Cirillo, M. Curcio, and N. Picci. 2008. Polymer in agriculture: a review. Am. J. Agric. Biol. Sci. 3:299–314.

Romero, P., C. Dodd Ian, and A. Martinez-Cutillas. 2012. Contrasting physiological effects of partial root zone drying in field-grown grapevine (Vitis vinifera L. cv. Monastrell) according to total soil water availability. J. Exp. Bot. 63:4071–4083.

Sadras, V. O. 2009. Does partial root-zone drying improve irrigation water productivity in the field? A meta-analysis. Irrig. Sci. 27:183–190.

Song, Y., Y. Zhao, and C. Wang. 2011. Changes of accumulated temperature, growing season and precipitation in the North China Plain from 1961 to 2009. Acta Meteorol. Sin. 25:534–543.

Tang, L., Y. Li, and J. Zhang. 2005. Physiological and yield responses of cotton under partial rootzone irrigation. Field Crops Res. 94:214–223.

Tang, L., Y. Li, and J. Zhang. 2010. Partial rootzone irrigation increases water use efficiency, maintains yield and enhances economic profit of cotton in arid area. Agric. Water Manag. 97:1527–1533.

UNIS. 2000. Secretary general address to developing countries 'South Summit'. UN Information Service Press Release, 13 April 2000, Vienna (see www.unis/unvienna.org/unis/pressrels/2000/sg2543.html).

Von Braun, J. 2007. Pp. 1–5 in Food policy report: the world food situation-new driving forces and required actions. International Food Policy Research Institute, Washington, DC.

Wang, S., X. Song, Q. Wang, G. Xiao, C. Liu, and J. Liu. 2009. Shallow groundwater dynamics in North China Plain. J. Geog. Sci. 19:175–188.

Water Resource Bureau of Gansu Province. 2010. Pp. 250–289 in Statistical yearbook of Gansu water conservation. China Statistics Press, China.

Watson, N. 2004. Integrated river basin management: a case for collaboration. Int. J. River Basin Manag. 2:243–257.

Wu, Q., G. Wang, W. Lin, and F. Zhang. 2012. Estimating groundwater recharge of Taihang Mountain piedmont in Luancheng County, Hebei Province China. Geol. Sci. Technol. Inf. 31:99–105.

Yellow River Conservancy Commission, Ministry of Water Resources. 2013. Yellow river water resources bulletin. Available at http://www.yellowriver.gov.cn/other/hhgb/ (accessed 10 September 2013).

Yun, Z., and D. Wang. 1997. Land salinization in China and the prevention countermeasures. Rural Eco-Environ. 13:1–5 (in Chinese with English abstract).

Zhang, J. 2011. China's success in increasing per capita food production. J. Exp. Bot. 62:3707–3711.

Zhang, Z., and Y. Lu. 2002. Pp. 1–10 in Water resources exploitation in Northwest China. China WaterPower Press, Beijing, China.

Zhang, J., X. Sui, B. Li, B. Su, J. Li, and D. Zhou. 1998. An improved water-use efficiency for winter wheat grown under reduced irrigation. Field Crops Res. 59:91–98.

Zhang, Y., E. Kendy, Q. Yu, C. Liu, Y. Shen, and H. Sun. 2004. Effect of soil water deficit on evapotranspiration, crop yield, and water use efficiency in the North China Plain. Agric. Water Manag. 64:107–122.

Zhang, G., Y. Fei, H. Wang, M. Yan, and Z. Liu. 2009. Impact of farmland production increasing under irrigation water saving on groundwater exploitation in Hebei plain, China. Geol. Bull. China 28:645–650.

Zhang, X., S. Chen, H. Sun, L. Shao, and Y. Wang. 2011. Changes in evapotranspiration over irrigated winter wheat and maize in North China Plain over three decades. Agric. Water Manag. 98:1097–1104.

Zhang, G., Y. Fei, J. Wang, and M. Yan, eds. 2012. P. 190 *in* Agricultural irrigation and groundwater adaptability in North China. China Science Press, Beijing, China.

Zhang, M., J. He, B. Wang, S. Wang, S. Li, W. Liu, et al. 2013a. Extreme drought changes in Southwest China from 1960 to 2009. J. Geog. Sci. 23:3–16.

Zhang, X., S. Wang, H. Sun, S. Chen, L. Shao, and X. Liu. 2013b. Contribution of cultivar, fertilizer and weather to yield variation of winter wheat over three decades: a case study in the North China Plain. Eur. J. Agron. 50:52–59.

16

Sweet sorghum ideotypes: genetic improvement of stress tolerance

Sylvester Elikana Anami[1,2], Li-Min Zhang[1], Yan Xia[1], Yu-Miao Zhang[1], Zhi-Quan Liu[1] & Hai-Chun Jing[1]

[1]Key Laboratory of Plant Resources, Institute of Botany, Chinese Academy of Sciences, Beijing 100093, China
[2]Institute of Biotechnology Research, Jomo Kenyatta University of Agriculture and Technology, Nairobi, Kenya

Keywords
Abiotic stress, biotic stress, Herbicides, Sweet sorghum

Correspondence
Hai-Chun Jing, Key Laboratory of Plant Resources, Institute of Botany, Chinese Academy of Sciences, Beijing 100093, China.

E-mail: hcjing@ibcas.ac.cn

Funding Information
No funding information provided.

Abstract

Stress tolerance is a prerequisite for the success of biofuel production, which normally requires the use of marginal lands and nonfood biofuel feedstocks. Sorghum is known for its ability to withstand stress conditions, however, terminal stresses threaten its growth and development negatively impacting yield and sugar accumulation. It is crucial, therefore, that research aimed at developing sorghum resistance to stress factors should be pursued to expand the range of its growth to marginal and barren soils to meet the needs of a growing population, changing diets, and biofuel production. In this context, the leaf architectural trait of stay-green drought tolerance, in addition to salinity, cold, and aluminium toxicity and biotic stress tolerance and their genetic basis discussed in this review are expected to be available in future sweet sorghum ideotypes. Also highlighted is the key role of efficient management of farming systems, in particular the use of herbicides to control weeds, to ensure the sustainability of the sweet sorghum biomass productions.

Introduction

Abiotic and biotic stress limits plant growth, crop productivity, and biofuel production. Changes in precipitation patterns due to climate change and meteo-climatic variability have become a critical issue and a limiting factor for the crops under rain-fed systems and for the water resource when irrigation is applied. In arid and semiarid areas where water is limiting, the cultivation of irrigated energy crops could exacerbate the problem of competitiveness with food crops for the water use. Therefore, drought-tolerant energy crops should be the preferred choice in terms of both adaptation and environmental sustainability.

Sorghum originated from Africa and is currently the fifth most important cereal crop in the world and a staple crop for humans and other animals for food, feed, fodder, fiber, and fuel. Cultivation of sweet sorghum for the production of bioenergy is an attractive option to cope with the challenges of climate change to which adaptation is necessary in order to maintain good levels of productions (Berndes et al. 2003; Sims et al. 2006; Orlandini et al. 2007; Dalla Marta et al. 2014).

Sweet sorghum accumulate soluble sugars in the stalk at the expense of panicle production, these sugars could be mechanically extracted and directly fermented to obtain first generation bioethanol. Abiotic stress is a serious environmental obstacle for sugar production in sweet sorghum and strongly threatens biomass production and biofuel yields (Zegada-Lizarazu and Monti 2013). For instance, the structural carbohydrates (cellulose, hemicelluloses, and lignin) and biomass yields in sweet sorghum are significantly affected by drought stress (Zegada-Lizarazu and Monti 2013). In addition, during the 2010–2011 La Niña period in East Africa, drought led to sharp declines in the production of sorghum, a staple food in this semiarid region. For Somalia, the total sorghum production in 2011 was 25 kilotons, more than 80% below normal and the lowest for the last decade (Anyamba et al. 2014). However, sweet sorghum is considered water stress-resistant and suitable for arid and semiarid marginal areas (Staggenborg et al. 2008), due to morphophysiological characteristics which confer the drought tolerance (Zegada-Lizarazu and Monti 2013), and to the C_4 photosynthetic system which allows efficient CO_2 fixation and an outstanding dry matter accumulation (Mastrorilli et al. 1999).

The high sugar content, and therefore the sweetness in sweet sorghum attracts about 150 insect pests throughout its life cycle, which negatively impacts on biomass production (Guo et al. 2011). Common examples of pests that can severely damage a sorghum crop include the Miridae and Lygaeidae (Kruger et al. 2008), sorghum midge, Greenbug, fall armyworm, corn borers (Munson et al. 1993; Wu and Huang 2008; Damte et al. 2009), grasshoppers, sorghum shoot fly, corn rootworm, and sorghum aphid. Sorghum may also be affected by a number of diseases including anthracnose, down mildew, and Fusarium.

Climate change is associated with an increase in the frequency of heat stress, droughts, and floods (Kim et al. 2014) that negatively affect crop yields and biomass production and the ability of the climate smart sorghum to adapt and yield under such harsh environment. Resistance to abiotic and biotic stress is crucial in determining the sustainability of food and biofuel production in future (http://dialogues.cgiar.org/blog/millets-sorghum-climate-smart-grains-warmer-world/).

Here, we review the abiotic and biotic stress resistance, key traits expected to become available in new sweet sorghum ideotypes dedicated for biofuel production in the future and the control of these traits at the genetic level. We highlight the key role of a proper management of the farming systems, in particular, the use of herbicides to control weeds, to ensure the sustainability of the sweet sorghum biomass and biofuel productions.

QTLs related to abiotic and biotic stress tolerance in sorghum

Traditional breeding and QTL analysis have been applied for the identification of genes responsible for biotic and abiotic stress tolerance in crops plants (Collins et al. 2008; Takeda and Matsuoka 2008). Thus, 350 QTLs related to abiotic and biotic stress tolerance in sorghum (Table 1) were identified. These genetic loci have the potential to be utilized in developing superior sorghum ideotypes for various agroecological climates through marker-assisted breeding and genetic engineering once candidate genes have been fine mapped. In total, 51 and 182 loci were found to have known physical and genetic map positions, respectively. To have an idea of their chromosomal distribution in relation to biotic and abiotic stress traits they

control, an atlas map (Fig. 1) was generated. The outer rectangular marks indicate QTLs with known genetic position and the inner circular marks shows QTLs with known physical map positions on sorghum chromosomes. The next-generation sequencing and advanced metabolic profiling might impact the field of QTL analysis and facilitate the cloning of more genes responsible for tolerance to abiotic and biotic stresses. The ability to resequence a large number of F2 or recombinant inbred lines coupled with statistical linkage analysis could open the way for a very rapid and new type of marker-assisted mapping at the genome or metabolome level (Zheng et al. 2011).

Leaf stay-green drought resistance trait in sorghum

Sorghum is a drought-tolerant crop due to its ability to display morphological changes such as dense and deep root system, reduction in transpiration through leaf rolling and stomatal closure, and lowering of metabolic processes to near dormancy in response to terminal stress (Schittenhelm and Schroetter 2014). In fact, sorghum can survive prolonged dry periods, then 'resurrect' and resume growth once the soil moisture becomes available. However, depending on the severity, sorghum still suffers yield and biomass losses of up to 90% (House 1985). Recently, drought led to a sharp decline in agricultural production of sorghum up to 80% below the normal in Somalia during the 2010–2011 La Nina period (Anyamba et al. 2014). The impact of drought stress is greater during the grain-filling stage and causes premature leaf death, plant senescence, stalk lodging, and charcoal rot with poor yield of seed and stover. Cultivars tolerant to drought stress at postflowering stage are referred to as stay-green cultivars.

Stay-green is an integrated heritable drought adaptation trait characterized by a distinct green leaf phenotype during grain filling or postflowering under drought (Borrell et al. 2014a), Indeed, genetic studies show that QTLs for temperature and drought responses coincide with loci for leaf senescence, and in numerous examples of improvements in stress tolerance achieved by simultaneous selection for stay-green (Ougham et al. 2008; Vijayalakshmi et al. 2010; Jordan et al. 2012; Emebiri 2013). Stay-green trait is characterized either as cosmetic in which a lesion interferes with an early step in chlorophyll catabolism or functional in which the transition from the carbon capture period to the nitrogen mobilization (senescence) phase of canopy development is delayed, and/or the senescence syndrome proceeds slowly. Alteration in hormone metabolism and signaling, particularly affecting the networks involving cytokinins and ethylene, could contribute to the stay-green phenotype.

Table 1. The biofuel-associated traits of stay-green drought resistance trait, resistance to abiotic and biotic stresses, and their genetic determinants.

Traits	Trait category	No. of QTLs	QTL names	References
Abiotic stress tolerance	Stay-green/leaf senescence	1	Stg1 (green leaf area retention)	Subudhi et al. (2000), Xu et al. (2000), Kebede et al. (2001), Sanchez et al. (2002), Harris et al. (2007), Sabadin et al. (2012)
		1	Stg2 (green leaf area retention)	
		1	Stg3 (green leaf area retention)	
		1	Stg4 (green leaf area retention)	
		1	St2-1 (Stay green 2-1)	Sabadin et al. (2012)
		1	St2-2(Stay green 2-2)	
		1	St3 (Stay green 3)	
		1	St4 (Stay green 4)	
		1	St6 (Stay green 6)	
		1	St8 (Stay green 8)	
		1	St9 (Stay green 9)	
		1	St10 (Stay green 10)	
		1	Stg C.2 (Stay green C.2)	Kebede et al. (2001)
		1	Stg C.1 (Stay green C.1)	
		1	Stg B (Stay green B)	
		1	Stg E (Stay green E)	
		1	Stg D (Stay green D)	
		1	Stg A (Stay green A)	
		3	Ldg G (Lodging G), Ldg F (Lodging F), Ldg J (Lodging J)	
		4	Prf C(Preflowering drought tolerance C), Prf F(Preflowering drought tolerance F), Prf E(Preflowering drought tolerance E), Prf G(Preflowering drought tolerance G)	
		1	Stg F(Stay green F)	
		9	(% GL15) (green leaf area percentages)	Haussmann et al. (2002)
		14	(% GL30)green leaf area percentages)	
		13	(% GL45)green leaf area percentages)	
		6	t-E8/102, th19/50, tD9/103, t329/132, bB20/205, umc84	Tuinstra et al. (1997)
		1	SGA (Stay-green A)	Crasta et al. (1999)
		1	SGD (Stay-green D)	
		1	SGG (Stay-green G)	
		1	SGB (Stay-green B)	
		1	SG1.1 (Stay-green 1.1)	
		1	SG1.2 (Stay-green 1.2)	
		1	SGJ (Stay-green J)	
		1	MB6-84-TS136	Tao et al. (2000)
		1	TXS654-TXS943	
		1	ST1668-TXS558	
		1	CDO460-SSCIR165	
		2	QLsn.txs-B, QLsn.txs-Ea/Eb (Leaf senescence)	Feltus et al. (2006)
	SPAD at booting	3	QSpadb-dsr09-1, QSpadb-dsr06-1, QSpadb-dsr03	Reddy et al. (2014)
	SPAD values at maturity	1	QSpadm-dsr09-1	
	Green leaves at booting	4	QGlb-dsr01-1a, QGlb-dsr04-1, QGlb-dsr02-1, QGlb-dsr04-3	
	Green leaves at maturity	2	QGlm-dsr04-1, QGlm-dsr09-2	
	Per cent green leaves retained at maturity	2	QPglm-dsr04-2, QPglm-dsr09-2	
	Green leaf area at booting	2	QGlab-dsr10-1, QGlab-dsr02-1,	
	Green leaf area at maturity	1	QGlab-dsr02-2	

(Continued)

Table 1. Continued.

Traits	Trait category	No. of QTLs	QTL names	References
	Rate of leaf senescence	1	*QRls-dsr10-1*	
	Cold germinability/field emergence and early seedling vigor	16	*Germ30-1.1, Germ30-1.2, Germ30-2.1, Germ12-2.1, Germ12-9.2; Fearlygerm-1.2, Fearlygerm-7.1, Fearlygerm-9.2, Fearlygerm-9.3, Fearlygerm-1.1, Fearlygerm-4.1, Fearlygerm-9.1, Fearlygerm-9.3; Xtxp43, Xtxp51, Xtxp211*	Burow et al. (2011a,b),
	Aluminium tolerance	1	*Alt(SB)*	Magalhaes et al. (2007)
	Salinity stress (germination vigor, germination percentage, shoot height, root length, shoot fresh weight, root fresh weight, total fresh weight, shoot dry weight, root dry weight, total dry weight)	38	*qGV2-1, qGV2-2, qGV3; qGV1-1, qGV1-2, qGV4; qGP1, qGP2,qGP7-1, qGP7-2; qSH8, qSH1,qSH2, qSH4, qSH10;qRL1, qRL8, qRL3, qRL10-1, qRL10-2; qSFW8, qSFW9-1, qSFW4,qSFW9-2; qRFW6-1, qRFW2, qRFW6-2; qTFW6, qTFW9-1, qTFW1, qTFW4, qTFW9-2; qSDW4, qSDW9, qSDW6;qRDW3, qRDW6;qTDW6, qTDW8*	Wang et al. (2014a,b)
	Number of rhizomes/ subterranean rhizomes/ number of rhizome-derived shoots/ Overwintering	7	*pSB300a-pSBO88, pSB195-SH068, pSBJ02-pSB158; Overwintering2011A, Overwintering2011B; Ln2010RDS, Ln2010Dist*	Paterson et al. (1995), Washburn et al. (2013)
Subtotal		159		
Biotic stress tolerance	Midge disease egg count	2	*Flanking markers (ST698- RZ543, ST1017 -SG14)*	Tao et al. (2003)
	Midge disease pupal infestation	3	*Flanking markers (ST698- RZ543, ST1017 -SG14, TXS1931-SG37)*	
	Target leaf spot	1	*tls* (Target leaf spot)	Mohan et al. (2009)
	Zonate leaf spot	1	*Zls* (Zonate leaf spot)	
	Drechstera leaf plight	1	*Dls* (Drechstera leaf plight)	
	Rust resistance	8	*BNL5.09, TXS1625, RZ323, ISU102, ISU102, TXS2042, PSB47, TXS422*	Tao et al. (2003)
	Anthracnose resistance	14	*7 QTLs not named, Cg1, Locus 1-8, QAnt1, QAnt4, SC326-6, SCA 12, OPJ 0₁₁₄₃₇*	Boora et al. (1998), Klein et al. (2001), Singh et al. (2006), Singh et al. (2006), Perumal et al. (2009), Mohan et al. (2010), Upadhyaya et al. (2013)
	Percentage ergot infection	9	*Not named*	Parh et al. (2008)
	Pollen quantity	5	*Not named*	
	Pollen viability	4	*Not named*	
	Greenbug resistance to biotype I and K, C, and K/ greenbug feeding	34	*B18-885, OPC01-880, Sb5-214, Sb1-10, SbAGB03, Sb6-84, SbAGA01, OPA08-1150, OPB12-795; Ssg1, Ssg2, Ssg3, Ssg4, Ssg5, Ssg6,Ssg7, Ssg8, Ssg9; (8 unnamed QTLs);QSsgr-09-01, QSsgr-09-02;Qstsgr-sbi09ii, Qstsgrsbi09iii, Qstsgr-sbi09i, Qstsgr-sbi09iv;Xtxp16–Starssbem162, Starssbem162–Starssbem265*	Agrama et al. (2002), Katsar et al. (2002), Nagaraj et al. (2005), Wu and Huang (2008), Punnuri et al. (2013)
	Head bug resistance/ damage	10	*SbRPG943-RZ630, RZ476-SbRPG872, SbRPG667-CDO580, BNL5.37-SbRPG749, BNL5.37-SbRPG749, BNL5.37-SbRPG749, CDO20-C223, RZ630-SbRPG826, RZ244b-SbRPG852, mAGB03-UMC139*	Deu et al. (2005)
	Leaf scotch	1	*QLsc.txs-B (Leaf scotch)*	Feltus et al. (2006)

(Continued)

Table 1. Continued.

Traits	Trait category	No. of QTLs	QTL names	References
	Stalk rot resistance(No of internode, length of infection, percent lodging)	8	*xtxp297, xtxp213, AC13, xtxp343, xtxp176, (3 unnamed QTL for lodging resitance*	Reddy et al. (2008), Felderhoff et al. (2012)
	Shoot fly leaf glossiness	8	*QGs.dsr-3, QGs.dsr-5, QGs.dsr-6, QGs.dsr-10; QGs.dsr-1, QGs.dsr-4.1, QGs.dsr-2, QGs. dsr-4.2*	Satish et al. (2009), Aruna et al. (2011)
	Shoot fly seedling vigor	8	*QSv.dsr-3, QSv.dsr-6.1, QSv.dsr-6.2, QSv. dsr-10; QSv.dsr-1.1, QSv.dsr-1.2, QSv.dsr-2, QSv.dsr-9*	Satish et al. (2009), Aruna et al. (2011)
	Shoot fly oviposition	7	*QEg21.dsr-1, QEg21.dsr-7, QEg21.dsr-9, QEg21.dsr-10, QEg28.dsr-5, QEg28.dsr-7, QEg28.dsr-10*	Satish et al. (2009)
	Shoot fly deadheart	13	*QDh.dsr-5, QDh.dsr-10.3, QDh.dsr-10.4; QDh. dsr-1.1, QDh.dsr-1.2, QDh.dsr-2, QDh.dsr-6.1, QDh.dsr-6.2, QDh.dsr-7.1, QDh.dsr-7.2, QDh. dsr-9, QDh.dsr-10.1, QDh.dsr-10.2*	Satish et al. (2009), Aruna et al. (2011)
	Shoot fly adaxial trichome density	4	*QTdu.dsr-10.1, QTdu.dsr-10.2; QTdu.dsr-7, QTdu.dsr-10*	Satish et al. (2009), Aruna et al. (2011)
	Shoot fly abaxial trichome density	7	*QTdl.dsr-1.1, QTdl.dsr-1.2, QTdl.dsr-4, QTdl. dsr-6, QTdl.dsr-10.1, QTdl.dsr-10.2; QTdl.dsr-3*	Satish et al. (2009), Aruna et al. (2011)
	Root and crown rot resistance (*Pc* locus)	1	*PC*	Nagy et al. (2007)
	Striga resistance	39	*38 QTLs Not named; lgs*	Haussmann et al. (2004), Satish et al. (2012)
	Resistance to down mildew	3	*bin 2.04/05, bin 3.04/05, bin 6.05*	Nair et al. (2005)
Subtotal		191		
Total QTLs		350		

Cytokinin production increases growth and yield by improving foliar stay-green indices under drought conditions and improving processes that impact grain filling and grain number (Wilkinson et al. 2012). This suggests that the stay-green phenotype could be achieved by the biotechnological expression of isopentenyltransferase (*IPT*) gene whose protein catalyzes the rate-limiting step in cytokinin biosynthesis. Indeed, transgenic tobacco plants overexpressing *IPT* gene produced more trans-zeatin, did not senesce, had water content of 86%, maintained photosynthetic activity, and resurrected upon rewatering (Rivero et al. 2010). In addition, members of the WRKY and NAC families, and an ever-expanding cast of additional senescence-associated transcription factors, are identified by mutations that result in stay-green phenotype (Thomas and Ougham 2014).

Retention of chlorophyll in leaves of stay-green genotypes is associated with enhanced capacity to continue normal grain fill and maintenance of the ability to undergo photosynthesis for longer periods under drought conditions, reduced lodging, high stem carbohydrate content and grain weight, and resistance to charcoal stem rot (McBee et al. 1983; Borrell et al. 2000b; Burgess et al.

2002; Jordan et al. 2012). Hybrids involving stay-green sources produced close to 47% more biomass between anthesis and maturity in comparison with senescent checks (Borrell et al. 2000a). Moreover, the stay-green cultivar Sorcoll-141/07 had high plant height and green leaf number that contribute to its high biomass (Yemata et al., 2014). Leaves of stay-green cultivars have higher nutritional quality (Jordan et al. 2012). The high leaf nitrogen content and the simultaneous prolonged photosynthesis associated with stay-green trait (Borrell et al. 2000b) are correlated with higher sugar production in sorghum (Serrão et al. 2012) suggesting that leaf nitrogen concentrations and increased photosynthetic capacity are indicators for predicting sugar production in sweet sorghum. Therefore, breeding for stay-green trait specifically associated with high stem carbohydrates and leaf nitrogen will undoubtedly boost biofuel production in sweet sorghum cultivars.

The sources of stay-green trait used in most of the genetic studies and associated breeding programmes are lines BTx642, formally (B35), SC 56, E36-1, and KS19 (Haussmann et al. 2002; Mahalakshmi and Bidinger 2002; Hash et al. 2003) and have been reported to have greater

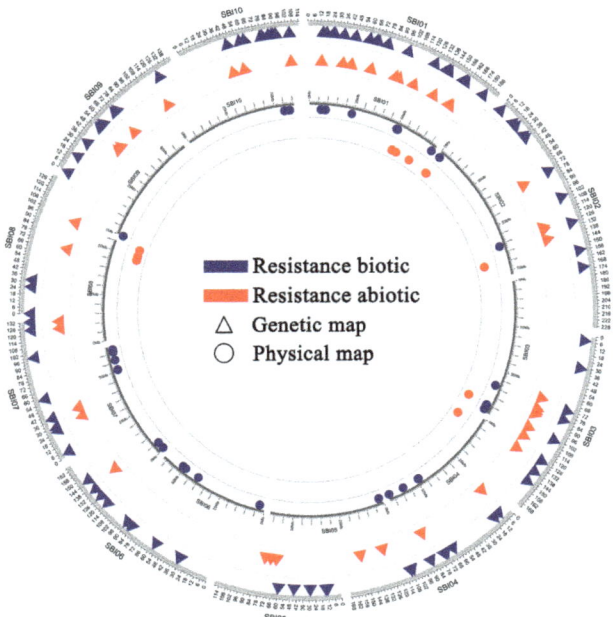

Figure 1. QTLs atlas map for biofuel-associated abiotic and biotic resistance traits in sorghum.

adaptation to drought stress through osmotic adjustment (Zhou et al. 2013). Indeed, the leaf relative water content of stay-green lines is much higher than those in nonstay-green lines, indicating that the stay-green lines keep the stalk transportation system functioning under severe drought conditions (Xu et al. 2000). Four loci *Stg1*, *Stg2*, *Stg3*, and *Stg4* (Table 1) controlling the stay-green trait in sorghum have consistently been identified across several environments in different mapping populations derived from crosses with BTx642, SC 56, E36-1, and KS19 lines, though each study reported varying phenotypic variation contributed by each QTL (Crasta et al. 1999; Subudhi et al. 2000; Tao et al. 2000; Xu et al. 2000; Kebede et al. 2001; Sanchez et al. 2002; Feltus et al. 2006; Harris et al. 2007; Sabadin et al. 2012). This suggests that although the ability of leaves to delay senescence has a genetic basis in sorghum, the expression of the character is strongly influenced by environmental cues (van Oosterom et al. 1996). QTLs controlling chlorophyll content colocated with the *Stg1*, *Stg2*, *Stg3*, and *Stg4* loci controlling stay-green (Subudhi et al. 2000) and (Xu et al. 2000). Therefore, chlorophyll content or loss of chlorophyll is a marker for the stay-green trait in sorghum during grain filling under drought.

The consistency of the four QTLs in various mapping populations across various environments and their combined phenotypic variation contribution of 54% to the stay-green drought trait suggest that they are stable and major QTLs for stay-green drought trait in sorghum

(Tanksley 1993; Xu et al. 2000; Sanchez et al. 2002). Fifteen (15) novel QTLs (Table 1) for various measures of stay-green trait have been identified using a genetic linkage map based on 245 F_9 Recombinant Inbred Lines (RILs) derived from a cross between M35-1 (more senescent) and B35 (less senescent) (Reddy et al. 2014). The phenotypic variation explained by each QTL ranged from 3.8 to 18.7%. Several other stay-green loci have also been reported (Table 1) but are generally unstable across environments (Crasta et al. 1999) with the potential to spontaneously reverts back to their parental phenotypes. The sorghum line E36-1, for instance, displays stay-green phenotype when grown under drought conditions in the field (van Oosterom et al. 1996), but not under well-watered conditions. (Tuinstra et al. 1997; Kebede et al. 2001; Thomas and Ougham 2014) identified six genetic loci controlling preflowering drought stress tolerance in sorghum from RILs derived from the crosses, SC 56 × Tx7000 and Tx7078 × B35, respectively with phenotypic variation contributed ranging between 15 to 40% under different environments, suggesting a strong genotype × environment (G × E) interaction at these loci.

Under water limited conditions, the stay-green alleles have been implicated to individually enhance grain and biomass yields in sorghum by modifying canopy development and water uptake patterns (Borrell et al. 2014b), indicating that stay-green phenotype and biomass yield could be achieved via the modification of root architecture (Mace et al. 2012), canopy development through reduction in tillering via increased size of lower leaves (Borrell et al. 2000a), or both. Thus far, breeders have transferred through marker-assisted backcrossing the stay-green trait into elite cultivars (Hash et al. 2003). The approach has been compromised because stress-related QTLs are dependent on the environmental conditions to which they were characterized (high G × E interaction) (Collins et al. 2008). In addition, different QTLs associated with stress-related traits can explain only a low percentage of the variation in the phenotype and that the effects of a favorable allele could not be transferred due to epistatic interactions (Peleg et al. 2009). Therefore, identifying QTLs of major effect that are independent of the particular genetic background and cloning the genes in the QTL could enhance breeding through biotechnology.

Functional analysis of the genes can be significantly aided through the application of reverse genetics approaches such as RNA interference (RNAi) and the type II CRISPR/ Cas system (Jiang et al. 2013) in order to characterize the individual gene function(s). Emphasis should be given to forward genetics studies where the identified genes can be expressed in genotypes that have been already selected for their adaptation to stressful environmental conditions. The availability of the sorghum genome may aid in fine mapping of the candidate genes responsible for the stay-green trait through increasing marker density within the target chromosomal region in addition to increasing the number of segregating population for which phenotypic information linked to the QTL can be obtained.

Proline accumulates to high levels in many plant species in response to environmental stresses and its role has been extensively investigated under stress conditions (Verbruggen and Hermans 2008). Recently, the Arabidopsis proline dehydrogenase (*AtProDH2*) was found to be strongly expressed in senescent leaves and in roots (Funck et al. 2010), suggesting that proline could have a new role during plant developmental processes. Similar findings in oilseed rape (*Brassica napus*), demonstrated that the *BnaProDH2* genes are specifically expressed in vasculature in an age-dependent manner in roots and senescent leaves (Faës et al. 2014). Thus, indicating that such expression could be related to the provision of reducing power for cell degradation mechanisms when chloroplasts become dysfunctional, an early process in autophagy (Avila-Ospina et al. 2014). The catabolism of proline could also contribute to recycling of metabolites in senescent leaves and provides glutamate then glutamine, principal forms of nitrogen compounds transported in the phloem from senescing leaves to sink organs (Tilsner et al. 2005). Similar studies could be carried out in sorghum in order to link the role of proline in abiotic stress conditions and in the developmental process of leaf senescence and biomass production.

Molecular mechanisms involved in the response of sorghum to abiotic stress

MiRNA expression

MicroRNAs (miRNAs) are a recently discovered class of gene expression regulators that have also been linked to several plant stress responses (Sunkar et al. 2007; Rajwanshi et al. 2014; Zhai et al. 2014) and in the biosynthesis pathways of carbon, glucose, starch, fatty acid, and lignin and in xylem formation, which could aid in designing next-generation sweet sorghum for biomass and biofuel. Differentially expressed miRNAs involved in the regulation of transcription (*bZIPs*, *MYBs*, *HOXs*), signal transduction (phosphoesterases, kinases, phosphatases), carbon metabolism (*NADP-ME*), detoxification (*CYPs*, *GST*, *AKRs*), osmoprotection mechanisms (*P5CS*), and stability of protein membranes (*DHN1*, *LEA*, *HSPs*) were upregulated upon imposition of drought stress in a four-leaf-old sorghum genotype IS1945 (Pasini et al. 2014), indicating that these drought-related genes could be used to screen for potential drought tolerance in other sorghum genotypes including sweet sorghum. Indeed, rice miRNA 169 g,

upregulated during drought stress (Zhao et al. 2007), has five sorghum homologs (*sbi-MIR169c*, *sbi-MIR169d*, *sbi-MIR169.p2*, *sbi-MIR169.p6*, and *sbi-MIR169.p7*), suggesting that the miRNA may be involved in many different processes related to drought stress resistance. The computationally predicted targets of the sbi-MIR169 subfamily comprise members of the plant nuclear factor Y (*NF-Y*) B transcription factor family, linked to improved performance in Arabidopsis, and maize under drought stress (Nelson et al. 2007). *GmNFYA3* gene, a target of miR169, is a positive regulator of plant tolerance to drought stress and has potential applications in molecular breeding to enhance drought tolerance in crops (Ni et al. 2013).

Using a deep sequencing approach to generate a genome-wide transcriptome of foxtail millet after exposure to simulated drought stress, one long noncoding RNAs (lncRNAs) was found to share sequence conservation and colinearity with its counterpart in sorghum suggesting that the lncRNAs in sorghum have the potential to have an impact on drought-regulated gene expression (Qi et al. 2013). The analysis of *cis*-elements of miRNA targets including transcription factors, genes for chaperonins, and metabolic enzymes and other genes necessary for proper plant development provides molecular evidence for the possible involvement of miRNAs in the process of abiotic stress tolerance in sorghum, indicating that miRNAs could play an important role in water stress tolerance in future sorghum studies (Ram and Sharma 2013). Therefore, miRNA 169 could be an excellent target for the generation of drought-resistant sweet sorghum genotypes through genetic engineering.

Auxin-related genes

Auxin-related gene families in *Sorghum bicolor* have also been implicated in abiotic stress response. The Gretchen Hagen3 (GH3) *SbGH3* and lateral organ boundaries (LBD) (*SbLBD*) genes, expressed at low levels under natural condition, were highly induced by salt and drought stress consistent with their products being involved in both abiotic stresses. Three genes, *SbIAA1*, *SbGH3-13*, and *SbLBD32*, were highly induced under all the four treatments, Indole-3-Acetic Acid, brassinosteroids, salt, and drought. The analysis provided new evidence for role of auxin in stress response, implying there are cross talk between auxin, brassinosteroids, and abiotic stresses (Wang et al. 2010).

Moisture stress triggered the upregulation of more transcription factor genes of MADS-box, Auxin Responsive Factors, Heme Activator Protein 2, Multiprotein Bridging Factors, and Homeobox families in root tissues compared to shoot tissues in sorghum (Aglawe et al. 2012). Under ABA, salt and drought treatments, sorghum auxin transporters *SbPIN4*, *-5*, *-8*, *-9*, and *-11* were highly increased, whereas *SbPIN1-3*, *-6*, *-7*, and *-10* were almost inhibited by all three treatments (Shen et al. 2010). The expression levels of *SbLAX1*, *-2*, *-4*, and *-5* compared with *SbLAX3* in leaves were lower than those in roots when treated with ABA. However, the response of *SbLAX* genes to salt and drought stresses was irregular, with *SbLAX4* expression downregulated dramatically under the stresses. Interestingly, transcription of the *SbPGP* gene family was almost inhibited in roots under salt treatment. *SbPGP1*, *-2*, *-5*, *-13*, *-14*, and *-15* were induced in roots under ABA treatment, whereas *SbPGP2*, *-3*, *-4*, *-7*, *-12*, and *-23* were induced in leaves under salt or drought stress. Under salt and drought treatment, *SbPGP13*, *-15*, *-17*, *-18*, *-20*, *-21*, and *-24* were all downregulated in both leaves and roots. Exploiting RNA-Seq technology in combination with the sorghum genome sequence (Paterson et al. 2009) and the SorghumCyc metabolic pathways database, (Dugas et al. 2011) characterized the sorghum transcriptome and reexamined the differential expression of sorghum genes in response to exogenous ABA and osmotic stress (Buchanan et al. 2005). Fifty differentially expressed drought-responsive gene orthologs specific to sorghum were identified for which no function had been previously assigned either in maize, rice, or Arabidopsis and were enriched for ABREs and CGTCA-motifs, or motifs that are involved in responses to ABA.

Transcription factors

The ethylene response factor family, members of the APETALA2 (AP2)/ERF transcription factor superfamily, is known to play an important role in plant adaptation to biotic and abiotic stress (Lata et al. 2014) and 105 sorghum ERF (*SbERF*) genes, categorized into 12 groups (A-1 to A-6 and B-1 to B-6) based on their sequence similarity have been identified in sorghum (Yan et al. 2013) Glutathione reductases (GRs) are important components of the antioxidant machinery that plants use to respond against abiotic stresses. Phylogenetic analysis identified two chloroplast GRs in sorghum that could possibly have a role in the modulation of abiotic stress. Since chloroplasts GR are also targeted to mitochondria suggest a combined antioxidant mechanism in both chloroplasts and mitochondria (Wu et al. 2013). In addition, phylogenetic analysis of the rice heterotrimeric G-protein complexes Gα subunit revealed high homology with sorghum. The promoter sequence analysis of RGA1(I) confirms the presence of stress-related *cis*-regulatory elements viz. ABA, MeJAE, ARE, GT-1 boxes, and LTR suggesting its active and possible independent roles in abiotic stress signaling. Furthermore, transcript profiling of RGA1(I) showed upregulation following NaCl, cold and drought stress, but under an elevated temperature, its transcript was downregulated. Heavy metal(loid)s stress showed rhythmic response in ABA stress and strong upregulation. These findings provide critical evidence for the active role of G-protein complexes in regulation of abiotic stresses in rice and possibly in sorghum and suggest that the Gα subunit of the heterotrimeric G-protein complexes could be exploited in the development of abiotic stress tolerance in sorghum. Genes coding for drought response element-binding (*DREB*) proteins regulate transcription of a large number of downstream genes involved in the plant response to abiotic stresses. An integration of abscisic acid, ethylene, auxin, and methyl jasmonate signaling was probably involved in regulating expression of the drought response through *DREB* transcription factors. The *SbEST8* gene was implicated to have a role in abiotic stress tolerance since imposition of drought resulted in rapid accumulation *SbEST8* mRNA in the germinating seeds of drought susceptible cultivar ICSV-272 (Dev Sharma et al. 2006).

Compatible solutes

The introduction of compatible solute synthesis pathway has emerged as a potential strategy for enhancing abiotic stress tolerance in crop plants (Rathinasabapathi 2000). The ability to synthesize and accumulate glycine betaine is widespread among angiosperms and is thought to contribute to salt and drought tolerance. Betaine aldehyde dehydrogenase *BADH1* and *BADH15* mRNA in sorghum were both induced by water deficit and their expression coincided with glycine betaine accumulation. The leaf water potential in stressed sorghum plants reached −2.3 MPa in the course of 17 days of water stress. Water deficit induced a 26-fold increase in glycine betaine levels and proline levels increased 108-fold (Wood et al. 1996). The upregulation of *Sorghum bicolor* glycine-rich RNA-binding protein designated as sbGR-RNP was induced by salinity and ABA and regulated by blue and red light, suggesting that there exists a cross talk between abiotic stress and light signaling in sorghum (Aneeta et al. 2002).

The mannitol biosynthetic pathway was engineered into *Sorghum bicolor* L. Moench cv. SPV462 with the *mtlD* gene encoding for mannitol-1-phosphate dehydrogenase from *E. coli*. The transgenic leaf segments were found to retain higher leaf water content when exposed to polyethylene glycol 8000 (−2.0 MPa) and maintained a 1.7- to 2.8-fold higher shoot and root growth, respectively, under NaCl stress (200 mmol/L) when compared to untransformed controls (Maheswari et al. 2010). These studies establish a role for a number of genes in modulating drought stress tolerance in sorghum. Therefore, functional characterization of these genes in sorghum including their overexpression or down regulation using genetic engineering could provide additional information as to their roles in broad abiotic stress tolerance.

Expression analysis of key stress inducible regulatory genes that play crucial roles in proline biosynthesis, *SbP5CS1* and *SbP5CS2*, revealed that the transcripts were upregulated after treatment of 10-day-old seedlings of sweet sorghum with drought, salt (250 mmol/L NaCl) and MeJA (10 μmol/L) indicating that the two genes could have the potential to be used in improving stress tolerance of sweet sorghum and other bioenergy feedstocks (Su et al. 2011).

Cold tolerance in sorghum

Soil temperatures below 15°C limit germination and seedling establishment for sorghum during early-season planting in temperate areas. Developing fast-growing sorghum seedlings is an important breeding goal for temperate climates since low spring time temperatures result in a prolonged juvenile development. In addition, this would allow expansion of sorghum to temperate region and for earlier planting in areas where it is being grown (Singh, 1985). In China, sorghum landrace, *kaoliang*, has poor agronomic characteristics though it exhibits higher seedling emergence and greater seedling vigor under cold conditions. The genetic basis of early-season cold tolerance in sorghum associated with germination, emergence, and vigor has been investigated and 15 QTLs have been identified (Table 1, (Burow et al. 2011a; Knoll and Ejeta 2008; Knoll et al. 2008). Using marker-assisted selection these desirable genomic regions can be introgressed into elite lines to improve early-season performance in sorghum. The quality of the messenger RNAs stored during embryo maturation on the mother plant, proteostasis, and DNA integrity play a major role in the germination phenotype. In addition, the sulfur amino acid metabolism pathway represents a key biochemical determinant of the commitment of the seed to initiate its development toward germination (Rajjou et al. 2012). Therefore, the characterization of molecular variables for germination and seed vigor under cold stress is expected to deliver new markers of seed quality that can be used in breeding programs and/or in biotechnological approaches to improve biomass yield in sweet sorghum. Further, a higher respiration rate is positively correlated with a higher germination rate and cultivars with higher respiration rate are likely to be resistant to early-season cold (Balota et al. 2010). Therefore, selection for a higher respiration rate can improve early-season vigor (germination, elongation, and growth rate in sorghum).

Rhizome formation trait is correlated and genetically linked to overwinter survival in sorghum (Washburn et al. 2013). The understanding of the genetic mechanisms

controlling overwintering has the potential to create perennial sorghums that can overwinter in climates where they previously could not. These overwintering sorghum types could be used for improvements in biofuel sorghum production by extending the period of biomass production and reducing production costs. In sorghum, rhizomatousness and overwintering are controlled by seven QTLs (Paterson et al. 1995; Washburn et al. 2013) Table 1). The QTLs were identified from a mapping population of a cross between BTx623 and *S. propinquum* and that regrowth after overwintering was associated with both rhizomatousness and tillering.

Salinity tolerance in sorghum

Salinity stress affects plant growth and productivity in many parts of the world and plants have developed adaptive responses to this external stress at the genetic level. For example, under sodium stress, *SbHKT1;4*, a member of the high-affinity potassium transporter gene family from *Sorghum bicolor*, functions to maintain optimal Na$^+$/K$^+$ balance (Wang et al. 2014b). Upon Na$^+$ stress *SbHKT1;4* expression was more strongly upregulated in salt-tolerant sorghum accessions, correlating with better balanced Na$^+$/K$^+$ ratio and enhanced plant growth. To gain insight into the genetic mechanism of salt tolerance at germination and seedling stage as a basis for improving salt tolerance in sorghum, (Wang et al. 2014a) identified 38 QTLs underlying salt tolerance (Table 1) from a 181 recombinant inbred lines derived from Shihong 137 and L-Tian. Six major QTLs with more than 10% phenotypic variation were detected at seedling stage under salt stress. These data indicate that the genetic mechanism for salt tolerance at germination and seedling stage in sorghum is different and that further research need to be done to identify genetic loci determining salt tolerance at different growth stages of sorghum during development.

Sorghum tolerance to Aluminium toxicity

Aluminum (Al) toxicity is an important limitation to food security in tropical and subtropical regions. In acidic soils, aluminum is solubilized into ionic forms (Al^{3+}), especially when the soil pH falls to lower than 5. This ionic form of Al is very toxic to plants, limiting the growth of roots either by inhibition of cell division, cell elongation, or by both. In this way, water and nutrient uptake by the roots is affected and as a consequence, plant growth and development is seriously hindered (Foy et al. 1993). Aluminum toxicity is, therefore, a major constraints for sorghum production in tropical and subtropical regions of the world (Doumbia et al. 1993, 1998). In addition to naturally occurring acid soils, agricultural practices may decrease soil pH, leading to yield losses due to Al toxicity. Elucidating the genetic and molecular mechanisms underlying sorghum Al tolerance is expected to accelerate the development of Al-tolerant cultivars. Using positional cloning, a gene encoding a member of the multidrug and toxic compound extrusion (MATE) family, an aluminum-activated citrate transporter, was identified as responsible for the major sorghum aluminum tolerance locus, Alt(Sb), on sorghum chromosome 3 (Magalhaes et al. 2007). These markers have been used by breeders to introgress rapidly the most favorable *SbMATE* alleles into sorghum germplasm, which is currently being field-tested in acid soils. Similar results have recently been demonstrated in maize where *ZmMATE1* expression, controlled either by three copies of the target gene or by an unknown molecular mechanism, is responsible for Al tolerance mediated by QTL mapped on chromosome 6 (*qALT6*) (Guimaraes et al. 2014). Polymorphisms in regulatory regions of Alt(Sb) are likely to contribute to large allelic effects, acting to increase Alt(Sb) expression in the root apex of tolerant genotypes. Furthermore, aluminum-inducible Alt(Sb) expression is associated with induction of aluminum tolerance via enhanced root citrate exudation (Magalhaes et al. 2007). These information could allow scientist to identify superior Alt(Sb) haplotypes that can be incorporated via molecular breeding and biotechnology into acid soil breeding programs, thus helping to increase crop yields in developing countries where acidic soils predominate.

Resistance to biotic stresses

Insect pests

Sorghum biomass and sugar yield are severely affected by biotic stresses including about 150 insect pests with more than 100 of them occurring in Africa (Guo et al. 2011), and new parental lines having genes for various biotic stress tolerances have the potential to mitigate this negative effect. The most destructive pests are the lepidopteran stem borer (*Chilo partellus*) and the dipterans, midge (*Stenodiplosis sorghicola*) and shoot fly (*Atherigona soccata*). Given the wide host range of some of the insect pests, and low level of resistance in the cultivated germplasm against major sorghum pests such as stem borers, head bugs, and armyworms, it will be highly desirable to invoke molecular plant breeding approaches combining conventional plant resistance with novel genes from other sources such as *Bacillus thuringiensis* (*Bt*) toxic protein. Insecticidal crystal proteins (CRY) from *Bacillus thuringiensis* are very effective against the lepidopterans and dipterans. *Bt* and other genes including protease inhibitors, enzymes, secondary plant metabolites, and plant lectins

with insecticidal activities are being evaluated for eventual use in transforming cotton, maize, rice, sorghum, grain legumes, tobacco, potato, sugarcane, groundnuts and tomatoes and reducing losses due to these pests(Sharma et al. 2004; Visarada and Kishore 2007). A transgenic sorghum plant was generated carrying a synthetic gene, *Bt cry*1Ac, under the control of a wound-inducible promoter from a maize protease inhibitor gene (*mpi*) (Girijashankar et al. 2005). The transgenic sorghum had low levels of *Bt* protein of 1–8 ng/g of fresh leaf tissue. A moderate level of tolerance was reported, which in turn conferred partial protection against neonate larvae of the spotted stem borer (*Chilo partellus*). Transgenic sorghum plants expressing *Bt cry*1Ab gene displayed insect-resistance to pink rice borer (*Sesamina inferens*) (Zhang et al. 2009).

Sorghum midge is the most damaging pest of grain sorghum worldwide (Young and Teetes 1977). Though sweet sorghum accumulates sugar at the expense of grain, the damage caused on the grain of sweet sorghum by sorghum midge could impact biomass and sugar accumulations. At flowering, female midges oviposit into spikelets, and the larvae feed on the ovary during the following 2 weeks, resulting in the failure of kernel development. Using classical approach, over 40 sorghum cultivars resistant to midge have been identified (Sharma et al. 1999) and could be useful for use in resistance breeding programs and to mitigate against biomass and sugar lose in sweet sorghum. Two genetic mechanisms of midge resistance, antixenosis and antibiosis, have been resolved in a recombinant inbred population from the cross of sorghum lines ICSV745 × 90,562 (Tao et al. 2003). Two genetic regions (between loci *ST698* and *RZ543* of linkage group A and loci *ST1017* and *SG14* of linkage group G, respectively, were significantly associated with egg counts (antixenosis) and the degree of phenotypic variation explained by each region was 12% and 15%, respectively. Three genetic regions located on linkage group A, linkage group G, and linkage group J, respectively, were found to be associated with pupal infestation. The levels of phenotypic variations explained by each region are 8.8% and 15%, respectively. The other region associated with pupal counts is the interval between loci *TXS1931* and *SG37* on linkage group J. explained 33.9% of total variation in pupal counts (antibiosis). The identification of genes for different mechanisms of midge resistance will be particularly useful for exploring new sources of midge resistance and for gene pyramiding of different mechanisms for increased security in sorghum breeding through marker-assisted selection and for the development of agronomically superior sorghum hybrids (Tao et al. 2003). Indeed, a putative candidate gene (*gm3*) for the recessive gall midge resistance gene (*gm3*) in rice was identified using a mapping population consisting of 302 F_{10} recombinant inbred lines derived from the cross TN1 (susceptible)/RP2068-18-3-5 (Sama et al. 2014). Comparative genomics could, therefore, identify similar syntenic genomic regions in sorghum for incorporation into midge sorghum resistance breeding programmes.

Greenbug, *Schizaphis graminum* (Rondani) is one of the major insect pests of sorghum and can cause serious damage to sorghum plants, particularly in the US Great Plains. Identification of chromosomal regions responsible for greenbug resistance will facilitate both map-based cloning and marker-assisted breeding. A total of 36 QTLs have been identified affecting both resistance and tolerance to greenbug insect pest (Agrama et al. 2002; Katsar et al. 2002; Nagaraj et al. 2005; Wu and Huang 2008; Punnuri et al. 2013).

Mirid panicle-feeding bugs (head bugs), particularly *Eurystylus oldi* Poppius, are major pests of sorghum in sub-Saharan Africa (Ajayi et al. 2001) and could also affect biomass and sugar accumulation upon infecting sweet sorghum panicles. Three significant QTLs on linkage group C accounted for 13% of the phenotypic variation for reduction in thousands kernel weight trait. Nine additional genomic regions in sorghum were identified to have a role in controlling head bug resistance in sorghum (Deu et al. 2005) and one leaf scorch QTL, *QLsc.txs-B*, explained 8.5% of the genetic variance (Feltus et al. 2006).

The shoot fly is a pest of sorghum, especially in America and Australia, and the larvae of this insect cut the growing point of the growing apical shoot resulting in a deadheart symptom. Genetic variations in sorghum resistance to shoot fly have been detected and this polymorphism has been used to identify genetic loci-controlling resistance to shoot fly. Nine QTLs associated with the resistance to leaf glossiness with phenotypic variation explained by individual QTL ranging from 7.6 to 14.0% were identified (Satish et al. 2009; Aruna et al. 2011). Seven QTLs distributed on five chromosomes, two each on SBI-07 and SBI-10, one each on SBI-01, SBI-05, and SBI-09, controlling oviposition were identified and the phenotypic variation explained by individual QTL ranged from 5.0 to 19.0%. A major QTL for this trait was detected on chromosome SBI-10 near the marker Xnhsbm 1044, explaining 19.0 and 16.1% of the phenotypic variation for mean eggs on 21 days after seedling emergence and mean eggs on 28 days after seedling emergence. Six QTLs for deadheart trait, which is a direct measure of resistance, were distributed on three chromosomes with one each on SBI-05 and SBI-09, and four on SBI-10 were identified in (Satish et al. 2009) study. The phenotypic variation explained by the individual trait ranged from 5.5 to 15.0%. Two major QTLs, *QDh.dsr-10.2* (explaining 11.4% of the phenotypic variation) and *QDh.dsr-10.3*, explaining 15.0%

of the phenotypic variation, were located on chromosome SBI-10. However, (Aruna et al. 2011) identified 10 QTLs on six chromosomes, SBI-02, SBI-09, SBI-01, SBI-06, SBI-07, and SBI-10) controlling deadheart trait with individual QTL explaining 4.5 to 12.8% phenotypic variation. Two major QTLs, *QTdu.dsr-10.1* and *QTdu.dsr-10.2*, were detected for adaxial trichome density on chromosome SBI-10, explaining 15.7 and 33.0% of the phenotypic variation, while six QTLs (*QTdl.dsr-1.1*, *QTdl.dsr-1.2*, *QTdl.dsr-4*, *QTdl.dsr-6*, *QTdl.dsr-10.1*, *QTdl.dsr-10.2*) were detected for abaxial trichome density distributed on four chromosomes with two on SBI-01, one each on SBI-04 and SBI-06, and two on SBI-10. The phenotypic variation explained by individual QTL ranged from 5.2 to 22.7% (Satish et al. 2009). In (Aruna et al. 2011) study, two QTLs (*QTdu.dsr-7*, *QTdu.dsr-10*) distributed on two chromosomes (one each on SBI-07 and SBI-10) were identified for adaxial trichome density, explaining 4.3-44.1% of the phenotypic variation with a QTL on chromosome SBI-10 being a major QTL contributing for 44.1% of phenotypic variation. In addition, three QTLs controlling abaxial trichome density were identified on chromosomes SBI-03 and SBI-10 accounting for 5.0 to 24.1% of the phenotypic variation. Cloning of genetic these genomic loci underlying resistance to sorghum diseases and the understanding of the mechanisms of how pathogens circumvent the genetic resistance will contribute toward sustainable intensification of biomass production.

Foliar diseases

Sorghum is also negatively affected by foliar diseases, viz. anthracnose, target leaf spot, zonate leaf spot, Drechstera leaf blight, and rust. Sorghum anthracnose is caused by *Colletotrichum sublineolum* and is characterized by weakening the plant, severely reducing grain yield, and quality and biomass production. The disease is more prevalent and severe in warm and humid environments, where it causes substantial economic losses. The pathogen causes seedling blight, leaf blight, stalk rot, head blight, and grain molding, and thus limits both forage and grain production. Among these, foliar anthracnose is the most pronounced and devastating, especially on sweet sorghum cultivars directly impacting sugar production (Dalianis 1997). The *Cg1* anthracnose resistance dominant gene located at the distal region of linkage group SBI-05 has been mapped in sorghum cultivar SC748-5 using four AFLP markers (Perumal et al. 2009; Ramasamy et al. 2009). In planta and ex planta *C. sublineolum*, infection assays were carried out using 1-week-old seedlings and it was observed that transgenic line, KOSA-1, was found to be significantly more tolerant to anthracnose than the parent wild-type, KAT 412 (Akosambo-Ayoo et al. 2013).

Association analysis of a sorghum mini-core collection consisting of 242 diverse accessions identified eight loci (loci 1-8) linked to anthracnose resistance in sorghum (Upadhyaya et al. 2013) and found genes associated with anthracnose resistance. They include NB-ARC class of R genes (*Sb10 g021850, Sb10 g021860*) in locus 7 that share 20% homology to *Pib* (accession number BAA76281) which confers resistance to rice blast disease (Wang et al. 1999). Autophagy-related protein 3 (*Sb01 g029070*) in locus 6 coding for *SbATG3* gene is 77% identical and 85% similar to the tobacco homolog *ATG3* (*AAW80629*). Silencing *ATG3* in tobacco resulted in unrestricted TMV-induced hypersensitive cell death due to increased pathogen propagation (Liu et al. 2005). The sorghum loci *Sb08 g003690, Sb08 g003705, Sb08 g003710*, and *Sb08 g003720* on locus 4 code for harpin-induced *Hin 1* and is a well-known hypertensive response marker gene (Pontier et al. 1999). Overexpression of the Arabidopsis *Hin 1* homolog, *AtNHL3*, enhances resistance to infection by *Pseudomonas syringae* pv.tomato DC3000 in Arabidopsis (Varet et al. 2003). RAV transcription factor (*Sb01 g049150*) in locus 3 is also associated with anthracnose resistance. Silencing RAV homolog in tomato abolished the resistance to bacterial wilt caused by *Ralstonia solanacearum* (Li et al. 2011). In addition, overexpression in Arabidopsis enhanced resistance to infection by *Pseudomonas syringae* pv.tomato DC3000 and to osmotic stresses by high salinity and dehydration (Sohn et al. 2006). The oxysterol-binding protein *Sb01 g010720* in locus 5 was also found to have a role in the disease resistance pathway. It is homolog in tomato (*StOBP1*) was found to be induced rapidly by *Phytophthora infestans* (Avrova et al. 2004). In addition, four homologs of menthone:neomenthol reductase 1 (*MNR*) in locus 1 potentially were found and silencing the *MNR* in pepper (*Capsicum annuum*) significantly increased its susceptibility to *Xanthomonas campestris* pv *vesicatoria* and *Colletotrichum coccodes* infection (Choi et al. 2008). Overexpressing rice PR-5 enhances resistance to *Rhizoctonia solani*, the causal agent of sheath blight (Datta et al. 1999). In sorghum, protein expression level of one TLP, sormatin, correlates with resistance to grain mold (Bueso et al. 2000). Taken together it suggests that modulation of genes with a role in resistance pathway has the potential to provide simultaneous resistance to multiple biotic and abiotic stresses in sorghum. These genes potentially play a role in countering pathogen attack in sorghum through the hypersensitive response, the rapid death of plant cells at the site of pathogen infection. Therefore, these genes and markers may be developed into molecular tools for the genetic improvement of anthracnose resistance in sorghum. (Mohan et al. 2010) mapped four (*QAnt1 QAnt2, QAnt3, and QAnt4)* anthracnose resistance loci. *QAnt3* was also mapped by (Klein et al. 2001) and locus 8 from the (Upadhyaya

et al. 2013) study was most likely *QAnt3* and locus 1 was close to *QAnt2* (Upadhyaya et al. 2013). A recessive anthracnose resistance gene in SC326-6 sorghum cultivar was mapped with a RAPD marker (Boora et al. 1998). Another recessive anthracnose resistance gene was mapped in G 73 sorghum cultivar with RAPD markers OPJ 0_{11437} at the same loci with SCAR marker SCJ 01 at 3.26 cM (Singh et al. 2006) and a RAPD-based SCAR marker SCA 12 at 6.03 cM (Singh et al. 2006). Anthracnose can be avoided by growing the sorghum in arid and semiarid environment. Additional genetic loci responsible for resistance to folia diseases in sorghum have been identified. Using 168 F7 recombinant inbred lines derived from a cross between 296 B (resistant) and IS18551 (susceptible) parents one major QTL with significant effects for each disease that colocated on SBI-06 was identified (Mohan et al. 2009). The variance explained by each QTL ranged from 12% to 50% with the QTL (*tls*) for target leaf spot explaining 50% of the total phenotypic variance. Similarly, one QTL each for zonate leaf spot (*zls*) and Drechstera leaf blight (*dls*) was identified as colocating with the QTL for target leaf spot disease. The QTL for Drechstera leaf blight explained 12%, while QTL for zonate leaf spot explained 16% of the phenotypic variance. The draft genome sequence of *Colletotrichum sublineola* has been presented and represents a new resource that will be useful for further research into the biology, ecology, and evolution of this key pathogen to find ways to mitigate its destructive ability on cultivated sorghum (Baroncelli et al. 2014).

Sorghum rust disease caused by *Puccinia purpurea* is important because its presence predisposes sorghum to other major disease problems like stalk rot and charcoal rot. Eight loci with significant effect on rust resistance have been identified (Tao et al. 1998) (Table 1). The percentage of the total phenotypic variation explained by each of these genomic regions varied from 6.8% to 42.6%.

Disease of the panicle

Sorghum ergot, caused predominantly by *Claviceps africana* is a significant threat to the sorghum industry worldwide and impacts juice and brix content in sweet sorghum (http://fenalce.org/archivos/SorFee.pdf). Ergot resistance in sorghum is controlled by many genes and that the pollen traits, pollen quantity, and pollen viability have moderate genetic correlation with percentage ergot infection. Nine genetic loci (Table 1) control percentage ergot infection in sorghum (Parh et al. 2008).

Stalk rot

Stalk rot caused by *Macrophomina phaseolina*, is also an economically important, soil-borne disease in major sorghum-growing areas across the world. It is associated with premature stem lodging and pith disintegration leading to inferior grain and fodder quality. Five QTLs were identified at Dharwad location and four QTLs at Bijapur locations for the component traits of stalk rot disease resistance (Reddy et al. 2008). Two QTLs associated with marker xtxp297, xtxp213 for number of internodes crossed on linkage group B, one QTL associated with marker AC13 for length of infection on linkage group D, and two QTLs associated with markers xtxp343, xtxp176 for per cent lodging on linkage group I accounted for 31.83, 10.76, and 18.90 per cent at Dharwad location and 14.87%, 10.47%, and 26.44% phenotypic variability at Bijapur location, respectively. The root and crown rot of sorghum known as milo disease is caused by the peritoxin produced by the saprophytic fungus *Periconia circinata* (Leukel 1948). The *PC* locus of sorghum (*Sorghum bicolor*) determines dominant sensitivity to a host-selective peritoxin. The *Pc* region was cloned by a map-based approach and found to contain three tandemly repeated genes with the structures of nucleotide-binding site-leucine-rich repeat (NBS-LRR) disease resistance genes (Nagy et al. 2007). The agronomically important gene *chi II*, encoding rice chitinase under the constitutive CaMV 35S promoter, has been transferred to sorghum for resistance to stalk rot (*Fusarium thapsinum*) (Krishnaveni et al. 2001). In addition, particle bombardment was used to genetically transform a sorghum genotype, KAT 412, with chitinase (harchit) and chitosanase (harcho) genes isolated from *Trichoderma harzianum*.

Resistance to *Striga* parasitism

Witchweed (*Striga* spp.) infestations are the greatest obstacle to sorghum [*Sorghum bicolor* (L.) Moench] grain and biomass production in many areas in Africa and Asia where they have a 20-100% yield reduction in any given season (Ejeta and Gressel 2007; Parker 2009). Sorghum coevolved with *Striga* in Africa and thus possesses intrinsic modicums of resistance that could be combined. Seeds of an acetolactate synthase (ALS) herbicide-tolerant sorghum hybrid mutant were treated with ALS-inhibiting herbicides before planting and the results showed that seeds treated with the highest herbicide rates had the fewest *Striga* attachments and the greatest delay in attachment (Tuinstra et al. 2009). Once the necessary sorghum genes are isolated and cloned, they could be transformed in a single, dominantly inherited construct containing a group of clustered genes, which would be a very effective strategy (Gressel 2010). Such resistance could easily be backcrossed into local varieties and land races preserving crop biodiversity, because it is inherited as a single dominant gene and not four separate recessive genes. Perhaps the resistance

genes from sorghum, once isolated, could be stacked with those responsible for Desmodium allelochemical production, along with resistance genes being found in cowpea and rice (Tuinstra et al. 2009), all into mini-chromosomes or into the genome at one locus. It would be very hard for the parasitic weeds to overcome such resistance and many crop species could be engineered with the same gene cluster. RNAi constructs encoding genes that suppress parasite-only metabolic pathways have been engineered into tomatoes (Aly et al. 2009) and the same strategy could be attempted in sorghum. In the field, drought stress and Striga infestation are rarely presented individually and sorghum plants are often subjected to a combination of stress types limiting its productivity. Identification of pathways and genetic loci directing specificity and crosstalk of sorghum responses combined with functional characterization of theses genetic signatures could lead to new targets for the enhancement of sorghum stress tolerance and identification of sorghum ideotypes specific to Africa.

Nevertheless, major QTLs for resistance of sorghum to the hemi-parasitic weed *Striga hermonthica* have been mapped in two recombinant inbred populations of $F_{3:5}$ lines developed from the crosses IS9830 × E36-1 and N13 × E36-1 (Haussmann et al. 2004) (Table 1). Sorghum cultivars resistant to Striga are known to produce low levels of strigolactone, a Striga germination stimulant. An in vitro assay for germination stimulant activity toward *Striga asiatica* in 354 recombinant inbred lines derived from SRN39 (low stimulant) × Shanqui Red (high stimulant), a single recessive gene *lgs* was precisely tagged and mapped (Satish et al. 2012) explaining about 40% of the phenotypic variance for area under the Striga number progress curve (Haussmann et al. 2004). So far, no QTL has been found to direct multiple disease resistance in sorghum. Given the selection pressure that many pathogens exert directly on natural plant populations and indirectly via variety improvement programs on crop plants, it is proposed that research should be focused on finding genetic loci responsible for multiple disease resistance as this has important implications for plant fitness. In maize, evidence of a locus conditioning resistance to multiple pathogens was found in bin 1.06 of the maize genome with the allele from inbred line 'Tx303' conditioning quantitative resistance to northern leaf blight (NLB) and qualitative resistance to Stewart's wilt and that *pan1* a gene conditioning susceptibility for NLB and Stewart's wilt was cloned (Jamann et al. 2014). Therefore, to reduce the risk of resistance breakdown and increase the levels of disease resistance in sorghum, new sources of disease resistance need to be explored to isolate and incorporate alternative mechanisms of resistance and to pyramid different resistance genes into commercial hybrids.

Weed control

Weeds cause a host of problems in agriculture, competing with crops for light, water, and nutrients, providing a reservoir for insects and diseases, and contaminating seedlots. Vegetative dispersal by rhizomes (underground stems) and seed dispersal by disarticulation of the mature inflorescence ("shattering") cause perennial monocots such as "johnsongrass" to rank among the world's most noxious weeds. Improvements in agricultural production have correlated well with the use of herbicides in controlling weeds. Transgenic glyphosate-resistant crops overexpressing 5-enolpyruvylshikimate-3-phosphate synthase (*cp4 epsps)* gene accelerated widespread use of glyphosate becoming the most widely used herbicide in world agriculture (Duke and Powles 2008) for its effectiveness in controlling recalcitrant weeds such as Johnsongrass. However, the increased utilization of these herbicides over a long period of time exerts selective pressure leading to widespread evolution of resistance in several weed species (Busi et al. 2013). For instance, metabolic resistance (enhanced metabolic capacity to detoxify herbicides) can be endowed by the increased activity of the endogenous cytochrome P450 mono-oxygenases, glucosyl transferases (GTs), glutathione S-transferases (GSTs), and/or other enzyme systems such as aryl acylamidase (Carey et al. 1995) that can metabolize herbicides (Yu and Powles 2014). Combined with lack of novel herbicides being brought to the market over the last 30 years and tougher registration and environmental regulations on herbicides have resulted in a loss of some herbicides, particularly in Europe, threatening crop production worldwide (Heap 1997).

Integrated weed management approach to control weed populations has been hailed as an effective tool in addition, reduce the environmental impact of individual weed management practices, increase cropping system sustainability, and reduce selection pressure for weed resistance to herbicides (Harker and O'Donovan 2013). Maize plants transformed with an aryloxyalkanoate dioxygenase (*AAD-1*) gene showed robust crop resistance to aryloxyphenoxypropionate herbicides over four generations and were also not injured by 2,4-dichlorophenoxyacetic acid (2,4-D) applications at any growth stage. Arabidopsis plants expressing *AAD-12* were resistant to 2,4-D as well as triclopyr and fluroxypyr, and transgenic soybean plants expressing *AAD-12* maintained field resistance to 2,4-D over five generations indicating that single *AAD* transgenes can provide simultaneous resistance to a broad repertoire of agronomically important classes of herbicides, including 2,4-D, with utility in both monocot and dicot crops which can help transgenes preserve the productivity and environmental benefits of herbicide-resistant crops (Wright et al. 2010). Recently, the use of multicopy transposons

bearing unfitness genes has been proposed in the management of weeds. Multicopy transposons rapidly disseminate through populations, appearing in ~100% of progeny unlike nuclear transgenes, which appear in a proportion of segregating populations (Gressel and Levy 2014). Here, weed populations could be generated that contain the unfitness gene under chemically or environmentally inducible promoters, activated after gene dissemination, or under constitutive promoters where the gene function is utilized only at special times (e.g., sensitivity to a herbicide), and thus are easily controllable. Efforts need to be accelerated to understand the genetic basis of weed resistance for employment of an RNAi approach to interfere with the expression of herbicide resistance genes in weeds (Sammons et al. 2012) and restore sensitivity of weeds to glyphosate (Green 2014). Enhancing crop competitiveness, for example, by genetic engineering with genes encoding phosphates, and the application of fertilizers with phosphites as the main source of phosphates is a key strategy for crops to outcompete weeds for essential nutrients (López-Arredondo and Herrera-Estrella 2012).

Perspectives: Sweet sorghum ideotypes with enhanced resistance to abiotic and biotic stresses

In the field, multiple abiotic stresses (drought, salinity, heat, cold, chilling, freezing, nutrient, high light intensity, ozone, and anaerobic stresses) and biotic stresses (insect pests and diseases) are presented. The performance of the plant, therefore, is affected by the degree of heterogeneity between stress levels, simultaneous occurrence of different stresses, the timing of the stress event with respect to the developmental stage of the plant and the intensity and duration of the stress (Mittler and Blumwald 2010). Plants respond differently to combined stresses as compared to their response to individual stress as it happens under laboratory conditions. The later response activates a specific program of gene expression relating to the exact environmental condition encountered. The responses are complex and could involve changes at transcriptomic, cellular, and physiological level. Genetic and genomic resources for sorghum breeding are available (Carpita and McCann 2008), and they offer an opportunity to employ multidisciplinary approaches involving traditional breeding and biotechnology to contribute to future improvements of sweet sorghum to adapt to both biotic and abiotic stresses. For instance, in tropical climate, drought stress is ranked most important followed by striga parasitism, and fungal and bacterial diseases in terms of limiting sorghum potential for growth and reproduction. Therefore, breeding for sorghum ideotypes tolerant to combined drought and Striga parasitism will be ideal for this region.

As discussed previously, resistance to abiotic and biotic stresses has been demonstrated through genetic engineering and classical breeding. Tolerance to both abiotic and biotic stresses has also been achieved. In maize, breeding programs have developed plants tolerant to drought and have additional resistance to the parasitic weed *Striga hermonthica* (Bänziger et al. 2006; Badu-Apraku and Yallou 2009). In Sudan, sorghum cultivars resistant to drought and striga infectivity have been developed through classical breeding (Nair Suliman personal communication), this suggests that hormone signaling pathways orchestrating the interaction between abiotic and biotic stresses are altered and in particular abscisic acid. This alteration could be interesting to breeders to further design sorghum ideotypes that can withstand a combination of stresses as presented in the field. In addition, research programs need to focus on developing tolerance to multiple stresses in order for the improved varieties to respond predictably under field condition.

In temperate environment, cool temperatures below 15°C during the early growing season limits optimal growth of sorghum, it is a key agronomic trait for warm season cereal crops such as sorghum. Breeding for sorghum ideotypes with improved early-season cold tolerance would be appropriate for temperate environments (Yu and Tuinstra 2001). So far, Chinese sorghum kaoliang, Shanqui Red (Knoll et al. 2008) and and F7 RIL population of RTx403xPI567946 (Burow et al. 2011b) have been found to exhibit higher emergence and greater seedling vigor under cool temperature than most breeding lines currently available. However, they lack desirable agronomic characteristics. Sorghum ideotypes' resistance to cold could be developed by introgression of desirable genes from Chinese landraces into elite lines through marker-assisted selection (Knoll et al. 2008; Burow et al. 2011a).

The use of herbicides is increasing in global crop production. Improved weed control with herbicides promotes fertilizer use and has the potential to improve crop yields in many developing countries in the near future (Gianessi 2013). Shattercane and *Sorghum halepense* (johnsongrass) are natural weeds for sorghum and resistant to most other herbicides used for their control (Heap 2014). In addition, they outcross with cultivated sorghum (Morrell et al. 2005; Muraya et al. 2011). This suggests that transgenes introduced to sorghum would readily introgress and be retained in these wild species, which often occur sympatrically with cultivated sorghum in Africa (Mutegi et al. 2010). Compared to other weed control strategies including manual hand weeding, herbicides are the key to sustainable crop production throughout the world, and, will remain the mainstay for weed control in the foreseeable future. Therefore, when developing sweet sorghum ideotypes for different ecological regions, the use of herbicides to control weeds should be considered. The rhizome formation trait is correlated and genetically linked to overwinter survival in sorghum. Genetic mechanisms controlling overwintering have the potential to minimize the risk of weediness and create perennial sorghums that can overwinter in climates where they previously could not (Paterson et al. 1995; Washburn et al. 2013). These perennial overwintering sorghum ideotypes could be used for improvements in biofuel production in sweet sorghum by extending the period of biomass production and reducing production costs.

Conflict of Interest

None declared.

References

Aglawe, S., B. Fakrudin, C. Patole, S. Bhairappanavar, R. Koti, and P. Krishnaraj. 2012. Quantitative RT-PCR analysis of 20 transcription factor genes of MADS, ARF, HAP2, MBF and HB families in moisture stressed shoot and root tissues of sorghum. Physiol. Mol. Biol. Plants 18:287–300.

Agrama, H., G. Widle, J. Reese, L. Campbell, and M. Tuinstra. 2002. Genetic mapping of QTLs associated with greenbug resistance and tolerance in *Sorghum bicolor*. Theor. Appl. Genet. 104:1373–1378.

Ajayi, O., H. Sharma, R. Tabo, A. Ratnadass, and Y. Doumbia. 2001. Incidence and distribution of the sorghum head bug, *Eurystylus oldi* Poppius (Heteroptera: Miridae) and other panicle pests of sorghum in West and Central Africa. Int. J. Trop. Insect Sci. 21:103–111.

Akosambo-Ayoo, L., M. Bader, H. Loerz, and D. Becker. 2013. Transgenic sorghum (*Sorghum bicolor* L. Moench) developed by transformation with chitinase and chitosanase genes from *Trichoderma harzianum* expresses tolerance to anthracnose. Afr. J. Biotechnol. 10:3659–3670.

Aly, R., H. Cholakh, D. M. Joel, D. Leibman, B. Steinitz, A. Zelcer, et al. 2009. Gene silencing of mannose 6-phosphate reductase in the parasitic weed *Orobanche aegyptiaca* through the production of homologous dsRNA sequences in the host plant. Plant Biotechnol. J. 7:487–498.

Aneeta, S. N. N. Tuteja, and S. Kumar Sopory. 2002. Salinity- and ABA-induced up-regulation and light-mediated modulation of mRNA encoding glycine-rich RNA-binding protein from *Sorghum bicolor*. Biochem. Biophys. Res. Commun., 296:1063–1068.

Anyamba, A., J. L. Small, S. C. Britch, C. J. Tucker, E. W. Pak, C. A. Reynolds, et al. 2014. Recent weather extremes and impacts on agricultural production and vector-borne disease outbreak patterns. PLoS ONE 9:e92538.

Aruna, C., V. Bhagwat, R. Madhusudhana, V. Sharma, T. Hussain, R. Ghorade, et al. 2011. Identification and validation of genomic regions that affect shoot fly resistance in sorghum [Sorghum bicolor (L.) Moench]. Theor. Appl. Genet. 122:1617–1630.

Avila-Ospina, L., M. Moison, K. Yoshimoto, and C. Masclaux-Daubresse. 2014. Autophagy, plant senescence, and nutrient recycling. J. Exp. Bot. doi:10.1093/jxb/eru039.

Avrova, A. O., N. Taleb, V. M. Rokka, J. Heilbronn, E. Campbell, I. Hein, et al. 2004. Potato oxysterol binding protein and cathepsin B are rapidly up-regulated in independent defence pathways that distinguish R gene-mediated and field resistances to Phytophthora infestans. Mol. Plant Pathol. 5:45–56.

Badu-Apraku, B., and C. Yallou. 2009. Registration of-resistant and drought-tolerant tropical early maize populations TZE-W Pop DT STR C and TZE-Y Pop DT STR C. J. Plant Regist. 3:86–90.

Balota, M., W. Payne, S. Veeragoni, B. Stewart, and D. Rosenow. 2010. Respiration and its relationship to germination, emergence, and early growth under cool temperatures in sorghum. Crop Sci. 50:1414–1422.

Bänziger, M., P. S. Setimela, D. Hodson, and B. Vivek. 2006. Breeding for improved abiotic stress tolerance in maize adapted to southern Africa. Agric. Water Manag. 80:212–224.

Baroncelli, R., J. M. Sanz-Martín, G. E. Rech, S. A. Sukno, and M. R. Thon. 2014. Draft genome sequence of Colletotrichum sublineola, a destructive pathogen of cultivated sorghum. Genome Announc. 2:e00540-14.

Berndes, G., M. Hoogwijk, and R. van den Broek. 2003. The contribution of biomass in the future global energy supply: a review of 17 studies. Biomass Bioenergy 25:1–28.

Boora, K. S., R. Frederiksen, and C. Magill. 1998. DNA-based markers for a recessive gene conferring anthracnose resistance in sorghum. Crop Sci. 38:1708–1709.

Borrell, A. K., G. L. Hammer, and A. C. Douglas. 2000a. Does maintaining green leaf area in sorghum improve yield under drought? I. Leaf growth and senescence. Crop Sci. 40:1026–1037.

Borrell, A. K., G. L. Hammer, and R. G. Henzell. 2000b. Does maintaining green leaf area in sorghum improve yield under drought? II. Dry matter production and yield. Crop Sci. 40:1037–1048.

Borrell, A. K., J. E. Mullet, B. George-Jaeggli, E. J. Van Oosterom, G. L. Hammer, P. E. Klein, et al. 2014a. Drought adaptation of stay-green sorghum is associated with canopy development, leaf anatomy, root growth, and water uptake. J. Exp. Bot., 65:6137–6139

Borrell, A. K., E. J. Oosterom, J. E. Mullet, B. George-Jaeggli, D. R. Jordan, P. E. Klein, et al. 2014b. Stay-green alleles individually enhance grain yield in sorghum under drought by modifying canopy development and water uptake patterns. New Phytol. 203:817–30.

Buchanan, C. D., S. Lim, R. A. Salzman, I. Kagiampakis, D. T. Morishige, B. D. Weers, et al. 2005. Sorghum bicolor's transcriptome response to dehydration, high salinity and ABA. Plant Mol. Biol. 58:699–720.

Bueso, F. J., R. D. Waniska, W. L. Rooney, and F. P. Bejosano. 2000. Activity of antifungal proteins against mold in sorghum caryopses in the field. J. Agric. Food Chem. 48:810–816.

Burgess, M. G., C. Rush, G. Piccinni, and G. Schuster. 2002. Relationship between charcoal rot, the stay-green trait, and irrigation in grain sorghum. Phytopathology 92:S10.

Burow, G., J. J. Burke, Z. Xin, and C. D. Franks. 2011a. Genetic dissection of early-season cold tolerance in sorghum (Sorghum bicolor (L.) Moench). Mol. Breeding 28:391–402.

Burow, G., Z. Xin, C. Franks, J. Burke, and P. Hi. 2011b. Genetic enhancement of cold tolerance to overcome a major limitation in sorghum. American Seed Trade Association Conference Proceedings.

Busi, R., M. M. Vila-Aiub, H. J. Beckie, T. A. Gaines, D. E. Goggin, S. S. Kaundun, et al. 2013. Herbicide-resistant weeds: from research and knowledge to future needs. Evol. Appl. 6:1218–1221.

Carey, V., S. O. Duke, R. E. Hoagland, and R. E. Talbert. 1995. Resistance Mechanism of Propanil-Resistant Barnyardgrass 1. Absorption, Translocation, and Site of Action Studies. Pestic. Biochem. Physiol. 52:182–189.

Carpita, N. C., and M. C. McCann. 2008. Maize and sorghum: genetic resources for bioenergy grasses. Trends Plant Sci. 13:415–420.

Choi, H. W., B. G. Lee, N. H. Kim, Y. Park, C. W. Lim, H. K. Song, et al. 2008. A role for a menthone reductase in resistance against microbial pathogens in plants. Plant Physiol. 148:383–401.

Collins, N. C., F. Tardieu, and R. Tuberosa. 2008. Quantitative trait loci and crop performance under abiotic stress: where do we stand? Plant Physiol. 147:469–486.

Crasta, O., W. Xu, D. Rosenow, J. Mullet, and H. Nguyen. 1999. Mapping of post-flowering drought resistance traits in grain sorghum: association between QTLs influencing premature senescence and maturity. Mol. Gen. Genet. 262:579–588.

Dalianis, C.1997. Productivity, sugar yields, ethanol potential and bottlenecks of sweet sorghum in European Union. Pp. 65–79 in D. Li, ed. Proceedings of the 1st International sweet sorghum conference. Ed: Dajue Li, 1997. 65–79

Dalla Marta, A., M. Mancini, F. Orlando, F. Natali, L. Capecchi, and S. Orlandini. 2014. Sweet sorghum for bioethanol production: crop responses to different water stress levels. Biomass Bioenergy, 64:211–219.

Damte, T., B. B. Pendleton, and L. K. Almas. 2009. Cost-benefit analysis of sorghum midge, Stenodiplosis sorghicola 1 (Coquillett)-resistant sorghum hybrid research and development in Texas. Southwest. Entomol. 34:395–405.

Datta, K., R. Velazhahan, N. Oliva, I. Ona, T. Mew, G. Khush, et al. 1999. Over-expression of the cloned rice thaumatin-like protein (PR-5) gene in transgenic rice plants enhances environmental friendly resistance to Rhizoctonia solani causing sheath blight disease. Theor. Appl. Genet. 98:1138–1145.

Deu, M., A. Ratnadass, M. Hamada, J. Noyer, M. Diabatedagger, and J. Chantereau. 2005. Quantitative trait loci for head-bug resistance in sorghum. Afr. J. Biotechnol. 4:247–250.

Dev Sharma, A., S. Kumar, and P. Singh. 2006. Expression analysis of a stress-modulated transcript in drought tolerant and susceptible cultivars of sorghum (Sorghum bicolor). J. Plant Physiol., 163:570–576.

Doumbia, M., L. Hossner, and A. Onken. 1993. Variable sorghum growth in acid soils of subhumid West Africa. Arid Land Res. Manag. 7:335–346.

Doumbia, M., L. Hossner, and A. Onken. 1998. Sorghum growth in acid soils of West Africa: variations in soil chemical properties. Arid Land Res. Manag. 12:179–190.

Dugas, D. V., M. K. Monaco, A. Olson, R. R. Klein, S. Kumari, D. Ware, et al. 2011. Functional annotation of the transcriptome of Sorghum bicolor in response to osmotic stress and abscisic acid. BMC Genom. 12:514.

Duke, S. O., and S. B. Powles. 2008. Glyphosate: a once-in-a-century herbicide. Pest Manag. Sci. 64:319–325.

Ejeta, G., and J. Gressel. 2007. Integrating new technologies for Striga control: towards ending the witch-hunt. World Scientific, Publishing Co. Pte Ltd, 5 Tol Tuck Link, Singapore, pp. 3–16.

Emebiri, L. C. 2013. QTL dissection of the loss of green colour during post-anthesis grain maturation in two-rowed barley. Theor. Appl. Genet. 126:1873–1884.

Faës, P., C. Deleu, A. Aïnouche, F. Le Cahérec, E. Montes, V. Clouet, et al. 2014. Molecular evolution and transcriptional regulation of the oilseed rape proline dehydrogenase genes suggest distinct roles of proline catabolism during development. Planta 241: 403-419

Feltus, F., G. Hart, K. Schertz, A. Casa, S. Kresovich, S. Abraham, et al. 2006. Alignment of genetic maps and QTLs between inter-and intra-specific sorghum populations. Theor. Appl. Genet. 112:1295–1305.

Felderhoff, T. J., et al. "QTLs for Energy-related Traits in a Sweet× Grain Sorghum [(L.) Moench] Mapping Population." Crop Science 52.5 (2012):2040–2049.

Foy, C., T. Jr Carter, J. Duke, and T. Devine. 1993. Correlation of shoot and root growth and its role in selecting for aluminum tolerance in soybean. J. Plant Nutr., 16:305–325.

Funck, D., S. Eckard, and G. Müller. 2010. Non-redundant functions of two proline dehydrogenase isoforms in Arabidopsis. BMC Plant Biol. 10:70.

Gianessi, L. P. 2013. The increasing importance of herbicides in worldwide crop production. Pest Manag. Sci. 69:1099–1105.

Girijashankar, V., H. Sharma, K. K. Sharma, V. Swathisree, L. S. Prasad, B. Bhat, et al. 2005. Development of transgenic sorghum for insect resistance against the spotted stem borer (Chilo partellus). Plant Cell Rep., 24:513–522.

Green, J. M.. 2014. Current state of herbicides in herbicide-resistant crops. Pest Manag. Sci. 70:1351–1357.

Gressel, J. 2010. Needs for and environmental risks from transgenic crops in the developing world. New Biotechnol. 27:522–527.

Gressel, J., and A. A. Levy. 2014. Use of multi-copy transposons bearing unfitness genes in weed control: four example scenarios. Plant Physiol. 166:1221–31.

Guimaraes, C. T., C. C. Simoes, M. M. Pastina, L. G. Maron, J. V. Magalhaes, R. C. Vasconcellos, et al. 2014. Genetic dissection of Al tolerance QTLs in the maize genome by high density SNP scan. BMC Genom. 15:153.

Guo, C., W. Cui, X. Feng, J. Zhao, and G. Lu. 2011. Sorghum insect problems and Managementf. J. Integr. Plant Biol. 53:178–192.

Harker, K. N., and J. T. O'Donovan. 2013. Recent weed control, weed management, and integrated weed management. Weed Technol. 27:1–11.

Harris, K., P. Subudhi, A. Borrell, D. Jordan, D. Rosenow, H. Nguyen, et al. 2007. Sorghum stay-green QTL individually reduce post-flowering drought-induced leaf senescence. J. Exp. Bot. 58:327–338.

Hash, C., A. Bhasker Raj, S. Lindup, A. Sharma, C. Beniwal, R. Folkertsma, et al. 2003. Opportunities for marker-assisted selection (MAS) to improve the feed quality of crop residues in pearl millet and sorghum. Field. Crop. Res., 84:79–88.

Haussmann, B., V. Mahalakshmi, B. Reddy, N. Seetharama, C. Hash, and H. Geiger. 2002. QTL mapping of stay-green in two sorghum recombinant inbred populations. Theor. Appl. Genet. 106:133–142.

Haussmann, B., D. Hess, G. Omanya, R. Folkertsma, B. Reddy, M. Kayentao, et al. 2004. Genomic regions influencing resistance to the parasitic weed Striga hermonthica in two recombinant inbred populations of sorghum. Theor. Appl. Genet. 109:1005–1016.

Heap, I. 1997. International survey of herbicide-resistant weeds. Western Society of Weed Science (USA), 1997.

Heap, I. 2014. Global perspective of herbicide-resistant weeds. Pest Manag. Sci. 70:1306–1315.

House, L. R. 1985. A guide to sorghum breeding. International Crops Research Institute for the Semi-Arid Tropics Patancheru, India.

Jamann, T. M., J. A. Poland, J. M. Kolkman, L. G. Smith, and R. J. Nelson. 2014. Unraveling genomic complexity at a quantitative disease resistance locus in Maize. Genetics 198:333–344.

Jiang, W., H. Zhou, H. Bi, M. Fromm, B. Yang, and D. P. Weeks. 2013. Demonstration of CRISPR/Cas9/sgRNA-mediated targeted gene modification in Arabidopsis, tobacco, sorghum and rice. Nucleic Acids Res., 41, e188

Jordan, D., C. Hunt, A. Cruickshank, A. Borrell, and R. Henzell. 2012. The relationship between the stay-green trait and grain yield in elite sorghum hybrids grown in a range of environments. Crop Sci. 52:1153–1161.

Katsar, C. S., A. H. Paterson, G. L. Teetes, and G. C. Peterson. 2002. Molecular analysis of sorghum resistance to the greenbug (Homoptera: Aphididae). J. Econ. Entomol. 95:448–457.

Kebede, H., P. Subudhi, D. Rosenow, and H. Nguyen. 2001. Quantitative trait loci influencing drought tolerance in grain sorghum (Sorghum bicolor L. Moench). Theor. Appl. Genet. 103:266–276.

Kim, K.-H., E. Kabir, and S. Ara Jahan. 2014. A review of the consequences of global climate change on human health. J. Environ. Sci. Health Part C 32:299–318.

Klein, R., R. Rodriguez-Herrera, J. Schlueter, P. Klein, Z. Yu, and W. Rooney. 2001. Identification of genomic regions that affect grain-mould incidence and other traits of agronomic importance in sorghum. Theor. Appl. Genet. 102:307–319.

Knoll, J., and G. Ejeta. 2008. Marker-assisted selection for early-season cold tolerance in sorghum: QTL validation across populations and environments. Theor. Appl. Genet. 116:541–553.

Knoll, J., N. Gunaratna, and G. Ejeta. 2008. QTL analysis of early-season cold tolerance in sorghum. Theor. Appl. Genet. 116:577–587.

Krishnaveni, S., J. Joeung, S. Muthukrishnan, and G. Liang. 2001. Transgenic sorghum plants constitutively expressing a rice chitinase gene show improved resistance to stalk rot. J. Genet. Breed. 55:151–158.

Kruger, M., J. van den Berg, and H. du Plessis. 2008. Diversity and seasonal abundance of sorghum panicle-feeding Hemiptera in South Africa. Crop Prot. 27:444–451.

Lata, C., A. K. Mishra, M. Muthamilarasan, V. S. Bonthala, Y. Khan, and M. Prasad. 2014. Genome-wide investigation and expression profiling of AP2/ERF transcription factor superfamily in Foxtail Millet (Setaria italica L.). PLoS ONE 9:e113092.

Leukel, R. 1948. Periconia circinata and its relation to Milo disease. J. Agric. Res. 77:201–222.

Li, J.-G., J. Cao, F.-F. Sun, D.-D. Niu, F. Yan, H.-X. Liu, et al. 2011. Control of tobacco mosaic virus by PopW as a result of induced resistance in tobacco under greenhouse and field conditions. Phytopathology 101:1202–1208.

Liu, Y., M. Schiff, K. Czymmek, Z. Tallóczy, B. Levine, and S. Dinesh-Kumar. 2005. Autophagy regulates programmed cell death during the plant innate immune response. Cell 121:567–577.

López-Arredondo, D. L., and L. Herrera-Estrella. 2012. Engineering phosphorus metabolism in plants to produce a dual fertilization and weed control system. Nat. Biotechnol. 30:889–893.

Mace, E., V. Singh, E. van Oosterom, G. Hammer, C. Hunt, and D. Jordan. 2012. QTL for nodal root angle in sorghum (Sorghum bicolor L. Moench) co-locate with QTL for traits associated with drought adaptation. Theor. Appl. Genet. 124:97–109.

Magalhaes, J. V., J. Liu, C. T. Guimaraes, U. G. Lana, V. M. Alves, Y.-H. Wang, et al. 2007. A gene in the multidrug and toxic compound extrusion (MATE) family confers aluminum tolerance in sorghum. Nat. Genet. 39:1156–1161.

Mahalakshmi, V., and F. R. Bidinger. 2002. Evaluation of stay-green sorghum germplasm lines at ICRISAT. Crop Sci. 42:965–974.

Maheswari, M., Y. Varalaxmi, A. Vijayalakshmi, S. Yadav, P. Sharmila, B. Venkateswarlu, et al. 2010. Metabolic engineering using mtlD gene enhances tolerance to water deficit and salinity in sorghum. Biol. Plant. 54:647–652.

Mastrorilli, M., N. Katerji, and G. Rana. 1999. Productivity and water use efficiency of sweet sorghum as affected by soil water deficit occurring at different vegetative growth stages. Eur. J. Agron. 11:207–215.

McBee, G., R. Waskom, and R. Creelman. 1983. Effect of senescene on carbohydrates in sorghum during late Kernel Maturity states. Crop Sci. 23:372–376.

Mittler, R., and E. Blumwald. 2010. Genetic engineering for modern agriculture: challenges and perspectives. Annu. Rev. Plant Biol. 61:443–462.

Mohan, S., R. Madhusudhana, K. Mathur, C. Howarth, G. Srinivas, K. Satish, et al. 2009. Co-localization of quantitative trait loci for foliar disease resistance in sorghum. Plant Breeding 128:532–535.

Mohan, S. M., R. Madhusudhana, K. Mathur, D. Chakravarthi, S. Rathore, R. N. Reddy, et al. 2010. Identification of quantitative trait loci associated with resistance to foliar diseases in sorghum [Sorghum bicolor (L.) Moench]. Euphytica 176:199–211.

Morrell, P., T. Williams-Coplin, A. Lattu, J. Bowers, J. Chandler, and A. Paterson. 2005. Crop-to-weed introgression has impacted allelic composition of johnsongrass populations with and without recent exposure to cultivated sorghum. Mol. Ecol. 14:2143–2154.

Munson, R. E., J. A. Schaffer, and E. W. Palm. 1993. Sorghum aphid pest management. University of Missouri-Extension.

Muraya, M. M., E. Mutegi, H. H. Geiger, S. M. de Villiers, F. Sagnard, B. M. Kanyenji, et al. 2011. Wild sorghum from different eco-geographic regions of Kenya display a mixed mating system. Theor. Appl. Genet. 122:1631–1639.

Mutegi, E., F. Sagnard, M. Muraya, B. Kanyenji, B. Rono, C. Mwongera, et al. 2010. Ecogeographical distribution of wild, weedy and cultivated Sorghum bicolor (L.) Moench in Kenya: implications for conservation and crop-to-wild gene flow. Genet. Resour. Crop Evol. 57:243–253.

Nagaraj, N., J. C. Reese, M. R. Tuinstra, C. M. Smith, P. St. Amand, M. Kirkham, et al. 2005. Molecular mapping of sorghum genes expressing tolerance to damage by greenbug (Homoptera: Aphididae). J. Econ. Entomol., 98:595–602.

Nagy, E. D., T.-C. Lee, W. Ramakrishna, Z. Xu, P. E. Klein, P. Sanmiguel, et al. 2007. Fine mapping of the Pc locus of Sorghum bicolor, a gene controlling the reaction to a fungal pathogen and its host-selective toxin. Theor. Appl. Genet. 114:961–970.

Nair, S. K., et al. "Identification and validation of QTLs conferring resistance to sorghum downy mildew (Peronosclerospora sorghi) and Rajasthan downy mildew (P. heteropogoni) in maize." Theoretical and applied genetics 110.8 (2005):1384–1392.

Nelson, D. E., P. P. Repetti, T. R. Adams, R. A. Creelman, J. Wu, D. C. Warner, et al. 2007. Plant nuclear factor Y (NF-Y) B subunits confer drought tolerance and lead to improved corn yields on water-limited acres. Proc. Natl Acad. Sci. 104:16450–16455.

Ni, Z., Z. Hu, Q. Jiang, and H. Zhang. 2013. GmNFYA3, a target gene of miR169, is a positive regulator of plant tolerance to drought stress. Plant Mol. Biol. 82:113–129.

van Oosterom, E., R. Jayachandran, and F. Bidinger. 1996. Diallel analysis of the stay-green trait and its components in sorghum. Crop Sci. 36:549–555.

Orlandini, S., M. Mancini, and A. Dalla Marta. 2007. Sistema per la realizzazione di una filiera corta per la produzione di energia da biomasse agricole. Proceedings of the XXXVII Convegno Nazionale della Società Italiana di Agronomia, Catania. 13-14.

Ougham, H., I. Armstead, C. Howarth, I. Galyuon, I. Donnison, and H. Thomas. 2008. The genetic control of senescence revealed by mapping quantitative trait loci. Ann. Plant Rev. Senes. Proc. Plants 26:171.

Parh, D., D. Jordan, E. Aitken, E. Mace, P. Jun-Ai, C. McIntyre, et al. 2008. QTL analysis of ergot resistance in sorghum. Theor. Appl. Genet. 117:369–382.

Parker, C. 2009. Observations on the current status of Orobanche and Striga problems worldwide. Pest Manag. Sci. 65:453–459.

Pasini, L., M. Bergonti, A. Fracasso, A. Marocco, and S. Amaducci. 2014. Microarray analysis of differentially expressed mRNAs and miRNAs in young leaves of sorghum under dry-down conditions. J. Plant Physiol. 171:537–548.

Paterson, A. H., K. F. Schertz, Y.-R. Lin, S.-C. Liu, and Y.-L. Chang. 1995. The weediness of wild plants: molecular analysis of genes influencing dispersal and persistence of johnsongrass, Sorghum halepense (L.) Pers. Proc. Natl Acad. Sci. 92:6127–6131.

Paterson, A. H., J. E. Bowers, R. Bruggmann, I. Dubchak, J. Grimwood, H. Gundlach, et al. 2009. The Sorghum bicolor genome and the diversification of grasses. Nature 457:551–556.

Peleg, Z., T. Fahima, T. Krugman, S. Abbo, D. Yakir, A. B. Korol, et al. 2009. Genomic dissection of drought resistance in durum wheat× wild emmer wheat recombinant inbreed line population. Plant, Cell Environ. 32:758–779.

Perumal, R., M. A. Menz, P. J. Mehta, S. Katile, L. A. Gutierrez-Rojas, R. R. Klein, et al. 2009. Molecular mapping of Cg1, a gene for resistance to anthracnose (Colletotrichum sublineolum) in sorghum. Euphytica 165:597–606.

Pontier, D., S. Gan, R. M. Amasino, D. Roby, and E. Lam. 1999. Markers for hypersensitive response and senescence show distinct patterns of expression. Plant Mol. Biol. 39:1243–1255.

Punnuri, S., Y. Huang, J. Steets, and Y. Wu. 2013. Developing new markers and QTL mapping for greenbug resistance in sorghum [Sorghum bicolor (L.) Moench]. Euphytica 191:191–203.

Qi, X., S. Xie, Y. Liu, F. Yi, and J. Yu. 2013. Genome-wide annotation of genes and noncoding RNAs of foxtail millet in response to simulated drought stress by deep sequencing. Plant Mol. Biol. 83:459–473.

Rajjou, L., M. Duval, K. Gallardo, J. Catusse, J. Bally, C. Job, et al. 2012. Seed germination and vigor. Annu. Rev. Plant Biol. 63:507–533.

Rajwanshi, R., S. Chakraborty, K. Jayanandi, B. Deb, and D. A. Lightfoot. 2014. Orthologous plant microRNAs: microregulators with great potential for improving stress tolerance in plants. Theor. Appl. Genet. 127:2525–2543.

Ram, G., and A. D. Sharma. 2013. In silico analysis of putative miRNAs and their target genes in sorghum (Sorghum bicolor). Int. J. Bioinform. Res. Appl. 9:349–364.

Ramasamy, P., M. Menz, P. Mehta, S. Katilé, L. Gutierrez-Rojas, R. Klein, et al. 2009. Molecular mapping of Cg1, a gene for resistance to anthracnose (Colletotrichum sublineolum) in sorghum. Euphytica 165:597–606.

Rathinasabapathi, B. 2000. Metabolic engineering for stress tolerance: installing osmoprotectant synthesis pathways. Ann. Bot. 86:709–716.

Reddy, P. S., B. Fakrudin, S. Punnuri, S. Arun, M. Kuruvinashetti, I. Das, et al. 2008. Molecular mapping of genomic regions harboring QTLs for stalk rot resistance in sorghum. Euphytica 159:191–198.

Reddy, N. R. R., M. Ragimasalawada, M. M. Sabbavarapu, S. Nadoor, and J. V. Patil. 2014. Detection and validation of stay-green QTL in post-rainy sorghum involving widely adapted cultivar, M35-1 and a popular stay-green genotype B35. BMC Genom. 15:909.

Rivero, R. M., J. Gimeno, A. van Deynze, H. Walia, and E. Blumwald. 2010. Enhanced cytokinin synthesis in tobacco plants expressing PSARK: IPT prevents the degradation of photosynthetic protein complexes during drought. Plant Cell Physiol. 51:1929–1941.

Sabadin, P., M. Malosetti, M. Boer, F. Tardin, F. Santos, C. Guimaraes, et al. 2012. Studying the genetic basis of drought tolerance in sorghum by managed stress trials and adjustments for phenological and plant height differences. Theor. Appl. Genet. 124:1389–1402.

Sama, V., N. Rawat, R. Sundaram, K. Himabindu, B. S. Naik, B. Viraktamath, et al. 2014. A putative candidate for the recessive gall midge resistance gene gm3 in rice identified and validated. Theor. Appl. Genet. 127:113–124.

Sammons, D. R., D. Wang, P. Morris, B. Duncan, G. Griffith, and D. Findley. 2012 Strategies for countering herbicide resistance. Abstracts of papers of the American Chemical Society. Amer Chemical Soc 1155 16TH ST, NW, Washington, DC 20036 USA.

Sanchez, A., P. Subudhi, D. Rosenow, and H. Nguyen. 2002. Mapping QTLs associated with drought resistance in sorghum (Sorghum bicolor L. Moench). Plant Mol. Biol. 48:713–726.

Satish, K., G. Srinivas, R. Madhusudhana, P. Padmaja, R. N. Reddy, S. M. Mohan, et al. 2009. Identification of quantitative trait loci for resistance to shoot fly in sorghum [Sorghum bicolor (L.) Moench]. Theor. Appl. Genet. 119:1425–1439.

Satish, K., Z. Gutema, C. Grenier, P. J. Rich, and G. Ejeta. 2012. Molecular tagging and validation of microsatellite markers linked to the low germination stimulant gene (lgs) for Striga resistance in sorghum [Sorghum bicolor (L.) Moench]. Theor. Appl. Genet. 124:989–1003.

Schittenhelm, S., and S. Schroetter. 2014. Comparison of drought tolerance of maize, sweet sorghum and sorghum-sudangrass hybrids. J. Agron. Crop Sci. 200:46–53.

Serrão, M., M. Menino, J. Martins, N. Castanheira, M. Lourenço, I. Januário, et al. 2012. Mineral leaf composition of sweet sorghum in relation to biomass and sugar yields under different nitrogen and salinity conditions. Commun. Soil Sci. Plant Anal. 43:2376–2388.

Sharma, H., S. Mukuru, K. Hari Prasad, E. Manyasa, and S. Pande. 1999. Identification of stable sources of resistance in sorghum to midge and their reaction to leaf diseases. Crop Prot., 18:29–37.

Sharma, H. C., K. K. Sharma, and J. H. Crouch. 2004. Genetic transformation of crops for insect resistance: potential and limitations. Crit. Rev. Plant Sci. 23:47–72.

Shen, C., Y. Bai, S. Wang, S. Zhang, Y. Wu, M. Chen, et al. 2010. Expression profile of PIN, AUX/LAX and PGP auxin transporter gene families in Sorghum bicolor under phytohormone and abiotic stress. FEBS J. 277:2954–2969.

Sims, R. E., A. Hastings, B. Schlamadinger, G. Taylor, and P. Smith. 2006. Energy crops: current status and future prospects. Glob. Change Biol. 12:2054–2076.

Singh SP. 1985. Sources of cold tolerance in grain sorghum. Can J Plant Sci. 65:251–257

Singh, M., K. Chaudhary, H. Singal, C. Magill, and K. Boora. 2006. Identification and characterization of RAPD and SCAR markers linked to anthracnose resistance gene in sorghum [Sorghum bicolor (L.) Moench]. Euphytica 149:179–187.

Sohn, K. H., S. C. Lee, H. W. Jung, J. K. Hong, and B. K. Hwang. 2006. Expression and functional roles of the pepper pathogen-induced transcription factor RAV1 in bacterial disease resistance, and drought and salt stress tolerance. Plant Mol. Biol. 61:897–915.

Staggenborg, S. A., K. C. Dhuyvetter, and W. Gordon. 2008. Grain sorghum and corn comparisons: yield, economic, and environmental responses. Agron. J. 100:1600–1604.

Su, M., X.-F. Li, X.-Y. Ma, X.-J. Peng, A.-G. Zhao, L.-Q. Cheng, et al. 2011. Cloning two P5CS genes from bioenergy sorghum and their expression profiles under abiotic stresses and MeJA treatment. Plant Sci. 181:652–659.

Subudhi, P., D. Rosenow, and H. Nguyen. 2000. Quantitative trait loci for the stay green trait in sorghum (Sorghum bicolor L. Moench): consistency across genetic backgrounds and environments. Theor. Appl. Genet. 101:733–741.

Sunkar, R., V. Chinnusamy, J. Zhu, and J.-K. Zhu. 2007. Small RNAs as big players in plant abiotic stress responses and nutrient deprivation. Trends Plant Sci. 12:301–309.

Takeda, S., and M. Matsuoka. 2008. Genetic approaches to crop improvement: responding to environmental and population changes. Nat. Rev. Genet. 9:444–457.

Tanksley, S. D. 1993. Mapping polygenes. Annu. Rev. Genet. 27:205–233.

Tao, Y., D. Jordan, R. Henzell, and C. McIntyre. 1998. Identification of genomic regions for rust resistance in sorghum. Euphytica 103:287–292.

Tao, Y., R. Henzell, D. Jordan, D. Butler, A. Kelly, and C. McIntyre. 2000. Identification of genomic regions associated with stay green in sorghum by testing RILs in multiple environments. Theor. Appl. Genet. 100:1225–1232.

Tao, Y., A. Hardy, J. Drenth, R. Henzell, B. Franzmann, D. Jordan, et al. 2003. Identifications of two different mechanisms for sorghum midge resistance through QTL mapping. Theor. Appl. Genet. 107:116–122.

Thomas, H., and H. Ougham. 2014. The stay-green trait. J. Exp. Bot., 65:3889–3900.

Tilsner, J., N. Kassner, C. Struck, and G. Lohaus. 2005. Amino acid contents and transport in oilseed rape (*Brassica napus* L.) under different nitrogen conditions. Planta 221:328–338.

Tuinstra, M. R., E. M. Grote, P. B. Goldsbrough, and G. Ejeta. 1997. Genetic analysis of post-flowering drought tolerance and components of grain development in *Sorghum bicolor* (L.) Moench. Mol. Breeding 3:439–448.

Tuinstra, M. R., S. Soumana, K. Al-Khatib, I. Kapran, A. Toure, A. van Ast, et al. 2009. Efficacy of herbicide seed treatments for controlling infestation of sorghum. Crop Sci. 49:923–929.

Upadhyaya, H. D., Y.-H. Wang, R. Sharma, and S. Sharma. 2013. Identification of genetic markers linked to anthracnose resistance in sorghum using association analysis. Theor. Appl. Genet. 126:1649–1657.

Varet, A., B. Hause, G. Hause, D. Scheel, and J. Lee. 2003. The Arabidopsis NHL3 gene encodes a plasma membrane protein and its overexpression correlates with increased resistance to *Pseudomonas syringae* pv. tomato DC3000. Plant Physiol. 132:2023–2033.

Verbruggen, N., and C. Hermans. 2008. Proline accumulation in plants: a review. Amino Acids 35:753–759.

Vijayalakshmi, K., A. K. Fritz, G. M. Paulsen, G. Bai, S. Pandravada, and B. S. Gill. 2010. Modeling and mapping QTL for senescence-related traits in winter wheat under high temperature. Mol. Breeding 26:163–175.

Visarada, K., and N. Kishore 2007. Improvement of Sorghum through transgenic technology. *Information System for Biotechnology News Report (Virginia tech, US)* pp, 1-3.

Wang, Z. X., M. Yano, U. Yamanouchi, M. Iwamoto, L. Monna, H. Hayasaka, et al. 1999. The Pib gene for rice blast resistance belongs to the nucleotide binding and leucine-rich repeat class of plant disease resistance genes. Plant J. 19:55–64.

Wang, S., Y. Bai, C. Shen, Y. Wu, S. Zhang, D. Jiang, et al. 2010. Auxin-related gene families in abiotic stress response in *Sorghum bicolor*. Funct. Integr. Genomics 10:533–546.

Wang, H., G. Chen, H. Zhang, B. Liu, Y. Yang, L. Qin, et al. 2014a. Identification of QTLs for salt tolerance at germination and seedling stage of *Sorghum bicolor* L Moench. Euphytica 196:117–127.

Wang, T. T., Z. J. Ren, Z. Q. Liu, X. Feng, R. Q. Guo, B. G. Li, et al. 2014b. SbHKT1; 4, a member of the high-affinity potassium transporter gene family from Sorghum bicolor, functions to maintain optimal Na^+/K^+ balance under Na^+ stress. J. Integr. Plant Biol. 56:315–332.

Washburn, J. D., S. C. Murray, B. L. Burson, R. R. Klein, and R. W. Jessup. 2013. Targeted mapping of quantitative trait locus regions for rhizomatousness in chromosome SBI-01 and analysis of overwintering in a *Sorghum bicolor*× *S. propinquum* population. Mol. Breeding 31:153–162.

Wilkinson, S., G. R. Kudoyarova, D. S. Veselov, T. N. Arkhipova, and W. J. Davies. 2012. Plant hormone interactions: innovative targets for crop breeding and management. J. Exp. Bot. 63:3499–3509.

Wood, A. J., H. Saneoka, D. Rhodes, R. J. Joly, and P. B. Goldsbrough. 1996. Betaine aldehyde dehydrogenase in sorghum (molecular cloning and expression of two related genes). Plant Physiol. 110:1301–1308.

Wright, T. R., G. Shan, T. A. Walsh, J. M. Lira, C. Cui, P. Song, et al. 2010. Robust crop resistance to broadleaf and grass herbicides provided by aryloxyalkanoate dioxygenase transgenes. Proc. Natl Acad. Sci. 107:20240–20245.

Wu, Y., and Y. Huang. 2008. Molecular mapping of QTLs for resistance to the greenbug *Schizaphis graminum* (Rondani) in *Sorghum bicolor* (Moench). Theor. Appl. Genet. 117:117–124.

Wu, T.-M., W.-R. Lin, Y.-T. Kao, Y.-T. Hsu, C.-H. Yeh, C.-Y. Hong, et al. 2013. Identification and characterization of a novel chloroplast/mitochondria co-localized glutathione reductase 3 involved in salt stress response in rice. Plant Mol. Biol. 83:379–390.

Xu, W., P. K. Subudhi, O. R. Crasta, D. T. Rosenow, J. E. Mullet, and H. T. Nguyen. 2000. Molecular mapping of QTLs conferring stay-green in grain sorghum (*Sorghum bicolor* L. Moench). Genome 43:461–469.

Yan, H., L. Hong, Y. Zhou, H. Jiang, S. Zhu, J. Fan, et al. 2013. A genome-wide analysis of the ERF gene family in sorghum. Genet. Mol. Res. 12:2038–2055.

Yemata, G., Fetene1, M., Assefa, A and Tesfaye, K (2014) Evaluation of the agronomic performance of stay green and farmer preferred sorghum (Sorghum bicolor (L) Moench) varieties at Kobo North Wello zone, Ethiopia. Sky Journal of Agricultural Research Vol. 3 240–248

Young, W., and G. Teetes. 1977. Sorghum entomology. Annu. Rev. Entomol. 22:193–218.

Yu, Q., and S. B. Powles. 2014. Metabolism-based herbicide resistance and cross-resistance in crop weeds: a threat to herbicide sustainability and global crop production. Plant Physiol. 166:1106–1118.

Yu, J., and M. R. Tuinstra. 2001. Genetic analysis of seedling growth under cold temperature stress in grain sorghum. Crop Sci. 41:1438–1443.

Zegada-Lizarazu, W., and A. Monti. 2013. Photosynthetic response of sweet sorghum to drought and re-watering at different growth stages. Physiol. Plant. 149:56–66.

Zhai, J., Y. Dong, Y. Sun, Q. Wang, N. Wang, F. Wang, et al. 2014. Discovery and analysis of microRNAs in Leymus chinensis under saline-alkali and drought stress using high-throughput sequencing. PLoS ONE 9:e105417.

Zhang, M., Q. Tang, Z. Chen, J. Liu, H. Cui, Q. Shu, et al. 2009. [Genetic transformation of Bt gene into sorghum

(*Sorghum bicolor* L.) mediated by *Agrobacterium tumefaciens*]. Chin. J. Biotechnol. 25:418–423.

Zhao, B., R. Liang, L. Ge, W. Li, H. Xiao, H. Lin, et al. 2007. Identification of drought-induced microRNAs in rice. Biochem. Biophys. Res. Commun. 354:585–590.

Zheng, L.-Y., X.-S. Guo, B. He, L.-J. Sun, Y. Peng, S.-S. Dong, et al. 2011. Genome-wide patterns of genetic variation in sweet and grain sorghum (*Sorghum bicolor*). Genome Biol. 12:R114.

Zhou, Y., D. Wang, Z. Lu, N. Wang, Y. Wang, F. Li, et al. 2013. [Impacts of drought stress on leaf osmotic adjustment and chloroplast ultrastructure of stay-green sorghum]. Ying Yong Sheng Tai Xue Bao 24:2545–2550.

PERMISSIONS

LIST OF CONTRIBUTORS

Kevin N. Lindegaard
Crops for Energy Ltd, 15 Sylvia Avenue, Knowle, Bristol BS3 5BX, UK

Paul W. R. Adams
Department of Mechanical Engineering, Faculty of Engineering and Design, Institute for Sustainable Energy and Environment (I·SEE), University of Bath, Claverton Down, Bath BA2 7AY, UK

Martin Holley and Annette Lamley
Centre for Sustainable Energy, 3 St Peter's Court, Bedminster Parade, Bristol BS3 4AQ, UK

Annika Henriksson
Henriksson Salix AB, Almhög, 241 92 Eslöv, Sweden

Stig Larsson
EWB, Spannmalsgatan 28, SE-268 32 SVALÖV, Sweden

Hans-Georg von Engelbrechten
Agraligna GmbH, Oststrasse 7, 38315 Schladen/OT Beuchte, Germany

Gonzalo Esteban Lopez
Agencia Provincial de la Energía de Granada, Edificio CIE - 1oPlanta. Avda. Andalucía s/n., 18015 Granada, Spain

Marcin Pisarek
PGNiG TERMIKA SA, Siedziba główna - Elektrociepłownia Żerań, ul. Modlińska 15, 03-216 Warszawa, Poland

Ian Mackay and Eric Ober
John Bingham Laboratory, NIAB, Huntingdon Road, Cambridge CB3 0LE, UK

John Hickey
The Roslin Institute and Royal (Dick) School of Veterinary Studies, University of Edinburgh, Easter Bush Research Centre, Midlothian EH25 9RG, UK

Athole H. Marshall, Rosemary P. Collins, Mike W. Humphreys and John Scullion
Institute of Biological and Environmental Research, Aberystwyth University, Gogerddan, Aberystwyth SY233EE, UK

Katrin Drastig, Hilde Klauss and Werner Berg
Leibniz-Institute for Agricultural Engineering Potsdam-Bornim, Max-Eyth-Allee 100, 14469 Potsdam, Germany

Annette Prochnow
Faculty of Agriculture and Horticulture, Humboldt-University of Berlin, Hinter der Reinhardtstr. 8-18, 10115 Berlin, Germany

Michael Robertson
CSIRO Agriculture, Wembley, Western Australia, Australia

John Kirkegaard and Greg Rebetzke
CSIRO Agriculture, Wembley, Western Australia, Australia
CSIRO Agriculture, Canberra, Australian Capital Territory 2601, Australia

Rick Llewellyn
CSIRO Agriculture, Wembley, Western Australia, Australia
CSIRO Agriculture, Urrbrae, South Australia, Australia

Tim Wark
CSIRO Data 61, Pullenvale, Queensland, Australia

Fabiani da Rocha, Caio Canella Vieira, Mônica Christina Ferreira, Kênia Carvalho de Oliveira, Fabiana Freitas Moreira and José Baldin Pinheiro
Departamento de Genética, Universidade de São Paulo, Escola Superior de Agricultura "Luiz de Queiroz", Avenida Pádua Dias, 11, 13.418-900 Piracicaba, SP, Brazil

Sandra J. Hey
Rothamsted Research, Harpenden, Hertfordshire
AL5 2JQ, UK

Peter R. Shewry
Rothamsted Research, Harpenden, Hertfordshire
AL5 2JQ, UK
University of Reading, Whiteknights, Reading
Berkshire RG6 6AH, UK

Qingfeng Song and Xin-Guang Zhu
CAS Key Laboratory for Computational Biology
and State Key Laboratory of Hybrid Rice,
Partner Institute for Computational Biology,
Chinese Academy of Sciences, Shanghai 200031,
China

Chengcai Chu
The State Key Laboratory of Plant Genomics and
National Center of Plant Gene Research (Beijing),
Institute of Genetics and Developmental Biology,
CAS, Beijing 100101, China

Martin A. J. Parry
Lancaster Environment Centre, Lancaster
University, Lancaster LA1 4YQ, UK

**Mike W. Humphreys, Sally A. O'Donovan,
Markku S. Farrell, Alan P. Gay and Alison H.
Kingston-Smith**
Institute of Biological, Environmental and Rural
Sciences, Aberystwyth University, Aberystwyth,
Wales, SY23 3EE, United Kingdom

Bishwajit Ghose
Institute of Nutrition and Food Science, University
of Dhaka, Dhaka, Bangladesh

Fernando P. Carvalho
Laboratório de Protecção e Segurança Radiológica,
Instituto Superior Técnico/Universidade de Lisboa,
Estrada Nacional 10, km 139, 2695-066 Bobadela
LRS, Portugal

**Simão Branco-Neves, Cristiano Soares, Alexandra
de Sousa and Fernanda Fidalgo**
BioISI – Biosystems and Integrative Sciences
Institute, Departamento de Biologia, Faculdade
de Ciências, Universidade do Porto, Rua Campo
Alegre s/n, 4169-007 Porto, Portugal

Viviana Martins
CITAB-UM – Centre for the Research and
Technology of Agro-Environmenal and Biological
Sciences, Universidade do Minho, Campus de
Gualtar, 4710-057 Braga, Portugal

Hernâni Gerós
CITAB-UM – Centre for the Research and
Technology of Agro-Environmenal and Biological
Sciences, Universidade do Minho, Campus de
Gualtar, 4710-057 Braga, Portugal
CBMA – Centre of Molecular and Environmental
Biology, Universidade do Minho, Campus de
Gualtar, 4710-057 Braga, Portugal
CEB – Centre of Biological Engineering, Department
of Biological Engineering, Universidade do Minho,
Campus de Gualtar, 4710-057 Braga, Portugal

Manuel Azenha
CIQ-UP, Departamento de Química e Bioquímica,
Faculdade de Ciências, Universidade do Porto, Rua
Campo Alegre 687, 4169-007 Porto, Portugal

William J. Davies
The Lancaster Environment Centre, Lancaster
University, Bailrigg, Lancaster LA1 4YQ, UK

Jean-Marcel Ribaut
Generation Challenge Programme (GCP) c/o
CIMMYT, Carretera Mexico-Veracruz, El Batan,
Texcoco, Estado de Mexico, Mexico

Phillip A. Davis and Claire Burns
Stockbridge Technology Centre, Cawood, Selby,
North Yorkshire YO8 3TZ, UK

**P. John Andralojc, Szilvia Bencze, Pippa J.
Madgwick, Hélène Philippe and Martin A. J. Parry**
Plant Biology and Crop Science Department,
Rothamsted Research, Harpenden, Hertfordshire,
AL5 2JQ, United Kingdom

Stephen J. Powers
Computational and Systems Biology Department,
Rothamsted Research, Harpenden, Hertfordshire,
AL5 2JQ, United Kingdom

Ian Shield and Angela Karp
AgroEcology Department, Rothamsted Research,
Harpenden, Hertfordshire, AL5 2JQ, United
Kingdom

Taisheng Du and Shaozhong Kang
Center for Agricultural Water Research in China, China Agricultural University, Beijing 100083, China

Xiying Zhang
Center for Agricultural Resources Research, Institute of Genetics and Developmental Biology, Chinese Academy of Sciences, Shijiazhuang 050021, China

Jianhua Zhang
School of Life Sciences and State Key Laboratory of Agrobiotechnology, The Chinese University of Hong Kong, Hong Kong, China

Sylvester Elikana Anami
Key Laboratory of Plant Resources, Institute of Botany, Chinese Academy of Sciences, Beijing 100093, China
Institute of Biotechnology Research, Jomo Kenyatta University of Agriculture and Technology, Nairobi, Kenya

Li-Min Zhang, Yan Xia, Yu-Miao Zhang, Zhi-Quan Liu and Hai-Chun Jing
Key Laboratory of Plant Resources, Institute of Botany, Chinese Academy of Sciences, Beijing 100093, China

Index

9 781639 872336